Molecular Biology of the Neuron

Second edition

(Molecular and Cellular Neurobiology)

OXFORD
UNIVERSITY PRESS

Great Clarendon Street, Oxford OX2 6DP

Oxford University Press is a department of the University of Oxford.
It furthers the University's objective of excellence in research, scholarship,
and education by publishing worldwide in

Oxford New York
Auckland Bangkok Buenos Aires Cape Town Chennai
Dar es Salaam Delhi Hong Kong Istanbul Karachi Kolkata
Kuala Lumpur Madrid Melbourne Mexico City Mumbai Nairobi
Sao Paulo Shanghai Taipei Tokyo Toronto

Oxford is a registered trade mark of Oxford University Press
in the UK and in certain other countries

Published in the United States
by Oxford University Press Inc., New York

First Edition published 1997
Second Edition published 2004

A catalogue record for this title is available from the British Library

ISBN 0-19-850998-7 (Hbk)

10 9 8 7 6 5 4 3 2 1

Typeset by Cepha Imaging Pvt. Ltd.
Printed in Great Britain
on acid-free paper by Biddles Ltd, King's Lynn

Preface

Neurones are arguably the most complex of all cells. At the current time, our knowledge of the molecular repertoire of cells is increasing at enormous speed, in the aftermath of the determination of the entire sequence of the human genome, and the genomes of a growing list of other species. It is, therefore, a challenging task to keep abreast of the many diverse advances in our understanding of how neurones function at a molecular level. The first edition of *Molecular Biology of the Neuron* was possibly unique in providing up-to-date and thorough reviews of the major classes of molecules which contribute to neuronal function. The chapters provided detailed molecular information as a valuable source of reference, while also keeping the text broad in coverage and accessible to non-specialists. This second edition provides an essential update to the original text, encompassing the many major advances in knowledge that have been made in the intervening years. As before, each chapter is written by leading researchers in the field and all provide comprehensive, yet comprehensible, coverage of the key molecules expressed in neurones, and how they act and interact.

All chapters have been rewritten to incorporate the immense amount of new molecular data about neurons that has been discovered in the past six years, since the publication of the first edition. Thus, significant new material has been added to the coverage of the regulation of neuronal gene expression, neuronal ion channels, ligand-gated ion channel receptors and G-protein linked receptors, the neuronal cytoskeleton, molecules involved in neurotransmitter release, and the molecular basis of neuronal plasticity and learning. The new review of the genetic and molecular basis of human neurological disease is perhaps the most complete and up-to-date summary of the rapidly expanding information available. New chapters have been added where appropriate, to cover fields where there was little or no molecular information six years ago. Thus there are new chapters on protein trafficking, post-synaptic density and neuronal ageing, plus additional chapters to cover signalling within neurons in greater depth: specifically, synapse to nucleus calcium signalling, signalling by tyrosine phosphorylation, and signalling by serine-threonine phosphorylation. In addition, new chapters presenting the genetic technologies of the Drosophila and mouse systems as applied to the study of neurons are included because of the ever-increasing importance of these systems to functional studies of neuronal genes and proteins.

These are exciting times for molecular studies of the nervous system. Knowing all the genes as we do, and with the pace of advance quickening daily, it is easy for students and researchers to lose track of progress outside their immediate subject area.

This book provides a platform of knowledge, which allows new advances to be put easily into context, without being an expert. Taken as a whole, we hope that this second edition of *Molecular Biology of the Neuron* will provide an authoritative and extremely useful overview of the molecular structure of neurones, of value to students and researchers alike.

R. Wayne Davies | *Glasgow*
Brian J. Morris | *February 2004*

Contents

Contributors

Armstrong, J. Douglas
Institute for Adaptive and Neural Computation
School of Informatics
University of Edinburgh
5 Forrest Hill
Edinburgh EH1 2QL
jda@inf.ed.ac.uk

Avila, Jesús
Centro de Biología Molecular 'Severo Ochoa'
Facultad de Ciencias
Universidad Autónoma de Madrid
Madrid E-28049
Spain
jesus.avila@cbm.uam.es

Bailey, Mark E.S.
Institute for Biomedical and Life Sciences
Division of Molecular Genetics
University of Glasgow
Anderson College
56 Dumbarton Road
Glasgow G11 6NU
U.K.
M.Bailey@bio.gla.ac.uk

Baldassa, Simona
Department of Biomolecular Sciences and Biotechnology
University of Milan
Via Celoria 26
20133 Milano
Italy
simona.baldessa@unimi.it

Catsicas, Stefan
Institut de Biologie Cellulaire et de Morphologie
Université de Lausanne
Rue du Bugnon 9
CH-1005 Lausanne
Switzerland

Cattabeni, Flaminio
Department of Pharmacological Sciences
Center of Excellence on Neurodegenerative Diseases
University of Milano
via Balzaretti 9
20133 Milano, Italy
flaminio.cattabeni@unimi.it

Connolly, Christopher N.
Division of Pathology and Neuroscience
Ninewells Medical School
University of Dundee
Dundee DDI 9SY
c.n.connolly@dundee.ac.uk

Davies, R. Wayne
Institute for Biomedical and Life Sciences
Division of Molecular Genetics
University of Glasgow
Robertson Building
54 Dumbarton Road
Glasgow G11 6NU
wayne.davies@bio.gla.ac.uk

Díaz-Nido, Javier
Centro de Biologia Molecular 'Severo Ochoa'
Facultad de Ciencias
Universidad Autónoma de Madrid
Madrid E-28049
Spain
javier.diazmido@cbm.uam.es

Di Luca, Monica
Department of Pharmacological Sciences
Center of Excellence on Neurodegenerative Diseases
University of Milano
via Balzaretti 9
20133 Milano, Italy
monica.diluca@unimi.it

Ernsberger, Uwe
Institut für Anatomie und Zellbiologie
Interdisziplinares Zentrum für Neurowissenschaften
Ruprecht-Karls-Universität Heidelberg
Im Neuenheimer Feld 307
69120 Heidelberg
Germany
uwe.ernsberger@urz.uni-heidelberg.de

Gardoni, Fabrizio
Department of Pharmacological Sciences
Center of Excellence on Neurodegenerative Diseases
University of Milano
via Balzaretti 9
20133 Milano, Italy
fabrizio.gardoni@unimi.it

Girault, Jean-Antoine
INSERM U 536
Institut du Fer à Moulin
17, rue du Fer à Moulin
75005 Paris
France
Girault@ifm.inserm.fr

Glencorse, Thora A.
Institute for Biomedical and Life Sciences
Division of Molecular Genetics
University of Glasgow
Robertson Building
54 Dumbarton Road
Glasgow G11 6NU
t_glencorse@yahoo.co.uk

Goodwin, Stephen F.
Institute for Biomedical and Life Sciences
Division of Molecular Genetics
University of Glasgow
Robertson Building
54 Dumbarton Road
Glasgow G11 6NU
s.goodwin@bio.gla.ac.uk

Hardingham, Giles E.
Department of Preclinical Veterinary Sciences
Royal (Dick) School of Veterinary Studies
University of Edinburgh
Summerhall
Edinburgh EH9 1QH
Giles.Hardingham@ed.ac.uk

Jockusch, Harald
Developmental Biology and Molecular Pathology
University of Bielefeld
D-33501 Bielefeld
Germany
h.jockusch@uni-bielefeld.de

Koenig, Jennifer A.
Department of Pharmacology
University of Cambridge
Tennis Court Road
Cambridge CB2 1PD
jk111@cam.ac.uk

Morris, Brian J.
Institute for Biomedical and
Life Sciences
Division of Neuroscience and Biomedical
Systems
University of Glasgow
West Medical Building
Glasgow G12 8QQ
B. Morris@bio.gla.ac.uk

Pessia, Mauro
Università degli Studi di Perugia
Facoltà di Medicina e Chirurgia
Dipartimento di Medicina Interna
Sezione di Fisiologia Umana
Via del Giochetto
1-06126 Perugia
Italy
pessia@unipg.it

Schmitt-John, Thomas
Developmental Biology and Moleular
Pathology
University of Bielefeld
D-33501 Bielefeld
Germany
thomas.john@uni-bielefeld.de

Shiels, Paul
Division of Cancer Sciences and
Molecular Pathology
Department of Surgery
Western Infirmary Glasgow
44 Church Street
Glasgow G11 6NT
P.Shiels@clinmed.gla.ac.uk

Staple, Julie
Institut de Biologie Cellulaire et de
Morphologie
Université de Lausanne
Rue du Bugnon 9
CH-1005 Lausanne
Switzerland

Sturami, Emmapaola
Department of Biomolecular Sciences
and Biotechnology
University of Milan
Via Celoria 26
20133 Milano
Italy
emmapaola.sturami@unimi.it

Zippel, Renata
Department of Biomolecular Sciences
and Biotechnology
University of Milan
Via Celoria 26
20133 Milano
Italy
renata.zippel@unimi.it

Abbreviations

Aβ	protein fragment found in β-amyloid and neuritic plaques
A–P	anterior–posterior
ABP	actin-binding protein
ACh	acetylcholine
AChR	acetylcholine receptor
AChT	acetylcholine transporter
AD	Alzheimer's disease
ADF	actin depolymerizing factor
ADNFE	autosomal dominant nocturnal frontal lobe epilepsy
ALS	amyotrophic lateral sclerosis
AMPA	D,L-amino-3-hydroxy-5-methyl-4-isoxalone propionic acid
ANP	atrial natriuretic peptide
APOE	apoprotein E locus
APP	amyloid precursor protein
AR	androgen receptor
ARF	ADP-ribosylation factor
AVED	selective vitamin E deficiency
βARK	β_2-adrenergic receptor kinase
BDNF	brain-derived neurotrophic factor
bHLH	basic helix–loop–helix
BMP	bone morphogenetic protein
BoNT	botulinum neurotoxin
CAM	cell adhesion molecule
CaMKII	Ca^{2+} calmodulin kinase II
CCK	cholecystokinin
ChAT	choline acetyltransferase
ClC	voltage-gated chloride channel
CJD	Creutzfeldt–Jakob disease
CLIP	cytoplasmic linker protein
CMT1A	Charcot–Marie–Tooth disease 1A
CNS	central nervous system
CNTF	ciliary neurotrophic factor
CNTFR	ciliary neurotrophic factor receptor
COMP	cartilage oligomeric matrix protein
cox-2	cyclooxygenase 2
$cPLA_2$	cytosolic phospholipase A_2

CRE	cAMP-response element
CREB	cAMP-response element binding protein
CT-1	cardiotropin-1
CTX	conotoxin
D–V	dorsal–ventral
DAG	diacylglycerol
DBH	dopamine-β-hydroxylase
DCC	deleted in colorectal cancer (gene)
DHP-R	dihydropyridine receptor
DM	myotonic dystrophy (dystrophia muscularis)
DMPK	myotonic dystrophy kinase/myotonin
DRG	dorsal root ganglion
DRPLA	dentatorubral-pallidoluysian atrophy
ECM	extracellular matrix
EGF	epidermal growth factor
EMSA	electrophoresis mobility shift assay
ER	endoplasmic reticulum
ES cells	embryonic stem cells
FALS	familial amyotrophic lateral sclerosis
FFI	fatal familial insomnia
FGF	fibroblast growth factor
FN	fibronectin
FRAXA	fragile X syndrome
FRDA	Friedreich's ataxia
GABA	γ-aminobutyric acid
GDNF	glia cell line-derived neurotrophic factor
GDNFR	glia cell line-derived neurotrophic factor receptor
GluR	glutamate receptor
GlyR	glycine receptor
GPI	glycosyl-phosphatidylinositol
GRK	G protein receptor kinase
GSS	Gerstmann–Sträussler–Scheinker syndrome
HD	Huntington's disease
HH	hedgehog
HOX	homeobox
HSV	herpes simplex virus
5-HT	5-hydroxytryptamine
hyperYPP	hyperkalemic periodic paralysis
hypoYPP	hypokalemic periodic paralysis
ICE	interleukin-1β converting enzyme
IEG	immediate-early genes

IFP	intermediate filament protein
IgSF	immunoglobulin superfamily
IL-6	interleukin-6
IL-6R	interleukin-6 receptor
$Ins(1,4,5)P_3$	inositol 1,4,5-triphosphate
IPL	intraperiod line (of myelin)
KAIN	kainate
lacZ	β-galactosidase
LAMP	limbic system-associated protein
LCR	locus control region
LDL	low density lipoprotein
LIF	leukemia inhibitory factor
LTD	long-term depression
LTP	long-term potentiation
LV	large vesicles
mAChR	muscarinic acetylcholine receptor
MAG	myelin-associated glycoprotein
MAOA	monoamine oxidase A
MAOB	monoamine oxidase B
MAP	microtubule-associated protein
MAPK	mitogen-activated protein kinase
MARCKS	myristoylated alanine-rich C kinase substrate
MBP	myelin basic protein
MCDP	mast cell degranulating peptide
MDL	major dense line (of myelin)
MDLS	Miller–Dieker lissencephaly syndrome
MED	motor endplate disease
mGluR	metabotropic glutamate receptors
MJD	Machado–Joseph disease
MUSK	muscle-specific kinase
NAIP	neuronal apoptosis inhibitory protein
NCAM	neural cell adhesion molecule
ND	Norrie disease
NDP	Norrie disease protein
NF-H	neurofilament triplet protein H
NF-L	neurofilament triplet protein L
NF-M	neurofilament triplet protein M
NFT	neurofibrillary tangles
NgCAM	neuron–glial cell adhesion molecule
NGF	nerve growth factor
NGICR	neurotransmitter-gated ion channel receptors

NMDA	*N*-methyl-D-aspartate
NO	nitric oxide
NrCAM	NgCAM-related cell adhesion molecule
NRSE	neuron restrictive silencer element
NRSF	neuron restrictive silencer factor
NSF	*N*-ethylmaleimide-sensitive factor
NT	neurotrophin
NT3	neurotrophin 3
OBCAM	opioid-binding cell adhesion molecule
OSM	oncostatin-M
P0	protein zero
PAF	platelet-activating factor
PHF	paired helical filaments
PKA	protein kinase A
PKC	protein kinase C
PKG	protein kinase G
PLA	phospholipase A
PLC	phospholipase C
PLD	phospholipase D
PLP	proteolipid protein
PMP-22	peripheral myelin protein-22
PNS	peripheral nervous system
PPV	plasmalemmal precursor vesicles
PSA	polysialic acid
RAGS	repulsive axonal guidance signal
RNP	ribonucleoprotein particle
RTK	receptor tyrosine kinase
RTP	receptor-like tyrosine phosphatases
RT-PCR	reverse transcriptase–polymerase chain reaction
SBMA	spinal and bulbar muscular atrophy
SCA	spinocerebellar ataxia
SCG	superior cervical sympathetic ganglion
ScTX	scorpion toxin
SDS-PAGE	sodium dodecyl sulfate–polyacrylamide gel electrophoresis
SHH	sonic hedgehog
SHH-C	C-terminal cleavage product of hedgehog protein
SHH-N	N-terminal cleavage product of hedgehog protein
SMA	spinal muscular atrophy
SMN	survival motor neuron (gene)
SNAP	synaptosomal-associated protein
SNARES	SNAP receptors

SOD	superoxide dismutase
$sPLA_2$	secreted phospholipase A_2
SR	splicing regulatory protein
SV	small vesicles
TeNT	tetanus neurotoxin
TF	transcription factor
TGF	transforming growth factor
TH	tyrosine hydroxylase
TN-C	tenascin-C
TN-R	tenascin-R
TN-X	tenascin-X
TNF	tumor necrosis factor
tPA	tissue plasminogen activator
Trk	receptor tyrosine kinases
TSP	thrombospondin
TSS	transcription start site
α-TTP	α-tocopherol transfer protein
TTX	tetrodotoxin
VAChT	vesicular acetylcholine transporter
VAMP	vesicle-associated membrane protein
VMAT	vesicular monoamine transporters

Chapter 1

Studying neuronal function using the *Drosophila* genetic system

J. Douglas Armstrong and Stephen F. Goodwin

1.1 Introduction

The detailed study of neuronal function, in particular the function of neurons involved in complex processing and behavioural roles, requires several factors to be controlled, measured or exploited. Of relevance to the study of neuronal function are:

1. *Drosophila*: the short life cycle and simple culture conditions make it easy to control external environmental variables which are often more problematic in other species. In particular, the 'sensory' and 'social' interactions of the organism can easily be monitored and/or manipulated directly.
2. Behaviour: in order to unravel the neuronal pathways underlying behaviour, it is essential to study behaviours that can be assayed both qualitatively and quantitatively. There is a range of assays that have been developed over the years for both larval and adult stages of *Drosophila*. These include the use of visual, olfactory, tactile and auditory cues and assessment of responses with respect to reflex reactions as well as more complex learned behaviour.
3. Neurogenetics: *Drosophila* is one of the longest established genetic models with a formidable array of genetic and molecular tools that can be brought to bear upon any biological research area. This organism has given us valuable insights into the molecular, cellular and evolutionary bases of behaviour. In this chapter we will restrict our attention to those particularly relevant to neuroscience research in the organism.
4. Nervous system: the detailed neuronal architecture of the nervous system needs to be known in order for pathways involving multiple neurons to be investigated; i.e. neurons act as parts in circuits and context is critical. For *Drosophila*, traditional neuroanatomical techniques have produced gross maps of the nervous system through development and have also contributed valuable information about single neuronal projection patterns. The use of gene expression profiles (via enhancer-traps) to unravel neuroanatomy was pioneered in *Drosophila* and recent techniques have refined such approaches to the extent that brain maps for *Drosophila* will soon be available to single neuron resolution.

5. Neurophysiology: sadly the small size of neurons in *Drosophila* makes traditional neurophysiology difficult. However, there have been advances recently with several emerging techniques that allow non-invasive recording of neuronal activity.
6. Neuroinformatics and high-throughput technologies, supporting informatics analysis toolkits and public database systems. Historically, *Drosophila* has had one of the most up-to-date and comprehensive databases describing the genetics of the organism (FlyBase: http://flybase.bio. indiana.edu) and this has been supplemented by a range of genome databases describing the sequences generated by the public and private genome sequencing projects and their annotation. In addition there are additional databases and resources of specific interest to *Drosophila* neuroscience that will be described towards the end of this chapter.

1.2 *Drosophila melanogaster*

Drosophila melanogaster (referred to here as *Drosophila*) is generally the first multicellular experimental organism that biologists are introduced to in their training. Unlike most other experimental systems in biology, their simple culture requirements are so simple and cost-effective that they can be maintained in high-school biology labs with ease. In a modern laboratory situation this makes getting started with *Drosophila* particularly easy, although as for any organism large-scale operations will require some dedicated equipment and support staff.

For the newcomer, there is a wealth of literature describing the culture requirements and the basic (and advanced) techniques possible. To begin with, we would recommend '*Drosophila* protocols' (Sullivan *et al.* 2000), which is a laboratory manual (in format and weight). It describes in detail many of the common laboratory techniques including many of the recent molecular/transgenic manipulations used in *Drosophila* neuroscience, some of which we will describe in more detail here.

One of the major rate-limiting steps in modern neuroscience, particularly with regard to the adult nervous system is quite simply the life-cycle of the organism. This is one of the key advantages when working with *Drosophila*. The life-cycle is temperature dependent and varies from around 21 days at 16°C to under 10 days at temperatures over 25°C (for details, see Ashburner 1989). Moreover, their proliferation under standard conditions allows small cultures to be rapidly scaled into the thousands within two generations (2 flies → 1000s in 20 days). Experiments investigating neuronal senescence in *Drosophila* (e.g. Lin *et al.* 1998; Rogina *et al.* 2000) take somewhat longer with the average life expectancy around 100 days under optimal conditions. This still compares favourably with more complex models.

1.3 Behaviour

The output of a complex nervous system is manifest as the behaviour of the animal. While neuroscience research often focuses on systems at much lower levels (e.g. small groups of neurons, single neurons, molecular complexes) for a holistic understanding of how the nervous system works it is important that we are able, at least as a research community, to bridge the system levels. This means that for a good model system, we need to make the connection between molecules, cells, tissues and ultimately behaviour and environment (Greenspan and Tully 1994).

Assayable behaviour in *Drosophila* starts around 24 hours after fertilization. At this stage, the larva starts moving and pushes its way out of the egg casing. From this stage onwards (apart from the pupal phase) *Drosophila* interacts with its environment and displays a wide range of initially instinctive behaviours, which in common with other organisms develop biologically as the animal matures and are refined by experience.

At the simple larval stage where much of the nervous system has yet to be deployed, assays have been developed to investigate movement, exploratory behaviour, phototaxis, olfactory responses, pain responses and learning amongst others (reviewed by Sokolowski 2001). In the mature adult, there is a much wider range of behaviour assays that reflect the relative complexity of the mature nervous system, its wider range of sensory inputs and wider options for motor response. Assays can be grouped according to the sensory modality they primarily use and these cover all primary senses: visual, olfactory, gustatory, tactile, gravitational and auditory.

Fruit flies can exhibit several different types of learning and memory. These include two general types of learning, non-associative and associative. Habituation (a decrement in behavioural response) and sensitization (an increment in behavioural response) are forms of non-associative learning that result from exposure only to one environmental stimulus. In contrast, associative learning results from the temporal association of two stimuli, one of which acts as a natural 'reinforcer' of the behavioural response. A variety of experimental paradigms, using a range of sensory modalities, have been developed that can assess the flies ability to learn and remember (reviewed by Connolly and Tully 1998).

1.4 Neurogenetics

1.4.1 Overview

From the genetic perspective, *Drosophila* has been one of the most intensely studied organisms using both classical and modern transgenic approaches. There is an extensive literature on the general techniques available and we would recommend 'Fly pushing' (Greenspan 2000), which describes basic culture conditions and classical genetic approaches routinely employed in *Drosophila*. This is a small and easy-to-read book — excellent for a train journey and as a primer for a new staff member or grad-student.

Among the advantages that have facilitated neuroscience research in *Drosophila* is the range of targeted expression methods available to researchers. For most, if not all organisms, known gene promoter sequences can be fused to reporter constructs and inserted into the genome at the embryonic stage. For those unfamiliar with *Drosophila*, we can take this much further. Stable transgenic lines can be generated that use random insertions close to genomic enhancers to drive subsequent transgenic constructs. The most widely used of these systems is the P{GAL4} system developed by Brand and Perrimon (1993). This is a binary system comprising an enhancer-trap element, in which expression of the yeast transcription factor GAL4 is activated by local genomic enhancers, and a secondary construct, which contains the GAL4 recognition sequence (upstream activation sequence UAS) adjacent to a cloning site upstream of any genetic construct of interest. When the enhancer is active, GAL4 is expressed and activates transcription of the construct downstream of the UAS (see Table 1.1 for some examples germane to this review). Researchers routinely use a UAS-lacZ or UAS-GFP reporter to characterize the temporal and spatial activation pattern of a new P{GAL4} insert (for review, see Brand and Dormand 1995; for examples in the adult brain, see www.fly-trap.org). Once characterized, a known P{GAL4} strain can then be crossed to fly strains

Table 1.1 Fly stocks carrying GAL4-responsive genes.

Fly Line	Description	Use	Localization
Reporter Genes			
UAS-*nuclear LacZ*	*E.coli* β-Galactosidase with a nuclear localization signal	Nuclear reporter protein	Nucleus
UAS-*eGFPnuclear*	Enhanced Green Fluorescent Protein (eGFP) with a nuclear localization signal	Fluorescent nuclear reporter protein	Nucleus
UAS-*mCD8::GFP*	Fusion protein between mouse lymphocyte marker CD8 and GFP	Outstanding labelling of neuronal processes	Membrane
Cell Death Inducer			
UAS-*hid*	*head involution defective* expression	Induces cell death	
UAS-*rpr*	*reaper* expression	Induces cell death	
UAS-*p35*	P35 expression	Rescues cell death induced by *hid*	
Targeted Suppression of Neuronal Activity			
UAS-*TeTxLC*	Tetanus Toxin light chain	Block synaptic vesicle release	Synapse
UAS-*Shi*^{1st}	Temperature sensitive mutation in *Shibire* (Dynamin)	Deplete synaptic vesicles Conditional block of neurotransmission	Site of endo-cytosis Cell membrane
UAS-*dORKΔ-C*	Constitutively open K^+-specific rectifier channel	Electrically silences neurons	Membrane
UAS-*dORKΔ-NC*	Inactivated K^+-specific channel	Does not conduct K^+ Used as control	Membrane

containing any or multiple UAS constructs. Collections of P{GAL4} lines (for example, an adult brain collection on-line at www.fly-trap.org) can thus be reused in multiple studies.

1.4.2 Using Enhancer-trap reporters for uncovering neuronal substructures

Given that enhancer-trap constructs including the P{GAL4} elements described above are reporting gene-expression, they reflect functional properties of the neurons they are active in. This has been used to suggest novel functional subdivision within neural structures that is not outwardly apparent on morphological examination alone (e.g. Yang *et al.* 1995; Han *et al.* 1996). These constructs have also been used to visualize neuronal development, but in this type of study the consequences of developmental changes in gene regulation have to be taken into consideration (e.g. Tettamanti *et al.* 1997; Ito *et al.* 1997). More recently, the Luo lab developed an elegant system they call MARCM, Mosaic Analysis with a Repressible Cell Marker, to examine fundamental issues of neuronal development (for review see Lee and Luo 2001). Induction of single-cell and two-cell clones at various time points during development allows the researcher to determine the projection patterns of any neuron or group of neurons of interest, that are generated at different stages during CNS development. This strategy has been elegantly employed in analyzing the development of the antennal lobes and mushroom bodies in *Drosophila* (Lee *et al.* 1999; Jeffris *et al.* 2001; Marin *et al.* 2002).

The intrinsic biological value of a defined set of GAL4 lines is enormous. The next step is functional analysis aimed at manipulating the functioning of subsets of relevant neurons and assaying the behavioural and developmental consequences.

1.4.3 Functional analysis of neurons in the CNS

Can various steps in a behavioural pathway be assigned to specific subsets of cells in the CNS that are expressing a gene of interest? Again in *Drosophila* the GAL4\UAS system can be exploited to ask whether particular steps in a behaviour can be associated with particular groups of neurons expressing candidate proteins. Approaches include:

- Selectively perturbing the molecular and cellular components of neurons of interest and determine how these perturbations affect its function in a given behaviour.
- Directed expression of the gene of interest in subsets of the cells in specific mutant backgrounds. Can distinct features of the mutant phenotype be rescued by *wild-type* expression in particular subsets of cells?

Control and experimental animals can then be assayed for the full spectrum of behavioural consequences that you are interested in *e.g.* circadian rhythms, learning and memory.

1.4.3.1 Targeted suppression of neuronal activity

In *Drosophila* it is possible to suppress neuronal activity in any neuron by inhibiting synaptic and electrical activity (review, White *et al.* 2001). All of these methodologies exploit the GAL4/UAS system. Firstly, the tetanus neurotoxin light chain (TeTxLC) can block synaptobrevin-dependent neurotransmitter release (Martin *et al.* 2002, Keller *et al.* 2002) and has been used in a variety of functional studies of specific neurons and behavioural roles (Reddy *et al.* 1997, Tissot *et al.* 1998, Heimbeck *et al.* 1999, Kaneko *et al.* 2000, Blanchardon *et al.* 2001, Suster *et al.* 2003). Secondly, the *shibire*^{ts1} mutation in *Drosophila*, which has a temperature-sensitive block in vesicle recycling due to a defective GTPase, dynamin (Kitamoto 2001; 2002*b*). These are potent tools for suppressing synaptic activity. However since the latter method allows perturbation of the neuronal activities rapidly and reversibly in a spatially and temporally restricted manner, it would seem the most effective method to study the functional significance of particular neuronal subsets in the behaviour of intact flies. Indeed Waddell *et al.* (2000) used it to demonstrate the role of two specific neurons in memory processes. Further studies by McGuire *et al.* (2001) and Dubnau *et al.* (2001) used the same system in Kenyon Cells and confirmed their link to learning processes. Kitamoto (2002*a*) later applied the same techniques to the control of the male courtship ritual. However, the *shibire* system is not ideal under all circumstances. Firstly, we may want temporal control over other processes, not just synaptic activity. Secondly, the use of temperature as a controlling factor is not ideal in some situations. Notably, at least two *Drosophila* behavioural assays use temperature: the flight simulator (Wolf and Heisenberg 1990) and the 'heat box' spatial learning assay (Wustmann and Heisenberg 1997) both frequently use high temperature in their training modes. An alternative is provided by recent developments in enhancer-trap systems that require co-factors before becoming active (see 1.4.4).

To address the role of electrical activity of a given set of neurons in a specific behaviour, one can employ a method for neuronal electrical silencing based upon UAS/GAL4-mediated expression of either of two distinct K^+ channels, UAS-*dORK-DC* and UAS-*dORK-DNC* (Nitabach *et al.* 2002). Such manipulations of membrane properties have been shown to be highly effective at shunting synaptic inputs and silencing activity in *Drosophila* excitable cells (Baines *et al.* 2001, Paradis *et al.* 2001, White *et al.* 2001). This approach allows one to suppress (or enhance) not only neurotransmission, but also other electrical processes involved in the modulation and integration of inputs or the encoding of outputs.

1.4.3.2 Phenocopying by RNAi

RNA interference (RNAi) can specifically inactivate a subset of proteins synthesized from genes encoding alternatively spliced mRNAs (Celotto and Graveley 2002). In theory it is possible to use an RNAi strategy to selectively degrade specific alternatively spliced mRNA isoforms. 'Heritable'-RNAi, utilizing the GAL4/UAS system (Kennerdell and Carthew 2000, Fortier and Belote 2000), has been successfully used to investigate larval and prepupal development and more recently adult behavioural rhythms (Martinek and Young 2000, Piccin *et al.* 2000, Kalidas and Smith 2002). One can generate a UAS–gene-of-interest–RNAi transgenic, to further elucidate the role of your gene in a given behaviour. More recently, Dzitoyeva *et al.* (2003) have demonstrated that injecting adult *Drosophila* intra-abdominally with dsRNA against a γ-aminobutyric acid (GABA) B receptor gene, resulted in cell-nonautonomous RNAi.

1.4.3.3 Selective ablation of *Drosophila* neurons

In comparison to inactivation of a neuron, cell ablation may reveal different behavioural and developmental defects. To address this issue, one can use a GAL4 line in conjunction with the UAS-cell death genes *reaper* (*rpr*) and *head involution defective* (*hid*) to ablate your neuron(s) of choice (McNabb *et al.* 1997, Renn *et al.* 1999, Rulifson *et al.* 2002, Park *et al.* 2003). In addition, p35 encodes a caspase inhibitor that can rescue *rpr*- or *hid*-mediated cell death (Zhou *et al.* 1997).

1.4.4 Targeted expression systems requiring co-factors

A major problem associated with the GAL4/UAS system is that often early dominant effects of mis- or overexpressed transgenes can preclude behavioural analysis in adult animals. Fortunately, there are now several methods for achieving temporal control of transgene activity beyond that inherited from the donor promoter/enhancer sequences.

Roman *et al.* (2001) have developed a conditional system, P{Switch}, which permits temporal as well as spatial control over a given UAS-transgene (Roman *et al.* 2001, Osterwalder *et al.* 2001). For temporal control they exploit Gene-Switch; the DNA binding domain encoding region of the GAL4 gene is fused to the ligand-binding domain encoding region of the human progesterone receptor and the p65 transcriptional domain (Burcin *et al.* 1999). Combined with *P*-element enhancer detection, for spatial control, this results in a system that is inactive without the progesterone analogue RU486. Once active the expression is presumed to be efficient, since it is amplified through the UAS steps of the transgene expression process. A further alternative is the use of the GAL4 inhibitor protein, GAL80. Expression of GAL80 can be used to block GAL4 activity (Lee and Luo 1999).

1.5 Neurophysiology

One of the drawbacks of neuroscience research in *Drosophila* is the difficulty in performing traditional electrophysiological techniques due to the small neuron size. These techniques can be performed in *Drosophila,* but research is largely restricted to larger neurons and is performed by a few specialized research groups (for a more comprehensive review, see Rohrbough *et al.* 2003). Of course, one can use fluorescent dyes such as fura-2 that are sensitive to cellular events such as changes in calcium concentration or voltage, but these are not ideal since loading into nervous tissue is technically challenging as the dye is non selective (for some examples, see Karunanithi *et al.* 1997, Wang *et al.* 2001, Berke and Wu, 2002).

The first use of a transgenic reporter for neuronal activity in *Drosophila* was described by Rosay *et al.* (2001). They used a UAS-aequorin construct that could be targeted into neuron groups using the P{GAL4} system discussed above. Aequorin is a luminescent protein that when combined with a luminophore co-factor (coelenterazine) emits photons in response to increases in calcium concentration. The technique is ideally suited to looking at slow changes (over seconds rather than milliseconds) taking place across groups or populations of neurons such as the Kenyon cells in the mushroom bodies.

Although sensitive to a relatively wide window of calcium concentration, the luminescent nature makes signal detection technically challenging, thus limiting the temporal and spatial resolution of the technique. Spatial resolution can be discarded if the GAL4 driver used is specific to the neurons of interest, and Rosay *et al.* (2001) describe temporal resolution of around 10Hz. With modified equipment, higher resolution (estimated around 50Hz) should be possible. If spatial resolution is required, very high sensitivity optics and imaging devices are essential and the temporal resolution drops dramatically. The technique also requires the addition of a co-factor (coelenterazine) for luminescence to occur. This was achieved in the study of Rosay *et al.* by adding the co-factor directly to the culture media and bathing a dissected brain, or by infusion through a small incision in the head capsule. An entirely non-invasive approach would be preferable and would potentially allow us to monitor neuronal activity in a behaving animal.

Several transgenic reporters based on modified GFP constructs that either use calcium-dependent fluorescent resonance energy transfer (FRET) or where the fluorescence of a single, modified GFP molecule is directly affected by calcium concentration, have been described in *Drosophila*. This technology has started to expand rapidly with the emergence of real-time calcium imaging in a variety of regions of the nervous system using transgenic chameleon constructs targeted using GAL4 to the tissue of choice in the developing and adult fly (e.g. Fiala *et al.* 2002, Reiff *et al.* 2002, Liu *et al.* 2003). Additional tools with altered sensitivity and alternative technical requirements (direct fluorescence measurement rather than via FRET) are also now available such as Camgaroo or G-CaMP (Yu *et al.* 2003, Wang *et al.* 2003). Key advantages over the aequorin system include the ability to report on several different molecules and a higher degree of temporal and spatial resolution as these can be viewed using fluorescent microscopy (commonly using multi-photon confocal microscopy).

1.6 Neuroinformatics and high-throughput technologies

1.6.1 Resources and Tools

One of the many advantages of using *Drosophila* is the wealth of informatics resources available to support established and new research projects. This is a constantly developing field but some of these resources are of particular interest to research groups in neuroscience:

1. FlyBase http://www.flybase.net/
2. BDGP http://fruitfly.org/
3. EDGP http://edgp.ebi.ac.uk/

These resources, and FlyBase in particular, also have extensive supporting documentation, data collections and links to other on-line resources. Several collections of fly strains useful to neuroscientists are available and FlyBase maintains an up-to-date set of links to these resources.

It should be noted that there is a large amount of neuroscience relevant data and resources embedded in the core genetic databases mentioned above. More specific collections and databases of enhancer-trap strains pre-screened for expression in the *Drosophila* nervous system exist (see www.fly-trap.org/ and Hayashi *et al.* 2002).

There have been several attempts at producing atlases of the *Drosophila* adult and pre-adult nervous systems and databases of gene expression patterns. The main on-line atlas for the *Drosophila* nervous system is Flybrain (www.flybrain.org), which contains an exemplary set of annotated reduced silver and autofluorescent head sections from the adult fly. Recent developments have the potential to revolutionize this aspect of research.

The Standard Brain system developed by the Heisenberg lab (Rein *et al.* 2002; Fig. 1.1) uses an anti-synaptic antibody to visualize the general regions of the brain. Using this information as a counter-stain for brain structure allowed them to manually segment the main brain regions in a number of genetic backgrounds. Their result was a statistical volumetric map of the key regions of the adult brain. This system facilitates the quantitative analysis of introducing mutations into the development of the brain.

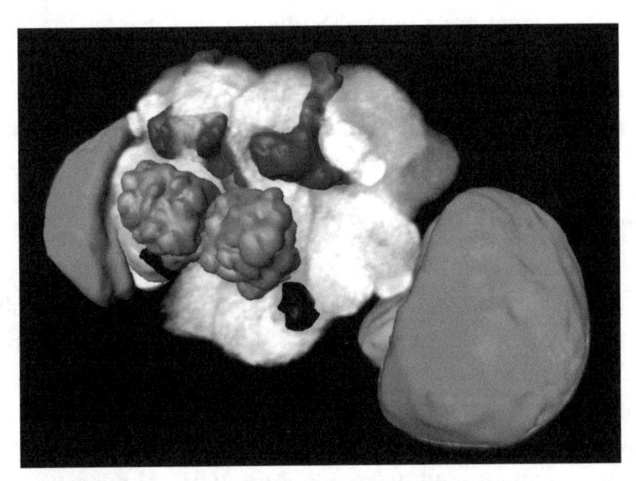

Fig. 1.1 StandardBrain representation of the adult *Drosophila* Brain.

An example image produced by the StandardBrain package (Rein *et al.* 2002) in greyscale. An anti-synaptic antibody (nc82) is used to get a general overview of the nervous system, which is then mapped onto a reference volume. Key brain structures are semi-automatically segmented and statistical information displayed. Image courtesy of Arnim Jenett and Martin Heisenberg (University of Würzburg, Germany).

It also allows for a degree of automatic annotation of double-labelled brain structure. Although the current system was developed in the adult central brain and optic lobes, it is presumably applicable to other regions and stages of the *Drosophila* nervous system and the general technique ultimately to other species.

The standard brain technique is good for the synaptic neurophil (brain regions formed from dendritic and axonal processes) but is less useful for the pericarya (cell body layers). In *Drosophila*, the neuronal cell bodies form a rind on the outer surface of the brain tissue and the neurites project into the brain mass before splitting to form dendritic and axonal processes. As gene expression is generally measured at the cell body/nucleus level, methods for identifying neuronal cell bodies would clearly be very useful in this post-genomic age. A new imaging technique that combines multiple 3D image stacks taken at high resolution from a single brain by multi-photon microscopy has been developed (Ponomarev and Davis 2003, *in press*). This approach allows individual, neuronal cell bodies to be resolved at very high resolution.

The two techniques described above, particularly if combined, represent an unparalleled opportunity to map an entire nervous system in a relatively complex behaving animal for the first time.

1.6.2 Functional Genomics and Proteomics

The promise of high throughput functional genomic and proteomic approaches is just starting to materialize in *Drosophila* neuroscience. Several recent studies demonstrate the ability of microarray technology to identify new candidate genes in neuron function (McDonald and Rosbash 2001; Toma *et al.* 2002; Dubnau *et al.* 2003; Johnson *et al.* 2003). The inherent noise in this technology makes inferring gene networks/ cascades directly from the data a particularly difficult challenge (e.g. Friedman *et al.* 2000). However, the long history of genetic analysis in *Drosophila* provides us with a few clues (around 10k known gene interactions — www.flybase.net) that can be used to help build, refine, and test gene network predictions.

1.6.3 Molecular modelling.

One of the key challenges in functional genomics and in high throughput proteomics is to understand the data produced — it is noisy and complex. The analysis tools for reconstructing gene networks given a series of microarray experiments are as yet in their infancy. Whilst some success has been achieved using probabilistic (e.g. Friedman *et al.* 2000) approaches, the assessment of derived networks is problematic. In this respect, research in *Drosophila* has an advantage. FlyBase catalogues an extensive list of known gene interactions (around 10,000 as of April 2003). Although the gene networks predicted by inference techniques are not directly comparable with the gene interactions derived from analysis of mutants, a degree of commonality should exist. Thus, the genetic interaction map for *Drosophila* could be used to assess the validity of predicted gene networks or, it could be used as prior information for supervised learning techniques.

1.7 **Conclusions**

Neuroscientists study the nervous system from the level of molecular genetics, to single cells (neurons), to systems, to behaviours. Molecular genetics, electrophysiology, pharmacology, anatomical techniques, microscopy, and behavioural tests are among the many approaches possible used to study the nervous system. The choice of model system in neuroscience requires a detailed analysis of the system itself, the tools available and appropriateness of the model for the study in mind. We have presented this overview as an introduction to the use of *Drosophila* as a system for studying neuroscience across multiple levels. The simple organism has some distinct advantages over many others yet at the molecular and genetic level, the nervous system shows a high degree of evolutionary conservation. There are a few technical limitations, largely in neurophysiological techniques but these are rapidly being addressed and recent transgenic developments have the potential to more than address the balance. *Drosophila* may not be ideal for every study, but where appropriate it is a fast, cost-effective, and extremely powerful model system.

References

Allen, M. J., O'Kane, C. J., and Moffat, K. G. (2002) Cell ablation using wild-type and cold-sensitive ricin-A chain in Drosophila embryonic mesoderm. *Genesis*, **34**, 132–4.

Armstrong, J. D., de Belle, J. S., Wang, Z., and Kaiser, K. (1998) Metamorphosis of the mushroom bodies; large-scale rearrangements of the neural substrates for associative learning and memory in Drosophila. *Learn Mem.*, **5**, 102–14.

Ashburner, M. (1989) Drosophila. A laboratory manual. Cold Spring Harbor Laboratory Press, New York.

Berke, B. and Wu, C. F. (2002) Regional calcium regulation within cultured *Drosophila* neurons: effects of altered cAMP metabolism by the learning mutations *dunce* and *rutabaga*. *J. Neurosci.*, **22**, 4437–47.

Blanchardon, E., Grima, B., Klarsfeld, A., Chelot, E., Hardin, P. E., Preat, T., *et al.* (2001) Defining the role of *Drosophila* lateral neurons in the control of circadian rhythms in motor activity and eclosion by targeted genetic ablation and PERIOD protein overexpression. *Eur. J. Neurosci.*, **13**, 871–88.

Brand, A. H. and Dormand, E. L. (1995) The GAL4 system as a tool for unravelling the mysteries of the *Drosophila* nervous system. *Curr. Opin. Neurobiol.*, **5**, 572–8.

Brand, A. H. and Perrimon, N. (1993) Targeted gene expression as a means of altering cell fates and generating dominant phenotypes. *Development*, **118**, 401–15.

Celotto, A. M. and Graveley, B. R. (2002) Exon-specific RNAi: a tool for dissecting the functional relevance of alternative splicing. *RNA*, **8**, 718–24.

Dubnau, J., Grady, L., Kitamoto, T., and Tully, T. (2001) Disruption of neurotransmission in Drosophila mushroom body blocks retrieval but not acquisition of memory. *Nature*, **411** 476–80.

Dubnau, J., Chiang, A. S, Grady, L., Barditch, J., Gossweiler, S., McNeil, J., *et al.* (2003) The staufen/pumilio pathway is involved in *Drosophila* long-term memory. *Curr. Biol.*, **13**, 286–96.

Dzitoyeva, S., Dimitrijevic, N., and Manev, H. (2003) Gamma-aminobutyric acid B receptor 1 mediates behavior-impairing actions of alcohol in *Drosophila*: adult RNA interference and pharmacological evidence. *Proc. Natl. Acad. Sci. USA*, **100**, 5485–90.

Fiala, A., Spall, T., Diegelmann, S., Eisermann, B., Sachse, S., Devaud, J. M., *et al.* (2002) Genetically expressed chameleon in *Drosophila melanogaster* is used to visualize olfactory information in projection neurons. *Curr. Biol.*, **12**, 1877–84.

Fortier, E. and Belote, J. M. (2000) Temperature-dependent gene silencing by an expressed inverted repeat in *Drosophila. Genesis*, **26**, 240–4.

Friedman, N., Linial, M., Nachman, I., and Pe'er D. (2000) Using Bayesian networks to analyze expression data. *J. Comput. Biol.*, **7**, 601–20.

Greenspan, R. J. and Tully, T. (1994) In *Flexibility and Constraint in Behavioral Systems* (ed. Greenspan, R. J. and Kyriacou, C. P.), pp. 65–80. Dahlem Konferenzen, Berlin.

Greenspan, R. J. (1997). *Fly Pushing: The Theory and Practice of Drosophila Genetics.* Cold Spring Harbor Laboratory Press, New York.

Han, P. L., Meller, V., and Davis, R. L. (1996) The *Drosophila* brain revisited by enhancer detection. *J. Neurobiol.*, **31**, 88–102.

Hayashi, S., Ito, K., Sado, Y., Taniguchi, M., Akimoto, A., Takeuchi, H., *et al.* (2002) GETDB, a database compiling expression patterns and molecular locations of a collection of Gal4 enhancer traps. *Genesis*, **34**, 58–61

Heimbeck, G., Bugnon, V., Gendre, N., Haberlin, C., and Stocker, R. F. (1999) Smell and taste perception in Drosophila melanogaster larva: toxin expression studies in chemosensory neurons. *J. Neurosci.*, **19**, 6599–609.

Ito, K., Awano, W., Suzuki, K., Hiromi, Y., and Yamamoto, D. (1997) The *Drosophila* mushroom body is a quadruple structure of clonal units each of which contains a virtually identical set of neurones and glial cells. *Development*, **124**, 761–71.

Jefferis, G. S., Marin, E. C., Stocker RF, and Luo L. (2001) Target neuron prespecification in the olfactory map of *Drosophila. Nature*, **414**, 204–8.

Johnson, E. C., Garczynski, S. F., Park, D., Crim, J. W., Nassel, D. R., and Taghert, P. H. (2003) Identification and characterization of a G protein-coupled receptor for the neuropeptide proctolin in Drosophila melanogaster. *Proc. Natl. Acad. Sci. USA*, **100**, 6198–203.

Kalidas, S. and Smith, D. P. (2002). Novel genomic cDNA hybrids produce effective RNA interference in adult *Drosophila. Neuron*, **33**, 177–84.

Kaneko, M., Park, J. H., Cheng, Y., Hardin, P. E., and Hall, J. C. (2000) Disruption of synaptic transmission or clock-gene-product oscillations in circadian pacemaker cells of *Drosophila* cause abnormal behavioral rhythms. *J. Neurobiol.*, **43**, 207–33.

Karunanithi, S., Georgiou, J., Charlton, M. P., and Atwood, H. L. (1997) Imaging of calcium in Drosophila larval motor nerve terminals. *J. Neurophysiol.*, **78**, 3465–7.

Keller, A., Sweeney, S. T., Zars, T., O'Kane. C. J., and Heisenberg, M. (2002) Targeted expression of tetanus neurotoxin interferes with behavioral responses to sensory input in Drosophila. *J. Neurobiol.*, **50**, 221–33.

Kennerdell, J. R. and Carthew, R. W. (2000) Heritable gene silencing in *Drosophila* using double-stranded RNA. *Nature Biotechnol.*, **18**, 896–8.

Kitamoto, T. (2001) Conditional modification of behavior in Drosophila by targeted expression of a temperature-sensitive shibire allele in defined neurons. *J. Neurobiol.*, **47**, 81–92.

Kitamoto, T. (2002*a*) Conditional disruption of synaptic transmission induces male-male courtship behavior in Drosophila. *Proc. Natl. Acad. Sci. USA*, **99**, 13232–7.

Kitamoto, T. (2002*b*) Targeted expression of temperature-sensitive dynamin to study neural mechanisms of complex behavior in *Drosophila. J. Neurogenet.*, **4**, 205–28.

Lee, T. and Luo, L. (2001) Mosaic analysis with a repressible cell marker (MARCM) for *Drosophila* neural development. *Trends Neurosci.*, **24**, 251–4.

Lee, T., Lee, A., and Luo, L. (1999) Development of the *Drosophila* mushroom bodies: sequential generation of three distinct types of neurons from a neuroblast. *Development,* **126,** 4065–76.

Lin, Y. J., Seroude, L., and Benzer, S. (1998) Extended life-span and stress resistance in the *Drosophila* mutant methuselah. *Science,* **282,** 943–6.

Liu, L., Yermolaieva, O., Johnson, W. A, Abboud, F. M, and Welsh, M. J. (2003) Identification and function of thermosensory neurons in *Drosophila* larvae. *Nat. Neurosci.,* **6,** 267–73.

Lukacsovich, T. and Yamamoto, D. (2001) Trap a gene and find out its function: toward functional genomics in *Drosophila. J. Neurogenet.,* **15,** 147–68.

Marin, E., Jefferis, G. S., Komiyama, T., Zhu, H., and Luo, L. (2002) Representation of the glomerular olfactory map in the *Drosophila* brain. *Cell,* **109,** 243–55.

Martin, J. R., Keller, A., and Sweeney, S. T. (2002) Targeted expression of tetanus toxin—a new tool to study the neurobiology of behavior. *Adv. Genet.,* **47,** 1–47.

Martinek, S. and Young, M. W. (2000) Specific genetic interference with behavioral rhythms in *Drosophila* by expression of inverted repeats. *Genetics,* **156,** 1717–25.

McDonald, M. J. and Rosbash, M. (2001) Microarray analysis and organization of circadian gene expression in *Drosophila. Cell,* **107,** 567–78.

McGuire, S. E., Le, P. T., and Davis, R. L. (2001) The role of *Drosophila* mushroom body signaling in olfactory memory. *Science,* **293,** 1330–3.

McNabb, S. L., Baker, J. D., Agapite, J., Steller, H., Riddiford, L. M., and Truman, J. W. (1997) Disruption of a behavioral sequence by targeted death of peptidergic neurons in *Drosophila. Neuron,* **19,** 813–23.

Moffat, K. G., Gould, J. H., Smith, H. K., and O'Kane, C. J. (1992) Inducible cell ablation in *Drosophila* by cold-sensitive ricin A chain. *Development,* **114,** 681–7.

Osterwalder, T., Yoon, K. S., White, B. H., and Keshishian, H. (2001) A conditional tissue-specific transgene expression system using inducible GAL4. *Proc. Natl. Acad. Sci. USA,* **98,** 12596–601.

Park, J. H., Schroeder, A. J., Helfrich-Forster, C., Jackson F. R., and Ewer, J. (2003) Targeted ablation of CCAP neuropeptide-containing neurons of *Drosophila* causes specific defects in execution and circadian timing of ecdysis behavior. *Development,* **130,** 2645–56.

Piccin, A., Salameh, A., Benna, C., Sandrelli, F., Mazzotta, G., Zordan, M., *et al.* (2001) Efficient and heritable functional knock-out of an adult phenotype in *Drosophila* using a GAL4-driven hairpin RNA incorporating a heterologous spacer. *Nucleic Acids Res.,* **29,** 55–65.

Ponomarev, A. and Davis, R. L. (2003) An adjustable-threshold algorithm for the identification of objects in three-dimensional images. *Bioinformatics,* **19,** 1431–5.

Reddy, S., Jin, P., Trimarchi, J., Caruccio, P., Phillis, R., and Murphey, R. K. (1997) Mutant molecular motors disrupt neural circuits in *Drosophila. J. Neurobiol.,* **33,** 711–23.

Reiff, D. F., Thiel, P. R., and Schuster, C. M. (2003) Differential regulation of active zone density during long-term strengthening of *Drosophila* neuromuscular junctions. *J. Neurosci.,* **22,** 9399–409.

Rein, K., Zockler, M., Mader, M. T., Grubel, C., and Heisenberg, M. (2002) The *Drosophila* standard brain. *Curr. Biol.,* **12,** 227–31.

Renn, S. C., Park, J. H., Rosbash, M., Hall, J. C., and Taghert, P. H. (1999) A *pdf* neuropeptide gene mutation and ablation of PDF neurons each cause severe abnormalities of behavioral circadian rhythms in *Drosophila. Cell,* **99,** 791–802.

Rogina, B., Reenan, R. A., Nilsen, S. P., and Helfand, S. L. (2000) Extended life-span conferred by co-transporter gene mutations in Drosophila. *Science,* **290,** 2137–40.

Rohrbough, J., O'Dowd, D. K., Baines, R. A., and Broadie, K. (2003) Cellular bases of behavioral plasticity: establishing and modifying synaptic circuits in the *Drosophila* genetic system. *J. Neurobiol.,* **54,** 254–71.

Roman, G., Endo, K., Zong, L., and Davis R. L. (2001) P[Switch], a system for spatial and temporal control of gene expression in Drosophila melanogaster. *Proc. Natl. Acad. Sci. USA*, **98**, 12602–7.

Rosay, P., Armstrong, J. D, Wang, Z., and Kaiser, K. (2001) Synchronized neural activity in the Drosophila memory centers and its modulation by amnesiac. *Neuron*, **30**, 759–70.

Rulifson E. J., Kim S. K., and Nusse R. (2002) Ablation of insulin-producing neurons in flies: growth & diabetic phenotypes. *Science*, **296**, 1118–20.

Sokolowski, M. B. (2001) Drosophila: genetics meets behaviour. *Nat. Rev. Genet.*, **11**, 879–90.

Sullivan, W., Ashburner, M., and Hawley R. S. (ed.) (2000) *Drosophila Protocols*. Cold Spring Harbor Laboratory Press, New York.

Suster, M. L., Martin, J. R., Sung, C., and Robinow, S. (2003) Targeted expression of tetanus toxin reveals sets of neurons involved in larval locomotion in *Drosophila. J. Neurobiol.*, **55**, 233–46.

Tettamanti, M., Armstrong, J. D., Endo, K., Yang, M. Y., Furukubo-Tokunaga, K., Kaiser, K., *et al.* (1997) Early development of the *Drosophila* mushroom bodies, brain centres for associative learning and memory. *Dev. Genes Evol.*, **207**, 242–52

Tissot, M., Gendre, N., and Stocker, R. F. (1998) *Drosophila* P[Gal4] lines reveal that motor neurons involved in feeding persist through metamorphosis. *J. Neurobiol.*, **37**, 237–50.

Toma, D. P., White, K. P., Hirsch, J., and Greenspan, R. J. (2002) Identification of genes involved in *Drosophila* melanogaster geotaxis, a complex behavioral trait. *Nat. Genet.*, **31**, 349–53.

Waddell, S., Armstrong, J. D., Kitamoto, T., Kaiser, K., and Quinn, W. G. (2000) The amnesiac gene product is expressed in two neurons in the *Drosophila* brain that are critical for memory. *Cell*, **103**, 805–13.

Wang, J. W., Wong, A. M., Flores, J., Vosshall, L. B., and Axel, R. (2003) Two-photon calcium imaging reveals an odor-evoked map of activity in the fly brain. *Cell*, **112**, 271–82.

Wang, Y., Wright, N. J., Guo, H., Xie, Z., Svoboda, K., Malinow, R., *et al.* (2001) Genetic manipulation of the odor-evoked distributed neural activity in the *Drosophila* mushroom body. *Neuron*, **29**, 267–76.

White, B., Osterwalder, T., and Keshishian, H. (2001) Molecular genetic approaches to the targeted suppression of neuronal activity. *Curr. Biol.*, **11**, 1041–53.

Wolf, R. and Heisenberg, M. (1990) Visual control of straight flight in *Drosophila melanogaster. J. Comp. Physiol. [A]*, **167**, 269–83.

Wustmann, G. and Heisenberg, M. (1997) Behavioral manipulation of retrieval in a spatial memory task for *Drosophila melanogaster. Learn Mem.*, **4**, 328–36.

Yang, M. Y., Armstrong, J. D., Vilinsky, I., Strausfeld, N. J., and Kaiser, K. (1995) Subdivision of the *Drosophila* mushroom bodies by enhancer-trap expression patterns. *Neuron*, **15**, 45–54.

Yu, D., Baird, G. S., Tsien, R. Y., and Davis, R. L. (2003) Detection of calcium transients in *Drosophila* mushroom body neurons with camgaroo reporters. *J. Neurosci.*, **23**, 64–72.

Zhou, L., Schnitzler, A., Agapite, J., Schwartz, L. M., Steller, H., and Nambu, J. R. (1997) Cooperative functions of the *reaper* and *head involution defective* genes in the programmed cell death of *Drosophila* central nervous system midline cells. *Proc. Natl. Acad. Sci. USA*, **94**, 5131–6.

Chapter 2

Using mouse genetics to study neuronal development and function

Harald Jockusch and Thomas Schmitt-John

2.1 Why the mouse?

Historically, the mouse has achieved the status of a genetic model for biomedical research through immunogenetics (cf. Green 1975). The role of genetics in elucidating the development and function of the nervous system seemed less obvious. Accordingly, the initial decades of neurobiology were dominated by pharmacological and surgical methods. The role of genetics in the function and long-term maintenance of the central nervous system (CNS) became evident in neuropathology, due to hereditary neurological diseases. Presently, there appear to be four main areas in which the mouse is used as a genetic system in neurobiology: (1) development of the nervous system, especially compartmentation and axon pathfinding, (2) glia–neuron interaction including myelination, (3) ion channels and signal transduction in neurological disease and (4) cognition and memory formation. All of these have important medical implications, so that a mammal seems to be the model of choice. Within the higher mammals (*Eutheria*), the mouse as a rodent is not particularly closely related to man, the last common ancestor being estimated to have lived 70 million years ago. However, in contrast to more closely related mammals like insectivores or primates, the mouse has unsurpassed qualities as a laboratory animal, especially for genetic analysis. There is little seasonal influence on mating, the intrauterine development takes only 19 days, litters are fairly large (6 to 14 babies), and, despite being born at a fetus-like stage, mice are sexually mature eight weeks postnatally. Thus, the life cycle can be completed within 11 weeks, and one can achieve four generations within one year. It is remarkable for a mammal, and mostly due to a continuous refinement of techniques, how well the mouse is suited for developmental and genetic manipulation. Furthermore, about 200 inbred laboratory strains are available (of which only a subset is commonly used), so that natural gene polymorphisms, the role of genetic background in phenotype expression and the influence of modifier genes can be readily studied. Lastly, sequencing of the human and mouse genomes has shown that the two species share an overwhelming fraction of genes so that functional genomics of both species support each other.

In this chapter, the methods available for the genetic analysis and manipulation of the mouse will be briefly outlined and their application to neurobiology will be illustrated by selected examples.

2.2 Methods useful for mouse neurogenetics

2.2.1 Comparative Genomics

The human nuclear genome comprises 23 chromosomes; 22 autosomes and the sex chromosomes, X and Y. The mouse has 20 chromosomes with 19 autosomes. The former has been sequenced to over 90% (as of June 2002, http://genome.ucsc.edu/), and the draft sequence for the mouse has been published (Waterston *et al.* 2002); http://www.ensembl.org/Mus_musculus/). The presence of several pairs of homologous genes on one given chromosome defines conserved synteny. The most striking example of conserved synteny is the conservation of gene content of mammalian X chromosomes. In any chromosomal segment of conserved synteny the local gene order is usually conserved to a high degree, making it possible to use mapping information from one species to identify a disease gene in the other species.

2.2.2 Classical mouse mutants and positional cloning

Classical mouse genetics relied on spontaneous or radiation-induced mutations (Lyon and Searle 1989) which in most cases lead to recessive 'loss-of-function' phenotypes, but also included cases of 'gain of function', i.e. dominant expression of the mutant phenotype. Radiation-induced mutations are often deletions whereas spontaneous mutations include point mutations, retroposon insertions and chromosomal rearrangements. These 'classical' mutations certainly represent a selective bias: lethal phenotypes early in life were likely to be overlooked, as were phenotypes which were too mild. Mutant loci were mapped using meiotic segregation in relation to markers of known position. In the early days, these were genes affecting the external appearance like coat color, or allelic forms of enzymes; in recent years, due to rapid PCR techniques, these have been replaced by DNA microsatellite polymorphisms. In order to identify the gene the tedious procedure of positional cloning, i.e. high resolution mapping followed by sequencing of sets of overlapping genomic DNA clones is necessary unless there is a physiological hint that allows one to focus on a candidate gene. The availability of a full mouse genome sequence (Waterston *et al.* 2002) will make this procedure much faster and easier.

In human neurological diseases such as *nervus opticus* atrophy and several muscle diseases, defects in mitochondrial genes play an important role. We are not aware of reports on spontaneously arising mitochondrial mutations that would cause neurological diseases in the mouse. However, methods to specifically introduce mutations into the mitochondrial genome have recently been developed (Wallace *et al.* 2001).

Methods to artificially induce mutations, either random or directed (Fig. 2.1) will be described in the following sections.

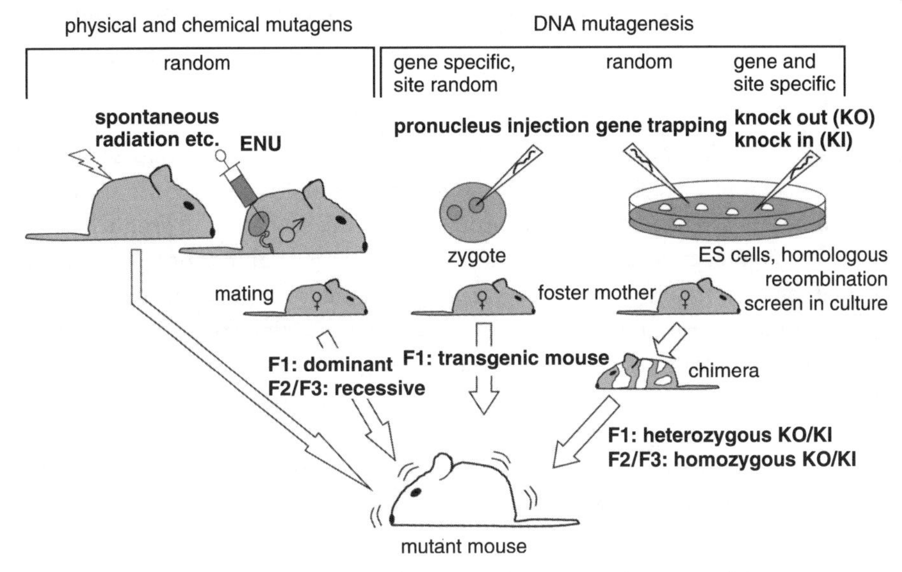

Fig. 2.1 Origin of mutant mice: Spontaneous and induced mutations. From left to right the methods used to obtain mouse mutants become increasingly sophisticated. In radiation or chemically induced mutagenesis whole animals are subjected to the treatment by shotgun methods. Using DNA as a mutagen, one chooses the gene to be introduced, but its insertion may be random (as in transgenes) and its nature may be that of a reporter (as in the 'gene trap'). Whereas transgenes are produced in zygotes, the other manipulations use embryonic stem (ES) cell lines as recipients. In knock-out (KO) and knock-in (KI) technologies, a specific gene is being functionally eliminated or replaced by a variant of interest (usually a human ortholog with a pathogenic mutation). Tissue specific and temporally controlled gene elimination is achieved by crossing mice with 'sensitized' ('floxed') genes to those carrying a 'destructive' Cre recombinase transgene with a cell type specific and/or inducible promoter.

2.2.3 **Induced random mutations: The ENU screening projects**

In the course of recent genome projects the chemical mutagenesis of the mouse had an unexpected comeback. Male mice are treated with a sublethal dosage of ethylnitrosurea (ENU), one of the most powerful chemical mutagens. ENU predominantly causes single basepair exchanges (Popp *et al.* 1983). The offspring of ENU treated males carry multiple paternally inherited point mutations (Balling 2001). To detect neurological mutations in subsequent generations several screening protocols have been set up, such as behavioural, anxiety and pain tests. The common aim of these mutagenesis programs is to saturate the genome with an unbiased spectrum of mutants. Mutant phenotypes are described in publicly available databases (Germany: www.gsf.de/ieg/groups/enu-mouse.html; UK: www.mgu.har.mrc.ac.uk/mutabase/;

Japan: www.gsc.riken.go.jp/Mouse/; USA: www.jax.org/nmf/documents/about.html; www.tnmouse.org/).

2.2.4 Transgenes

In transgenic animals genes of interest are added to the genome of a recipient animal. Expression plasmids (capacity < 20 Kb), which in the simplest case consist of a cloned cDNA downstream to a non-specific promoter, are injected into the pronucleus of a zygote. The zygotes are subsequently transferred into the genital tract of foster mothers. In about ten percent of the injected zygotes the plasmid is integrated into the genome and transgenic offspring are born. The site of transgene insertion is more or less random and usually multiple copies of the transgene are integrated in tandem arrangement. The expression of a transgene depends not only on the promoter used for the construct, but also on the copy number of insertions and on the location of the insertion. For this reason several transgenic lines usually have to be analyzed in parallel. To minimize the influence of the genetic environment on a given transgene it is preferable to insert the transgene including its normal chromosomal environment, in the form of a large genomic DNA-fragment (up to several hundred Kb). Yeast artificial chromosomes (YACs) and bacterial artificial chromosomes (BACs) have been successfully used for this purpose.

2.2.5 Gene trapping—random mutagenesis of embryonic stem cells

Gene trapping is a random insertional mutagenesis method applied to embryonic stem (ES) cells. The approach is based on the insertion of a promoterless reporter gene, usually the *E. coli LacZ* gene coding for β-galactosidase. The expression of the reporter, being under transcriptional control of the unknown gene into which it has been inserted, mimicks and allows visualization of the expression pattern of the 'trapped' gene. This facilitates the screening for genes of interest. Screening can be performed on the ES cell level for genes active at that stage or — with more relevance to neurobiology — after induced differentiation of ES cells in culture. Selected ES cell clones are transferred into blastocysts, and may be screened in the resulting chimeras. If they participate in the germ line heterozygous offspring may be used for the analysis of the expression pattern *in situ* and homozygous mutants obtained in the following generation, in which the insertion would destroy the function of both alleles, are tested for aberrant phenotypes (review: Hill and Wurst 1993).

Skarnes *et al.* (1995) developed a gene-trapping approach (secretion gene trap) that allows a pre-screen of gene loci encoding membrane and secreted proteins. The vector pGT1.8tm carries, from 5′ to 3′, the *engrailed 2* splice acceptor, the CD4 transmembrane domain and a β-geo (*lacZ-neo*^{R} fusion) sequence (Skarnes *et al.*, 1995). When insertions occur into genes, the transcripts of which lack a 5′ secretion signal sequence, the CD4 transmembrane domain acts as a secretion signal, leading to translocation of the carboxy-terminal end of the translation product into the endoplasmic reticulum. In the lumen of the ER, the fusion protein is exposed to conditions, including glycosylation enzymes, which lead to inactivation of β-galactosidase activity. However, insertions into a gene encoding a signal sequence, when they occur 3′ to the signal sequence, lead to the CD4 domain acting as a simple Type I transmembrane domain, terminating further translocation into the ER. This leaves the 3′ transgene product outside of the lumen of the ER in the cytoplasm, providing an active β-galactosidase marker.

2.2.6 **Site-specific deletion mutagenesis**

Any point mutation, spontaneous or ENU induced, may be combined with targeted deletion mutagenesis. Two different strategies have been developed for targeted deletion mutagenesis in mouse ES cells. The first is based on the targeted insertion of two loxP sites (see 2.2.9) into a specific mouse chromosome by homologous recombination in ES cells, followed by Cre recombinase mediated deletion of the DNA segment between the loxP sites (Ramirez-Solis *et al.* 1995). ES cell clones transmitted to the germ line yield hemizygous 'deletion mice'. The second approach utilizes a viral thymidine kinase as a negative selection marker, which is introduced into the genomic area of interest by homologous recombination in ES cells. Thereafter ES cells are subjected to X-irradiation induced deletion mutagenesis followed by selection for the loss of thymidine kinase (You *et al.* 1997). Deletions within the target region are mapped and hemizygous deletion mice are generated.

Hemizygous deletion females are mated with ENU-treated males and the F1 offspring are screened for aberrant phenotypes. Above a background of dominant gain-of-function mutations located anywhere in the mouse genome this approach allows the F1 screening for recessive mutations located within the hemizygous 'deletion window' (www.mouse-genome.bcm.tmc.edu/ENU/MutagenesisProj.asp).

2.2.7 **Knock-outs: Artificial loss-of-function mutations of known genes**

The targeted generation of null-mutations in mice (Thomas and Capecchi 1987) is based on the alteration of a known gene locus by homologous recombination in ES cells. The homologous recombination is driven by a targeting vector which comprises a selection marker (drug resistance) flanked by genomic gene-specific targeting fragments. The targeting vector is transfected into the ES cells via electroporation, resulting clones are selected for drug resistance and screened for the gene-specific homologous recombination event by genomic PCR or Southern blotting. Selected ES cells are injected into blastocysts or aggregated with morulae to yield chimeric blastocysts in culture that are subjected to the procedures described for gene trapping (Fig. 2.1). Fifty percent of the ES cell offspring should be heterozygous for the knock-out. Intercrossing yields 25% of homozygous 'KO mice' for phenotype analysis. Several thousands of genes have been knocked out and their phenotypes have been characterized (tbase.jax.org/). The knock-out strategy is not restricted to proper genes, but might also be useful for the functional characterization of regulatory sequence elements and conserved non-coding sequences, the latter of which have been identified via interspecies sequence comparison (Loots *et al.* 2000).

2.2.8 Knock-ins of human 'pathogenes'

In most cases of knock-out experiments a reporter and/or a selection gene is introduced concomitantly with disruption of the target gene. The knock-in of a 'reporter gene' like *LacZ* or jellyfish green fluorescent protein (GFP) and its artificial variants allows, if the promoterless reporter is correctly controlled by the target gene regulatory sequences, the analysis of the expression pattern of the target gene *in situ*. Apart from this somewhat artificial analytical tool, knock-in technology is used to produce accurate models of human disease. A mouse gene can be replaced by a pathogenic allele of the orthologous human gene. Alternatively an equivalent mutation may be introduced into the mouse gene by *in vitro* mutagenesis and knock-in technology.

2.2.9 Conditional mutations

The value of the knock-out approach for the analysis of complex physiological and behavioural capabilities of an organism has been questioned (Routtenberg 1995). A gene function represents just a node in a complicated network to which other gene functions contribute, modulated by their regulation as well as exogenous influences. If one gene function is missing during the development of an organism, the whole network might react and even compensate for the loss. Thus, the case is certainly different from a 'minus one' music recording. This thought was already implicit in Richardt Goldschmidt's 'theory of gene physiology' (Goldschmidt 1927) long before DNA had been identified as the hereditary material. It not only implies that the loss of a seemingly important gene function may cause no overt abnormalities ("no phenotype") but also that a single gene defect may change the expression of a host of other genes. Another complication of constitutive gene disruption is the fact that during development one and the same gene may be used at different times in different tissues in a different context. The phenotype resulting from its loss may thus be a superposition of physiologically unrelated events.

In order to avoid these complications, more specific gene targetting methods have been developed. They would either eliminate a gene function specifically in one organ or cell type or at a chosen time, or both.

For tissue specific knock-out, targeted mutagenesis in ES cells has to be silent and should not impair gene function. Usually essential exons are 'floxed' by flanking them with loxP sites (short oriented recognition sequences for the phage P1-derived Cre recombinase). Using homologous recombination in ES cells, the floxed exon including a floxed selection marker is introduced into the gene of interest. Targeted ES cell clones are selected and transiently transfected with a Cre-recombinase expression plasmid. Cre recombinase deletes the sequences between two tandemly arranged loxP sites and

ES cell clones have to be selected which have lost the selection cassette but not the essential exon. ES cell clones with the floxed essential exon are used to obtain homozygous floxed mice as described for conventional knock-outs. These mice are bred with Cre recombinase transgenic mice, the transgene of which is regulated by a tissue specific promoter. In Cre-transgenic homozygously floxed mice the essential exon is deleted exclusively in cells which express the Cre and the effects of the gene knock-out in specific tissues and organs can be analyzed.

Inducible knock-outs utilize "on/off" promoter systems for the expression of Cre-recombinase transgenes, which can be regulated externally e.g. by application of effector molecules like tetracyclin or steroid hormones, depending on the artificially introduced regulator system. In this case the Cre-mediated deletion of an essential exon of a gene can be induced at a specific timepoint so that critical developmental stages are overcome and acute effects of the loss of function may be studied: see e.g. the experiments of Malleret *et al.* (2001) and Gross *et al.* (2002) using tetracycline regulated systems.

2.3 Selected neurological mouse mutants

The genetic map positions of a number of spontaneous and induced neurological mutations in the mouse and their human homologs are shown in Fig. 2.2.

2.3.1 Development and Pathfinding

Neural development in all organisms studied is governed by an interplay of transcription factors, cell surface and matrix molecules and signal transduction systems (Yu and Bargmann 2001, Lee and Pfaff 2001, Monuki and Walsh 2001). A number of genes affecting the development and compartmentation of the mammalian CNS were discovered by the knock-out of genes that had been identified as homologs to *Drosophila* developmental genes.

A specific gene trap strategy for the mouse has been developed to identify genes involved in axonal pathfinding, based on the secretion gene trap approach (see 2.2.5; Skarnes *et al.* 1995). In a bicistronic construct, coupled to the LacZ and selection genes, a gene coding for an alkaline phosphatase with a GPI anchor was introduced (Leighton *et al.* 2001). Thus in transgenic animals, there was a double reporter system: the cell bodies of neurons expressed the β-galactosidase and the neurites of the same cells were labelled with the membrane-anchored phosphatase. In heterozygous animals the wild-type axon wiring pattern was visualized whereas in homozygotes, due to a complete loss of the wildtype gene, aberrant neurite orientations could be directly visualized by the phosphatase stain. Thus, among a number of surface and extracellular proteins the transmembrane protein semaphorin 6A was found to be necessary for the correct pathfinding of thalamocortical axons, and the receptor tyrosine kinase EphA4 for pathfinding and chiasma formation in the corticospinal tract.

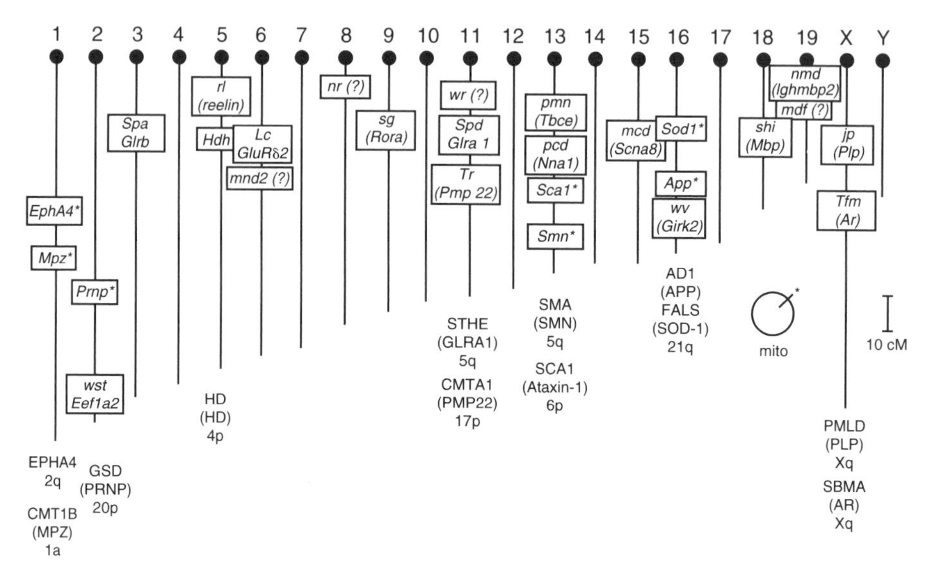

Fig. 2.2 Mutations that cause neurological symptoms in the mouse and orthologous disease loci in humans. The 19 telocentric autosomes and two sex chromosomes of the mouse are shown (the black balls are the centromeres). The circle 'mito' symbolizes the mitochondrial genome (16 Kb), enlarged about 500 times in comparison to the chromosomes. Loci of neurological mutations (with asterisks indicating gene manipulations like KO´s and KI´s) are shown for the mouse. Below the mouse chromosomes the orthologous human neurological diseases are shown with their locations, with chromosome numbers, and p for short, q for long arms of chromosomes. cM, centimorgans (recombination units). For detailed information see http://www.informatics.jax.org/.

2.3.2 Mutations affecting myelin formation and stability

Myelin, the structural prerequisite for rapid saltatory impulse conduction in vertebrates, is formed by glial cells, oligodendrocytes in the central nervous system (CNS) and Schwann cells in the peripheral nervous system (PNS). In humans and rodents myelination takes place postnatally. Myelin deficiency may cause striking behavioural symptoms, such as ataxia and tremor, which reflect the loss of motor control. This late-onset phenotype has allowed the identification of a number of spontaneous mutations with both CNS and PNS dysmyelinations. Most of these have now been molecularly defined as mutations in genes coding for myelin-associated proteins (review: Nave 1994).

Myelin basic proteins (MBPs) are a group of small positively-charged proteins derived from an autosomal gene by alternative mRNA splicing. The *shiverer* mutation of the mouse is a loss-of-function deletion in the MBP gene. The *shiverer* defect, thin and largely uncompacted myelin, has been complemented by a wildtype MBP transgene.

The *jimpy* mouse (*jp*) is affected in the gene for proteolipid protein (PLP), which is the major integral membrane protein of CNS myelin. There is a smaller isoform

DM-20 which is derived by alternative RNA splicing. *jp* is a point mutation leading to a fatal error of mRNA splicing and consequently to a severe CNS-specific dysmyelination (Nave 1994). Other murine alleles are point mutations causing single amino acid substitutions. Over 30 different mutations of the PLP gene have been found in human patients with Pelizaeus-Merzbacher disease. The PLP gene is X-linked in both species.

Mutations of the small glycoprotein PMP22 were first identified in the *Trembler* mouse and later in human patients with a form of *Charcot-Marie-Tooth disease,* type 1A (CMT1A). Homozygous *Trembler* mice have virtually no peripheral myelin (but are fully viable). CMT1A in humans is associated with a duplication of the PMP22 gene.

P0 is a highly abundant glycosylated membrane protein of 30 kD of the PNS where it fulfills functions analogous to those of MBP in the CNS. Elimination of one P0 allele in mice leads to ultrastructural abnormalities reminiscent of myelin in *Charcot-Marie-Tooth* disease type 1B (CMT1B), previously associated with point mutations in the human P0 gene.

The myelin mutants of the mouse thus provide a collection of genocopies for a number of human neurological diseases.

2.3.3 Genetics of excitability and ion channels

The cerebellar mutation 'weaver' (*wv*) had been classified as a developmental mutation because granule cells fail to properly migrate and subsequently degenerate; however, it turned out to be a mutation in a GTP-binding protein regulated potassium channel, GIRK-2, the human gene for which, GIRK2, was already known (Patil *et al.* 1995). This is a case where conserved synteny between mouse and human genomes (cf. Gregory *et al.* 2002, Waterston *et al.* 2002) was helpful to identify the disease gene, in this case in the mouse.

In the case of the neuromuscular mutation 'motor endplate disease' (*med*), several spontaneous alleles were available but finally serendipity was helpful in that an insertional mutation produced the same symptoms, mapped to the same chromosome (Chr 15) and did not complement the authentic *med* mutation. In addition, the expression of the candidate gene specifically in the brain and spinal cord supported the notion of a CNS disease. The *med* locus was cloned, was found to code for a new sodium channel subunit protein, and was therefore renamed *Scna8* for 'sodium channel α subunit 8' (Burgess et al. 1995).

Loss-of-function of inhibitory ion channels may lead to hyperexcitability, as evidenced by seizures. Mutant excitatory ion channels may also produce hyperexcitability, in a gain-of-function mode if their inactivation is delayed (review: Meisler *et al.* 2001). More than 40 genes in humans and mice have been found to be associated with CNS hyperexcitability, i.e. the symptoms of epilepsy. The ion channels found to be responsible by positional cloning range from sodium, potassium and calcium channels (Fig. 2.2) to acetylcholine and GABA receptors.

The glycine receptor of brain stem and spinal cord is a ligand-gated chloride channel consisting of three α and two β subunits, with the glycine binding sites on the α subunits (Davies and Glencorse, this volume chapter 8). Mouse mutations in the genes for either the adult α subunit, *spasmodic,* or the β subunits, *spastic,* result in symptoms similar to those after strychnine poisoning as this alkaloid is an inhibitory analogue of glycine. A hereditary human neurological disease, startle disease (Hyperekplexia, stiff baby syndrome) is due to mutations in the gene for the glycine receptor α subunit (Fig. 2.2).

2.4 **Hereditary neurodegenerative diseases: natural and transgenic models**

In recent years it became evident that there is no common denominator for the genetic basis of neurodegenerative diseases: They may be dominant or recessive, may affect ubiquitously expressed genes, and these may code for enzymes, ion channels or hitherto unknown proteins (review: Bailey, this volume, chapter 15).

In a number of neurodegenerative diseases, e.g. Huntington's disease (HD), spinocerebellar ataxia type I (SCA1), and spinobulbar muscular atrophy (SBMA), CAG triplet repeat expansions are found within the coding region of the mutated genes where they translate into polyglutamine (poly Q) tracts (Brooks 1995). The complete sequence of the human SCA1 gene product 'ataxin-1' is known and antibodies directed against it are available. Although ataxin-1 does not resemble known proteins and its cellular function remains elusive, recent experiments have contributed to the understanding of the etiology of SCA. Transgenic mice have been produced for different alleles (differing in the lengths of their CAG repeats) of the SCA1 gene, conclusively showing that the severity of the neurological symptoms in the transgenic mouse, as in human patients, correlates with the length n of the trinucleotide repeat. In the mouse *Sca1* gene, there are only two adjacent CAGs, $n = 2$. Yet, the repeat length found in normal humans, $n = 30$, is still not pathogenic in mice, but an allele with $n = 82$ that is harmful in humans also causes neurodegeneration in the mouse. Non-neural cells express ataxin-1 in their cytoplasm, and most neurons express it in the nucleus. Purkinje cells express ataxin-1 both in their cytoplasm and nuclei and their specific susceptibility to toxic influences of overlength polyglutamine tracts may be related to the localization of ataxin-1. The present evidence suggests that poly Q localized in the nucleus is toxic due to interaction with other nuclear proteins (Zoghbi and Orr 2000).

The most common human neurodegenerative disease, Werdnig–Hoffmann spinal muscular atrophy (SMA), does not have a homologous counterpart in the mouse. The human SMA gene is located on the short arm of Chr 5 (Chr 5q), and has been termed 'SMN' for 'survival motor neuron (gene)' (Lefebvre *et al.* 1995). The SMN protein is a ubiquitously expressed component of the spliceosome. There is a difference between mouse and man at this locus: whereas the human SMN locus is duplicated and contains SMN1 and SMN2, there is only one SMN gene on Chr 13 of the mouse. Its knock-out leads to early embryonic lethality (Schrank *et al.* 1997), and therefore does not provide a model for the human disease in which SMN2 usually remains intact, whereas SMN1 is subject to deleterious mutations. However, after the introduction of human SMN2 into the mouse, *Smn* KO mice survived and developed symptoms similar to human SMA (Hsieh-Li *et al.* 2000). Recently, tissue-specific knockouts in the mouse have led to surprising results: not only does an SMN KO lead to neurodegeneration in the CNS and atrophic muscle (Frugier *et al.* 2000) but also the muscle specific KO has deleterious effects, confined to muscle (Cifuendes-Diaz *et al.* 2001).

For another human neurodegenerative disease, familial amyotrophic lateral sclerosis (FALS), an affected gene locus has been shown to code for the well-known enzyme cytosolic copper/zinc superoxide dismutase, SOD-1, an ubiquitously expressed enzyme thought to protect cells against oxidative stress. However, the etiology was not clear, as in most patients the catalytic activity of SOD-1 was not affected, and, again no homologous mouse mutants were available to clarify the issue. Recently, a SOD-1 KO-mouse has been produced and found to display no neurological symptoms under standard conditions. Therefore, transgenic mice bearing FALS alleles of the human *SOD1* gene (i.e. a situation analogous to the SCA1 transgenic mouse) were produced. In humans the amino acid replacement glycine to alanine at position 93 does not impair enzyme activity but, in a gain-of-function mechanism, causes neurodegeneration. The same symptom was observed in mice made transgenic for this allele but not in mice transgenic for normal human SOD1 (Gurney *et al.* 1994, review: Julien 2001).

A number of the classical mouse disease genes that affect motoneurons such as 'motoneuron degeneration 2' (*mnd2*, Chr 6[*]) 'wobbler' (*wr*, Chr 11), and 'muscle deficient' (*mdf*, Chr 19) still await positional cloning.

2.5 **Modifier genes and quantitative trait loci (QTL)**

An important issue of current genetics is that of inherited genetic risk factors which may have a dramatic effect on the severity and progression of disease symptoms. For a wide variety of complex human diseases, modifying genetically inherited risk factors have been described (Mackay 2001). These modifiers have been called 'quantitative trait loci' (QTL) because of their quantitative influence on certain diseases or other phenotypes. For many diseases, both inherited and acquired, modifying genes are known. Alzheimer's disease and epilepsy (Legare *et al.* 2000) are modified in the severity of their symptoms by QTLs. In the mouse, the neurodegenerative disease, *wobbler*, can be ameliorated (Kaupmann *et al.* 1992) or aggravated (Ulbrich *et al.* 2002) by modifiers.

2.6 **Genetic control of behaviour, cognitive functions and memory formation**

Considering the criticisms cited above (Section 2.2.9) it might seem hopeless to unravel complex functions of the CNS with the toolbox of mouse molecular genetics.

[*]The *mnd2* gene has recently been identified. It codes for a serine protease (Omi) which is present in all tissues and is localized in mitochondria (Jones *et al.* 2003).

However, from twin studies in humans and the comparison of inbred mouse strains it seems clear that there is a strong genetic component to behaviour and cognitive functions. We know that these capabilities are the result of prenatal and postnatal elaboration of neuronal wiring, the interaction with external influences (plasticity) and acute enzyme, transmitter and hormone levels. In some cases a behavioural trait may be altered by a single biochemical change, as in the case of the increased aggressiveness of NO synthase KO mice (Nelson *et al.* 1995).

To distinguish between developmental influences in the history of the CNS to be studied and the acute balance of biochemical parameters in neurons, a site and time controlled genetic manipulation seems necessary. An example is the study of the influence of the kinase/phosphatase balance on memory formation in mice. The protein kinase A (PKA)/calcineurin (CN) balance was studied to test the hypothesis that the phosphatase CN would counteract protein phosphorylation by PKA, and thus suppress early steps of memory formation. This was achieved by pharmacological interference using a genetic method: a neuron-expressed transgene for a CN inhibitory peptide was introduced, which could be positively regulated by the tetracyclin-related inducer doxycyclin. When 'switched on' the inhibitory transgene would reversibly suppress CN activity by about 40%. It was found that long-term potentiation (LTP), and short and long-term memory, were reversibly enhanced by regulated expression of the phosphatase inhibitor (Malleret *et al.* 2001).

Another example for the application of a region-specific and time-controlled transgene expression is an analysis of the role of the serotonin$_{1A}$ receptor in anxiety-like behaviour (Gross *et al.* 2002). It was known that agonists of this receptor act as anxiolytics and conversely, absence of the serotonin$_{1A}$ receptor in knockout mice increases anxiety-like behaviour. This receptor has a restricted expression pattern in the brain, comprising the hippocampus, septum, and cortex, and the serotonergic neurons of the brainstem raphe nuclei. It was not known where and during what developmental stage of the brain the receptor exerts its function to ensure a normal anxiety level. To answer this question a tissue-specific and time-controlled transgenic rescue (as measured by a normalized anxiety level) of the knockout mouse using a tetracycline regulated construct was established. The effect of the receptor seemed to be focussed in the hippocampus and cortex but not in the raphe nuclei , and during the early postnatal period, but not in the adult.

2.7 Outlook

The future use of the mouse as a model organism for neurogenetics, both for basic research and for the understanding of neurological diseases, will be based on the progress of comparative functional genetics of mouse and man and on evermore refined technologies of genetic manipulation (Mayford et al. 1997). The only hope of elucidating brain functions would be to combine spatially controlled changes in the neuronal wiring pattern with time-controlled changes in relevant biochemical parameters, i.e. transmitter levels, activities of ion channels and receptors, and of components of intracellular signal transduction. In the near future, transgenic or knock-in mice and the identification of modifier loci will be important for drug development and testing as well as for cell-mediated therapies.

References

Balling, R., (2001) ENU mutagenesis: analyzing gene function in mice. *Annu. Rev. Genomics Hum. Genet.*, **2**, 463–92.

Brooks, B. P. F. and Fischbeck, K. H. (1995) Spinal and bulbar muscular atrophy: a trinucleotide-repeat expansion neurodegenerative disease. *Trends Neurosci.*, **18**, 459–61.

Burgess, D. L., Kohrman, D. C., Galt, J., Plummer, N. W., Jones, J. M., Spear, B., *et al.* (1995) Mutation of a new sodium channel gene, Scn8a, in the mouse mutant 'motor endplate disease'. *Nat. Genet.*, **10**, 461–5.

Cifuentes-Diaz, C., Frugier, T., Tiziano, F. D., Lacene, E., Roblot, N., Joshi, V., *et al.* (2001) Deletion of murine SMN exon 7 directed to skeletal muscle leads to severe muscular dystrophy. *J. Cell Biol.*, **152**, 1107–14.

Frugier, T., Tiziano, F. D., Cifuentes-Diaz, C., Miniou, P., Roblot, N., Dierich, A., *et al.* (2000) Nuclear targeting defect of SMN lacking the C-terminus in a mouse model of spinal muscular atrophy. *Hum. Mol. Genet.*, **9**, 849–58.

Goldschmidt, R. (1927) *Physiologische Theorie der Vererbung*, Springer, Berlin.

Green, E. L. (1975) Biology of the laboratory mouse. Dover Publications, New York.

Gregory, S. G., Sekhon, M., Schein, J., Zhao, S., Osoegawa, K., Scott, C. E., *et al.* (2002) A physical map of the mouse genome. *Nature*, **418**, 743–50.

Gross, C., Zhuang, X., Stark, K., Ramboz, S., Oosting, R., Kirby, L., *et al.* (2002) Serotonin$_{1A}$ receptor acts during development to establish normal anxiety-like behaviour in the adult. *Nature*, **416**, 396–400.

Gurney, M. E., Pu, H., Chiu, A. Y., Dal Canto, M. C., Polchow, C. Y., Alexander, D. D., *et al.* (1994) Motor neuron degeneration in mice that express a human Cu, Zn superoxide dismutase mutation. *Science*, **264**, 1772–5.

Hill, D. P. and Wurst, W. (1993) Gene and enhancer trapping: mutagenic strategies for developmental studies. *Curr. Top. Dev. Biol.*, **28**, 181–206.

Hsieh-Li, H. M., Chang, J. G., Jong, Y. J., Wu, M. H., Wang, N. M., Tsai, C. H., *et al.* (2000) A mouse model for spinal muscular atrophy. *Nat. Genet.* **24**, 66–70.

Jones, J. M., Datta, P., Srinivasula, S. M., Ji, W., Gupta, S., Zhang, Z., *et al.* (2003) Loss of Omi mitochondrial protease activity causes the neuromuscular disorder of mnd2 mutant mice. *Nature*, **425**, 721–7.

Julien, J. P. (2001) Amyotrophic lateral sclerosis. unfolding the toxicity of the misfolded. *Cell*, **104**, 581–91.

Kaupmann, K., Simon-Chazottes, D., Guénet, J.-L., and Jockusch, H. (1992) Wobbler, a mutation affecting motoneuron survival and gonadal functions in the mouse, maps to proximal chromosome 11. *Genomics*, **13**, 39–43.

Lee, S. K. and Pfaff, S. L. (2001) Transcriptional networks regulating neuronal identity in the developing spinal cord. *Nat. Neurosci.*, **4**, 1183–91.

Lefêbvre, S., Burglen, L., Reboullet, S., Clermont, O., Burlet, P., Viollet, L., *et al.* (1995) Identification and characterization of a spinal muscular atrophy-determining gene. *Cell*, **80**, 155–65.

Legare, M. E., Bartlett, F. S., 2nd and Frankel, W. N. (2000) A major effect QTL determined by multiple genes in epileptic EL mice. *Genome Res.*, **10**, 42–8.

Leighton, P. A., Mitchell, K. J., Goodrich, L. V., Lu, X., Pinson, K., Scherz, P., *et al.* (2001) Defining brain wiring patterns and mechanisms through gene trapping in mice. *Nature*, **410**, 174–9.

Loots, G. G., Locksley, R. M., Blankespoor, C. M., Wang, Z. E., Miller, W., Rubin, E. M. *et al.* (2000) Identification of a coordinate regulator of interleukins 4, 13, and 5 by cross-species sequence comparisons. *Science*, **288**, 136–40.

Lyon, M. F. and Searle, A.G. (1989) *Genetic variants and strains of the laboratory mouse.* Oxford University Press, Oxford, UK.

Mackay, T. F. (2001) The genetic architecture of quantitative traits. *Annu. Rev. Genet.*, **35**, 303–39.

Malleret, G., Haditsch, U., Genoux, D., Jones, M. W., Bliss, T. V., Vanhoose, A. M., *et al.* (2001) Inducible and reversible enhancement of learning, memory, and long-term potentiation by genetic inhibition of calcineurin. *Cell*, **104**, 675–86.

Mayford, M., Mansuy, I. M., Muller, R. U., and Kandel, E. R. (1997) Memory and behavior: a second generation of genetically modified mice. *Curr. Biol.*, **7**, R580–9.

Meisler, M. H., Kearney, J., Ottman, R., and Escayg, A. (2001) Identification of epilepsy genes in human and mouse. *Annu. Rev. Genet.*, **35**, 567–88.

Monuki, E. S. and Walsh, C. A. (2001) Mechanisms of cerebral cortical patterning in mice and humans. *Nat. Neurosci.*, **4**, 1199–206.

Nave, K. A. (1994) Neurological mouse mutants and the genes of myelin. *J Neurosci Res*, **38**, 607–12.

Nelson, R. J., Demas, G. E., Huang, P. L., Fishman, M. C., Dawson, V. L., Dawson, T. M., *et al.* (1995) Behavioural abnormalities in male mice lacking neuronal nitric oxide synthase. *Nature*, **378**, 383–6.

Patil, N., Cox, D. R., Bhat, D., Faham, M., Myers, R. M., and Peterson, A. S. (1995) A potassium channel mutation in weaver mice implicates membrane excitability in granule cell differentiation. *Nat. Genet.*, **11**, 126–9.

Popp, R. A., Bailiff, E. G., Skow, L. C., Johnson, F. M., and Lewis, S. E. (1983) Analysis of a mouse alpha-globin gene mutation induced by ethylnitrosourea. *Genetics*, **105**, 157–67.

Ramirez-Solis, R., Liu, P., and Bradley, A. (1995) Chromosome engineering in mice. *Nature*, **378**, 720–4.

Routtenberg, A. (1995) Knockout mouse fault lines. *Nature*, **374**, 314–5.

Schrank, B., Gotz, R., Gunnersen, J. M., Ure, J. M., Toyka, K. V., Smith, A. G., *et al.* (1997) Inactivation of the survival motor neuron gene, a candidate gene for human spinal muscular atrophy, leads to massive cell death in early mouse embryos. *Proc. Natl. Acad. Sci. USA*, **94**, 9920–5.

Skarnes, W. C., Moss, J. E., Hurtley, S. M., and Beddington, R. S. (1995) Capturing genes encoding membrane and secreted proteins important for mouse development. *Proc. Natl. Acad. Sci. USA*, **92**, 6592–6.

Thomas, K. R. and Capecchi, M. R. (1987) Site-directed mutagenesis by gene targeting in mouse embryo-derived stem cells. *Cell*, **51**, 503–12.

Ulbrich, M., Schmidt, V. C., Ronsiek, M., Mußmann, A., Bartsch, J. W., Augustin, M., *et al.* (2002) Genetic modifiers that aggravate the neurological phenotype of the wobbler mouse. *NeuroReport*, **13**, 535–9.

Waterston, R. H., Lindblad-Toh, K., Birney, E., Rogers, J., Abril, J. F., Agarwal, P., *et al.* (2002) Initial sequencing and comparative analysis of the mouse genome. *Nature*, **420**, 520–62.

Wallace, D. C. (2001) Mouse models for mitochondrial disease. *Am. J. Med. Genet.*, **106**, 71–93.

You, Y., Bergstrom, R., Klemm, M., Lederman, B., Nelson, H., Ticknor, C., *et al.* (1997) Chromosomal deletion complexes in mice by radiation of embryonic stem cells. *Nat. Genet.*, **15**, 285–8.

Yu, T. W. and Bargmann, C. I. (2001) Dynamic regulation of axon guidance. *Nat. Neurosci.*, **4**, 1169–76.

Zoghbi, H. Y. and Orr, H. T. (2000) Glutamine repeats and neurodegeneration. *Annu. Rev. Neurosci.*, **23**, 217–47.

Chapter 3

Gene expression: from precursor to mature neuron

Uwe Ernsberger

3.1 Introduction

Gene expression specific to neurons and their precursors allows the formation of an interconnected set of cells able to process information encoded by electrical signals. In order to realize the appropriate part of the genetic information required to properly connect neurons and to process electrical information, a succession of gene regulatory events occurs. This includes the acquisition of neuron-specific gene regulatory competence during development, the resulting mature pattern of neuron-specific gene expression, and the plasticity of the mature expression pattern under changing physiological conditions.

Neurons are characterized in a simple and unifying manner by their ability to handle electrical signals due to ionic membrane conductances and the remarkable property of propagating changes in membrane potential along neuritic processes to transmit the encoded information onto target cells. Yet, the underlying molecular mechanisms show an astonishing diversity. The genes coding for the proteins involved in the very neuronal properties generally come in families which may include several to several dozen family members, the number increasing with ongoing search for homologues in the genomes whose sequences become available. This diversity of proteins available to perform a certain or similar function with slight differences in performance properties is the founding basis for neuronal diversity. The crude estimate of the number of different neuron classes ranges from several hundreds to several thousands, and depends on the complexity of the nervous system under investigation and the stringency of the classification. The degree of diversity provided by the existence of multigene families coding for related proteins is amplified by the combination of protein subunits in complexes such as many ion channels involved in electrical membrane processes. In addition, the generation of different protein isoforms by differential RNA splicing and the posttranslational modification of proteins by a range of biochemical processes make the nervous system an almost endless domain of diversity. And it makes the understanding of neurons-specific gene expression a formidable task where diversity becomes one of the most prominent features.

The challenge will be to first comprehend how, on a molecular basis, the expression of neuron-specific genes is regulated. As the genes come in families whose members are differentially expressed among different neuron populations (Fig. 3.1), the next problem is to understand why a specific type of neuron expresses a certain member of a class of neuron-specific genes and not one, two or several others. Moreover, as many

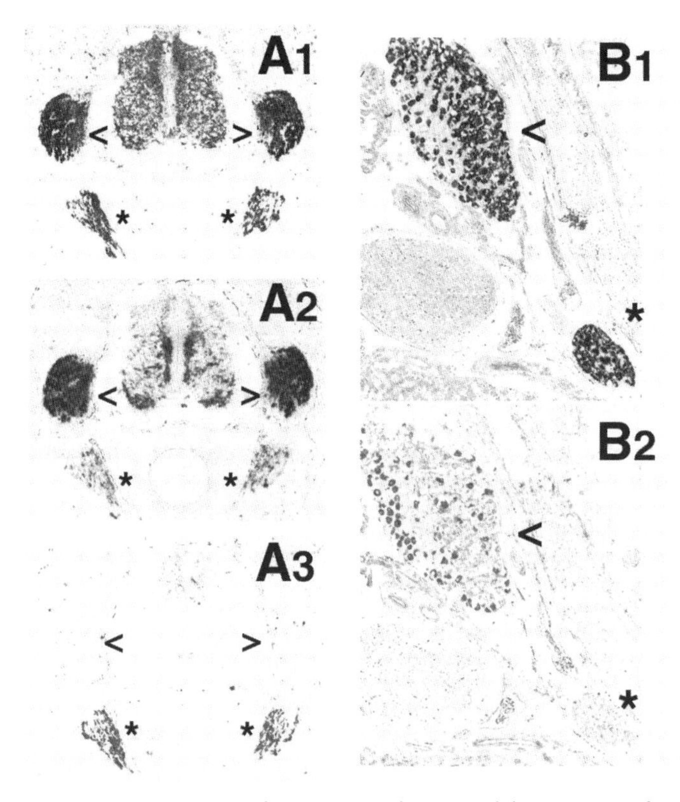

Fig. 3.1 Cellular expression patterns of neuron-specific genes. (A) Expression of general and population-specific neuronal mRNAs as detected by in situ hybridization. (A1) SCG10 is a general neuronal marker. In this cross section from the trunk region of a 7 day old chick embryo, SCG10 mRNA is observed in the grey matter of the spinal cord, dorsal root ganglia (arrowheads) and sympathetic ganglia (asterisks). (A2) Expression of mRNA for neurofilament middle subunit in an adjacent section. Expression levels in the spinal cord vary strongly depending on neuron population and developmental stage. (A3) Tyrosine hydroxylase mRNA is restricted to a small population of central and peripheral neurons such as the sympathetic neurons shown here. The enzyme is required for catecholamine biosynthesis and labels neurons using a certain class of neurotransmitters. (B) Differential expression of representatives of the synaptotagmin gene family in dorsal root (arrow head) and sympathetic (asterisk) ganglia. (B1) Synaptotagmin I mRNA is detectable in neurons from both ganglia. (B2) Synaptotagmin II is detectable in some dorsal root ganglion neurons which seem to be preferentially localized in a different part of the ganglion than synaptotagmin I expressing cells. Sympathetic ganglia do not express detectable synaptotagmin II levels.

physiologically, biochemically or structurally defined neuronal properties depend on the interaction of different proteins coming from different protein families, the issue arises as to how the coordinate expression of the different genes is achieved. And again why certain representatives of the respective gene or protein families are selected.

To meet these goals, the transcription apparatus needs to target genes in a highly ordered fashion. For this purpose, the clustering of chromatin domains is thought to provide a nuclear context beyond that provided by the DNA flanking sequences of the genes (Cockell and Gasser 1999). The importance of chromatin structure is emphasized by the statement that the 'transcription apparatus deals with chromatin, not DNA' (Felsenfeld 1996). Still, the precise link between chromatin structure and gene regulation has to be established (Belmont *et al.* 1999). The accessability of DNA to nucleases is increased in regions surrounding actively expressed genes. In addition, there is a correlation with histone hyperacetylation. The covalent modification of histones may affect nucleosome structure and can be directly linked to transcriptional activation (Strahl and Allis 2000). Consequently, a 'histone code' is recognized to be essential in regulating gene expression (Jenuwein and Allis 2001), and the activities of transcriptional activators and repressors in chromatin alteration by histone modification is becoming a central issue of current research.

Transcriptional activators and repressors are involved in a complex set of protein/DNA and protein/protein interactions. RNA polymerase and a group of general transcription factors (Orphanides *et al.* 1996) constitute the basal transcription machinery that assembles at the core promoter. This complex turns out to be insufficient to attain physiological levels of gene expression and additional activators and silencers are required which bind in a sequence-specific manner to DNA motifs which may be located in the 5′ upstream region as well as in exon or intron sequences of the gene of interest (Figs. 3.2 and 3.3). The communication between the basal transcription machinery and these activators or silencers is mediated by a third class of transcription factors, the co-activators or co-repressors (Näär *et al.* 2001). The combinatorial action of the transcription factors and the basal transcription machinery allows for the constitutive, inducible and tissue-specific regulation of gene expression and may mediate the functional autonomy of adjacent genes (Dillon and Sabbattini 2000). The specificity that will be of interest in the discussion of neuron-specific gene expression is expected to reside, to a large extent, in the sequence-specific activators and silencers. In addition, recent work indicates that the basal transcription machinery may also show cell type-specific properties (Verrijzer 2001), even though such a role for neuron-specific gene expression has not been demonstrated yet.

3.2 **Promoter analysis for the study of neuron-specific regulatory elements**

Important regulatory elements controlling gene expression located in the 5′ regions of the genes are analysed with the help of promoter/reporter DNA constructs (Fig. 3.2).

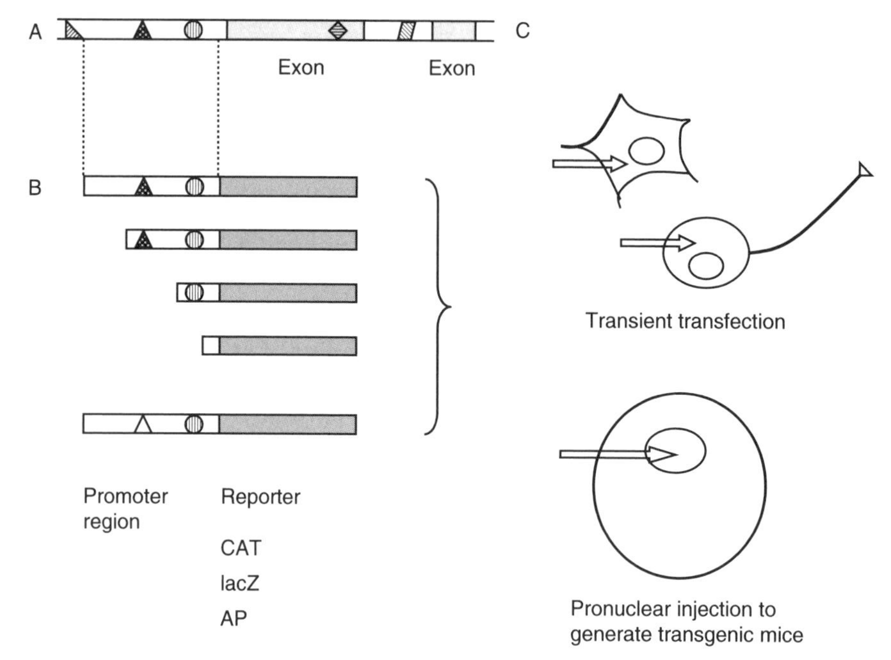

Fig. 3.2 Analysis of promoter elements regulating transcription. (A) Schematic arrangement of transcription factor binding sites (filled symbols) and exons (gray boxes). Binding sites may be located in the 5′ upstream region, introns, or exons. (B) Regulatory regions of the gene of interest are fused to the coding region of a reporter gene. Frequently used reporters are chloramphenicol acetyltransferase (CAT), β-galactosidase (lacZ), alkaline phosphatase (AP), and luciferase. Progressive truncation or mutation (white symbol) of the regulatory region allows the removal of transcription factor binding sites. (C) Introduction of the promoter/ reporter constructs into different cell types or transgenic animals allows the analysis of the efficacy of the promoter by analysing expression of the reporter. Transient transfection into cell lines or primary cells provides a means to compare promoter activity in neuronal versus non-neuronal cells or in different neuronal populations. Injection into oocyte pronuclei is used to generate transgenic mice in which the stage and cell type-specific activity of a promoter can be analysed via reporter detection.

They contain promoter fragments of differing length from the gene of interest, for example a gene encoding the subunit of a neuronal ion channel, in front of a reporter gene with enzymatic activity. Frequently used reporter genes are chloramphenicol acetyltransferase (CAT), β-galactosidase (lacZ), alkaline phosphatase (AP), and luciferase (luc). After introduction of such promoter/reporter constructs into primary cells or cell lines, the expression of the reporter gene depends on the ability of the cell's transcriptional apparatus to interact with the promoter and actively transcribe the reporter. The interaction of the cell's transcription factors with binding sites in the promoter of the construct is the limiting factor for reporter expression and allows the characterization of crucial promoter elements.

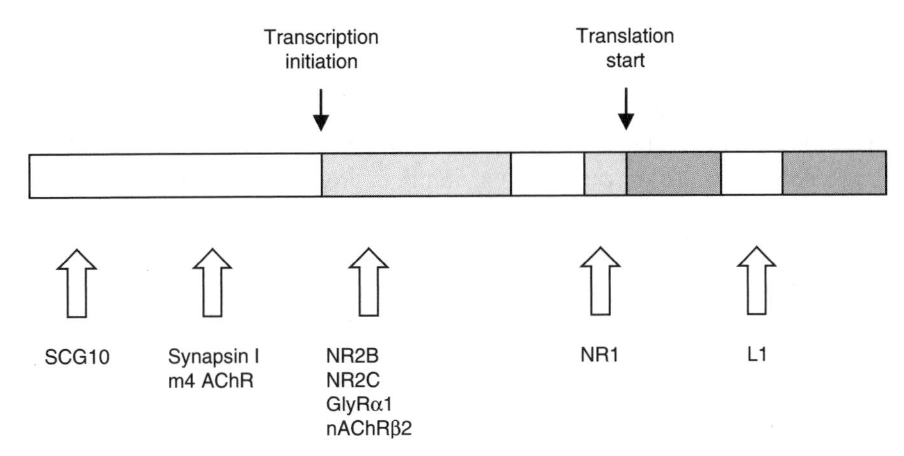

Fig. 3.3 Different location of RE1/NRSE motifs in different genes. Gene structure is shown schematically and exons are shown by grey boxes. Transcription and translation start sites are indicated. The position of the NRSE/RE1 in the 5′ upstream region, introns or untranslated exons is indicated for the genes shown. Modified from Thiel *et al.* (1999), and Myers *et al.* (1999).

Such promoter/reporter constructs can be introduced transiently into cell lines of different origin. Comparing the reporter expression in non-neuronally and neuronally derived cell lines is a powerful way to indicate the presence of transcription factors limited to one type of cells, as illustrated in the following chapter. Successively restricting the length of the promoter and mutation of promoter elements can be employed to characterize critical transription factor binding sites in the promoter. Promoter fragments including such sites can then be used to screen for the presence of transcription factors in different cell types and to clone these factors with expression strategies.

This approach can be powered up by introducing the promoter/reporter permanently into transgenic animals (Palmiter and Brinster 1986). Pronuclear injection of the construct into mouse oocytes is used to achieve chromosomal integration and the generation of transgenic mouse lines. This puts the reporter system in a close-to-normal developmental and physiological context. By standard histological techniques, the reporter expression can be monitored to determine whether a defined promoter fragment can drive expression in distinct cell types and at specific developmental stages. In this manner, it can be determined whether a particular promoter fragment, for example from a gene encoding a neurotransmitter-synthesizing enzyme, is sufficient to direct reporter expression to specifically those cells which express the gene in the normal animal. The limited length of the promoter fragments, usually below 10 kb, as well as the influence of the surroundings of the integration site which cannot be chosen in conventional transgenic approaches restrict the scope of this technology which, nevertheless, has produced many of the results dicussed below. The use of artificial minichromosomes (Heintz 2000; Giraldo and Montoliu 2001) into which an order of magnitude larger DNA inserts can be included will help to overcome the size limitation but has not yet been used extensively to analyse neuron-specific gene expression.

The combination of these techniques with the biochemical analysis of protein/DNA interactions and the characterization of promoter sequences suitable for transcription factor binding provides knowledge about a range of promoter elements and transcription factors that may be involved in the expression of neuronal genes and will be discussed below. Still, the emerging complexity of the protein complexes involved in transcriptional regulation is far from fully understood and results in statements such as 'surprisingly little is known about how individual genes are turned on or off' (Lemon and Tjian 2000). This is illustrated by the unfolding complexity of the interaction of coactivators and corepressors with the basal transcription machinery, and also by the considerable variation of the regulatory architecture of neuron-specific genes with respect to activating and silencing elements that has been appreciated already a decade ago (Mandel and McKinnon 1993; Grant and Wisden 1997).

3.3 **Regulation of neuronal gene expression by the neuron-restrictive silencer**

3.3.1 **The gene encoding the type II voltage-gated sodium channel — on the importance of transcriptional repression to divide neuronal from non-neuronal gene expression**

Since the characterization of the role of voltage-gated sodium conductances and the underlying channel proteins in the generation of the action potential, their expression may be considered one of the functional hallmarks of neurons. Thus, the analysis of regulatory elements involved in neuron-specific expression of the genes coding for voltage-gated sodium channels promises to unravel crucial players involved in the specification of neurons by differential expression of functionally relevant proteins. The analysis of the promoter region of the type II sodium channel held this promise.

Fusing approximately 1 kb of the 5′ flanking region of the sodium channel gene to bacterial chloramphenicol acetyltransferase as reporter demonstrates that, upon transient transfection of this construct into different cell types, expression of the reporter is detectable in cells with neuronal properties but not in non-neuronal cells (Maue *et al.* 1990). Deletion of sequence elements within this promoter region leads to a dramatic increase of the reporter expression in non-neuronal cells and suggested that a silencer element in the 5′ flanking region of the sodium channel gene is able to suppress expression in non-neuronal cells. The regulation is mediated by a silencer element called RE1/NRSE which is active in cell lines that do not express the type II sodium channel (Kraner *et al.* 1992; Mori *et al.* 1992). Expression cloning of a protein that binds to this silencer element resulted in the identification of the zinc finger transcription factor REST/NRSF which, in the developing mouse embryo, is expressed ubiquitously outside the nervous system (Chong *et al.* 1995; Schoenherr and Anderson 1995). Recombinant

REST/NRSF protein silences expression from a reporter construct carrying the RE1/NRSE domain in the promoter region. Conversely, a dominant-negative form of REST/NRSF that obstructs the activity of endogenous REST/NRSF relieves silencing mediated by the native protein (Chong *et al.* 1995; Eggen and Mandel 1997).

Importantly, similar silencer elements were identified in a large number of neuron-specific genes (Schoenherr *et al.* 1996) such as those coding for neurotransmitter receptors or transmitter release-associated proteins (Myers *et al.* 1999; Thiel *et al.* 1999). In many cases they are able to bind REST/NRSF and repress expression from a promoter/reporter construct. This exciting observation led to the conclusion that REST/NRSF is a silencing transcription factor expressed specifically in non-neuronal cells that prevents the expression of a large battery of neuron-specific genes in non-neuronal cells.

3.3.2 The β2 neuronal acetylcholine receptor subunit gene — turning repressors into activators?

Detection of REST/NRSF expression in adult neurons (Palm *et al.* 1998) strongly suggests that a silencing hypothesis assuming the exclusive action of the NRSE/RE1 in non-neuronal cells may be too simplistic. Also reporter gene expression mediated by the NRSE/RE1 motif in the β2 neuronal acetylcholine receptor (nAChR) subunit promoter casts doubt on the exclusive silencing action of the motif.

The β2 subunit of the nAChR is the most widely expressed subunit in neurons (Hill *et al.* 1993). A β-galactosidase reporter driven by the β2 nAChR promoter in transgenic mice is expressed in the peripheral nervous system and different regions of the central nervous system such as thalamus and colliculus. Mutation of the RE1/NRSE switches off expression in most of these structures but increases expression in some others such as the colliculus (Bessis *et al.* 1997). This suggests an intrinsic activating or repressing activity of RE1/NRSE in neurons depending on the neuron population. Similarly, in neuronal cell lines the RE1/NRSE element may act as a silencer or as enhancer in transfection experiments. Inserting spacers of different length between the RE1/NRSE and the TATA box of the promoter/reporter construct indicates that RE1/NRSE activity depends on its position in the promoter.

These conclusions remain controversial, however. REST/NRSF functions as repressor on model promoters containing strong promoter/enhancers in addition to RE1/NRSE motifs derived from various neuronal genes including the β2 neuronal acetylcholine receptor subunit (Thiel *et al.* 1998). Regardless of the position of the RE1/NRSE motifs, no activator function could be attributed to REST/NRSF in these experiments. Moreover, transcriptional activation mediated by activator domains from different transcription factors is blocked by REST/NRSF independent of the RE1/NRSE position (Lietz *et al.* 2001). Together with these observations, the existence of different repressor domains in REST/NRSF and their recruitment of histone deacetylases provoke the question how REST/NRSF may mediate an activating function.

3.3.3 The cholinergic gene locus — modulating the silencer function by alternative splicing?

The detection of REST/NRSF splice variants and their dynamic control in neural tissue (Palm *et al.* 1998) adds to the complexity of transcriptional regulation by this factor. Analysis of the action of the splice variants on the ChAT promoter demonstrates their possible antagonistic interaction.

Fragments of 5′ flanking region of the ChAT gene from different mammalian species are able to direct expression of a CAT reporter gene to neuronal but not to non-neuronal cell lines (Misawa *et al.* 1992; Li *et al.* 1993; Lönnerberg *et al.* 1995). In non-neuronal cell lines, the RE1/NRSE motif in the proximal part of the 5′ flanking region (Lönnerberg *et al.* 1996) shows repressing activity which is removed by specific deletion of this motif. Expression of the REST/NRSF splice variant REST 4 in a cell line expressing endogenous full-length REST/NRSF is able to transcriptionally activate the cholinergic gene locus (Shimojo *et al.* 1999). A direct interaction between REST/NRSF and REST4 could be demonstrated by co-immunoprecipitation. The relatively weak binding of REST4 to the cholinergic RE1/NRSE as compared to full-length REST/NRSF (Lee *et al.* 2000) suggests that the derepressing action of REST4 rests on a direct interaction with the full-length protein. Using a synapsin I promoter, the derepressing action of REST4 is not confirmed, however (Magin *et al.* 2002). Thus, the possible antagonistic action of splice variants remains controversial.

Regulation of REST/NRSF expression and splicing via neuronal activity (Palm *et al.* 1998) and protein kinase-mediated signal transduction (Shimojo *et al.* 1999) as well as the action of neurotrophin growth factors via the RE1/NRSE motif (Brene *et al.* 2000) indicate that this motif and the REST/NRSF splice variants may participate in the dynamic control of neuronal gene expression. The low level of expression of splice variants such as REST4 (Palm *et al.* 1998) leaves open the question for their significance *in vivo*.

3.3.4 REST/NRSF as master regulator?

The identification of the neuron-restrictive silencer element, RE1/NRSE, and the zinc finger protein REST/NRSF mark major progress in the quest to identify molecular players involved in the regulation of neuron-specific gene expression. The characterization of this element and its activity in regulating the expression of genes such as voltage-gated sodium channels, neuronal acetylcholine receptors, glutamate receptors, synapsins, choline acetyltransferase, and the like (Schoenherr *et al.* 1996) pointed out a role in the specification of neuronal properties. The quasi-ubiquitous expression in non-neuronal tissues as well as the expression in undifferentiated neuronal progenitors but not differentiated neurons, as it was initially described (Chong *et al.* 1995; Schoenherr and Anderson 1995), shaped the hypothesis that REST/NRSF could be a transcription factor crucial for differentiating gene expression between neurons and non-neuronal cells. Thus, REST/NRSF was considered to be 'a master negative regulator of neurogenesis' (Schoenherr and Anderson 1995) where the expression of neuron-specific genes marks a default pathway which is blocked in non-neuronal cells by the presence of REST/NRSF (Chong *et al.* 1995).

Several lines of evidence indicate that the situation is not that simple. In mice lacking functional REST/NRSF, *de novo* ectopic expression in non-neuronal tissue was observed for neuronal class III β tubulin but not for a range of other neuron-specific genes such as SCG 10, synapsin I, and neurofilament M all of which contain RE1/NRSE motifs in their promoters (Chen *et al.* 1998). The possibility that the mutation was hypomorphic rather than completely blocking the function of REST/NRSF seems low as no RE1/NRSE binding activity could be detected that was immunologically related to REST/NRSF. Analysis is compromised, however, by the early embryonic lethality of the mutant mice. Interestingly, derepression was achieved for more neuronal genes in the chick embryo when a dominant-negative REST/NRSF, which blocks the activity of endogenous REST/NRSF, was overexpressed (Chen *et al.* 1998). Irrespective of the reason for the low degree of derepression of neuronal genes in non-neuronal tissue of the mouse mutant embryos, the results demonstrate that there may be no simple unitary mechanism regulating the expression of neuronal genes, not even of those containing a NRSE/RE1 motif.

The comparatively low impact of the REST/NRSF mutation on neuronal gene expression in mice is in agreement with the expression of a lacZ reporter in transgenic mice that is driven by regulating regions of the L1 cell adhesion molecule gene (Kallunki *et al.* 1997) or the β2 nAChR subunit gene (Bessis *et al.* 1997). In particular the analysis of the β2 nAChR regulatory regions demonstrates that a mutation of the NRSE/RE1 motif does not lead to a general non-neuronal expression of the promoter except for a few oligodendrocytes. Together, the studies show that the expression of REST/NRSF by itself and the presence of the respective binding motif, RE1/NRSE, in the regulatory region of a gene may not suffice to correctly differentiate neuronal from non-neuronal gene expression.

Surprisingly, the mutation of the NRSE/RE1 motif in a β2 nACHR subunit transgene results in massive alterations in the neuronal expression of the transgene. The detection of REST/NRSF expression in neurons of different brain regions (Palm *et al.* 1998) strongly indicates that this factor may not only silence the expression of neuron-specific genes in non-neuronal cells but regulate, by activation or repression, gene expression in neurons. The different activity of the β2 RE1/NRSE motif in different brain regions suggests interaction with other transcriptional regulators. Alternatively different variants of REST/NRSF may be active in different brain regions.

An additional twist of complication lies in the observation that NRSE/RE1 may be found not only in neuronal genes, but also in non-neuronal genes (Schoenherr *et al.* 1996). Moreover, their intragenic position may be conserved between species and they may be functional in promoter/reporter constructs. This further suggests that the NRSE/RE1 motif must not necessarily result in repression of gene expression in cells expressing REST/NRSF.

3.3.5 Molecular mechanism of REST/NRSF action

Biochemical analysis begins to illuminate the mechanism by which such gene regulation may be accomplished. The important observation that REST/NRSF binds the corepressor Sin3 and recruits histone deacetylase (HDAC) into a complex (Naruse *et al.* 1999; Huang *et al.* 1999*a*; Roopra *et al.* 2000) suggests that REST/NRSF-mediated repression involves histone deacetylation. Indeed, REST/NRSF binding to RE1/NRSE is accompanied by a decrease of histone acetylation around the RE1/NRSE motif and inhibition of histone deacetylation leads to expression of neuron-specific genes in non-neuronal cells. Whereas the corepressor mSin3A/B interacts with the N terminus of REST/NRSF, another corepressor, CoRest, interacts with the C-terminal repressor domain of REST/NRSF (Andres *et al.* 1999). Significantly, CoRest is tightly associated with HDAC1/2 and the combination of REST/NRSF, CoRest and HDAC2 may repress the induction of type II sodium channel expression by nerve growth factor in PC12 cells (Ballas *et al.* 2001). These observations point out the recruitment of histone deacatylating activity by REST/NRSF and the role of corepressor proteins. On the other hand, due to the involvement of different corepressors binding to different repressor domains on REST/NRSF, the possibility arises that the successive recruitment of different corepressor complexes may help to explain dynamic versus stable, long-lasting regulation of neuronal gene expression (Griffith *et al.* 2001).

3.3.6 Conclusions

Currently, we have no complete picture of the functional role of the NRSE/RE1 motif and its binding factor REST/NRSF in neuron-specific gene expression. It may rather serve as a model to illustrate the complexity of protein–protein and protein–DNA interactions involved in gene regulation.

A plausible hypothesis for the role of REST/NRSF could be an early differentiation of neuronal and non-neuronal lineages and an additional, later action in differentiated neurons. Its role in non-neuronal cells may rest on an interaction with other transcription factors which differs between different genes and their regulatory regions. This may explain the non-neuronal expression of genes which despite containing the NRSE/RE1 motif are not neuron-specific.

While REST/NRSF has been analysed in vertebrates, tramtrack, another zinc finger transcription factor found in *Drosophila* represses neuroblast-specific genes and controls glial development (Badenhorst *et al.* 1996; Badenhorst 2001). This further underscores an early role of zinc finger transcription factors in lineage decisions. The comparison of vertebrate and invertebrate REST/NRSF and tramtrack homologues will show whether there is a conserved interaction or succession of such transcription factors during the early division of neuronal and non-neuronal development.

3.4 The regulation of neuronal gene expression by population-specific transcription factors

In addition to the quasi-ubiquitously expressed silencer factor REST/NRSF, analysis of promoter/reporter constructs in cell lines and transgenic animals provides evidence for a number of activating promoter elements and the respective binding factors. These include factors which are available in many different cell types, not only neurons. And, more interestingly, transcription factors expressed specifically in neurons, and, to be more precise, in neuronal subpopulations.

The characterization of Arix/Phox2 transcription factors will be discussed as a prototypic example for a neuron population-specific activator regulating the expression of the neurotransmitter synthesizing enzyme dopamine β-hydroxylase (DBH). It is particularly interesting as the characterization went all the way from the determination of critical promoter regions and expression cloning of binding factors to the analysis of effects of overexpression or mutational inactivation *in vivo*. As a result, a family of transcription factors is described which, across different vertebrate taxa, are involved in the induction of DBH expression as part of a synexpression group of transmitter-synthesizing enzymes.

3.4.1 Dopamine β-hydroxylase — expression cloning of a population-specific homeobox-containing transcription activator

Dopamine β-hydroxylase (DBH) is the enzyme required for norepinephrine synthesis in noradrenergic and adrenergic neurons and endocrine cells. Analysing the regulatory elements involved in the specific expression of this transmitter-synthesizing enzyme led to the characterization of a homeobox-containing transcription factor family required for the development of certain peripheral and central neuron populations (Table 3.1).

Expression of a lacZ reporter under the control of 5.8 kb 5′ flanking sequence of the human DBH gene in transgenic mice resulted in reporter gene activity in noradrenergic and adrenergic neurons in addition to few other sites (Mercer *et al.* 1991). In noradrenergic sympathetic ganglia, transgene expression was detectable at around the time when endogenous noradrenergic properties become induced (Kapur *et al.* 1991). This indicates that the promoter region tested is sufficient to generate a spatial and temporal expression pattern roughly equal to that of endogenous DBH. Analysing promoter fragments after transfection into cell lines demonstrates distinct activating and repressing elements involved in cell type-selective expression of the reporter (Shaskus *et al.* 1992, 1995). With expression cloning using the activating element DB1, a homeodomain-containing transcription factor called Arix was isolated (Zellmer *et al.* 1995). Earlier, the same transcription factor was isolated with an expression screen using a fragment of the NCAM promoter and called Phox2a (Valarché *et al.* 1993). A closely related family member called Phox2b (Pattyn *et al.* 1997) has been characterized and both Arix/Phox2a and 2b are expressed in adrenergic and noradrenergic neurons and endocrine cells. The DBH promoter contains multiple Arix/Phox2 binding sites and both Phox2a and 2b activate DBH transcription *in vitro* (Zellmer *et al.* 1995; Kim *et al.* 1998; Yang *et al.* 1998). The tightly coupled expression of Phox2 transcription factors and DBH in differentiating sympathetic neurons strongly suggests that these transcription factors are required for induction of the noradrenergic transmitter phenotype (Ernsberger *et al.* 2000). This is confirmed by the development of excess

Table 3.1 Characterization of Phox2 transcription factors and their role in the regulation of DBH expression

Analysis performed	**Results obtained**
DBH promoter/reporter analysis in transgenic mice	reproduction of the spatial and temporal expression pattern of the endogenous gene
DBH promoter/reporter analysis in cell lines	characterization of activating and repressing promoter elements
electrophoretic mobility shift analysis of an activating DBH enhancer fragment DB1	characterization of cell type-specific nuclear proteins binding to the DB1 fragment
expression cloning with the DB1 promoter fragment	characterization of Arix cDNAs and their identity with Phox 2a
homology screen	characterization of Phox 2b
comparison of DBH and Phox 2a and 2b expression patterns in adult nervous tissue	evidence for a role of Phox 2a and 2b in the specification of the noradrenergic phenotype
comparison of DBH and Phox 2a and 2b expression patterns during development	evidence for a role of Phox 2a and 2b in the induction of the noradrenergic phenotype
overexpression of Phox 2a and 2b in precursor cells in the chick embryo	evidence that Phox 2a and 2b are sufficient to induce noradrenergic differentiation
mutational inactivation of the Phox 2b gene in mice	demonstration of the necessity of Phox2b for DBH expression in various types of neurons
analysis of Phox2a mutations in mice and zebrafish	demonstration of the necessity of Phox2a for noradrenergic locus coeruleus neurons
comparison of observations in fish, chick and mouse embryos	evidence for an evolutionary conserved role of Phox 2a and 2b in the development of noradrenergic neurons

DBH-expressing cells after retrovirally mediated overexpression of Phox2a or 2b in chick embryos (Stanke *et al.* 1999) and the lack of DBH expression in developing sympathetic ganglia of Phox2b mutant mice (Pattyn *et al.* 1999). In addition, all noradrenergic centers in the brain are missing in these mice (Pattyn *et al.* 2000*a*). In the case of the Phox2a knockout, the noradrenergic cells in the locus coeruleus do not form (Morin *et al.* 1997).

These studies demonstrate that the homeodomain-containing Phox2 transcription factors can specifically activate DBH expression in noradrenergic and adrenergic neurons. The absence of these cells in the peripheral and the central nervous system of Phox2b mutant mice makes Phox2 transcription factors master regulators of neurons with these transmitter properties. Comparison of different species suggests that this developmental function is conserved throughout different classes of vertebrates (Guo *et al.* 1999; Ernsberger 2000).

The expression of Phox2 transcription factors also in cholinergic sympathetic neurons at later developmental stages (Ernsberger *et al.* 2000), the induction of cholinergic traits upon misexpression of Phox2b (Stanke *et al.* 1999; Dubreuil *et al.* 2000), and the lack of autonomic neurons irrespective of transmitter phenotype in Phox2b knockout mice (Pattyn *et al.* 1999) indicate that these transcription factors have other functions in addition to the regulation of DBH expression. Apart from a role in the development of cholinergic neurons which is not characterized in detail, Phox2 transcription factors are involved in the acquisition of general neuronal properties. Phox2b inactivation leads to disruption of the general neuronal differentiation program in certain populations of motoneurons (Pattyn *et al.* 2000*b*). Overexpression of Phox2a and 2b induces the development of ectopic neurons expressing different general neuronal genes such as neurofilament M or synaptotagmin I (Stanke *et al.* 1999; Patzke *et al.* 2001). As Phox2a binds to elements in the NCAM promoter (Valarché *et al.* 1993), the question arises whether corresponding binding elements in other neuronal genes such as those coding for neurofilaments or synaptotagmins may allow direct regulation by Phox2s. Thus, Phox2 transcription factors are not exclusively in charge of regulating a very modular aspect of neuronal differentiation, the expression of DBH. They also contribute to other aspects of differentiation in a restricted set of neuronal precursor populations. The molecular substrate of this coordinated expression of general neuronal properties and subpopulation-specific features remains to be worked out.

3.4.2 Transmitter release-related protein and neurotrophin receptor gene promoters — more evidence for regulation by subpopulation-specific transcription factors

The SNAP-25 gene encodes a protein involved in neurotransmitter release and is expressed in many neuron populations (Oyler *et al.* 1989). Interestingly, POU domain proteins of the Brn3 subfamily, which are restricted to certain neuronal subpopulations (Lillicrop *et al.* 1992; Xiang *et al.* 1993; Turner *et al.* 1994), can regulate SNAP-25 expression Cotransfection of a CAT reporter driven by some 2.1 kb of 5′ flanking region of the SNAP-25 gene with a Brn3a expression vector in ND7 cells stimulates reporter gene expression as compared to an expression vector without Brn3a coding sequence (Lakin *et al.* 1995). In addition to Brn3a, the closely related Brn3c is able to activate this promoter, whereas Brn3b represses activity (Morris *et al.* 1996). The promoter of the gene encoding the synaptic vesicle protein synapsin I is activated by all three Brn3 factors. The important topic highlighted by these studies is the role of neuronal subpopulation-specific transcription factors for the expression of genes not restricted to certain neuronal subpopulations. Evidence for a role of other transcription factors with restricted expression pattern, neurogenin1, Phox2a and Phox2b, for the regulation of SNAP-25, synaptotagmin I, and neurexin I expression has been obtained after overexpression of the transcription factors in Xenopus and chick embryos (Olson *et al.* 1998; Patzke *et al.* 2001). From these studies the question arises whether the expression of widely expressed neuron-specific genes is regulated by different transcription factors in distinct neuronal subpopulations *in vivo*.

Trk neurotrophin receptors are not expressed as widely as the transmitter release-related proteins but can be detected in a variety of neuron subpopulations. In Brn3a mutant mice, Trk expression is affected in a neuron subpopulation and stage-specific manner resulting in the death of certain populations of sensory neurons such as trigeminal neurons whereas dorsal root ganglion neurons are unaffected (Huang *et al.* 1999*b*). The analysis of the trkA/NGF receptor promoter demonstrates a region conserved between mammals and birds that is sufficient to direct reporter gene expression in transgenic mice with the appropriate timing (Ma *et al.* 2000). Mutation of distinct transcription factor binding motifs in this region demonstrates different binding sites to be required for expression in trigeminal, dorsal root and sympathetic ganglia.

These studies support the crucial role of neuronal subpopulation-specific transcription factors for neuron-specific genes widely expressed in different neuron populations. As the expression patterns of many neuronal genes do not respect the boundaries of expression of individual subpopulation-specific transcription factors, widely expressed neuron-specific genes may be regulated under the regime of different transcription factors in different neuron populations.

3.4.3 Conclusions

So far, the analysis of activating elements in the regulatory regions of neuron-specific genes has demonstrated transcriptional activation at a level specific for neuronal subpopulations and at a level which is not restricted to neurons at all. What has not been found to date is an activating counterpart to REST, namely a transactivating factor generally expressed in neurons but not non-neuronal cells. Such a factor may still appear with ongoing genomic analysis. If this will not be the case, a combination of neuron-restrictive silencing, general, not neuron-restrictive activation, and neuron subpopulation-specific activation is sufficient to drive the appropriate levels of expression for neuron-specific genes in the different neuron populations.

Importantly, neuron subpopulation-specific activation of gene expression is confirmed *in vivo* in mutant mice for different classes of homeobox-containing transcription factors and for neuron-specific genes as diverse as those coding for transmitter synthesis enzymes or neurotrophin receptors. These transcription factors belong to different families such as paired homeobox or POU proteins. Certain transcription factors or promoter elements may be important only during restricted periods in a neuron's lifespan. Consequently, the question arises what developmental cascades of sequential transcription factor action are required for neuronal differentiation and specification.

3.5 Developmental acquisition of a neuronal phenotype — population-specific transcription factor cascades

Neuronal differentiation involves the acquisition of neuron-specific morphological, biochemical, and functional properties in a precursor population by specific expression of neuronal genes. The time schedule of this differentiation process and the different stages on this path, characterized by the expression of distinct sets of neuron-specific genes, are beginning to be understood. Different transcription factors analysed in a number of neuron populations are involved and a pictures begins to unfold how the expression of distinct transcription factors at successive differentiation stages propels this developmental process.

3.5.1 Basic helix-loop-helix transcription factors — proneural function and neuronal subtype specification

The genes of the achaete-scute complex in *Drosophila melanogaster*, in particular the achaete and the scute genes, play a decisive role in the development of the fly peripheral nervous system (Ghysen and Dambly-Chaudiere 1988). Loss of function mutations obstruct sensory organ development such that cells normally destined to become neurons adopt epidermal fates. Gain of function mutations resulting in misexpression of achaete or scute lead to formation of sensory organs in 'false' positions, so called 'ectopic' structures. Due to their early function in singling out sensory organ precursor cells that give rise to sensory neurons, achaete and scute were called proneural genes. Importantly, they are involved in the development of specific types of sensory organs, the external sensory organs. For chordotonal organs or photoreceptors a different proneural gene, atonal, is required (Jarman *et al.* 1993, 1994). Genes in the achaete scute complex as well as atonal code for transcription factors of the basic helix-loop-helix (bHLH) family. Information specifying the type of sensory organ generated may reside in the bHLH domain of the protein and involve interactions with other cofactors (Chien *et al.* 1996). Thus, this class of transcription factors plays a crucial role in the very early decisions on the way from a precursor to a neuron and, at the same time, specifies the differentiation path taken.

The demonstration of bHLH transcription factors in vertebrates and their ability to convert non-neuronal to neuronal fate establishes these transcription factors as important regulators of neuronal development throughout the animal kingdom (Lee 1997). As in *Drosophila*, vertebrate bHLH transcription factors confer not only generic but also subtype-specific neuronal properties (Brunet and Ghysen 1999). The diversity of neuronal bHLH proteins in different neuronal subpopulations and the sequence of expression during neuronal differentiation indicate that they are involved in the determination of neuronal precursors and regulate consecutive steps of differentiation. The sensitivity of their expression to lateral inhibition as well as the timing of their expressing during differentiation indicate that only some of these transcription factors qualify as proneural genes. Others are involved in more advanced events during neurogenesis.

In mouse cranial sensory ganglia a cascade of bHLH gene expression including neurogenins 1 and 2, Math3, NeuroD, and Nscl1 is correlated with different steps of development (Fode *et al.* 1998; Ma *et al.* 1998). Since this differentiation cascade and neurogenesis is blocked in neurogenin mutant mice and neurogenins are the first bHLH factors detectable, it is concluded that these bHLH transcription factors function like *Drosophila* proneural genes. The later expressed bHLH proteins, such as neuro D, may in part be direct transcriptional targets of neurogenins and function during more downstream events of neurogenesis.

In the spinal cord, the bHLH factor olig2 is involved in motoneuron and at a later stage in oligodendrocyte development (Marquardt and Pfaff 2001). Interaction with neurogenin 2 and homeodomain proteins promotes neuronal differentiation and specification of neuron subtype identity (Mizuguchi *et al.* 2001; Novitch *et al.* 2001; Scardigli *et al.* 2001). As shown by gene mutation in mouse and overexpression in chick embryos, neurogenins together with mouse atonal homolog 1 (Math1), may specifiy neuron subpopulations by mutual cross-inhibitory actions and by regulation of expression of homeodomain proteins such as LIM transcription factors (Gowan *et al.* 2001) as will be discussed below.

In sympathetic ganglia, the bHLH gene Mash1 in mice and Cash1 in chick are expressed early during development (Lo *et al.* 1991; Ernsberger *et al.* 1995; Groves *et al.* 1995). Mutation in mice obstructs generation of sympathetic ganglia (Guillemot *et al.* 1993), which are populated by partially differentiating precursors expressing neurofilament but not the neuronal marker SCG10 (Sommer *et al.* 1995), indicating different requirements for the expression of distinct neuronal properties. At more advanced stages of sympathetic neuron development, the bHLH transcription factors dHAND and eHAND as well as the homeodomain proteins Phox2a and 2b are expressed in sympathetic ganglia and implicated in noradrenergic differentiation (Howard *et al.* 2000; Ernsberger *et al.* 2000). While the role of Phox2 proteins in noradrenergic induction may be direct as discussed below, the precise mechanism of the regulation of noradrenergic target genes by bHLH factors remains to be determined.

In the retina, the possibility of their direct regulation of target gene expression was demonstrated in addition to their importance for neuronal development. In the mouse retina, mutation of both bHLH factors Mash1 and Math3 leads to a complete loss of bipolar cells (Hatakeyama *et al.* 2001). Math5 is required for retina ganglion cell formation (Brown *et al.* 2001). Also in the zebrafish, ATH5 is required for retinal ganglion cell genesis (Kay *et al.* 2001) indicating that the involvement of bHLH factors of this class in eye development is conserved between vertebrate classes. As shown by promoter analysis in the chick embryo, ATH5 may regulate the expression of the β3 nAChR subunit gene which is specifically expressed in retina ganglion cells. The close correlation of β3 and ATH5 expression as well as the activation of expression from a β3 promoter by ATH5 suggest that this bHLH factors may directly regulate expression of transmitter receptor subunits as part of its role in retinal ganglion cell development. As ATH5 is expressed transiently in retinal ganglion cells, the question arises how ganglion cell-specific expression is maintained. Brn3 POU-domain transcription factors, which are also required for retinal development can be induced by chick and mouse ATH5 and may participate in transcriptional regulation in ganglion cells at more advanced stages of development (Liu *et al.* 2001).

3.5.2 LIM domain transcription factors — specifying neuronal subpopulations

In the vertebrate spinal cord, partitioning of motoneuron and interneuron populations is apparent along the antero-posterior and the dorso-ventral axes. In addition to the bHLH transcription factors discussed above, homeobox-containing transcription factors are essential for the specification of progenitor cell identity (Briscoe and Erickson 2001) and development of the differentiated neurons (Jurata *et al.* 2000).

The segregation of neuronal fates along the dorso-ventral and medio-lateral axes in the vertebrate spinal cord has been correlated with a combinatorial expression of LIM homeodomain proteins (Tanabe and Jessell 1996), the so-called LIM code (Lumsden 1995). In the chick spinal cord, the LIM factors Islet-1, Islet-2, Lim-1 and Lim-3 define subpopulations of motoneurons that segregate into distinct columns and select different axon pathways (Tsuchida *et al.* 1994). Also in zebrafish, primary motoneurons are characterized by particular LIM factor combinations (Appel *et al.* 1995). The lack of motor and certain interneuron populations in mice mutant for the LIM factor islet-1 demonstrates the crucial importance of these transcription factors for specific neuron subpopulations (Pfaff *et al.* 1996).

The homeodomain protein MNR2 is expressed in motoneuron progenitors and transiently in postmitotic motoneurons (Tanabe *et al.* 1998). Its expression precedes that of LIM3, Islet1 and Islet2. The induction by MNR2 overexpression of the motoneuron-specific LIM factors as well as markers of more advanced stages in motoneuron differentiation such as choline acetyltransferase demonstrates a motoneuron differentiation cascade where different transcription factors act sequentially.

Additional transcription factors such as the Nkx and Pax homeodomain proteins are involved in the patterning of the spinal cord (Briscoe *et al.* 1999, 2000; Sander *et al.* 2000). Due to mutual cross-inhibitory regulation, they help to establish sharp boundaries in transcription factor expression domains in the spinal cord (Briscoe and Ericson 2001). They are required for the commitment of progenitors to particular developmental fates and suppress genes involved in the development of neighbouring neuronal subtypes. By this means they help to segregate different neuron populations in the spinal cord. By regulating LIM factor expression, a sequence of homeodomain transcription factors together with bHLH proteins regulates the specification of neuron populations that differ in their location and connectivity. The regulation of the target genes involved in establishing axonal projections remains to be established.

3.5.3 POU domain transcription factors — regulating development and expression of function-specific genes

The family of POU domain transcription factors emerged when structural similarities between a transactivating factor involved in pituitary development and hormone expression in mammals and a protein affecting neuronal development and behaviour in *C. elegans* were observed (Schonemann *et al.* 1998). It was soon recognized that

POU domain proteins constitute a large protein family which is expressed in the developing brain (He *et al.* 1989) and plays crucial roles in pituitary and brain development (McEvilly and Rosenfeld 1999). Their action is analysed in greatest detail for Pit-1/GHF-1 and its importance for pituitary-specific gene expression.

Searching for activators of pituitary-specific growth hormone and prolactin expression led to the identification and cloning of Pit-1 also called GHF-1 (Bodner *et al.* 1988; Ingraham *et al.* 1988), a POU domain transcription factor required in somatotroph and lactotroph cells of the anterior pituitary. Pit-1 protein is able to bind cell type-specific cis-acting elements in the prolactin and growth hormone genes and to activate expression from the respective promoters (Nelson *et al.* 1988; Mangalam *et al.* 1989). The good correlation between the appearance of Pit-1/GHF-1 protein and of growth hormone expression indicates that this transcription factor is responsible for the differentiation of somatotrophic cells and hormone expression (Dollé *et al.* 1990). Direct proof of the Pit-1 function *in vivo* comes from dwarf mice which show disruptions of the Pit-1 gene and lack detectable expression of growth hormone and prolactin (Li *et al.* 1990). The additional loss of TSH expression extends the role of Pit-1 to the development of thyrotroph cells and indicates that additional factors confine prolactin and growth hormone expression to their respective cell types (Simmons *et al.* 1990).

Interestingly, using a Pit-1/lacZ transgene in Pit-1-defective dwarf mice demonstrates that Pit-1 expression is regulated by distinct enhancers for initial gene activation and subsequent autoregulation (DiMattia *et al.* 1997). This observation suggests a developmental sequence where distinct transcription factors initiate Pit-1 expression, which in turn initiates hormone production and maintains its own expression. The detection of the paired-like homeodomain transcription factor Prop-1 before the onset of Pit-1 expression and its requirement for Pit-1 gene activation illustrates such a cascade of tissue-specific regulators (Sornson *et al.* 1996). A range of additional transcription factors from different families are involved in different aspects of pituitary cell development (Dasen and Rosenfeld 1999). These include the LIM homeodomain factors Lhx3 and 4 as well as the Pitx homeodomain factors Pitx1 and 2 during early phases of pituitary development and nuclear hormone receptors involved in pituitary cell type-specific expression of hormone-encoding genes (Dasen and Rosenfeld 2001).

In the nervous system, POU domain proteins of the Brn3 subfamily are required for neuronal development. Targeted mutations in mice show that Brn3b and 3c are essential for differentiation and survival of retinal ganglion cells (Xiang 1998; Gan *et al.* 1999) and inner ear hair cells (Xiang *et al.* 1998), respectively. Axonal growth is affected by all Brn3 subfamily members as observed in Brn3a-deficient trigeminal neurons (Eng *et al.* 2001) or Brn3b and 3c-deficient retinal ganglion cells (Erkman *et al.* 2000; Wang *et al.* 2000, 2002). Whether direct regulation by Brn3 transcription factors of the SNAP-25 gene as discussed above and of GAP-43 expression as suggested by microarray analysis of cDNA from wild-type and Brn3b mutant retinas (Mu *et al.* 2001) are involved remains to be proven. A range of other transcription factors is affected by the

Brn3b mutation (Erkman *et al.* 2000) and may contribute to the altered expression of target genes and axonal outgrowth disruption.

3.5.4 Conclusions

The results from different parts of vertebrate and invertebrate nervous systems provide a first impression of transcription factor cascades involved in neuronal differentiation and specification. A general theme is the early role of bHLH proteins which may excert proneural functions and are involved in the specification of developing neurons. Cascades of bHLH proteins found in vertebrate neurogenesis indicate the diversification of their function along different stages during neurogenesis. A crucial role for the differentiation of the diverse neuronal lineages can be attributed to transcription factors of distinct homeobox protein families. There is currently no unifying concept for the relative roles played by the different homeobox factor families during differentiation of the diverse neuronal lineages.

A classical example for the combinatorial action of different homeobox transcription factors is the specification of neuron lineages in the nematode *C. elegans*. The POU domain protein unc-86 and the LIM domain protein mec-3, both homeobox-containing transcription factors, are necessary for development of distinct neurons (Finney *et al.* 1988; Way and Chalfie 1989). The sequential expression of unc-86 before mec-3 suggest hierarchical actions in the touch neuron differentiation cascade (Duggan *et al.* 1998). Their action must not be strictly sequential, however. unc-86/mec-3 heterodimers can form and bind to the mec-3 promoter (Xue *et al.* 1993; Rockelein *et al.* 2000) as well as presumed target gene promoters (Duggan *et al.* 1998). This illustrates how sequential activation of transcription factors may propel cell fate determination not only by successive production of distinct transcription factors but also by synergistic action of these transcription factors recruited into hetero-oligomere complexes.

Direct target genes of the respective transcription factors are known for only a limited number of examples such as Pit-1, Phox2, Ath-5, or unc-86. Moreover, the proof that the transcription factors bind to the presumed promoter target sequences *in vivo* is, in the majority of cases, not existent. Nevertheless, the methods to undertake this analysis are available. By the combination of these approaches with the increasing knowledge of model organism genomes, we can within the foreseeable future expect to have a profound look into the similarities and differences between transcription factor cascades specifying different classes of neurons.

3.6 Organizing neuronal function at the transcriptional level — synexpression groups

Neuronal properties typically depend on the interaction of several proteins. Ion channels may be composed of several subunits, transmitter synthesis may occur along a cascade of enzymes, and transmitter release requires an intricate apparatus composed of a multitude of proteins which belong to different protein families. Consequently, the coordinate expression of a range of proteins is necessary to enable a neuron to perform specific tasks. To achieve this goal, mechanisms exist to coordinate the expression of neuron-specific genes. Synexpression groups are groups of genes whose protein products are involved in a certain functional context and whose expression is regulated by common mechanisms (Niehrs and Pollet 1999).

3.6.1 Specifying neurotransmitter phenotypes as synexpression groups

Substantial evidence exists to indicate that neurotransmitter phenotypes are regulated as synexpression groups. The neurotransmitter phenotype is a term that describes the ability of a neuron to synthesize, store, and release a certain neurotransmitter. Neurons using glutamate as transmitter are called glutamatergic to indicate their phenotype; those which use GABA are referred to as GABA-ergic; cholinergic neurons use acetylcholine, dopaminergic neurons dopamine, noradrenergic neurons noradrenaline, and so forth. For transmitter synthesis they employ specific enzymes, and for the loading of the transmitter or its precursor into vesicles they use specific vesicular transport proteins. In addition, specific plasma membrane transporters may be expressed for reuptake of transmitter or breakdown products. Evidence has been obtained that the synthesizing enzymes and transport proteins for a certain transmitter can be expressed in coordinate manner, thus qualifying them as synexpression groups.

Coordinated transcriptional regulation has been observed for glutamic acid decarboxylase and the vesicular GABA transporter in *C. elegans* GABA-ergic neurons (Eastman *et al.* 1999). The homeodomain protein unc-30, which upon ectopic expression can induce a GABA-ergic phenotype (Jin *et al.* 1994), sequence-specifically binds to the promoters of both genes. Mutation in the unc-30 binding sites abolishes expression of reporter genes with promoters of glutamic acid decarboxylase and of the vesicular GABA transporter.

The ETS domain transcription factor Pet-1 colocalizes with serotonergic neurons in the mouse brain (Hendricks *et al.* 1999). Conserved Pet-binding sites are present in genes coding for tryptophan hydroxylase, aromatic L-amino acid decarboxylase, serotonin transporter, and 5HT-1a receptor, and are capable of supporting transcriptional activation through interaction with the Pet-1 ETS domain.

A particular interesting example is the cholinergic neurotransmitter phenotype encoded by the cholinergic gene locus (Mallet *et al.* 1998; Eiden 1998). The locus contains the genes coding for the enzyme for acetylcholine biosynthesis, choline acetyltransferase (ChAT), as well as the vesicular acetylcholine transporter (VAChT) gene (Fig. 3.4). The VAChT gene is located in the first intron of the ChAT gene, indicating that the expression of both genes may be driven by the same regulatory mechanism. Indeed, both genes can be coregulated by growth factor-treatment (Misawa *et al.* 1995, Lopez-Coviella *et al.* 2000) and by the common 5′ flanking region including the NRSE/RE1 motif (De Gois *et al.* 2000). Regulatory regions in the locus directing

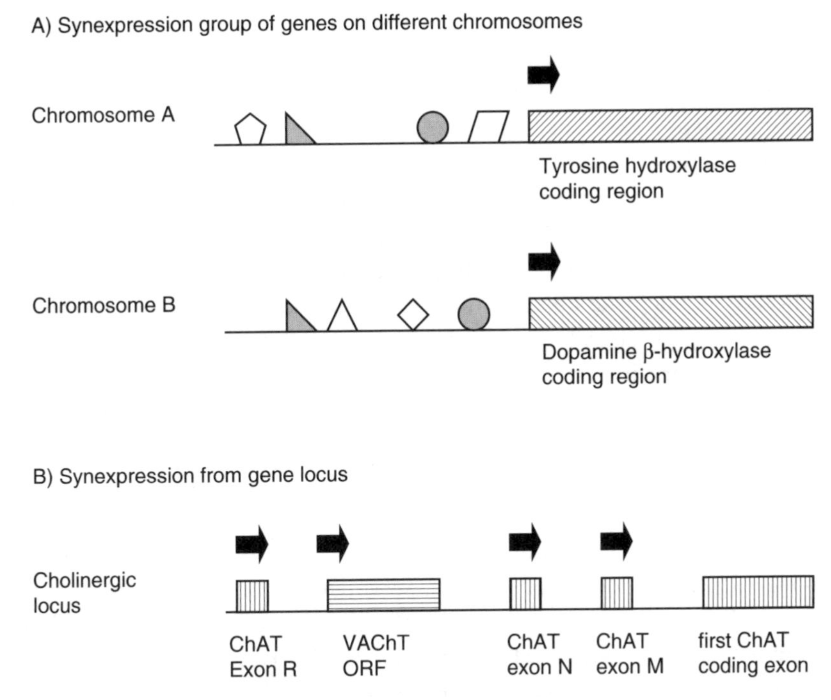

Fig. 3.4 Regulation of neuronal genes as synexpression groups. Genes regulated as synexpression groups can be located on different chromosomes (A). Gene structures are given schematically with the 5′ upstream region as line and the coding region as hatched box. Different symbols in the 5′ upstream region indicate distinct transcription factor binding sites. Common transcription factor binding sites in the regulatory regions of different genes are considered to mediate the coordinate regulation in certain classes of neurons as observed for tyrosine hydroxylase and dopamine β-hydroxylase in noradrenergic neurons. Different elements in the regulatory region may contribute to different regulation in other classes of neurons as observed for tyrosine hydroxylase and dopamine β-hydroxylase in noradrenergic versus dopaminergic neurons. Genes regulated as synexpression group can also be located within one locus (B) as shown for the cholinergic locus encoding choline acetyltrasferase (ChAT) and the vesicular acetylcholine transporter (VAChT). Exons are indicated by hatched boxes and transcription start sites by arrows. (B) Modified from Eiden (1998) and Mallet *et al.* (1998).

expression in cholinergic cells have been identified (Lönnerberg *et al.* 1995, 1996; Naciff *et al.* 1999) but the transcription factors interacting with these sites remain to be determined.

The close packing of genes into loci such as the cholinergic one constitutes an obvious way to coordinate gene expression in a synexpression group. In the case of the noradrenergic transmitter phenotype, genes encoding two different transmitter synthesizing enzymes may be coordinately regulated (Ernsberger 2000) and, at least in humans, are located on different chromosomes (www.ncbi.nlm.nih.gov:80/LocusLink). Tyrosine hydroxylase, the rate-limiting enzyme in the noradrenaline biosynthesis cascade, and dopamine β-hydroxylase, the final enzyme converting dopamine into noradrenaline, become detectable at the same time during sympathetic neuron development (Ernsberger *et al.* 2000). Coordinate induction by overexpression of the transcription factors Phox2a and 2b (Stanke *et al.* 1999) and lack of expression after functional inactivation of the Phox2b gene (Pattyn *et al.* 1999) strongly support the hypothesis that the two enzymes are regulated as a synexpression group in noradrenergic neuron populations. Dopaminergic neurons require TH but not DBH for transmitter synthesis. This indicates that coregulation of TH and DBH expression is not obligatory but depends on the neuron population and the respective transcription factor equipment. This may be mediated by transcription factor binding sites which in part are common and in part distinct between the genes (Fig. 3.4).

3.6.2 Coexpression of ion channel subunits — a question to be addressed

Ligand-gated ion channels such as the acetylcholine receptors or the glutamate receptors are composed of several subunits that contibute to the channel pore. In the case of the voltage-dependent ion channels, the same situation can be found for potassium-specific channels. In the case of voltage-dependent sodium and calcium channels, there is one subunit sufficient for pore formation and additional subunits may affect channel properties such as kinetic behavior. For this reason, the coordinate expression of channel subunits in individual neurons has important impact on the electrical properties of the neuronal plasmamembrane.

In the case of the neuronal AChR subunits alpha 3, alpha 5, and beta 4, the location of the genes in a cluster of 60 kb has been observed in the rat genome (Boulter *et al.* 1990). As there is extensive overlap of expression in the peripheral nervous system (Yang *et al.* 1997), this has sparked speculations on a common regulation of these clustered genes. Due to only partial overlap of the expression in the central nervous system, the issue may not have a simple solution. A reporter construct driven by a fragment spanning the region from the alpha 3 promoter to the beta 4 untranslated exon in transgenic mice results in CNS but not PNS expression (Yang *et al.* 1997). CNS expression is localized to a subset of CNS nuclei that expresses endogenous alpha 3, shows some overlap with beta 4 and none with alpha 5. The precise mechanism by which subunit expression is established in some neuron populations and prevented in others and the role of the clustering of genes therein still needs to be worked out.

3.6.3 Olfactory neurons — establishing neuronal identity by coordinately regulated gene expression

Analysing the regulatory region of a number of proteins specifically expressed in olfactory neurons, such as the olfactory marker protein (OMP) and components of olfactory signal transduction, identified a novel sequence motif binding specifically nuclear proteins present in olfactory neuroepithelium (Kudrycki *et al.* 1993; Wang *et al.* 1993). Olf-1, a helix-loop-helix transcription factor expressed exclusively in olfactory neurons and their precursor cells, binds to this sequence motif (Wang and Reed 1993). In transgenic mice, a 300 bp fragment of the OMP 5' flanking region containing the Olf-1 binding motif is sufficient for olfactory-specific expression of a reporter gene (Kudrycki *et al.* 1993). Surprisingly, a mutation that prevents the interaction of Olf-1 with its binding site does not alter the expression pattern of a reporter construct in the olfactory epithelium of transgenic mice (Kudrycki *et al.* 1998). As the mutated DNA sequence is still able to bind nuclear proteins from nervous tissue, the results are difficult to interpret.

The finding of related genes establishes Olf-1 as the founding member of the Olf-1/EBF-like HLH transcription factor family (Garel *et al.* 1997; Malgaretti *et al.* 1997; Wang *et al.* 1997). The expression of the Olf-1-relatives O/E2 and O/E3 in the olfactory epithelium indicates that there may be functional redundancy in their regulation of transcription (Wang *et al.* 1997). Their expression in other parts of the nervous system suggests that they may have additional functions in neuronal development.

Olf proteins represent transcription factors which may regulate an entire aspect of function in sensory neurons related to their stimulus transduction cascade. Still, the demonstration of their necessity in vivo is lacking due to technical difficulties. The expression of several family members in the olfactory epithelium and their possible redundant action compromises simple gene inactivation strategies. The mutation of promoter binding sites in transgene approaches is complicated by the creation of new, even more powerful binding sites due to the mutation. Thus, it still remains to be proven that a number of olfactory neuron-specific proteins involved in transduction of the sensory stimulus are regulated as a synexpression group *in vivo.*

3.7 **Specifying neuronal functions at the transcriptional level — the choice amongst the many**

The coordinate expression of neuronal genes is one important prerequisite for neuronal function in development and the mature state. As functionally relevant neuronal proteins and their respective genes typically come as families with a choice among two to dozens of homologous family members, another issue turns out to be important: the choice of expression of specific genes in neuronal subpopulations that will shape the precise function of a neuron in a neural network. Be it ion channels determining the kinetic properties of electrical membrane processes, transmitter synthesizing enzymes determining transmitter phenotype, or receptor expression determining the receptive features of a neuron, the specific function of a certain neuron not only rests on the expression of a certain set of genes but also on not expressing other, related genes. The activity of subpopulation-specific transactivating proteins may direct high level expression of defined target genes to specific subpopulations of neurons. Evidence exists that subpopulation-specific repression of neural genes may sharpen neuronal properties.

3.7.1 Specifying transmitter phenotypes may include transcriptional repression

In general neurons use either one of the transmitters glutamate, GABA, glycine, acetylcholine, noradrenaline, and some others, even though they may coexpress a range of neuropeptides and other mediators. In sympathetic neurons of birds and mammals, two populations of neurons exist which differ in their transmitter phenotype (Ernsberger and Rohrer 1999). A majority of neurons uses noradrenaline and is called 'noradrenergic', whereas a smaller population of neurons uses acetylcholine and is called 'cholinergic'. During development cholinergic neurons may be derived from noradrenergic ones (Landis 1990), but mature neurons are characterized by the usage of one of the transmitters and by a specific expression of the respective transmitter-synthesizing enzymes.

The analysis of the 5′ flanking regions of the genes for the transmitter-synthesizing enzymes choline acetyltransferase (ChAT) and dopamine β-hydroxylase (DBH) has provided some evidence that, in addition to inducing the 'right' enzyme in a neuron population, the expression of the 'wrong' enzyme is suppressed. A luciferase reporter driven by several kb of 5′ flanking sequences of the human ChAT gene expresses differently in cholinergic and non-cholinergic cell lines (Li *et al.* 1993). Elements with silencer activity were characterized which repress promoter activity in an adrenergic cell line to a much higher degree than in cholinergic cell lines. One of the silencer elements containes E-box motifs required for silencing activity (Li *et al.* 1995). Nuclear proteins from adrenergic cells specifically bind to these E-boxes. Using 2.3 kb of 5′ flanking region of the rat ChAT gene to drive a CAT reporter demonstrates enhancer activity in cholinergic and repressor activity in non-cholinergic cell lines and targets transgene expression to cholinergic sites (Lönnerberg *et al.* 1995). The activity of a RE1/NRSE motif does not discriminate between cholinergic and non-cholinergic neuronal cells but silences activity in non-neuronal cells (Lönnerberg *et al.* 1996). A different part of the flanking sequences demonstrates cholinergic-specific enhancer activity and is inactive in non-cholinergic neuronal and in non-neuronal cells. These data demonstrate that there is an interaction of positive and negative regulating elements which enhance ChAT expression in cholinergic cells and represses its expression in non-cholinergic and specifically adrenergic cells. They also suggest that the RE1/ NRSE may not be involved in this aspect of neuron subpopulation-specific silencing.

Analysis of 5′ flanking sequences of the DBH gene in transgenic mice complements this picture (Hoyle *et al.* 1994). Comparing reporter gene expression driven by 5′ flanking fragments of different length shows expression in noradrenergic and to different degrees in non-noradrenergic neuron populations. The comparison of a 1.1 and 1.5 kb fragment indicates the presence of sequences that may be responsible for the repression of expression in three sites which normally do not express DBH.

These studies illustrate the possible interaction of population-specific activation and silencing of gene expression. The proteins involved in this antagonistic regulation still have to be determined. The Phox2 transcription factors are implicated in noradrenergic induction (Ernsberger 2000). One has to bear in mind, however, that they are also expressed in other autonomic neurons (Ernsberger *et al.* 2000) and may induce the expression of both noradrenergic and cholinergic properties (Stanke *et al.* 1999).

3.7.2 Specifying receptive properties in sensory cells — locus control regions

Red–green color vision is based on different visual pigments encoded by an array of genes on the human X chromosome (Neitz and Neitz 1995). Individual photoreceptor cells express typically one gene. Analysis of the DNA sequences upstream of the visual pigment genes provides evidence that a locus control region is involved in the decision as to which of the genes from the cluster to express.

5′ flanking regions of visual pigment genes fused to a lacZ reporter direct expression to cone cells in transgenic mouse retina (Wang *et al.* 1992). A 600 bp fragment 5′ to the human red pigment gene is essential for expression and contains sequences highly conserved between vertebrate species. In an exciting set of experiments, a minimal human X chromosome visual pigment gene array was constructed, where the red and green pigment gene transcription units were replaced, respectively, by alkaline phosphatase and β-galactosidase reporters (Wang *et al.* 1999). In transgenic mice, the two enzymes were in the large majority of cases expressed in different cones. More than 70% of the expressing cones showed either one or the other enzyme activity. 30% expressed both activities. A fascinating aspect of this result is that the mutually exclusive expression of the two transgenes in different cones is achieved in a dichromat mammal.

An even more demanding task is the control of odorant receptor gene expression, where some thousand receptor encoding genes are expressed in unique patterns in the sensory neurons of the olfactory epithelium (Buck and Axel 1991; Buck 1992). The receptor genes are clustered across the genome (Ben-Arie *et al.* 1994; Sullivan *et al.* 1996) and expressed in a specific mode that has been hypothesized to occur via stochastic selection (Ressler *et al.* 1993). Comparing sequences surrounding the β-globin locus in mice and humans shows a high degree of conservation (Bulger *et al.* 1999). Surprisingly, the β-globin loci turned out to be embedded within an array of odorant receptor genes, suggesting a role of the β-globin locus control region in control of these odorant receptors. Deletion of the endogenous β-globin locus control region in mice did, however, not result in altered expression of neighbouring olfactory receptor genes (Bulger *et al.* 2000).

3.7.3 Conclusions

The hierarchical action of transcription factors in cascades during the rapid progression from neuronal precursors to immature neurons leads to the expression of a number of neuron-specific genes which in a large number of cases may be restricted to distinct neuronal subpopulations. A least in part, the expression of these genes may occur in a coordinate manner as synexpression groups. To sharpen the functional profile of a developing neuron, the induction of a certain set of genes may go along with the repression of others resulting in a mature phenotype or gene expression profile that distinguishes a certain neuronal subpopulation from others. Growth factors are involved in these induction processes. Neuronal activity may also shape gene expression profiles and contribute to plasticity in the mature nervous system.

3.8 **Regulation of gene expression by growth factors — a path from development to plasticity**

The alteration of gene expression by growth factors is an important way to regulate neuronal properties during development as well as in the mature organism. During development a multitude of growth factors is involved in different aspects of differentiation. The gene regulatory action of growth factors may be mediated by activation of transcription factors located in the cytoplasm as observed for the STAT (Bromberg and Chen 2001; Ihle 2001) and SMAD (Itoh *et al.* 2000; Massague and Wotton 2000) transcription factors after binding of neurokine and TGFβ family growth factors, respectively. In such cases receptor binding of the growth factors leads to direct phosphorylation of the respective transcription factors and sequestration of the activated transcription factor to the nucleus to elicit alterations in gene expression. Alternatively, the growth factor signal may be transduced by cytoplasmic signalling pathways which regulate the activity of protein kinases and only indirectly, via this protein kinases, the activity of nuclear transcription factors. The neurotrophin signalling pathway acts by this second mechanism to affect gene expression in the context of neuronal differentiation, survival, and plasticity.

Already twenty years ago, the importance of neurotrophin NGF (nerve growth factor)-induced changes in gene expression for neurite outgrowth in the phaeochromocytoma cell line PC12 was recognized (Greene *et al.* 1982). NGF induces large numbers of genes in PC12 cells which can be grouped into immediate early genes (IEG) regulated within minutes and late response genes (LRG) regulated within hours (Bonni and Greeneberg 1997). The IEGs include transcription factors such as c-fos and c-jun (Greenberg *et al.* 1985; Sheng and Greenberg 1990). NGF induction of c-fos in PC12 cells has been shown to be independent of new protein synthesis (Greenberg *et al.* 1986) and mediated by phosphorylation of transcription factors binding to c-fos promoter elements. Deletion analysis of the c-fos promoter in PC12 cells demonstrates the role of the serum response element (SRE) in the induction process (Sheng *et al.* 1988). In addition, the cAMP response element (CRE) in the c-fos promoter is required for NGF induction and NGF-induced phosphorylation of the CRE binding protein CREB at Ser-133 stimulates its ability to enhance transcription (Ginty *et al.* 1994). CRE and SRE binding factors interact to stimulate transcription (Bonni *et al.* 1995).

CREB, a stimulus-induced transcription factor (Shaywitz and Greenberg 1999), together with CREM and ATF-1 proteins forms a subfamily of leucine zipper-containing proteins. By means of the leucine zipper they may dimerize and, with the help of a basic domain, bind to DNA and regulate gene expression. These structural domains characterize them as members of the bZIP superfamily of transcriptional regulators where they are more distantly related to c-fos, c-jun, C/EBP and others. CREB proteins are particularly interesting in the context of neuronal gene expression as they mediate responses to growth factors and electrical activity and play a critical role in learning processes (Silva *et al.* 1998, and also Chapter 14).

For c-fos induction by the neurotrophin BDNF (brain-derived neurotrophic factor) in cortical neurons, the SRE is largely dispensable (Finkbeiner *et al.* 1997). BDNF activates CREB by calcium influx and CaMKIV as well as the Ras/ERK pathway to induce transcription through the CREB binding site (see also Chapter 10). Inhibition of the Ras/ERK pathway attenuates the survival promoting action of BDNF on granule cells in culture (Bonni *et al.* 1999). Importantly, CREB mutant overexpression blocks the survival effect whereas constitutively active CREB supports survival in the absence of the growth factor. Similarly, NGF-dependent survival of sympathetic neurons *in vitro* is blocked by CREB mutants, for example those that prevent phosphorylation at Ser-133 (Riccio *et al.* 1999). Again, a constitutively active CREB supports survival after NGF withdrawal. Important progress is made by the characterization of the pro-survival factor bcl-2 as neurotrophin target gene that is regulated by a CREB-dependent mechanism (Riccio *et al.* 1999).

The importance of CREB in activity and calcium-dependent gene regulation (see below, and also Chapter 10) as well as in growth factor-mediated survival and differentiation provokes the question to which extent the genes regulated in these diverse conditions overlap and what mechanisms may exist to select different groups of target genes. Important hints come from the bcl-2 and the BDNF gene itself which are both targets of CREB regulation (Shieh and Ghosh 1999; Finkbeiner 2000). In both cases, upstream regulating elements binding yet unknown transcription factors are important in addition to the CRE. The characterization of these upstream binding factors and their stage and cell type-specific expression may shed light on the issue of target gene selectivity. In addition selectivity may be achieved within a given cell via recruitment of different signalling pathways and different transcriptional coregulators. Neurotrophins applied to either distal axons or the region surrounding the neuronal cell body may recruit different kinases for intracellular signal transduction (Watson *et al.* 2001) potentially leading to the expression of diverging sets of target genes by specific phosphorylation of distinct transcription factors.

3.9 Ca^{2+}-dependent regulation of neural gene expression

After neuronal precursors go through a range of differentiation steps characterized by the expression of transcription factor cascades and the successive acquisition of neuronal properties regulated by different classes of growth factors, cells reach a quasi-stable condition, the mature neuronal state. This state is characterized by the cell's final position and connectivity as well as the expression of a certain set of neuron-specific genes whose protein products contribute to the neuron's functional properties. An essential feature of nervous tissue is that during this mature state, cellular properties and consequently network properties can be modified within certain limits to adjust to altering physiological conditions. The posttranslational modification of proteins plays a crucial role in this process, in particular on rapid timescales in the range of seconds to minutes. Adaptive alterations in the range of hours to weeks require transcriptional regulation of gene expression (Chapter 14).

Already more than 15 years ago, activation of ion channels was demonstrated to rapidly affect gene expression (Morgan and Curran 1986; Greenberg *et al.* 1986*b*). Ca^{2+} permeation through ion channels plays a crucial role in this process and voltage-gated calcium channels (Murphy *et al.* 1991) or NMDA receptors (Lerea *et al.* 1992; Bading *et al.* 1993, 1995) are the most prominent but not exclusive channel types involved. Activity-mediated calcium influx leads to a rapid and transient transcription of immediate early genes such as c-fos, c-jun, and others (Greenberg *et al.* 1986*b*; Bading *et al.* 1995) which does not depend on previous transcription or protein synthesis. Their gene products are transcription factors themselves which in the case of c-jun and c-fos may bind to AP1 motifs in downstream target genes to affect transcription.

The Ca^{2+} responsiveness is most thoroughly mapped in the c-fos gene (Finkbeiner and Greenberg 1998). Expression in transgenic mice of c-fos/lacZ fusion genes with mutations in distinct regulatory sequences of the c-fos gene such as the CRE, SRE, and AP-1 site demonstrates the requirement of multiple elements for full stimulus-induced activation (Robertson *et al.* 1995). Different signaling cascades can mediate the calcium control of expression from the c-fos promoter (Johnson *et al.* 1997) and may be specifically recruited by Ca^{2+} influx through NMDA receptors and L-type calcium channels (Bading *et al.* 1993). Interestingly, elevations in cytoplasmic and in nuclear calcium levels may act via different binding motifs, the SRE and CRE, respectively (Hardingham *et al.* 1997, and also Chapter 10). Blockade of nuclear but not cytoplasmic Ca^{2+} increase in the pituitary AtT20 cell line by microinjection of a non-diffusible calcium chelator demonstrates that increases in nuclear calcium concentration control calcium-activated gene expression mediated by the CRE motif.

Sustained phosphorylation of CRE-binding protein CREB is essential for regulation of c-fos expression in hippocampal and striatal neurons (Bito *et al.* 1996; Liu and Graybiel *et al.* 1996). Phosphorylated CREB interacts specifically with the nuclear protein CBP (Chrivia *et al.* 1993), which has intrinsic histone acetyltransferase activity (Bannister and Kouzarides 1996). In addition, CBP may associate with RNA polymerase II which recruits the polymerase to CREB in a phospho-Ser133-dependent manner as shown with nuclear extracts from Hela cells (Kee *et al.* 1996). Importantly, CREB phosphorylation and CREB-mediated transcription can be uncoupled by nuclear Ca^{2+} buffering or blockade of calcium/calmodulin-dependent protein kinases (CaMKs) (Chawla *et al.* 1998). This observation led to a 2-step model of transcriptional regulation by calcium-mediated CREB activation. The first step comprises the phosphorylation of CREB at Ser-133 and the recruitment of CBP. CBP recruitment may be differently controlled by Ca^{2+} influx through NMDA receptors and L-type calcium channels (Hardingham *et al.* 1999). The second step required for CREB-mediated transcription includes the additional activation of CBP by CaMKs.

Calcium-dependent nuclear CREB phosphorylation in hippocampal neurons can be induced by synaptic stimuli including those that alter synaptic strength (Deisseroth *et al.* 1996; Hardingham *et al.* 2001). The signal is initiated by a highly local rise of calcium concentration near the cell membrane and leads to the activation of CaMKIV (Bito *et al.* 1996). A cytoplasmic signaling pathway acting through ERK/RSK2 and a nuclear signaling pathway are involved. Initially, import of calcium/calmodulin has been proposed to deliver the nuclear signal (Deisseroth *et al.* 1998) whereas recent experiments with hippocampal neurons demonstrate that Ca^{2+} release from internal stores without protein import into the nucleus may be sufficient to mediate nuclear CREB phosphorylation (Hardingham *et al.* 2001). The kinetically distinct signaling pathways in combination with protein phosphatase action are considered to convey information about stimulus patterns to the nucleus (Wu *et al.* 2001). The stimulation frequency-coding of cytoplasmic Ca^{2+} signals is converted in the nucleus to amplitude-coded signals which result in transcriptional responses (Hardingham *et al.* 2001).

Ca^{2+} signaling to the nucleus enables the transcription apparatus to regulate neuronal gene expression as a function of the neuronal activity pattern. Electrical stimulation of hippocampal afferents results in the expression of different immediate early genes depending on the stimulation pattern (Worley *et al.* 1993). Electrical stimulation of mouse dorsal root ganglion neurons shows the importance of temporal dynamics on the degree of stimulation of c-fos expression (Fields *et al.* 1997). In addition, expression of cell adhesion molecules such as N-cadherin, NCAM, and L1 in cultured dorsal root ganglion neurons can be differentially regulated by distinct action potential patterns (Itoh *et al.* 1997).

In conclusion, the sensitivity of the gene regulatory response to changing activation patterns provides neurons with the ability to integrate the spatial and temporal aspects of synaptic activation at the level of gene expression. The degree of complexity of this regulation process still remains to be uncovered. This is nicely illustrated by the characterization of calcium-responsive elements (CaREs) different from CRE in the BDNF gene and the cloning of a new transcription factor binding to CaREs, CaRF (Tao *et al.* 2002). The fact that CaRF shows no apparent homology to known proteins as well as the interaction between the CaRF-binding CaRE and the CRE suggest that even more transcriptional regulators and elements involved in stimulus-mediated gene expression may be uncovered.

3.10 **Transcriptional regulation of learning processes**

A remarkable feature of nervous tissue is the storage of information induced by neural activity and synaptic plasticity is considered to be the fundamental underlying mechanism (Milner *et al.* 1998). Stimulation-induced long-lasting alterations in neuronal connectivity require RNA and protein synthesis as shown for sensitization and facilitation in Aplysia or long-term potentiation in mammals (Bailey *et al.* 1996). Key molecular players in this process have been recognised during the last decade even though the targets of the regulatory process may have to be determined in more detail.

The requirement for new protein synthesis in long-term memory formation is known since several decades (Davis and Squire 1984) and points to the importance of new gene expression in long-lasting learning processes. Comparison of learning processes in vertebrates and invertebrates strongly suggests that conserved gene regulatory mechanisms are involved (Kandel 2001). In particular CREB plays a central role in the formation of long-term memory in *Aplysia, Drosophila,* and mice (Yin and Tully 1996). More recent studies indicate that the bZIP transcription factor C/EBP is involved downstream of CREB in invertebrate and vertebrate learning cascades.

Sensitization, classical conditioning, and habituation in the marine snail *Aplysia* result from changes in the strength of synaptic connections between sensory and motor neurons (Kandel 2001). Experiments with cultured neurons, where long-term changes in synaptic connectivity may be induced by repeated puffs of the transmitter serotonin (Montarolo *et al.* 1986), show that the number of sensory neuron varicosities changes (Bailey *et al.* 1992) and that these processes depend on RNA and protein synthesis. The blockade of long-term facilitation by injection of DNA fragments containing CRE motifs (Dash *et al.* 1990) as well as the initiation of long-term memory processes by injection of phosphorylated CREB protein (Bartsch *et al.* 1998) suggests that binding of phosphorylated CREB to CRE motifs in target genes is a crucial step in these invertebrate learning processes. Interestingly, different CREB protein isoforms may affect learning processes differently (Bartsch *et al.* 1998). Downstream of CREB, C/EBP transcription factors may be essentially involved in learning as disruption of C/EBP function blocks long-term synaptic facilitation (Alberini *et al.* 1994). These observations seem to be of general importance as this cascade of transcription factors also seems to work in vertebrate long-term memory formation.

In vertebrates a number of different learning protocols are available for experimentation (Stork and Welzl 1999) and the regulation of immediate early gene expression such as c-fos is known from different brain regions studied in different species under different learning regimes (Tischmeyer and Grimm, 1999). In addition, late memory-related gene expression is under investigation with RNA fingerprinting and gene profiling (Cavallaro *et al.* 1997, 2001). Hippocampus-dependent learning increases CRE-dependent gene expression in CA1 and CA3 of transgenic mice carrying the lacZ gene under control of a promoter with a CRE motif (Impey *et al.* 1998). This indicates that the alterations in gene expression related to long-term memory may be mediated by CREB transcription factors which is corroborated by learning deficiencies observed in mice with mutation of CREB (Bourtchuladze *et al.* 1994). Disruption of hippocampal CREB levels by antisense oligonucleotides also may affect learning (Guzowski and McGaugh 1997). As in the learning processes studied in Aplysia, C/EBP transcription factors play a critical role downstream of CREB in long-term memory consolidation in the mammalian hippocampus (Taubenfeld *et al.* 2001*a*, *b*). Since the immediate induction of CREB and the delayed induction of C/EBP β are temporally discontinuous (Carew and Sutton 2001), the precise relationship between the induction of the two transcription factor families as well as their regulation of downstream target genes remains to be uncovered.

3.11 **Perspective**

Two decades ago, the number of neuronally expressed genes was estimated to be around 20 000 (Sutcliff *et al.* 1984). Since that time, a range of transcription factors involved in the regulation of neuronal gene expression have been characterized. They belong to a variety of families with distinct structural domains such as zinc fingers, basic helix-loop-helix domains, homeo domains, leucine zippers and others. These factors control the differentiation of precursor cells into mature neurons along a series of developmental stages which have not been fully characterized yet. The succession of different transcription factors during the differentiation process is being analysed in different neuronal lineages, but no unifying picture is currently available. The progression from an early action of zinc finger proteins during segregation of neuronal and non-neuronal lineages, via bHLH and HD protein function during specification of neuronal subpopulations, to the action of bZIP factors during maturation and plasticity (Fig. 3.5) is not more than a highly simplified working hypothesis. Still, this model emphasizes an important aspect of neuronal gene expression, namely its distinct regulation during development and in the mature state. This is nicely illustrated by the expression of a reporter driven by regulatory sequences of the tyrosine hydroxylase gene in transgenic mice (Trocmé *et al.* 1998). While the CRE motif in this promoter is dispensible for embryonic expression, no reporter expression is detectable in adult animals when the CRE motif is mutated. An important question to be solved is the molecular characterisation of the transition from embryonic to adult gene expression which may involve alterations in chromatine structure in addition to the expression of different transcription factor combinations.

Even though an increasing number of transcription factors has been shown to act during different periods of neurogenesis, it is fair to say that we do not understand the inner grammar of neuronal gene expression. Already upon induction of general neuronal properties, subpopulation-specific features become apparent (Ernsberger 2001), and it is not yet clear how this is coordinated with the segregation of neuronal and non-neuronal pathways. Likewise, the precise molecular mechanism resulting in the diversification of gene expression patterns characterizing different neuron classes remains to be determined. Comparison of transcription factors and their function in a given neuronal population of different species will demonstrate what is indispensible for the generation of a certain class of neurons. The comparison of different neuron populations will show the prerequisites for diversification. Forced expression of

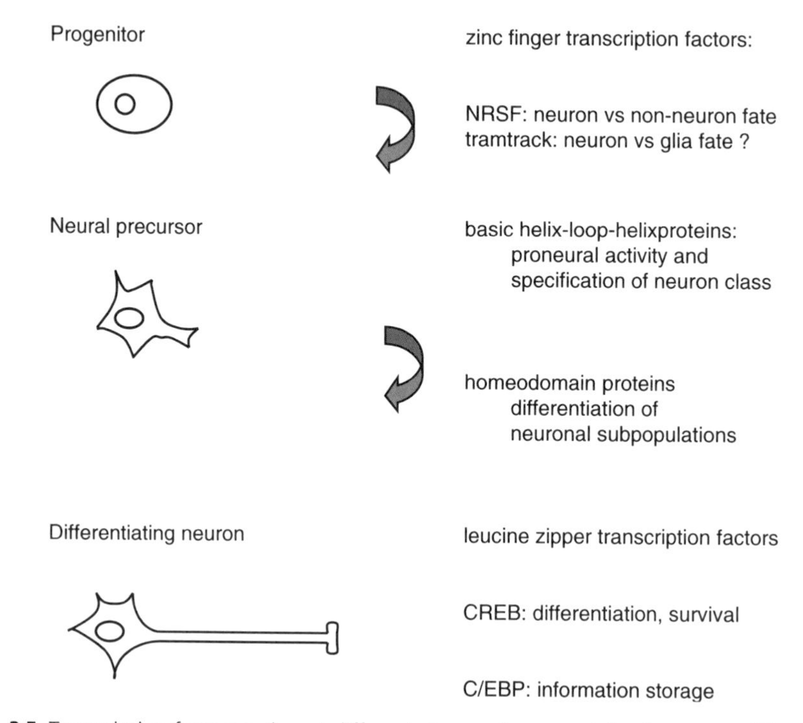

Fig. 3.5 Transcription factors acting at different stages of neuronal development. Different families of transcription factors are observed to affect gene expression in neurons and their precursors at different developmental stages. Down-regulation of REST/NRSF and tramtrack may clear the path for a progenitor to neuronal differentiation. bHLH and HD proteins may initiate neuronal differentiation proper by regulating the expression of general and subpopulation-specific neuronal properties. During final neuronal maturation as well as for survival and plasticity, CREB and C/EBP are important.

transcriptional regulators in stem cells or neuronal precursors should provide experimental proof and will be of importance in clinical application.

The complete sequencing of genomes in different species including humans will eventually enable us to paint a full picture of regulators involved in gene expression. Currently, we cannot be sure how many transcription factors in addition to those characterized are involved in the regulation of a certain gene at a certain developmental stage in the cell type under investigation. We also do not know the full account of population-specific gene expression. cDNA arrays will provide a means to screen the expression of large numbers of genes (Luo and Geschwind 2001) and thus may bridge the gap between the molecular and the systems approach in neurosciences (Geschwind 2000). This technique is being applied to developing and adult nervous tissue (Geschwind *et al.* 2001; Miki *et al.* 2001; Mody *et al.* 2001). Its full potential will show when purified population of primary neural cells are analysed and the sensitivity is sufficient to detect gene expression of low abundance. The combination of such approaches with conditional knockout strategies for transcriptional regulators can be expected to considerably further our understanding of basic principles and population-specific details in neuronal gene expression.

3.12 Acknowledgements

I wish to thank Hilmar Bading (Heidelberg), Jean-Francois Brunet (Marseille), and Gerald Thiel (Homburg) for their critical reading and helpful suggestions on parts of the manuscript.

References

Alberini, C. M., Ghirardi, M., Metz, R., and Kandel, E. R. (1994) C/EBP is an immediate-early gene required for the consolidation of long-term facilitation in Aplysia. *Cell*, **76**, 1099–114.

Andres, M. E., Burger, C., Peral-Rubio, M. J., Battaglioli, E., Anderson, M. E., Grimes, J., *et al.* (1999) CoREST: a functional corepressor required for regulation of neural-specific gene expression. *Proc. Natl. Acad. Sci. USA*, **96**, 9873–7.

Appel, B., Korzh, V., Glasgow, E., Thor, S., Edlund, T., Dawid, I. B., *et al.* (1995) Motoneuron fate specification revealed by patterned LIM homeobox gene expression in embryonic zebrafish. *Development*, **121**, 4117–25.

Badenhorst, P. (2001) Tramtrack controls glial number and identity in the Drosophila embryonic CNS. *Development*, **128**, 4093–101.

Badenhorst, P., Harrison, S., and Travers, A. (1996) End of the line? Tramtrack and cell fate determination in Drosophila. *Genes Cells*, **1**, 707–16

Bading, H., Ginty, D. D., and Greenberg, M. E. (1993) Regulation of gene expression in hippocampal neurons by distinct calcium signaling pathways. *Science*, **260**, 181–6.

Bading, H., Segal, M. M., Sucher, N. J., Dudek, H., Lipton, S. A., and Greenberg, M. E. (1995) N-methyl-D-aspartate receptors are critical for mediating the effects of glutamate on intracellular calcium concentration and immediate early gene expression in cultured hippocampal neurons. *Neuroscience*, **64**, 653–64.

Bailey, C. H., Montarolo, P., Chen, M., Kandel, E. R., and Schacher, S. (1992) Inhibitors of protein and RNA synthesis block structural changes that accompany long-term heterosynaptic plasticity in Aplysia. *Neuron*, **9**, 749–58.

Bailey, C. H., Bartsch, D., and Kandel, E. R. (1996) Toward a molecular definition of long-term memory storage. *Proc. Natl. Acad. Sci. USA*, **93**, 13445–52.

Ballas, N., Battaglioli, E., Atouf, F., Andres, M. E., Chenoweth, J., Anderson, M. E., *et al.* (2001) Regulation of neuronal traits by a novel transcriptional complex. *Neuron*, **31**, 353–65.

Bannister, A. J. and Kouzaridis, T. (1996) The CBP co-activator is a histone acetyltransferase. *Nature*, **384**, 641–3.

Bartsch, D., Casadio, A., Karl, K. A., Serodio, P., and Kandel, E. R. (1998) CREB1 encodes a nuclear activator, a repressor, and a cytoplasmic modulator that form a regulatory unit critical for long-term facilitation. *Cell*, **95**, 211–23.

Belmont, A. S., Dietzel, S., Nye, A. C., Strukov, Y. G., and Tumbar, T. (1999) Large-scale chromatin structure and function. *Curr. Opin. Cell Biol.*, **11**, 307–11.

Ben-Arie, N., Lancet, D., Taylor, C., Khen, M., Walker, N., Ledbetter, D. H., *et al.* (1994) Olfactory receptor gene cluster on human chromosome 17: possible duplication of an ancestral receptor repertoire. *Hum. Mol. Genet.*, **3**, 229–35.

Bessis, A., Champtiaux, N., Chatelin, L., and Changeux, J. P. (1997) The neuron-restrictive silencer element: a dual enhancer/silencer crucial for patterned expression of a nicotinic receptor gene in the brain. *Proc. Natl. Acad. Sci. USA*, **94**, 5906–11.

Bito, H., Deisseroth, K., and Tsien, R. W. (1996) CREB phosphorylation and dephosphorylation: a Ca(2+)- and stimulus duration-dependent switch for hippocampal gene expression. *Cell*, **87**, 1203–14.

Bodner, M., Castrillo, J. L., Theill, L. E., Deerinck, T., Ellisman, M., and Karin, M. (1988) The pituitary-specific transcription factor GHF-1 is a homeobox-containing protein. *Cell*, **55**, 505–18.

Bonni, A. and Greeneberg, M. E. (1997) Neurotrophin regulation of gene expression. *Can. J. Neurol. Sci.*, **24**, 272–83

Bonni, A., Ginty, D. D., Dudek, H., and Greenberg, M. E. (1995) Serine 133-phosphorylated CREB induces transcription via a cooperative mechanism that may confer specificity to neurotrophin signals. *Mol. Cell Neurosci.*, **6**, 168–83.

Bonni, A., Brunet, A., West, A. E., Datta, S. R., Takasu, M. A., and Greenberg, M. E. (1999) Cell survival promoted by the Ras-MAPK signaling pathway by transcription-dependent and -independent mechanisms. *Science*, **286**, 1358–62.

Boulter, J., O'Shea-Greenfield, A., Duvoisin, R. M., Conolly, J. G., Wada, E., Jensen, A., *et al.* (1990) Alpha 3, alpha 5, and beta 4: three members of the rat neuronal nicotinic acetylcholine receptor-related gene family form a gene cluster. *J. Biol. Chem.*, **265**, 4472–82.

Bourtchuladze, R., Frenguelli, B., Blendy, J., Cioffi, D., Schutz, G., and Silva, A. J. (1994) Deficient long-term memory in mice with a targeted mutation of the cAMP-responsive element binding protein. *Cell*, **79**, 59–68.

Brené, S., Messer, C., Okado, H., Hartley, M., Heinemann, S. F., and Nestler, E. J. (2000) Regulation of GluR2 promoter activity by neurotrophic factors via a neuron-restrictive silencer element. *Eur. J. Neurosci.*, **12**, 1525–33.

Briscoe, J. and Ericson, J. (2001) Specification of neuronal fates in the ventral neural tube. *Curr. Opin. Neurobiol.*, **11**, 43–9.

Briscoe, J., Sussel, L., Serup, P., Hartigan-O'Connor, D., Jessell, T. M., Rubenstein, J. L., *et al.* (1999) Homeobox gene Nkx2.2 and specification of neuronal identity by graded sonic hedgehog signalling. *Nature*, **398**, 622– 7.

Briscoe, J., Pierani, A., Jessell, T. M., and Ericson, J. (2000) A homeodomain protein code specifies progenitor cell identity and neuronal fate in the ventral neural tube. *Cell*, **101**, 435–45.

Bromberg, J. and Chen, X. (2001) STAT proteins: signal transducers and activators of transcription. *Meth. Enzymol.*, **333**, 138–51

Brown, N. L., Patel, S., Brzezinski, J., and Glaser, T. (2001) Math5 is required for retinal ganglion cell and optic nerve formation. *Development*, **128**, 2497–508.

Brunet, J. F. and Ghysen, A. (1999) Deconstructing cell determination: proneural genes and neuronal identity. *Bioessays*, **21**, 313–8.

Buck, L. B. (1992) The olfactory multigene family. *Curr. Opin. Neurobiol.*, **2**, 282–8.

Buck, L. and Axel, R. (1991) A novel multigene family may encode odorant receptors: a molecular basis for odor recognition. *Cell*, **65**, 175–87.

Bulger, M., van Doorninck, J. H., Saitoh, N., Telling, A., Farrell, C., Bender, M. A., *et al.* (1999) Conservation of sequence and structure flanking the mouse and human beta-globin loci: the beta-globin genes are embedded within an array of odorant receptor genes. *Proc. Natl. Acad. Sci. USA*, **96**, 5129–34.

Bulger, M., Bender, M. A., van Doorninck, J. H., Wertman, B., Farrell, C. M., Felsenfeld, G., *et al.* (2000) Comparative structural and functional analysis of the olfactory receptor genes flanking the human and mouse beta-globin gene clusters. *Proc. Natl. Acad. Sci. USA*, **97**, 14560–5.

Carew, T. J. and Sutton, M. A. (2001) Molecular stepping stones in memory consolidation. *Nature Neuroscience*, **4**, 769–71.

Cavallaro, S., Meiri, N., Yi, C. L., Musco, S., Ma, W., Goldberg, J., and Alkon, D. L. (1997) Late memory-related genes in the hippocampus revealed by RNA fingerprinting. *Proc. Natl. Acad. Sci. USA*, **94**, 9669–73.

Cavallaro, S., Schreurs, B. G., Zhao, W., D'Agata, V., *et al.* (2001) Gene expression profiles during long-term memory consolidation. *European J. Neurosci.*, **13**, 1809–15.

Chawla, S., Hardingham, G. E., Quinn, D. R., and Bading, H. (1998) CBP: a signal-regulated transcriptional coactivator controlled by nuclear calcium and CaM kinase IV. *Science*, **281**, 1505–9.

Chen, Z. F., Paquette, A. J., and Anderson, D. J. (1998) NRSF/REST is required in vivo for repression of multiple neuronal target genes during embryogenesis. *Nat. Genet.*, **20**, 136–42.

Chien, C. T., Hsiao, C. D., Jan, L. Y., and Jan. Y. N. (1996) Neuronal type information encoded in the basic-helix-loop-helix domain of proneural genes. *Proc. Natl. Acad. Sci. USA*, **93**, 13239–44.

Chong, J. A., Tapia-Ramirez, J., Kim, S., Toledo-Aral, J. J., Zheng, Y., Boutros, M. C., *et al.* (1995) REST: a mammalian silencer protein that restricts sodium channel gene expression to neurons. *Cell*, **80**, 949–57.

Chrivia, J. C., Kwok, R. P., Lamb, N., Hagiwara, M., Montminy, M. R., and Goodman, R. H. (1993) Phosphorylated CREB binds specifically to the nuclear protein CBP. *Nature*, **365**, 855–9.

Cockell, M. and Gasser, S. M (1999) Nuclear compartments and gene regulation. *Curr. Opin. Genet. Dev.*, **9**, 199–205.

Dasen, J. S. and Rosenfeld, M. G. (1999) Signaling mechanisms in pituitary morphogenesis and cell fate determination. *Curr. Opin. Cell Biol.*, **11**, 669–77.

Dasen, J. S. and Rosenfeld, M. G. (2001) Signaling and transcriptional mechanisms in pituitary development. *Annu. Rev. in Neurosci.*, **24**, 327–55.

Dash, P. K., Hochner, B., and Kandel, E. R. (1990) Injection of the cAMP-responsive element into the nucleus of Aplysia sensory neurons blocks long-term facilitation. *Nature*, **345**, 718–21.

Davis, H. P. and Squire, L. R. (1984) Protein synthesis and memory: a review. *Psychological Bulletin*, **96**, 518–59.

De Gois, S., Houhou, L., Oda, Y., Corbex, M., Pajak, F., Thévenot, E., *et al.* (2000) Is RE1/NRSE a common cis-regulatory sequence for ChAT and VAChT genes? *J. Biol. Chem.*, **275**, 36683–90.

Deisseroth, K., Bito, H., and Tsien, R. W. (1996) Signaling from synapse to nucleus: postsynaptic CREB phosphorylation during multiple forms of hippocampal synaptic plasticity. *Neuron*, **16**, 89–101.

Deisseroth, K., Heist, E. K., and Tsien, R. W. (1998) Translocation of calmodulin to the nucleus supports CREB phosphorylation in hippocampal neurons. *Nature*, **392**, 198–202.

Dillon, N. and Sabbattini, P. (2000) Functional gene expression domains: defining the functional unit of eukaryotic gene regulation. *BioEssays*, **22**, 657–65.

DiMattia, G. E., Rhodes, S. J., Krones, A., Carrière, C., O'Connell, S., Kalla, K., *et al.* (1997) The Pit-1 gene is regulated by distinct early and late pituitary-specific enhancers. *Developmental Biology*, **182**, 180–90.

Dollé, P., Castrillo, J. L., Theill, L. E., Deerinck, T., Ellisman, M., and Karin, M. (1990) Expression of GHF-1 protein in mouse pituitaries correlates both temporally and spatially with the onset of growth hormone gene activity. *Cell*, **60**, 809–20.

Dubreuil, V., Hirsch, M. R., Pattyn, A., Brunet, J. F., and Goridis, G. (2000) The Phox2b transcription factor coordinately regulates neuronal cell cycle exit and identity. *Development*, **127**, 5191–201.

Duggan, A., Ma, C., and Chalfie, M. (1998) Regulation of touch receptor differentiation by the Caenorhabditis elegans mec-3 and unc-86 genes. *Development*, **125**, 4107–19.

Eastman, C., Horvitz, H. R., and Jin, Y. (1999) Coordinated transcriptional regulation of the unc-25 glutamic acid decarboxylase and the unc-47 GABA vesicular transporter by the Caenorhabditis elegans UNC-30 homeodomain protein. *J. Neurosci.*, **19**, 6225–34.

Eiden, L. E. (1998) The cholinergic gene locus. *J. Neurochem.*, **70**, 2227–40.

Eggen, B. J. and Mandel, G. (1997) Regulation of sodium channel gene expression by transcriptional silencing. *Dev. Neurosci.*, **19**, 25–6.

Eng, S. R., Gratwick, K., Rhee, J. M., Fedtsova, N., Gan, L., and Turner, E. E. (2001) Defects in sensory axon growth precede neuronal death in Brn3a-deficient mice. *J. Neurosci.*, **21**, 541–9.

Erkman, L., Yates, P. A., McLaughlin, T., McEvilly, R. J., Whisenhunt, T., O'Connell, S. M., *et al.* (2000) A POU domain transcription factor-dependent program regulates axon pathfinding in the vertebrate visual system. *Neuron*, **28**, 779–92.

Ernsberger, U. (2000) Evidence for an evolutionary conserved role of BMP growth factors and Phox2 transcription factors during noradrenergic differentiation of sympathetic neurons: induction of a putative synexpression group of neurotransmitter-synthesizing enzymes. *Eur. J. Biochem.*, **267**, 6976–81.

Ernsberger, U. (2001) The development of postganglionic sympathetic neurons: coordinating neuronal differentiation and diversification. *Auton. Neurosci.*, **94**, 1–13.

Ernsberger, U. and Rohrer, H. (1999) Development of the cholinergic neurotransmitter phenotype in postganglionic sympathetic neurons. *Cell and Tissue Research*, **297**, 339–61.

Ernsberger, U., Patzke, H., Tissier-Seta, J. P., Reh, T., Goridis, C., and Rohrer, H. (1995) The expression of tyrosine hydroxylase and the transcription factors cPhox–2 and Cash-1: evidence for distinct inductive steps in the differentiation of chick sympathetic precursor cells. *Mech. Dev.*, **52**, 125–36.

Ernsberger, U., Reissmann, E., Mason, I., and Rohrer, H. (2000) The expression of dopamine β-hydroxylase, tyrosine hydroxylase, and Phox2 transcription factors in sympathetic neurons: evidence for common regulation during noradrenergic induction and diverging regulation later in development. *Mech. Dev.*, **92**, 169–77.

Felsenfeld, G. (1996) Chromatin unfolds. *Cell*, **86**, 13–19.

Fields, R. D., Eshete, F., Stevens, B., and Itoh, K. (1997) Action potential-dependent regulation of gene expression: temporal specificity in Ca2+, cAMP-responsive element binding proteins, and mitogen-activated protein kinase signaling. *J. Neurosci.*, **17**, 7252–66.

Finkbeiner, S. (2000) CREB couples neurotrophin signals to survival messages. *Neuron*, **25**, 11–14.

Finkbeiner, S. and Greenberg, M. E. (1998) Ca2+ channel-regulated neuronal gene expression. *J. Neurobiol.*, **37**, 171–89.

Finkbeiner, S., Tavazoie, S. F., Maloratsky, A., Jacobs, K. M., Harris, K. M., and Greenberg, M. E. (1997) CREB: a major mediator of neuronal neurotrophin responses. *Neuron*, **19**, 1031–47.

Finney, M., Ruvkun, G., and Horvitz, H. R. (1988) The C. elegans cell lineage and differentiation gene unc-86 encodes a protein with a homeodomain and extended similarity to transcription factors. *Cell*, **55**, 757–69.

Fode, C., Gradwohl, G., Morin, X., Dierich, A., LeMeur, M., Goridis, C., *et al.* (1998) The bHLH protein NEUROGENIN 2 is a determination factor for epibranchial placode-derived sensory neurons. *Neuron*, **20**, 483–94.

Gan, L., Wang, S. W., Huang, Z., and Klein, W. H. (1999) POU domain factor Brn-3b is essential for retinal ganglion cell differentiation and survival but not for initial cell fate specification. *Dev. Biol.*, **210**, 469–80.

Garel, S., Marin, F., Mattei, M. G., Vesque, C., Vincent, A., and Charnay, P. (1997) Family of Ebf/Olf-1-related genes potentially involved in neuronal differentiation and regional specification in the central nervous system. *Dev. Dyn.*, **210**, 191–205.

Geschwind, D. H. (2000) Mice, microarrays, and the genetic diversity of the brain. *Proc. Natl. Acad. Sci. USA*, **97**, 10676–8.

Geschwind, D. H., Ou, J., Easterday, M. C., Dougherty, J. D., Jackson, R. L., Chen, Z., *et al.* (2001) A genetic analysis of neural progenitor differentiation. *Neuron*, **29**, 325–39.

Ghysen, A. and Dambly-Chaudiere, C. (1988) From DNA to form: the achaete-scute complex. *Genes Dev.*, **2**, 495–501.

Ginty, D. D., Bonni, A., and Greenberg, M. E. (1994) Nerve growth factor activates a Ras-dependent protein kinase that stimulates c-fos transcription via phosphorylation of CREB. *Cell*, **77**, 713–25.

Giraldo, P. and Montoliu, L. (2001) Size matters: use of YACs, BACs and PACs in transgenic animals. *Transgenic Res.*, **10**, 83–103.

Gowan, K., Helms, A. W., Hunsaker, T. L., Collisson, T., Ebert, P. J., Odom, R., *et al.* (2001) Crossinhibitory activities of ngn1 and math1 allow specification of distinct dorsal interneurons. *Neuron*, **31**, 219–32.

Grant, A. L. and Wisden, W. (1997) Neuron-specific gene expression. In R. W. Davies and B. J. Morris, ed. *Molecular Biology of the Gene*, pp 67–93. Bios Scientific Publishers, Oxford.

Greene, L. A., Burstein, D. E., and Black, M. M. (1982) The role of transcription-dependent priming in nerve growth factor promoted neurite outgrowth. *Dev. Biol.*, **91**, 305–16.

Greenberg, M. E., Greene, L. A., and Ziff, E. B. (1985) Nerve growth factor and epidermal growth factor induce rapid transient changes in proto-oncogene transcription in PC12 cells. *J. Biol. Chem.*, **260**, 14101–10.

Greenberg, M. E., Hermanowski, A. L., and Ziff, E. B. (1986*a*) Effect of protein synthesis inhibitors on growth factor activation of c-fos, c-myc, and actin gene transcription. *Mol. Cell. Biol.*, **6**, 1050–7.

Greenberg, M. E., Ziff, E. B., and Greene, L. A. (1986*b*) Stimulation of neuronal acetylcholine receptors induces rapid gene transcription. *Science*, **234**, 80–3

Griffith, E. C., Cowan, C. W., and Greenberg, M. E. (2001) REST acts through multiple deacetylase complexes. *Neuron*, **31**, 339–44.

Groves, A. K., George, K. M., Tissier-Seta, J. P., Engel, J. D., Brunet, J. F., and Anderson, D. J. (1995) Differential regulation of transcription factor gene expression and phenotypic markers in developing sympathetic neurons. *Development*, **121**, 887–901.

Guillemot, F., Lo, L. C., Johnson, J. E., Auerbach, A., Anderson, D. J., and Joyner, A. L. (1993) Mammalian achaete-scute homologue 1 is required for the early development of olfactory and autonomic neurons. *Cell*, **75**, 463–76.

Guo, S., Brush, J., Teraoka, H., Goddard, A., Wilson, S. W., Mullins, M. C., *et al.* (1999) Development of noradrenergic neurons in the zebrafish hindbrain requires BMP, FGF8, and the homeodomain protein soulless/Phox2a. *Neuron*, **24**, 555–66.

Guzowski, J. F. and McGaugh, J. L. (1997) Antisense oligodeoxynucleotide-mediated disruption of hippocampal cAMP response element binding protein levels impairs consolidation of memory for water maze training. *Proc. Natl. Acad. Sci. USA*, **94**, 2693–8.

Hardingham, G. E., Chawla, S., Johnson, C. M., and Bading, H. (1997) Distinct functions of nuclear and cytoplasmic calcium in the control of gene expression. *Nature*, **385**, 260–5.

Hardingham, G. E., Chawla, S., Cruzalegui, F. H., and Bading, H. (1999) Control of recruitment and transcription-activating function of CBP determines gene regulation by NMDA receptors and L-type calcium channels. *Neuron*, **22**, 789–98.

Hardingham, G. E., Arnold F. J., and Bading, H. (2001) Nuclear calcium signaling controls CREB-mediated gene expression triggered by synaptic activity. *Nat. Neurosci.*, **4**, 261–7.

Hatakeyama, J., Tomita, K., Inoue T., and Kageyama, R. (2001) Roles of homeobox and bHLH genes in specification of a retinal cell type. *Development*, **128**, 1313–22.

He, X., Treacy, M. N., Simmons, D. M., Ingraham, H. A., Swanson, L.W., and Rosenfeld, M. G. (1989) Expression of a large family of POU-domain regulatory genes in mammalian brain development. *Nature*, **340**, 35–41.

Heintz, N. (2000) Analysis of mammalian central nervous system gene expression and function using bacterial artificial chromosome-mediated transgenesis. *Hum. Mol. Genet.*, **9**, 937–43.

Hendricks, T., Francis, N., Fyodorov D., and Deneris, E. S. (1999) The ETS domain factor Pet-1 is an early and precise marker of central serotonin neurons and interacts with a conserved element in serotonergic genes. *J. Neurosci.*, **19**, 10348–56.

Hill, J. A. Jr, Zoli, M., Bourgeois, J. P., and Changeux, J. P. (1993) Immunocytochemical localization of a neuronal nicotinic receptor: the beta-2 subunit. *J. Neurosci.*, **13**, 1551–68.

Howard, M. J., Stanke, M., Schneider, C., Wu X., and Rohrer, H. (2000) The transcription factor dHAND is a downstream effector of BMPs in sympathetic neuron specification. *Development*, **127**, 4073–81.

Hoyle, G. W., Mercer, E. H., Palmiter, R. D., and Brinster, R. L. (1994) Cell-specific expression from the human dopamine β-hydroxylase promoter in transgenic mice is controlled via a combination of positive and negative regulatory elements. *J. Neurosci.*, **14**, 2455–63.

Huang, Y., Myers, S. J., and Dingledine, R. (1999*a*) Transcriptional repression by REST: recruitment of Sin3A and histone deacetylase to neuronal genes. *Nat. Neurosci.*, **2**, 867–72.

Huang, E. J., Zang, K., Schmidt, A., Saulys, A., Xiang, M., and Reichardt, L. F. (1999*b*) POU domain factor Brn-3a controls the differentiation and survival of trigeminal neurons by regulating trk receptor expression. *Development*, **126**, 2869–82.

Ihle, J. N. (2001) The Stat family in cytokine signaling. *Curr. Opin. Cell Biol.*, **13**, 211–17.

Impey, S., Smith, D. M., Obrietan, K., Donahue, R., Wade C., and Storm, D. R. (1998) Stimulation of cAMP response element (CRE)-mediated transcription during contextual learning. *Nat. Neurosci.*, **1**, 595–601.

Ingraham, H. A., Chen, R. P., Mangalam, H. J., Elsholtz, H. P., Flynn, S. E., Lin, C. R., *et al.* (1988) A tissue-specific transcription factor containing a homeodomain specifies a pituitary phenotype. *Cell*, **55**, 519–29.

Itoh, K., Ozaki, M., Stevens, B., and Fields, R. D. (1997) Activity-dependent regulation of N-cadherin in DRG neurons: differential regulation of N-cadherin, NCAM, and L1 by distinct patterns of action potentials. *J. Neurobiol.*, **33**, 735–48.

Itoh, S., Itoh, F., Goumans, M. J., and Ten Dijke, P. (2000) Signaling of transforming growth factor-beta family members through Smad proteins. *Eur. J. Biochem.*, **267**, 6954–67.

Jarman, A. P., Grau, Y., Jan, L. Y., and Jan, Y. N. (1993) Atonal is a proneural gene that directs chordotonal organ formation in the Drosophila peripheral nervous system. *Cell*, **73**, 1307–21.

Jarman, A. P., Grell, E. H., Ackerman, L., Jan, L. Y., and Jan, Y. N. (1994) Atonal is the proneural gene for Drosophila photoreceptors. *Nature*, **369**, 398–400.

Jenuwein, T. and Allis, C. D. (2001) Translating the histone code. *Science*, **293**, 1074–80.

Jin, Y., Hoskins, R., and Horvitz, H. R. (1994) Control of type-D GABAergic neuron differentiation by C. elegans unc-30 homeodomain protein. *Nature*, **372**, 780–3.

Johnson, C. M., Hill, C. S., Chawla, S., Treisman, R., and Bading, H. (1997) Calcium controls gene expression via three distinct pathways that can function independently of the Ras/mitogen-activated protein kinases (ERKs) signaling cascade. *J. Neurosci.*, **17**, 6189–202.

Jurata, L. W., Thomas., J. B., and Pfaff, S. L. (2000) Transcriptional mechanisms in the development of motor control. *Curr. Opin. Neurobiol.*, **10**, 72–9.

Kallunki, P., Edelman, G. M., and Jones, F. S. (1997) Tissue-specific expression of the L1 adhesion molecule is modulated by the neural restrictive silencer element. *J. Cell Biol.*, **138**, 1343–54.

Kandel, E. R. (2001) The molecular biology of memory storage: a dialogue between genes and synapses. *Science*, **294**, 1030–8.

Kapur, R. P., Hoyle, G. W., Mercer, E. H., Brinster, R. L., and Palmiter, R. D. (1991) Some neuronal cell populations express human dopamine β-hydroxylase-lacZ transgenes transiently during embryonic development. *Neuron*, **7**, 717–27.

Kay, J. N., Finger-Baier, K. C., Roeser, T., Staub, W., and Baier, H. (2001) Retinal ganglion cell genesis requires lakritz, a zebrafish atonal homologue. *Neuron*, **30**, 725–36.

Kee, B. L., Arias, J., and Montminy, M. R. (1996) Adaptor-mediated recruitment of RNA polymerase II to a signal-dependent activator. *J. Biol. Chem.*, **271**, 2373–5.

Kim, H. S., Seo, H., Yang, C., Brunet, J. F., and Kim, K. S. (1998) Noradrenergic-specific transcription of the dopamine beta-hydroxylase gene requires synergy of multiple cis-acting elements including at least two Phox2a-binding sites. *J. Neurosci.*, **18**, 8247–60.

Kraner, S. D., Chong, J. A., Tsay, H. J., and Mandel, G. (1992) Silencing the type II sodium channel gene: a model for neural-specific gene regulation. *Neuron*, **9**, 37–44.

Kudrycki, K., Stein-Izsak, C., Behn, C., Grillo, M., Akeson, R., and Margolis, F. L. (1993) Olf-1-binding site: characterization of an olfactory neuron-specific promoter motif. *Mol. Cell. Biol.*, **13**, 3002–14.

Kudrycki, K. E., Buiakova, O., Tarozzo, G., Grillo, M., Walters, E., and Margolis, F. L. (1998) Effects of mutation of the Olf-1 motif on transgene expression in olfactory receptor neurons. *J. Neurosci. Res.*, **52**, 159–72.

Lakin, N. D., Morris, P. J., Theil, T., Sato, T. N., Moroy, T., Wilson, M. C., *et al.* (1995) Regulation of neurite outgrowth and SNAP-25 gene expression by the Brn-3a transcription factor. *J. Biol. Chem.*, **270**, 15858–63.

Landis, S. C. (1990) Target regulation of neurotransmitter phenotype. *Trends Neurosci.*, **13**, 344–50.

Lee, J. E. (1997) Basic helix-loop-helix genes in neural development. *Curr. Opin. Neurobiol.*, **7**, 13–20.

Lee, J. H., Shimojo, M., Chai, Y. G., and Hersh, L. B. (2000) Studies on the interaction of REST4 with the cholinergic repressor element-1/neuron restrictive silencer element. *Mol. Brain Res.*, **80**, 88–98.

Lemon, B. and Tjian, R. (2000) Orchestrated response: a symphony of transcription factors for gene control. *Genes Dev.*, **14**, 2551–69.

Lerea, L. S., Butler. L. S., and McNamara, J. O. (1992) NMDA and non-NMDA receptor-mediated increase of c-fos mRNA in dentate gyrus neurons involves calcium influx via different routes. *J. Neurosci.*, **12**, 2973–81.

Li, S., Crenshaw III, E. B., Rawson, E. J., Simmons, D. M., Swanson, L. W., and Rosenfeld, M. G. (1990) Dwarf locus mutants lacking three pituitary cell types result from mutations in the POU-domain gene pit-1. *Nature*, **347**, 528–33.

Li, Y. P., Baskin, F., Davis, R., and Hersh, L. B. (1993) Cholinergic neuron-specific expression of the human choline acetyltransferase gene is controlled by silencer elements. *J. Neurochem.*, **61**, 748–51.

Li, Y. P., Baskin, F., Davis, R., Wu, D., and Hersh, L. B. (1995) A cell type-specific silencer in the human choline acetyltransferase gene requiring two distinct and interactive E boxes. *Mol. Brain Res.*, **30**, 106–14.

Lietz, M., Bach, K., and Thiel, G. (2001) Biological activity of RE-1 silencing transcription factor (REST) towards distinct transcriptional activators. *Eur. J. Neurosci.*, **14**, 1303–12.

Lillicrop, K. A., Budrahan, V. S., Lakin, N. D., Terrenghi, G., Wood, J. N., Polak, J. M., *et al.* (1992) A novel POU family transcription factor is closely related to brn-3 but has a distinct expression pattern in neuronal cells. *Nucleic Acids Res.*, **20**, 5093–6.

Liu, F. C. and Graybiel, A. M. (1996) Spatiotemporal dynamics of CREB phosphorylation: transient versus sustained phosphorylation in the developing striatum. *Neuron*, **17**, 1133–44.

Liu, W., Mo, Z., and Xiang, M. (2001) The ath5 proneural genes function upstream of Brn3 POU domain transcription factor genes to promote retinal ganglion cell development. *Proc. Natl. Acad. Sci. USA*, **98**, 1649–54.

Lo, L. C., Johnson, J. E., Wuenschell, C. W., Saito, T., and Anderson, D. J. (1991) Mammalian achaete scute homologue I is transiently expressed by spatially restricted subsets of early neuroepithelial and neural crest cells. *Genes Dev.*, **5**, 1524–37.

Lönnerberg, P., Lehndahl, U., Funakoshi, H., Ärhlund-Richter, L., Persson, H., and Ibánez, C. F. (1995) Regulatory region in choline acetyltransferase gene directs developmental and tissue-specific expression in transgenic mice. *Proc. Natl. Acad. Sci. USA*, **92**, 4046–50.

Lönnerberg, P., Schoenherr, C. J., Anderson, D. J., and Ibánez, C. F. (1996) Cell type-specific regulation of choline acetyltransferase gene expression. *J. Biol. Chem.*, **271**, 33358–65.

Lopez-Coviella, I., Berse, B., Krauss, R., Thies, R. S., and Blusztajn, J. K. (2000) Induction and maintenance of the neuronal cholinergic phenotype in the central nervous system by BMP-9. *Science*, **289**, 313–16.

Luo, Z. and Geschwind, D. H. (2001) Microarray applications in neuroscience. *Neurobiol. Dis.*, **8**, 183–93.

Lumsden, A. (1995) A 'LIM code' for motor neurons? *Curr. Biol.*, **5**, 491–5.

Ma, Q., Chen, Z., del Barco Barrantes I., de la Pompa, J. L., and Anderson, D. J. (1998) Neurogenin1 is essential for the determination of neuronal precursors for proximal cranial sensory ganglia. *Neuron*, **20**, 469–82.

Ma, L., Merenmies J., and Parada, L. F. (2000) Molecular characterization of the TrkA/NGF receptor minimal enhancer reveals regulation by multiple cis elements to drive embryonic neuron expression. *Development*, **127**, 3777–88.

Magin, A., Lietz, M., Cibelli, G., and Thiel, G. (2002) RE-1 silencing transcription factor-4 (REST4) is neither a transcriptional repressor nor a de-repressor. *Neurochem. Int.*, **40**, 195–2002.

Malgaretti, N., Pozzoli, O., Bosetti, A., Corradi, A., Ciarmatori, S., Panigada, M., *et al.* (1997) Mmot1, a new helix-loop-helix transcription factor gene displaying a sharp expression boundary in the embryonic mous brain. *J. Biol. Chem.*, **272**, 17632–9.

Mallet, J., Houhou, L., Pajak, F., Oda, Y., Cervini, R., Bejanin, S., and Berrard, S. (1998) The cholinergic locus: ChAT and VAChT genes. *Journal de Physiologie (Paris)*, **92**, 145–7.

Mandel G. and McKinnon, D. (1993) Molecular basis of neural-specific gene expression. *Annu. Rev. Neurosci.*, **16**, 323–45.

Mangalam, H. J., Albert, V. R., Ingraham, H. A., Kapiloff, M., Wilson, L., Nelson, C., *et al.* (1989) A pituitary POU domain protein, Pit-1, activates both growth hormone and prolactin promoters transcriptionally. *Genes Dev.*, **3**, 946–58.

Marquardt, T., and Pfaff, S. L. (2001) Cracking the transcriptional code for cell specification in the neural tube. *Cell*, **106**, 651–4.

Massague, J. and Wotton, D. (2000) Transcriptional control by the TGF-beta/Smad signaling system. *EMBO J.*, **19**, 1745–54

Maue, R. A., Kraner, S. D., Goodman, R. H., and Mandel, G. (1990) Neuron-specific expression of the rat brain type II sodium channel gene is directed by upstream regulatory elements. *Neuron*, **4**, 223–31.

McEvilly, R. J. and Rosenfeld, M. G. (1999) The role of POU domain proteins in the regulation of mammalian pituitary and nervous system development. *Prog. Nucleic Acid Res. Mol. Biol.*, **63**, 223–55.

Mercer, E. H., Hoyle, G. W., Kapur, R. P., Brinster, R. L., and Palmiter, R. D. (1991) The dopamine β-hydroxylase gene promoter directs expression of E. coli lacZ to sympathetic and other neurons in adult transgenic mice. *Neuron*, **7**, 703–16.

Miki, R., Kadota, K., Bono, H., Mizuno, Y., Tomaru, Y., Carninci, P., *et al.* (2001) Delineating developmental and metabolic pathways in vivo by expression profiling using the RIKEN set of 18,816 full-length enriched mouse cDNA arrays. *Proc. Natl. Acad. Sci. USA*, **98**, 2199–204.

Milner, B., Squire, L. R., and Kandel, E. R. (1998) Cognitive neuroscience and the study of memory. *Neuron*, **20**, 445–68.

Misawa, H., Ishii, K., and Deguchi, T. (1992) Gene expression of mouse choline acetyltransferase: Alternative splicing and identification of a highly active promoter region. *J. Biol. Chem.*, **267**, 20392–9.

Misawa, H., Takahashi, R., and Deguchi, T. (1995) Coordinate expression of vesicular acetylcholine transporter and choline acetyltransferase in sympathetic superior cervical neurons. *Neuroreport*, **6**, 965–8.

Mizuguchi, R., Sugimori, M., Takebayashi, H., Kosako, H., Nagao, M., Yoshida, S., *et al.* (2001) Combinatorial roles of olig2 and neurogenin2 in the coordinated induction of pan-neuronal and subtype-specific properties of motoneurons. *Neuron*, **31**, 757–71.

Mody, M., Cao, Y., Cui, Z., Tay, K. Y., Shyong, A., Shimizu, E., *et al.* (2001) Genome-wide gene expression profiles of the developing mouse hippocampus. *Proc. Natl. Acad. Sci. USA*, **98**, 8862– 7.

Montarolo, P. G., Goelet, P., Castellucci, V. F., Morgan, J., Kandel, E. R., and Schacher, S. (1986) A critical period for macromolecular synthesis in long-term heterosynaptic facilitation in Aplysia. *Science*, **234**, 1249–54.

Morgan, J. I. and Curran, T. (1986) Role of ion flux in the control of c-fos expression. *Nature*, **322**, 552–5.

Mori, N., Schoenherr, C., Vandenbergh, D. J., and Anderson, D. J. (1992) A common silencer element in the SCG10 and type II Na+ channel genes binds a factor present in nonneuronal cells but not in neuronal cells. *Neuron*, **9**, 45–54.

Morin, X., Cremer, H., Hirsch, M. R., Kapur, R. P., Goridis, C., and Brunet, J. F. (1997) Defects in sensory and autonomic ganglia and absence of locus coeruleus in mice deficient for the homeobox gene Phox2a. *Neuron*, **18**, 411–23.

Morris, P. J., Lakin, N. D., Dawson, S. J., Ryabinin, A. E., Kilimann, M. W., Wilson, M. C., *et al.* (1996) Differential regulation of genes encoding synaptic proteins by members of the Brn-3 subfamily of POU transcription factors. *Mol. Brain Res.*, **43**, 279–85.

Mu, X., Zhao, S., Pershad, R., Hsieh, T. F., Scarpa, A., Wang, S. W., *et al.* (2001) Gene expression in the developing mouse retina by EST sequencing and microarray analysis. *Nucleic Acids Res.*, **29**, 4983–93.

Murphy, T. H., Worley, P. F., and Baraban, J. M. (1991) L-type voltage-sensitive calcium channels mediate synaptic activation of immediate early genes. *Neuron*, **7**, 625–35.

Myers, S. J., Dingledine, R., and Borges, K. (1999) Genetic regulation of glutamate receptor ion channels. *Ann. Rev. Pharmacol. Toxicol.*, **39**, 221–41.

Näär, A. M., Lemon, B. D., and Tjian, R. (2001) Transcriptional coactivator complexes. *Annu. Rev. Biochem.*, **70**, 475–501.

Naciff, J. M., Behbehani, M. M., Misawa, H., and Dedman, J. R. (1999) Identification and transgenic analysis of a murine promoter that targets cholinergic neuron expression. *J. Neurochem.*, **72**, 17–28.

Naruse, Y., Aoki, T., Kojima, T., and Mori, N. (1999) Neural restrictive silencer factor recruits mSin3 and histone deacetylase complex to repress neuron-specific target genes. *Proc. Natl. Acad. Sci. USA*, **96**, 13691–6.

Neitz, M. and Neitz, J. (1995) Numbers and ratios of visual pigment genes for normal red-green color vision. *Science*, **267**, 1013–16.

Nelson, C., Albert, V. R., Elsholtz, H. P., Lu, L. I., and Rosenfeld, M. G. (1988) Activation of cell-specific expression of rat growth hormone and prolactin genes by a common transcription factor. *Science*, **239**, 1400–5.

Niehrs, C. and Pollet, N. (1999) Synexpression groups in eukaryotes. *Nature*, **402**, 483–7.

Novitch, B. G., Chen, AI., and Jessell, T. M. (2001) Coordinate regulation of motor neuron subtype identity and pan-neuronal properties by the bHLH repressor olig2. *Neuron*, **31**, 773–89.

Olson, E. C., Schinder, A. F., Dantzker, J. L., Marcus, E. A., Spitzer, N. C., and Harris, W. A. (1998) Properties of ectopic neurons induced by Xenopus neurogenin1 misexpression. *Mol. Cell. Neurosci.*, **12**, 281–99.

Orphanides, G., Lagrange, T., and Reinberg, D. (1996) The general transcription factors of RNA polymerase II. *Genes Dev.*, **10**, 2657–83.

Oyler, G. A., Higgins, G. A., Hart, R. A., Battenberg, E., Billingsley, M., Bloom, F. E., *et al.* (1989) The identification of a novel synaptosomal-associated protein, SNAP-25, differentially expressed by neuronal subpopulations. *J. Cell Biol.*, **109**, 3039–52.

Palm, K., Belluardo, N., Metsis, M., and Timmusk, T. (1998) Neuronal expression of zinc finger transcription factor REST/NRSF/XBR gene. *J. Neurosci.*, **18**, 1280–96.

Palmiter, R. D. and Brinster, R. L. (1986) Germ-line transformation of mice. *Annu. Rev. Genet.*, **20**, 465–99.

Pattyn, A., Morin, X., Cremer, H., Goridis, C., and Brunet, J. F. (1997) Expression and interactions of the two closely related homeobox genes Phox2a and Phox2b during neurogenesis. *Development*, **124**, 4065–75.

Pattyn, A., Morin, X., Cremer, H., Goridis, C., and Brunet, J. F. (1999) The homeobox gene Phox2b is essential for the development of autonomic neural crest derivatives. *Nature*, **399**, 366–70.

Pattyn, A., Goridis, C., and Brunet, J. F. (2000*a*) Specification of the central noradrenergic phenotype by the homeobox gene Phox2b. *Mol. Cell. Neurosci.*, **15**, 235–43.

Pattyn, A., Hirsch, M. R., Goridis, C., and Brunet, J. F. (2000*b*) Control of hindbrain motor neuron differentiation by the homeobox gene Phox2b. *Development*, **127**, 1349–58.

Patzke, H., Reissmann, E., Stanke, M., Bixby, J. L., and Ernsberger, U. (2001) BMP growth factors and Phox2 transcription factors can induce synaptotagmin I and neurexin I during sympathetic neuron development. *Mech. Dev.*, **108**, 149–59.

Pfaff, S. L., Mendelsohn M., Stewart, C. L., Edlund, T., and Jessell, T. M. (1996) Requirement for LIM homeobox gene Isl1 in motor neuron generation reveals a motor neuron-dependent step in interneuron differentiation. *Cell*, **84**, 309–20.

Riccio, A., Ahn, S., Davenport, C. M., Blendy, J. A., and Ginty, D. D. (1999) Mediation by a CREB family transcription factor of NGF-dependent survival of sympathetic neurons. *Science*, **286**, 2358–61.

Ressler, K. J., Sullivan, S. L., and Buck, L. B. (1993) A zonal organization of odorant receptor gene expression in the olfactory epithelium. *Cell*, **73**, 597–609.

Robertson, L. M., Kerppola, T. K., Vendrell, M., Luk, D., Smeyne, R. J., Bocchiaro, C., *et al.* (1995) Regulation of c-fos expression in transgenic mice requires multiple interdependent transcription control elements. *Neuron*, **14**, 241–52.

Rockelein, I., Rohrig, S., Donhauser, R., Eimer, S., and Baumeister, R. (2000) Identification of amino acid residues in the Caenorhabditis elegans POU protein UNC-86 that mediate UNC-86-MEC-3-DNA ternary complex formation. *Mol. Cell. Biol.*, **20**, 4806–13.

Roopra, A., Sharling, L., Wood, I. C., Briggs, T., Bachfischer, U., Paquette, A. J., *et al.* (2000) Transcriptional repression by neuron-restrictive silencer factor is mediated via the Sin3-histone deacetylase complex. *Mol. Cell. Biol.*, **20**, 2147–57.

Sander, M., Paydar, S., Ericson, J., Briscoe, J., Berber, E., German, M., *et al.* (2000) Ventral neural patterning by Nkx homeobox genes: Nkx 6.1 controls somatic motor neuron and ventral interneuron fates. *Genes Dev.*, **14**, 2134–9.

Scardigli, R., Schuurmans, C., Gradwohl, G., and Guillemot, F. (2001) Crossregulation between neurogenin2 and pathways specifying neuronal identity in the spinal cord. *Neuron*, **31**, 203–17.

Schoenherr, C. J. and Anderson D. J. (1995) The neuron-restrictive silencer factor (NRSF): a coordinate repressor of multiple neuron–specific genes. *Science*, **267**, 1360–3.

Schoenherr, C. J., Paquette, A. J., and Anderson, D. J. (1996) Identification of potential target genes for the neuron-restrictive silencer factor. *Proc. Natl. Acad. Sci. USA*, **93**, 9881–6.

Schonemann, M. D., Ryan, A. K., Erkman, L., McEvilly, R. J., Bermingham J., and Rosenfeld, M. G. (1998) POU domain factors in neural development. *Adv. Exp. Med. Biol.*, **449**, 39–53.

Shaskus, J., Greco, D., Asnani, L. P., and Lewis, E. J. (1992) A bifunctional genetic regulatory element of the rat dopamine beta-hydroxylase gene influences cell type specificity and second messenger-mediated transcription. *J. Biol. Chem.*, **267**, 18821–30.

Shaskus, J., Zellmer, E., and Lewis, E. J. (1995) A negative regulatory element in the rat dopamine β-hydroxylase gene contributes to the cell type specificity of expression. *J. Neurochem.*, **64**, 52–60.

Shaywitz. A. J. and Greenberg. M. E. (1999) CREB: a stimulus-induced transcription factor activated by a diverse array of extracellular signals. *Annu. Rev. Biochem.*, **68**, 821–61.

Sheng, M., Dougan, S. T., McFadden, G., and Greenberg, M. E. (1988) Calcium and growth factor pathways of c-fos transcriptional activation require distinct upstream regulatory sequences. *Mol. Cell. Biol.*, **8**, 2787–96.

Sheng, M. and Greenberg, M. E. (1990) The regulation and function of c-fos and other immediate early genes in the nervous system. *Neuron*, **4**, 477–85.

Shieh, P. B. and Ghosh, A. (1999) Molecular mechanisms underlying activity-dependent regulation of BDNF expression. *J. Neurobiol.*, **41**, 127–34.

Shimojo, M., Paquette, A. J., Anderson, D. J., and Hersh, L. B. (1999) Protein kinase A regulates cholinergic gene expression in PC12 cells: REST4 silences the silencing activity of neuron-restrictive silencer factor/REST. *Mol. Cell. Biol.*, **19**, 6788–95.

Silva, A. J., Kogan, J. H., Frankland, P. W., and Kida, S. (1998) CREB and memory. *Annu. Rev. Neurosci.*, **21**, 127–48.

Simmons, D. M., Voss, J. W., Ingraham, H. A., Holloway, J. M., Broide, R. S., Rosenfeld, M. G., *et al.* (1990) Pituitary cell phenotypes involve cell-specific Pit-1 mRNA translation and synergistic interactions with other classes of transcription factors. *Genes Dev.*, **4**, 695–711.

Sommer, L., Shah, N., Rao, M., and Anderson D. J. (1995) The cellular function of MASH1 in autonomic neurogenesis. *Neuron*, **15**, 1245–58.

Sornson, M. W., Wu, W., Dasen, J. S., Flynn, S. E., Norman, D. J., O'Connell, S. M., *et al.* (1996) Pituitary lineage determination by the Prophet of Pit-1 homeodomain factor defective in Ames dwarfism. *Nature*, **384**, 327–33.

Stanke, M., Junghans, D., Geissen, M., Goridis, C., Ernsberger, U., and Rohrer, H. (1999) The Phox2 homeodomain proteins are sufficient to promote the development of sympathetic neurons. *Development*, **126**, 4087–94.

Stork, O. and Welzl, H. (1999) Memory formation and the regulation of gene expression. *Cell. Mol. Life Sci.*, **55**, 575–92.

Strahl, B. D. and Allis, C. D. (2000) The language of covalent histone modifications. *Nature*, **403**, 41–5.

Sullivan, S. L., Adamson, M. C., Ressler, K. J., Kozak, C. A., and Buck, L. B. (1996) The chromosomal distribution of mouse odorant receptor genes. *Proc. Natl. Acad. Sci. USA*, **93**, 884–8.

Sutcliffe, J. G., Milner, R. J., Gottesfeld, J. M., and Reynolds, W. (1984) Control of neuronal gene expression. *Science*, **225**, 1308–15.

Tanabe, Y. and Jessell, T. M. (1996) Diversity and pattern in the developing spinal cord. *Science*, **274**, 1115–23.

Tanabe, Y., William, C., and Jessell, T. M. (1998) Specification of motor neuron identity by the MNR2 homeodomain protein. *Cell*, **95**, 67–80.

Tao, X., West, A. E., Chen, W. G., Corfas, G., and Greenberg, M. E. (2002) A calcium-responsive transcription factor, CaRF, that regulates neuronal activity-dependent expression of BDNF. *Neuron*, **33**, 383–95.

Taubenfeld, S. M., Wiig, K. A., Monti, B., Dolan, B., Pollonini, G., and Alberini, C. M. (2001*a*) Fornix-dependent induction of hippocampal CCAAT enhancer-binding protein [beta] and [delta] Co-localizes with phosphorylated cAMP response element-binding protein and accompanies long-term memory consolidation. *J. Neurosci.*, **21**, 84–91.

Taubenfeld, S. M., Milekic, M. H., Monti, B., and Alberini, C. M. (2001*b*) The consolidation of new but not reactivated memory requires hippocampal C/EBPβ. *Nat. Neurosci.*, **4**, 813–18.

Tischmeyer, W. and Grimm, R. (1999) Activation of immediate early genes and memory formation. *Cell. Mol. Life Sci.*, **55**, 564–74.

Thiel, G., Lietz, M., and Cramer, M. (1998) Biological activity and modular structure of RE-1-silencing transcription factor (REST), a repressor of neuronal genes. *J. Biol. Chem.*, **273**, 26891–9.

Thiel, G., Lietz, M., and Leichter, M. (1999) Regulation of neuronal gene expression. *Naturwissenschaften*, **86**, 1–7.

Trocmé, C., Sarkis, C., Hermel, J. M., Duchateau, R., Harrison, S., Simonneau, M., *et al.* (1998) CRE and TRE sequences of the rat tyrosine hydroxylase promoter are required for TH basal expression in adult mice but not in the embryo. *Eur. J. Neurosci.*, **10**, 508–21.

Tsuchida, T., Ensini, M., Morton, S. B., Baldassare, M., Edlund, T., Jessell, T. M., *et al.* (1994) Topographic organization of embryonic motor neurons defined by expression of LIM homeobox genes. *Cell*, **79**, 957–70.

Turner, E. E., Jenne, K. J., and Rosenfeld, M. G. (1994) Brn-3.2: a brn-3-related transcription factor with distinctive central nervous system expression and regulation by retinoic acid. *Neuron*, **12**, 205–18.

Valarché, I., Tissier-Seta, J. P., Hirsch, M. R., Martinez, S., Goridis, C., and Brunet, J. F. (1993) The mouse homeodomain protein Phox2 regulates Ncam promoter activity in concert with Cux/CDP and is a putative determinant of neurotransmitter phenotype. *Development*, **119**, 881–96.

Verrijzer, C. P. (2001) Transcription factor IID — not so basal after all. *Science*, **293**, 2010–11.

Wang, M. M. and Reed, R. R. (1993) Molecular cloning of the olfactory neuronal transcription factor Olf-1 by genetic selection in yeast. *Nature*, **364**, 121–6.

Wang, M. M., Tsai, R. Y., Schrader, K. A., and Reed, R. R. (1993) Genes encoding components of the olfactory signal transduction cascade contain a DNA binding site that may direct neuronal expression. *Mol. Cell. Biol.*, **13**, 5805–13.

Wang, S. S., Tsai, R. Y., and Reed, R. R. (1997) The characterization of the Olf-1/EBF-like HLH transcription factor family: implications in olfactory gene regulation and neuronal development. *J. Neurosci.*, **17**, 4149–58.

Wang, S. W., Gan, L., Martin, S. E., and Klein, W. H. (2000) Abnormal polarization and axon outgrowth in retinal ganglion cells lacking the POU-domain transcription factor Brn-3b. *Mol. Cell. Neurosci.*, **16**, 141–56.

Wang, S. W., Mu, X., Bowers, W. J., Kim, D. S., Plas, D. J., Crair, M. C., *et al.* (2002) Brn3b/Brn3c double knockout mice reveal an unsuspected role for Brn3c in retinal ganglion cell axon outgrowth. *Development*, **129**, 467–77.

Wang, Y., Macke, J. P., Merbs, S. L., Zack, D. J., Klaunberg, B., Bennett, J., *et al.* (1992) A locus control region adjacent to the human red and green visual pigment genes. *Neuron*, **9**, 429–40.

Wang, Y., Smallwood, P. M., Cowan, M., Blesh, D., Lawler, A., and Nathans, J. (1999) Mutually exclusive expression of human red and green visual pigment-reporter transgenes occurs at high frequency in murine cone photoreceptors. *Proc. Natl. Acad. Sci. USA*, **96**, 5251–6.

Watson, F. L., Heerssen, H. M., Bhattacharyya, A., Klesse, L., Lin, M. Z., and Segal, R. A. (2001) Neurotrophins use the Erk5 pathway to mediate a retrograde survival response. *Nat. Neurosci.*, **4**, 981–7.

Way, J. C. and Chalfie, M. (1989) The mec-3 gene of Caenorhabditis elegans requires its own product for maintained expression and is expressed in three neuronal cell types. *Genes Dev.*, **3**, 1823–33.

Worley, P. F., Bhat, R. V., Baraban, J. M., Erickson, C. A., McNaughton, B. L., and Barnes, C. A. (1993) Thresholds for synaptic activation of transcription factors in hippocampus: correlation with long-term enhancement. *J. Neurosci.*, **13**, 4776–86.

Wu, G. Y., Deisseroth, K., and Tsien, R. W. (2001) Activity-dependent CREB phosphorylation: convergence of a fast, sensitive calmodulin kinase pathway and a slow, less sensitive mitogen-activated protein kinase pathway. *Proc. Natl. Acad. Sci. USA*, **98**, 2808–13.

Xiang, M. (1998) Requirement for Brn-3b in early differentiation of postmitotic retinal ganglion cell precursors. *Dev. Biol.*, **197**, 155–69.

Xiang, M., Zhou, L., Peng, Y. W., Eddy, R. L., Shows, T. B., and Nathans, J. (1993) Brn-3b: a POU domain gene expressed in a subset of retinal ganglion cells. *Neuron*, **11**, 689–701.

Xiang, M., Gao, W. Q., Hasson, T., and Shin, J. J. (1998) Requirement for Brn-3c in maturation and survival, but not in fate determination of inner ear hair cells. *Development*, **125**, 3935–46.

Xue, D., Tu, Y., and Chalfie, M. (1993) Cooperative interactions between the Caenorhabditis elegans homeoproteins UNC-86 and MEC-3. *Science*, **261**, 1324–8.

Yang, C., Kim, H. S., Seo, H., Kim, C. H., Brunet, J. F., and Kim, K. S. (1998) Paired-like homeodomain proteins, Phox2a and Phox2b, are responsible for noradrenergic cell-specific transcription of the dopamine β-hydroxylase gene. *J. Neurochem.*, **71**, 1813–26.

Yang, X., Yang, F., Fyodorov, D., Wang, F., McDonough, J., Herrup, K., *et al.* (1997) Elements between the protein-coding regions of the adjacent beta 4 and alpha 3 acetylcholine receptor genes direct neuron-specific expression in the central nervous system. *J. Neurobiol.*, **32**, 311–24.

Yin, J. C. and Tully, T. (1996) CREB and the formation of long-term memory. *Curr. Opin. Neurobiol.*, **6**, 264–8.

Zellmer, E., Zhang, Z., Greco, D., Rhodes, J., Cassel, S., and Lewis, E. J. (1995) A homeodomain protein selectively expressed in noradrenergic tissue regulates transcription of neurotransmitter biosynthetic genes. *J. Neurosci.*, **15**, 8109–20.

Chapter 4

Protein trafficking in neurons

Christopher N. Connolly

4.1 Introduction

Neurons are the basic cellular units that comprise neural networks that are responsible for information flow in the nervous system. Information transfer between neurons is carried by chemical signals, in the form of neurotransmitters, which are released specifically (from presynaptic sites on a donor neuron) onto discrete postsynaptic sites presenting the appropriate neurotransmitter receptors on the recipient neuron. Information flow through the neuron is conducted electrically to presynaptic neurotransmitter release sites where neurotransmitter is released onto the next neuron in the pathway. However, neurons are not passive electrical cables but integrative units, collating information from multiple sources. Information received may be excitatory or inhibitory and the cumulative response to these inputs is unequivocal, to fire an action potential or to remain silent.

These processing skills require a highly organized cellular structure and consequently, neurons exhibit a remarkably complex phenotype. From a typical neuron, multiple extensions (neurites) emanate from the cell body as either dendrites or axons (Fig. 4.1). Neurons may possess multiple (1–11), relatively short (<2 mm), and highly branched (0–440 branch points) dendrites that are covered in synapses. Dendritic postsynaptic sites are responsible for receiving and processing the vast majority of excitatory inputs, with inhibitory inputs being located on both dendrites and the cell body/axonal initial segment (AIS). A single excitatory input is insufficient to activate a neuron, however, upon multiple 'hits' a significant membrane depolarization is transferred to the proximal region of the axon, where voltage-gated sodium (Nav) channels are stimulated to produce an action potential along a single axon. Unlike dendrites, that capture and collate local information, axons can extend over a metre (e.g. motor neurons) to reach their target and transfer information. In neurons where axons transmit output over such large distances, axons are ensheathed by myelin-rich Schwann cell (PNS) or oligodendrocyte (CNS) membranes. This sheath serves to reduce decay of the electrical output as it travels long distances along the axon. Furthermore, signal amplification sites exist at gaps in the myelin sheath, called nodes of Ranvier, where Nav channels respond to reinforce the signal. The action potential travels to the end of

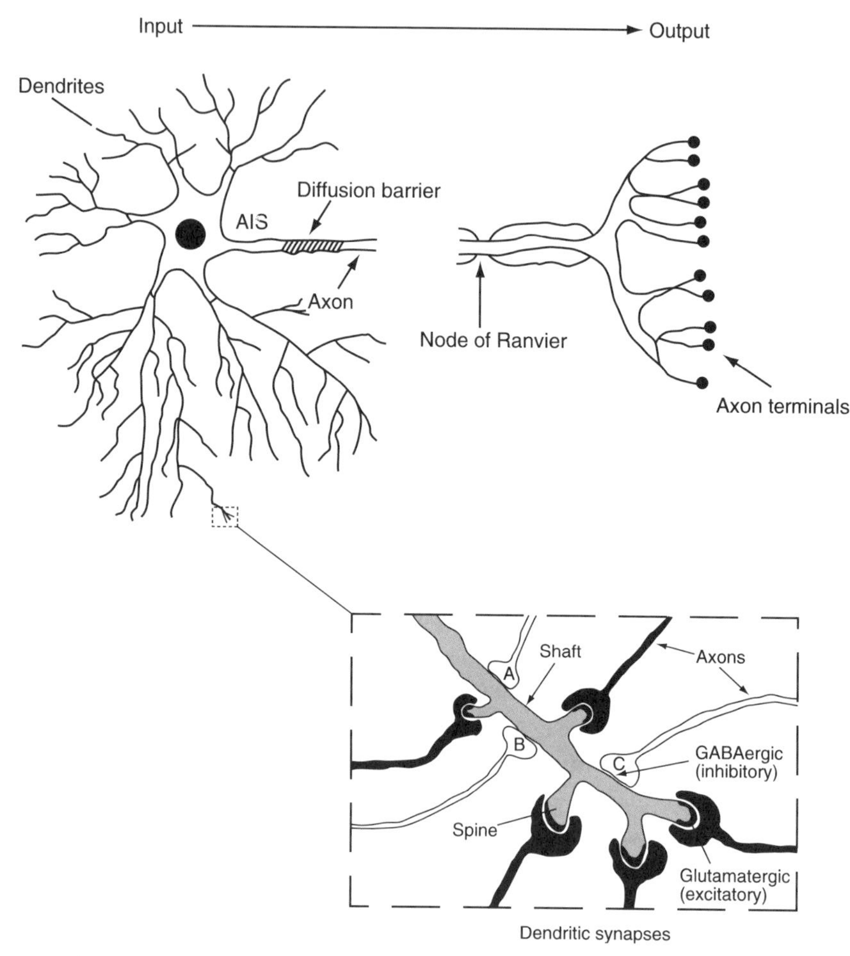

Fig. 4.1 The polarized phenotype of a typical neuron. Neuronal plasma membranes are segregated into the somatodendritic compartment, which includes the cell soma and dendrites, and axons. The diffusion barrier prevents the intermixing of proteins between the two compartments. The axonal plasma membrane may be divided into subcompartments; nodes of Ranvier (and paranodal regions) and presynaptic terminals. The somatodendritc compartment is studded with postsynaptic specializations. In particular, inhibitory synapses are found on both the cell soma and dendritic shafts, whereas excitatory synapses are commonly formed on dendritic spines emanating from dendritic shafts. All synapses are not equivalent, differentially expressing a number of proteins, most notably ion channels and clustering molecules. Furthermore, the composition of a synapses is not fixed but changes dynamically in response to activity. Thus the number of distinct compartments within neurons may be infinite.

the axon, terminating in presynaptic neurotransmitter release sites. At these locations, the action potential activates voltage-gated calcium channels (VGCCs), with the subsequent influx of calcium triggering the exocytosis of synaptic vesicles and thus neurotransmitter release (Staple and Catsicas, this volume Chapter 6).

Recent advances in molecular and cell biological technologies have enhanced our knowledge of protein trafficking in neurons. Techniques to isolate and culture many types of primary neurons have permitted long-term (several weeks) studies on neurite outgrowth and neuronal polarity. Although neurons were initially thought to be intractable to transfection, improvements in culture conditions and perseverance in the development of transfection protocols have paid off. It is now possible to culture polarized neurons and express recombinant proteins by a variety of methods including: intranuclear microinjection, DNA biolistics (DNA-gun), viral vectors, and more classical approaches such as calcium phosphate precipitation, electroporation, and liposome-based protocols. Furthermore, with the advent of optical tweezers and green fluorescent protein (and its multi-coloured derivatives) has arrived the opportunity to study protein trafficking in living neurons.

4.2 **Regulated transport to the trans-Golgi network**

Many ion channels are constructed from a variety of different subunits to form heteromeric quaternary structures. Recombinant expression studies have revealed that partial subunit combinations do not reach the cell surface, but are efficiently retained within the endoplasmic reticulum (ER). This mechanism is shared by metabotropic (e.g. mGluR and $GABA_B$) receptors and a wide variety of other multimeric proteins. For clarity, I will focus my discussion on the K_{ATP} channels, and the NMDA and $GABA_A$ receptors. K_{ATP} channels are constructed by the co-assembly of four Kir 6.1/6.2 subunits with four SUR1/2A/2B subunits (Schwappach *et al.* 2000), with individually expressed subunits being retained in the ER. The efficient assembly of NMDA receptors requires a combination of NR1, NR2A-D, or NR3A-B subunits (Meddows *et al.* 2001). Receptor assembly is particularly complex for the pentameric $GABA_A$ receptors, which are constructed from a selection of 16 different mammalian isoforms (α1–6, β1–3, γ1–3, δ, ε, π, and θ). Although the total number of possible receptor compositions exceeds one million (16^5), only 20–30 functionally distinct receptor types are thought to be expressed commonly. Although silent phenotypes inevitably exist to direct or regulate receptor function, it is clear that pathways exist to limit receptor diversity. Indeed, an examination of the assembly of $GABA_A$ receptors revealed that assembly occurred by defined pathways, restricting the cell surface expression to certain subunit combinations, with others being retained in the ER (Connolly *et al.* 1996). The identification of differential assembly signals (Bollan *et al.* 2003) are beginning to shed light on the basis of $GABA_A$ receptor and K_{ATP} channel assembly, but do not address the mechanism of efficient ER retention.

During $GABA_A$ receptor assembly, subunits interact with the ER chaperones, Immunoglobulin-binding protein (BiP), and calnexin (Connolly *et al.* 1996). Interactions with BiP are thought to result from the exposure of hydrophobic domains of incorrectly folded proteins, whereas calnexin shows specificity for glycoproteins containing partially glucose trimmed (monoglucosylated) carbohydrate side chains.

Association with these molecules results in ER retention, by virtue of specific ER retention signals (KDEL) within the chaperone proteins, until folding is complete or the subunits are degraded (Boyd *et al.* 2002). This quality control mechanism operates within the lumen of the ER and does not monitor the integrity of cytoplasmic domains.

However, although $GABA_A$ receptor γ2L is efficiently retained in the ER, γ2S (a cytoplasmic splice variant lacking 8 amino acids) can reach the cell surface as a monomeric subunit. Similarly, cytoplasmic splice variants of the NMDA receptor NR1 subunit differ in their ability to reach the cell surface. The ability to reach the cell surface suggests that these subunits are correctly folded and no longer interacting with the ER chaperones. Furthermore, the γ2S subunit did not require oligomerization to attain this transport competence. Thus, a mechanism operating cytoplasmically must also control the export of proteins from the ER.

Recent studies have identified the presence of cytoplasmically localized ER retention/export signals in a number of proteins (Ma *et al.* 2001; Scott *et al.* 2001; Zerangue *et al.* 2001; Votsmeier and Gallwitz 2001; Boyd *et al.* 2003 and references therein). The presence of a three amino acid signal (RxR) present in the cytoplasmic domain of K_{ATP} subunits, was identified as being responsible for the ER retention of Kir6.2 tetramers and SUR1 monomers. The presence of the RxR motif in all eight subunits comprising the K_{ATP} channel, suggests that the subunits may be sequentially masked during assembly. Partial complexes may mask many, but not all, of the ER retention signals and would be efficiently retained in the ER. An RxR motif has also been identified within the C1 cassette of the NR1 subunit. Interestingly, this ER retention signal appears to be regulated by the presence of a PDZ-binding domain and subunit phosphorylation by PKC. In addition, AMPA receptor surface expression is regulated by RNA editing at a single site leading to an amino acid change (Q→R) from the genetically encoded DNA blueprint. This substitution produces an arginine-based ER retention signal that requires masking by its assembly with a different subunit (Greger *et al.* 2002). Curiously, this arginine residue is bifunctional, rendering the AMPA receptors impermeable to calcium, a critical feature of AMPA receptors required to limit excitotoxicity (Davies and Glencorse, this volume Chapter 8). These findings suggest that ER retention/export may represent a dynamic regulatory process that is not restricted to quality control monitoring, but also controlled by trafficking molecules and second messenger pathways.

The control of ER transport appears to be bidirectional. In addition to the presence of retention/retrieval signals, anterograde signals required for ER export have also been identified, including dileucine, diphenylalanine, and diacidic residues. Furthermore, the observed functioning of Yxxθ (retrograde, where θ represents any hydrophobic residue) and dileucine (anterograde) signals is reminiscent of signals interacting with clathrin adaptor proteins (involved in endocytosis) and serving as basolateral (MDCK cells) or somatodendritic (neurons) targeting signals (see later). The alternative use of the same signals suggests that signals may function at multiple steps in the targeting of proteins to subdomains in the plasma membrane. There is no reason why a higher standard of quality control should be reserved for multimeric proteins. Monomeric proteins might be regulated in a similar way by intramolecular interactions.

The exact mechanism of ER retention may involve an interaction with an ER matrix, failure to be recruited for transport, or retrieval from the *cis* Golgi. Coatomer proteins (COPs) are involved in the selection of cargo for anterograde (COPII) and retrograde (COPI) transport steps between organelles. That COPI binds to ER retention signals (RK, KK but not RxR) (Zerangue *et al.* 2001) and COPII binds to ER forward signals (DxE and FF) (Votsmeier and Gallwitz 2001) favours active bidirectional transport to maintain proteins within the ER. That RxR signals do not appear to bind to individual components of COPI suggests a different mode of interaction or a distinct mechanism may be responsible for ER retention via this signal. Such a mechanism might involve direct interactions between RxR signals in associated subunits, as observed for the EGHRG motif in asialoglycoprotein receptor. Alternatively, RxR signals may interact with other trafficking proteins responsible for recruitment into COPI-independent vesicles for retrograde transport to the ER. In the studies discussed above, mutation of ER retention signals leads to cell surface expression, implying that there is no requirement for transport signals through the Golgi. On the other hand, active export signals have been shown to be required for the trafficking of Kir channels beyond the Golgi (Stockklausner and Klocker 2003) suggesting the presence of ubiquitously expressed Golgi export signals. Similarly, the mutation of an ER retention signal (RK) in the nicotinic acetylcholine receptor α subunit releases the subunit from ER retention, only to be captured within the Golgi. Although Golgi localization signals exist to retain resident proteins, they have not been reported to act transiently in the transport of cargo proteins to the cell surface.

4.3 **Morphology**

In polarized neurons, axons and dendrites perform different functions. These differences are reflected in their distinct molecular compositions and organization. Early studies on the polarized trafficking of neuronal proteins were performed using polarized epithelial cells as a model system. In epithelial cells, the two distinct membrane domains (apical and basolateral) are physically isolated by tight junctions that prevent the intermixing of proteins and lipids in the exoplasmic leaflet. Neurons do not exhibit an obvious physical barrier, despite the fact that they exhibit a polarized distinction between their somatodendritic and axonal membranes (Jareb and Banker 1998; Winckler *et al.* 1999). Indeed, when the artificial lipid DiI is incorporated into neuronal membranes, it does not encounter a diffusion barrier between the somatodendritic and axonal membranes.

To investigate the mechanism segregating somatodendritic and axonal proteins investigators concentrated on the region separating these two domains. As somatodendritic proteins such as GluR1 were found to access the proximal region of the axon, from which axonal L1 was excluded, it was concluded that any diffusion barrier must exist beyond the AIS (Winckler *et al.* 1999). These investigators used optical (laser) tweezers to apply antibody-coupled beads to axonal processes in order to study the lateral mobility of proteins of interest. Beads were positioned on axonal processes, allowed to bind, and physically dragged using the optical tweezers to determine tractability (and therefore mobility). Beads located on distal regions of axons (or dendrites) were freely mobile and could be dragged, on average, 3.7 μm. In contrast, beads located proximal to the AIS could be dragged only, on average, 0.3 μm. Interestingly, even GPI-anchored proteins that are exclusively within the exoplasmic leaflet appear to be restricted within the AIS, suggesting that the barrier can penetrate both leaflets to restrict proteins, but not lipids (Winckler *et al.* 1999).

Exogenous expression of neuron-glia adhesion molecule (NgCAM) in neurons infected with recombinant adenovirus revealed that both immobility in the diffusion barrier and NgCAM polarity required the existence of the actin cytoskeleton. An hypothesis of the physical tethering of proteins within the diffusion barrier was supported by their insolubility (non-ionic detergents) within this region, relative to other membrane or neurons with their actin cytoskeleton disrupted (Winckler *et al.* 1999). The specialized cytoskeletal structure formed between actin–spectrin–ankyrin at the diffusion barrier directly anchors membrane proteins such as Nav channels. Furthermore, the high concentration of proteins may itself induce a crowding effect to further reduce protein diffusion.

Beyond the AIS/diffusion barrier stretches the isolated axon (Fig. 4.1). In the myelinated axon, subdomains exist at the nodes of Ranvier, where Nav channels are clustered. Closely apposed (~5 μm) to these regions, separated by paranodes, are the juxtaparanodal regions at which voltage-gated potassium channels are clustered (Rasband and Trimmer 2001). At presynaptic terminals, synaptic vesicles and a complex exocytotic machinery, triggered by VGCCs (modulated by neurotransmitter receptors) exists.

Although the cell soma and the dendrites are thought to be continuous, they do not represent a single functional unit. On the contrary, this compartment is studded with multiple inhibitory and excitatory postsynaptic subcompartments. Inhibitory synapses found on the soma, AIS, and dendritic shafts are termed symmetrical synapses, as the electron density on both membranes (pre and post) in electron-micrographs are similar. Excitatory synapses are formed on dendritic spines of varying morphology and are called asymmetric due to their conspicuous postsynaptic density at the postsynaptic membrane. Glutamate (AMPA and NMDA) receptors are localized to excitatory synapses via PDZ-binding domains, found at receptor subunit carboxy-termini, interacting with a variety of PDZ proteins that comprise the postsynaptic density. The construction of inhibitory synapses is much less understood, particularly for $GABA_A$ receptors in the CNS. Candidate molecules for the clustering of $GABA_A$ receptors include: rapsyn, dystrophin, GABARAP, and gephyrin. It is at postsynaptic structures that the greatest potential for diversity exists, with distinct neurotransmitter receptors and activity-related proteins maintaining a dynamic profile. The detailed composition and organization of postsynaptic structures is discussed by Cattabeni in Chapter 7 of this volume.

4.4 Development of polarity

The process of neurite outgrowth during development defines neuronal shape, polarity, and the formation of synaptic connections to other neurons or target tissue. Materials required (proteins and membrane) are derived from the *trans*-Golgi network and delivered along microtubule tracks. In keeping with membrane transport in other cell types, both vesicular and target SNAREs and Rab proteins are required for neurite outgrowth. However, the identity of the individual components of these transport proteins is unknown (reviewed by Tang 2001).

Studies on the development of neuronal polarity have been limited due to the low level of polarized endogenous marker proteins. To circumvent this problem, Silverman *et al.* (2001) utilized replication-defective viruses (herpes simplex virus [HSV] and adenovirus [Adv]) to recombinantly express exogenous proteins as dendritic (transferrin receptor: TfR, and low-density lipoprotein receptor: LDLR) and axonal NgCAM markers. Using these molecular tools, the authors established multiple distinct developmental stages. Developmental stage 2 is initiated as multiple non-polarized (expressing both dendritic and axonal markers) neurites extend from the neuronal cell body. Neurons cultured *in vitro* undergo a remarkable intrinsic programme in which multiple, apparently identical, neurites compete to become the cell's single axon (Fukata *et al.* 2002; Banker 2003). At stage 3 of development (12–36 h), recombinantly expressed dendritic and axonal markers were correctly localized to the appropriate neurites. Developmental stage 4 occurred over the next few days, with the emergence of dendritic spines and the formation of functional synapses between days 5–16 (Ziv and Garner 2001).

4.4.1 Synapse formation

Previous studies have implicated a role of dendritic filopodia in synapse formation, possibly by probing the extracellular environment in search of a presynaptic partner. Using biolistic particle-mediated transfer of cDNA constructs into primary neurons, Prange and Murphy (2001) studied the role of green fluorescent protein (GFP)-tagged postsynaptic density protein (PSD-95), a major constituent of the excitatory postsynaptic densities, in synapse formation. To visualize all filopodia and developing spine synapses, the authors microinjected a second fluorescent dye (sulforhodamine) to highlight the entire cytoplasm of a transfected (GFP-labelled) neuron. Using 2-channel confocal microscopy, it was discovered that filopodia are highly motile and continually extend and retract. In the absence of GFP-PSD-95 expression, filopodia were relatively short-lived ($t_{1/2}$ ~1 h). In contrast, filopodia expressing GFP-PSD-95 appeared stable (>2.5 h), with no evidence of extension or retraction. Interestingly, GFP-PSD-95 appeared to be translocated to filopodia as preassembled clusters, rather than accumulating gradually. PSD-95 is central to the formation of the excitatory postsynaptic scaffold/signaling complex (Cattabeni, this volume Chapter 7) but cannot cluster when expressed alone (Imamura *et al.* 2002). Therefore, it is feasible that the observed PSD-95 clusters might represent preassembled, or partly so, postsynaptic densities. Within 45 min, the presence of AMPA and NMDA receptors can be found postsynaptically.

At the presynaptic site, the use of GFP-tagged synaptobrevin (a synaptic vesicle protein) has demonstrated that clusters of synaptic vesicles accumulate at new synapses and are capable of activity-induced exocytosis (and subsequent endocytotic retrieval) within one hour of cell–cell contact. Indeed, it appears that several components of the presynaptic active zone may be preassembled and transported to presynaptic terminals within 80 nm dense-core vesicles (Zhai *et al.* 2001). If neuroligin were preassembled with PSD-95 within postsynaptic filopodia and neurexins preassembled with the presynaptic active zones, it might be possible for the two modular units to be rapidly slotted together like lego bricks, via neuroligin–neurexin interactions, to form immature synapses to which AMPA and NMDA receptors could be recruited.

4.4.2 Axonal development

The segregation of ion channels into distinct subcellular domains within the axon occurs early in development and is responsible for a 50-fold increase in action potential velocities (Rasband and Trimmer 2001). Nav channels appear to be excluded from regions of axoglial contact. As these gaps shorten during myelination, Nav channels are compressed into what becomes ultimately, the nodes of Ranvier. Once the mature myelinated axon is formed, new Nav channels are targeted directly to the nodes of Ranvier. The mechanism of Nav clustering and maintenance is unknown, beyond the requirement for cell–cell contact between the axon and glia, and a possible involvement of an ezrin–radixin–moesin protein complex present on the microvilli of myelinating Schwann cells (Girault and Peles 2002). However, once established, Nav channels become independent and are stable in demyelinated axons, suggesting some form of direct receptor anchoring occurs, possibly via Ankyrin G within the axon. In contrast, juxtaparanodal Kv channels in the PNS appear to cluster initially within nodes prior to lateral diffusion to their final destination. In demyelinated axons juxtaparanodal Kv channels diffuse rapidly, suggesting a requirement for axoglial junctions in maintaining Kv channel localization. In the CNS, Kv channels appear to be directly targeted to the juxtaparanode, despite the fact that the Kv isoforms are identical to those in the PNS. These observations implicate the myelinating glia (PNS: Schwann cells; CNS: oligodendrocytes) as playing a pivotal role in the clustering of Kv channels.

4.5 Polarity in mature neurons

Preliminary studies (using the targeting of viral proteins in neurons infected with vesicular stomatitis virus or influenza virus) on the polarity of neurons determined that MDCK cells and neurons share common pathways for sorting. In support of this comparison, studies using replication-defective viruses (HSV and Adv) determined that TfR, LDLR, and pIgR (all transported to the basolateral domain of MDCK cells) were all targeted to the somatodendritic compartment in neurons (Jareb and Banker 1998; Silverman 2001). Furthermore, mutation of the basolateral targeting signals (Yxxθ or NPxY) in these proteins, led to a non-polarized distribution in neurons. In addition, after transport to the basolateral/somatodendritic domain, pIgR was transcytosed to the apical/axonal domain, suggesting that transcytotic routes are also conserved between the two cell types. However, clear differences appeared despite this apparent conservation. In MDCK cells, the removal of basolateral signals results in apical localization, whereas a non-polarized distribution occurs in neurons, suggesting that the signals may be conserved but the mechanisms may differ.

To examine axonal targeting, three representative apical proteins were studied, HA, CD8α, and P75/NGFR. In contrast to the similarities between the two cell types for somatodendritic sorting, all three apical proteins exhibited a non-polarized distribution in neurons. Thus, although polarized epithelial cells have provided an invaluable model for studying neuronal polarity, it is now evident that there is a clear need to focus future studies on neurons. As discussed earlier, recombinantly expressed NgCAM is restricted to the axonal membrane. However, if the neurons were permeabilized, intracellular NgCAM localization revealed a non-polarized distribution, suggesting that NgCAM is selectively polarized only at the level of the plasma membrane. In contrast, the intracellular distribution of TfR mirrors that of surface receptors and is restricted to the dendritic cytoplasm. Thus, NgCAM is either excluded from the plasma membrane of the dendrites and cell soma or rapidly removed after insertion.

The polarized distribution of the recombinantly expressed metabotropic receptors, mGluR2 and mGlur7, are similar to the intracellular polarity observed for TfR and NgCAM. The mGluR2 is restricted to the somatodendritic compartment, whereas mGluR7 is non-polarized (Stowell and Craig 1999). Like the TfR, the mGluR2 may fail to access the axon, due to the lack of an appropriate targeting signal or the presence of an exclusion signal. The expression of a truncation mutant of mGluR2 revealed that the recombinant mGluR2 could now access the axonal plasma membrane in 65 % of neurons, supporting the removal of an axonal exclusion signal. Although the presence of an axonal exclusion signal could explain the distinct localization of mGluR2 and mGluR7, it cannot result in the specific axonal targeting of molecules such as NgCAM, which are not excluded from dendrites. Curiously, the recombinant expression of a similarly truncated mGluR7 construct in neurons resulted in access to the axonal plasma membrane in only 8 % of neurons. This striking observation supports the existence of an axonal targeting signal within the carboxy-terminus of mGluR7. Furthermore, when this region was fused to telencephalin to produce a chimaeric hybrid protein, the chimaera was observed in the axonal plasma membranes of 77 % of neurons (0.7 % for wild-type telencephalin). Thus the axonal targeting signal residing in mGluR7 appears to be both necessary and sufficient for axonal targeting. When this region was added to the mGluR2, the axonal targeting signal dominated over the exclusion signal. However, the targeting of truncated mGluRs were not identical, as expected for the absence of both targeting and exclusion signals, suggesting that other targeting/exclusion signals may function as cryptic (that may not normally function) targeting signals.

Just as similarities and differences exist in the polarized transport of proteins between epithelial cells and neurons (Jareb and Banker 1998), the same appears to be true between different neuronal cell types as well as within an individual neuron, depending on previous activity (Agno *et al.* 2000). For example, the recombinant expression of mGluR5 in striatal or cerebellar granule neurons (CGNs) revealed that mGluR5 was restricted to the cell soma of CGNs, but was dendritically localized within striatal neurons. As these neurons expressed different homer proteins (mGluR-interacting proteins known to regulate ER-retention and cell surface delivery), it was hypothezised that the lack of homer 1 expression in CGNs might explain the failure to export mGluR5 from the cell soma. Consistent with this hypothesis, when homer 1b was transfected into CGNs it could be detected in dendrites. Furthermore, when co-transfected with mGluR5, both proteins were detected in dendrites, but not axons. Interestingly, when homer 1a (absent from both striatal and CGNs) was transfected into CGNs, it was found to gain access to both axons and dendrites. Upon co-transfection with mGluR5, both molecules could be detected in axons and dendrites. Similar results were observed for homer 1c in cortical neurons, which caused an increase in the dendritic trafficking of GluR1a. Homer 1a is an immediate early gene product, whose expression is induced after intense depolarization of neurons (including CGNs) and may deliver mGluRs to dendrites and axons during high-level neuronal activity, as occurs during convulsive seizures and the induction of long-term potentiation (LTP)(see Chapters 14 and 15). Homer proteins do not appear to be essential, as endogenous mGluR1a is dendritically expressed in CGNs, despite the lack of these proteins.

Two of the greatest advances in the study of neuronal trafficking must be the discovery of GFP and the ability to transfect neurons. When used together, it is possible to follow neuronal trafficking pathways, in real time, in living neurons. Apart from the breathtaking video images generated of protein trafficking within neurons (for example see: http:/www.neuron.org/cgi/content /full/26/2/465/DCI), significant scientific advances are being made. In the study of Burack *et al.* (2000), TfR-GFP and NgCAM were observed to traffic within discrete organelles. These organelles are highly pleomorphic and dynamic, ranging from small vesicles to long (>1 μm) tubules. Movement for both proteins occurred bi-directionally in dendrites (TfR and NgCAM) and axons (NgCAM).

The trafficking of transport vesicles in both dendrites and axons was inhibited by >80 % following treatment with the microtubule-disrupting agent, nocodazole, suggesting that transport occurs along microtubule tracks in both types of neurite. Microtubule tracks in dendrites exist in both orientations, with their plus and minus ends facing away from the cell body. In contrast, microtubules in axons exhibit a uniform polarity, with their plus ends facing away from the cell body (Fig. 4.2). If carrier vesicles containing dendritic proteins, such as TfR, exclusively associate with minus end-directed microtubule motors (e.g. dynein) then they would be excluded from axons, yet be capable of bi-directional transport in dendrites. Conversely, axonal vesicles, containing molecules such as NgCAM, could associate with plus end-directed motors (e.g. kinesins), gaining them bi-directional access to dendrites, but only uni-directional (anterograde) access to axons. Although this could explain the polarized distribution of proteins discussed above, it is not consistent with the real time studies identifying bi-directional trafficking of both dendritic and axonal transport organelles.

When the chimaeric molecules, amyloid precursor protein (APP)-YFP and synaptophysin (p38)-GFP, were recombinantly expressed in hippocampal neurons in culture and their transport along axons investigated by two-colour video microscopy, it was found that these proteins were segregated prior to transport along axons (Kaether 2000). In doubly transfected neurons, APP-YFP and p38-GFP exhibited mutually exclusive distributions. APP-YFP was transported rapidly (~4.5 μmS^{-1}) within elongated tubules up to 10 μm in length. In contrast, p38-GFP was restricted to slow (~1 μmS^{-1}) tubulovesicular carriers. The use of antisense oligonucleotides to block the translation of Kinesin mRNA, resulted in a disruption of APP-YFP carriers but did

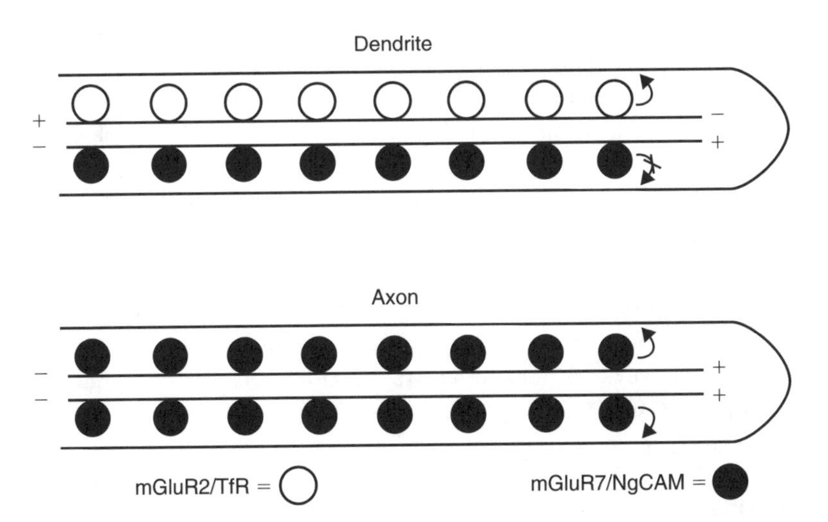

Fig. 4.2 Transport into dendrites and axons occurs along microtubules. Microtubules in dendrites are oriented in both directions, whereas the microtubules in axons are all polarized with their plus end distal to the cell body. Vesicles carrying cargo into dendrites are excluded from axons, but axonally targeted vesicles are transported into both axons and dendrites. Thus, dendritic versus axonal sorting may occur early in the biosynthetic pathway, with axonal targeting being achieved by a retention mechanism or the failure of vesicle to fuse with dendritic plasma membrane. Dendritic targeting may be achieved by exclusion from the axon.

not affect p38-GFP transport. Thus, other microtubule motors may be involved in some of these transport pathways, particularly within the axon (Diaz-Nido and Avila, this volume, Chapter 15).

4.6 Polarity signals

4.6.1 Dendrites

Basolateral targeting signals present in TfR, pIgR, and LDLR (Yxxθ or NPxY) have been shown to be required for the dendritic localization of these receptors in neurons (Jareb and Banker 1998). Paradoxically, a Yxxθ motif has also been implicated in the sorting of a neuronal form of the cell adhesion molecule L1 to axonal growth cones and the NPxY for the axonal localization of Megalin. Furthermore, these basolateral targeting signals are similar, and sometimes identical, to endocytosis motifs that are responsible for the recruitment of proteins into clathrin-coated vesicles. Interestingly, another endocytosis/ER export signal, a dileucine motif, has been implicated in the basolateral/somatodendritic targeting of a glycine transporter as well as the axonal localization of Nav1.2. The similarity between sorting signals operating at diverse locations (ER, TGN, endosomes, and plasma membrane) appears incongruous with the requirement for specificity. Thus, either other proximal (or distal?) residues may be required to provide specificity, or these common motifs are required to access a common sorting compartment from which other unidentified signals may operate, failure to do so might result in non-selective transport by default. Alternatively, these signals may operate within different environments, such as secondary sorting sites.

Protein recruitment within the TGN is driven by their direct interaction with the μ1 subunit of the AP1 clathrin adaptor complex. The role of the AP1 complex in protein targeting is further supported by the observation that most AP1-coated vesicles derived from the TGN fuse with endosomes, where their content may be resorted. Indeed, basolateral vesicles have been observed to arise from clathrin-adaptin-coated structures on endosomal tubules in MDCK cells. The μ1 subunit of AP1 has now been directly implicated in the dendritic targeting of odorant receptors in neurons of *Caenorhabditis elegans*. Distinct isoforms of the μ subunit exist, offering the potential for selectivity. However, other unidentified dendritic targeting signals, distinct from tyrosine and dileucine motifs, clearly exist and may function sequentially at multiple sorting sites.

4.6.2 Axons

As for dendrites, there appears to be little consensus on the identity and mechanism of axonal targeting. Cell adhesion molecule L1 requires the presence of a cytoplasmic Yxxθ signal (Kamiguchi and Lemmon 1998), Nav1.2 requires a dileucine motif (Rivera *et al.* 2003) and Megalin requires one of its three NPxY motifs (Takeda *et al.* 2003) to target to axons, all of which are also required for the somatodendritic targeting of other proteins (Jareb and Banker 1998). Perhaps a resolution to this apparent lack of specificity, is the environment in which these signals operate. In the axonally localized proteins, perhaps the somatodendritic/endocytic signals function to remove the proteins exclusively from the somatodendritic membrane, leading ultimately to axonal localization. Indeed this appears to be the case for the axonally localized VAMP2 (Sampo *et al.* 2003).

In contrast to the use of cytoplasmic targeting signals, APP and NgCAM require ectodomain axonal signals which, being on the lumenal side of the membrane, cannot associate with soluble cytosolic sorting proteins such as the adaptor complex. An apparent third mechanism for axonal targeting involves the formation of protein–lipid complexes. Both haemagglutinin and Thy1, but not APP, associate with CHAPS-insoluble sphingolipid–cholesterol lipid rafts and become mis-sorted in sphingolipid-depleted cells.

The construction of a synaptobrevin-TfR chimaera revealed that the cytoplasmic domain of synaptobrevin contains an axonal signal capable of transporting the chimaera to presynaptic terminals. The synaptic targeting signal could not be isolated, suggesting that either the signal is dispersed, multiple signals are present, or that the overall conformation of the cytoplasmic domain is necessary. This chimaera was not appropriately targeted to synaptic vesicles, but localized to a presynaptic endosome that could recycle alongside synaptic vesicles, but not intermix. Correct localization of the synaptobrevin-TfR chimaera was achieved upon the deletion of a 10 amino acid region from synaptobrevin. Thus, synaptobrevin must pass at least three sorting tests following exit from the TGN: axonal targeting, synaptic targeting, and synaptic vesicle targeting.

Sorting signals have recently been identified in peripheral membrane proteins (cytosolic proteins associated with membranes)(El-Husseini *et al.* 2001). This study examined the targeting of two palmitoylated proteins, PSD-95 and growth-associated protein-43 (GAP43). Palmitate is a saturated fatty acid, linked via thioester bonds, to specific cysteine residues and is responsible for membrane association. Both PSD-95 and GAP43 are dually palmitoylated, yet targeted to dendritic synapses or axons, respectively. Differential sorting appears to be specified by the spacing of the two cysteines and the presence of local basic residues and operated, in the case of GAP43, by the incorporation into lipid rafts mediating transport into axons. However, like many axonally localized integral membrane proteins, GAP43 is also transported to dendrites. Although palmitoylation is required for PSD localization to dendritic synapses, it is insufficient by itself. It is possible that unidentified dendritic targeting/axonal exclusion signals require membrane association by palmitoylation in order to function. In contrast to the dendritic localization observed in most neurons of the forebrain, PSD-95 is highly expressed in axons of cerebellar basket cells. The axons of these neurons are unusual in that they lack microtubules. This intriguing anomaly incriminates a role of minus end motors in the axonal exclusion of PSD-95. It is tempting to speculate that in forebrain neurons, dendritic transport vesicles with associated PSD-95 (palmitoylated) that approach the AIS (Fig. 4.1) might be efficiently captured and put on the first available minus-end motor and driven out along microtubules causing axonal exclusion. The same mechanism would serve to efficiently target the vesicles into dendrites.

4.7 **Postsynaptic targeting**

4.7.1 **mRNA targeting**

Until recently, it has been envisaged that protein localization is achieved exclusively by protein trafficking events preceding anchoring mechanisms. However, a completely different strategy, involving the targeted transport of mRNA molecules is gaining popularity (reviewed by Job and Eberwine 2001). A particular attraction of this mechanism is its economical use of cellular components, since one mRNA molecule has the capacity to produce many molecules of the protein product. The indication of localized dendritic translation of mRNA was obtained by electron microscopy by the observation of polysomes at the base of dendritic spines. Individual mRNA species (encoding MAP2, CaMKII, and glycine receptors) were subsequently identified by *in situ* hybridization.

The full potential for localized translation was realized by the extraction of mRNA from mechanically isolated dendrites using a patch pipette, followed by PCR amplification. Using this methodology, it was determined that, of the 10 000 mRNAs thought to be present in neuronal soma, approximately 400 were estimated to be present in dendrites. More significantly, differential display revealed that all dendrites do not contain the same complement of mRNAs, raising the possibility of differential targeting, or at least, specificity.

To date, mRNAs for structural proteins (e.g. MAP2), enzymes (e.g. CamKII), growth factors (e.g. BDNF), ligand-gated ion channels (e.g. glycine receptors), and transcription factors (e.g. CREB) have been found within dendrites. The significance of mRNA being present within dendrites was enhanced by the immunocytochemical colocalization of proteins involved in the translation process and present within the ER and Golgi, a prerequisite for the assembly and trafficking of integral membrane proteins such as ion channels (Gardiol *et al.* 1999).

Direct evidence that dendritically localized mRNA are actually translated locally was first obtained via the incorporation of radiolabelled amino acids into isolated dendrites. Synaptoneurosome preparations were used to detect and quantify protein synthesis. Intriguingly, a glutamate-responsive mRNA encoding an RNA-binding protein, FMR1 (Fragile-X mental retardation protein), which can regulate translation from polysomes was detected. Thus, synaptic activity may initiate local protein synthesis, ensuring a supply of required proteins exclusively to the site(s) requiring them.

Definitive proof that dendrites do possess the capacity to translate proteins came from an elegant molecular approach. Dendrites were isolated from neuronal cell bodies and transfected with a recombinantly transcribed mRNA. Consistent with dendritic translation being regulated, expression could only be detected following stimulation with growth factors (e.g. BDNF).

Given the almost infinite number of distinct synapses, it is unlikely that distinct targeting signals could exist for each. Alternatively, dendritically targeted mRNAs could be targeted to all synapses, but translated only at synaptically active sites. Combined with temporally regulated transcription in the nucleus, a simple, yet effective, mechanism for selective targeting could be achieved. As synaptic activity is known to regulate the transcription of certain genes in the nucleus, it is tempting to speculate that a mechanism could exist in which active synapses might specify the proteins that need to be recruited. Support for such an hypothesis has been provided by the discovery of the transcription factor (CREB) mRNA and protein within dendrites. More importantly, the CREB protein could be activated locally by phosphorylation (a requirement for transcription) and was capable of retrograde transport to the cell nucleus. Thus, restricted synaptic activity may initiate the local translation of proteins required within the synapse as well as those required in the nucleus for the transcription of new mRNAs. An exciting possibility is whether certain types of synaptic activity could lead to the transcription of specific genes, i.e. the provision of a shopping list, rather than just requesting 'food', for delivery. Of course, what is delivered by such a mechanism is the recipe, in the form of mRNA molecules.

4.7.2 **Protein targeting via lipid rafts**

It is becoming increasingly apparent that the segregation of specific proteins into sphingolipid/cholesterol-containing microdomains is a commonly used sorting mechanism operating within the exocytic and endocytic pathways (reviewed in Ikonen 2001). The role of lipid rafts in organizing the exocytotic machinery has been implicated by their association with SNARE proteins. The association of VAMP2 (a vesicular (*v*)-SNARE) with lubrol-insoluble lipid domains, such as are found on synaptic vesicles, suggests a role of lipid rafts in targeting. Similarly, the association of syntaxin 1A (a target (*t*)-SNARE) within a distinct subset of lipid rafts (based on detergent solubility characteristics) suggests that sites of exocytosis may be restricted to discrete subdomains in the plasma membrane. Although VAMP2 and syntaxin 1A are involved in the regulated exocytosis of presynaptic vesicles, syntaxin 1A and SNAP-25 also occur along the axonal plasma membrane, implicating a more general role for lipid rafts in exocytosis. Interestingly, different Kv channel isoforms show distinct localization profiles with respect to lipid rafts, with Kv1.5 in caveolar rafts and Kv2.1 in non-caveolar rafts, while Kv4.2 is not associated with any lipid rafts. Thus, distinct lipid rafts may play a critical role in determining the sites of exocytosis for different cargo proteins. Furthermore, these specialist sites may not be restricted to axonal destinations. AMPA receptors have been found to be equally distributed between lipid rafts and postsynaptic densities and may be recruited by an interaction with raft-associated GRIP, via the AMPA receptor GluR2 subunit. In addition, the $\alpha7$ nicotinic acetylcholine receptor localization and clustering in somatic spines of ciliary neurons requires lipid raft association. The full significance and potential of lipid rafts in protein sorting remains to be established (reviewed in Tsui-Pierchala *et al.* 2002).

4.7.3 Specific transport proteins and pathways

With the advent of 2-hybrid (yeast, bacterial and mammalian) screening, a molecular approach to the identification and cloning of interacting proteins, has come the identification of numerous clustering, anchoring, signaling, and transporting proteins involved in postsynaptic receptor function. This technology has been particularly fruitful in the identification of participants in the transport of $GABA_A$ and AMPA receptors to the cell surface (Kittler and Moss 2001; Sheng and Lee 2001; Malinow and Malenka 2002). Equally important has been the development of methods to transfect neurons and the application of GFP chimaeras to the study of trafficking pathways leading to synaptic targeting.

4.7.3.1 $GABA_A$ receptors

It is known, from transgenic studies, that $GABA_A$ receptor synaptic targeting at most inhibitory synapses requires both the γ subunit and gephyrin. GABARAP ($GABA_A$ receptor associated protein) was the first yeast two-hybrid hit for the $GABA_A$ receptors. GABARAP interacts with the γ2 subunit of $GABA_A$ receptors, but is rarely associated with clustered $GABA_A$ receptors or gephyrin. Instead, GABARAP appears to be concentrated at transport sites on the Golgi, where it interacts with *N*-ethylmaleimide-sensitive fusion protein (NSF), suggesting a role in $GABA_A$ receptor trafficking. This is supported by the close structural homology of GABARAP to GATE-16 (Golgi associated ATP enhancer) a known transport molecule. The crystal structure of GABARAP suggests the ability to bind the γ2 subunit and tubulin on opposite faces. Thus, GABARAP may select vesicles (possibly even sort the receptors into vesicles) with γ2-containing receptors and transport them along microtubules to inhibitory synapses. As GABARAP is not localized to these sites, it presumably hands over the receptors to microtubule-associated gephyrin clusters and returns to the Golgi (reviewed in Kneussel 2002).

Using the large intracellular loop of the α1 subunit of $GABA_A$ receptor as bait in a 2-hybrid screen revealed Plic-1 as an interacting protein capable of binding to all α and β subunits, but not γ or δ. Like GABARAP, Plic-1 is found on intracellular membranes and is not found significantly at synapses. Having identified the $GABA_A$ receptor-binding domain, the authors delivered (using the *antennapedia* internalization peptide) inhibitory peptide to cells. These experiments revealed a requirement for Plic-1 in the maintenance of $GABA_A$ receptors at the cell surface. As $GABA_A$ receptors constitutively recycle in neurons, Plic-1 may be required for recycling to the plasma membrane, or protection from degradation during recycling. Plic-1 is a ubiquitin-like protein and may be a negative regulator of the proteasome. It is not clear if the sole function of Plic-1 is to prevent $GABA_A$ receptor degradation, thus leaving more receptors capable of recycling back to the cell surface, or if Plic-1 may play a direct role in returning receptors. Plic-1 may also increase receptor number by protecting receptors from ubiquitin-dependent degradation within the ER during assembly. Unlike

GABARAP, Plic-1 would be capable of maintaining both synaptic and extrasynaptic surface $GABA_A$ receptors.

As mentioned above, multiple sorting stations may exist in neurons, serving to filter and direct proteins to their final destination. In addition to providing specificity, these compartments may also offer the opportunity to regulate the delivery of its cargo to the cell surface. Indeed, the exocytosis of synaptically targeted (γ subunit-containing) $GABA_A$ receptors appears to be regulated by a mechanism distinct from extrasynaptic (lacking γ subunits) $GABA_A$ receptors. In recombinant expression systems, both $\alpha\beta$ and $\alpha\beta\gamma$ receptors constitutively recycle between the cell surface and peripheral endosomes. However, $\alpha\beta\gamma$ receptors are subsequently diverted from peripheral early endosomes and routed into a perinuclear (in fibroblasts) late endosomal compartment. Upon PKC stimulation, exocytosis from this late compartment is blocked, leading to a reduction in the surface expression of $\alpha\beta\gamma$, but not $\alpha\beta$ receptors. Such a mechanism may provide the basis for the regulation of $GABA_A$ receptor synaptic targeting (involving GABARAP?) (Kittler and Moss 2001; Kneussel 2002). The presence of a subcellular compartment from which $GABA_A$ receptors may be recruited rapidly is supported by the findings that insulin (via tyrosine kinase activity) causes the fast (<10 min) translocation of $GABA_A$ receptors to the cell surface of transfected fibroblasts or synapses in neurons. The speed of surface delivery is consistent with receptor recruitment from an endosome. Thus, kinase activity is capable of modulating the cell surface delivery of $GABA_A$ receptors by either inhibiting (PKC) or promoting (tyrosine kinase) receptor targeting from endosomes.

4.7.3.2 NMDA receptors

The delivery of NMDA receptors to the plasma membrane is also subject to kinase regulation. In contrast to the findings for $GABA_A$ receptors, PKC activation was found to promote NMDA receptor targeting to the cell surface. Importantly, endogenous levels of PKC stimulation (via mGluR1 activation) are sufficient to increase surface targeting of NMDA receptors. This targeting can be inhibited by botulinum neurotoxin A and a dominant negative mutation of soluble NSF-associated protein (SNAP-25), implicating the involvement of SNAP-25 dependent exocytosis. This PKC-dependent pathway is conserved in *Xenopus* oocytes, fibroblasts, and hippocampal neurons. Intriguingly, NMDA receptors are also recruited to the cell surface of *Xenopus* oocytes in response to insulin treatment, via a SNAP-25 dependent process. It appears that the regulated exocytosis of NMDA receptors may also occur in response to synaptic activity. In the adult (but not neonate; see below) CA1 region of the rat hippocampus, an increase in synaptic strength leads to the rapid surface expression of NMDA receptors by a PKC and tyrosine kinase-dependent mechanism (Grosshans *et al.* 2001). Given the opposite nature, and distinct localization of $GABA_A$ and NMDA receptors in the adult, it seems likely that at least two distinct, highly homologous, sub-synaptic endosomal sorting compartments co-exist within neuronal dendrites.

4.7.3.3 AMPA receptors

In neonates, NMDA receptors are relatively fixed components of the postsynaptic densities, whereas AMPA receptors are more loosely associated. Long-term potentiation (LTP) and long-term depression (LTD) are modifications to synaptic strength, expressed as AMPA receptor responsiveness to glutamate, and are currently the best molecular correlate of learning and memory. The modulation of synaptic strength has been perceived traditionally to reflect alterations in neurotransmitter release. However, the modulation of AMPA receptor numbers in postsynaptic membranes might also provide a powerful mechanism to modulate synaptic strength. Indeed, a large body of evidence supporting this hypothesis has emerged over the last three years (reviewed in Sheng and Lee 2001; Malinow and Malenka 2002).

The identification of a postsynaptic compartment from which regulated exocytosis might occur was achieved by loading cells with the membrane dye FM1-43. Prolonged labelling was necessary for the dye to access a compartment from which regulated exocytosis could occur, consistent with its distribution within the TGN/late endosome. The exocytotic release of FM1-43 could be induced by glutamate, electrical stimulation (50Hz), calcium and CaMKII stimulation, conditions necessary for the expression of LTP. In primary hippocampal neurons, this calcium-evoked dendritic exocytosis is only evident after 7 days in culture, unless αCaMKII (absent until day 7) was expressed using recombinant viral constructs. The requirement for postsynaptic membrane fusion was confirmed by the postsynaptic introduction (via a microelectrode) of several agents (*N*-ethylmaleimide, NSF-SNAP binding inhibitory peptides and botulinum toxin) that block membrane fusion events. In all cases, LTP in the CA1 region of the hippocampus was blocked, confirming the importance of postsynaptic membrane fusion in the expression of LTP at these synapses.

NSF plays a critical role in vesicle fusion with target membranes in both constitutive and regulated exocytosis. NSF is recruited to v-SNARE/*t*-SNARE/SNAP-25 complexes via soluble NSF-attachment proteins. It was quite unexpected, therefore, when 2-hybrid screening identified a direct interaction between the AMPA receptor GluR2 subunit and NSF. Support for the validity of this interaction was obtained by the perturbation of NSF-GluR2 interactions. The injection of inhibitory peptides (corresponding to NSF-binding site on GluR2) or monoclonal antibodies (against NSF) into postsynaptic hippocampal neurons resulted in a rapid rundown of AMPA receptor currents. Although GluR2 is clearly not a SNARE protein, it is possible that NSF may mediate conformational changes in GluR2 exposing a plasma membrane targeting signal or dissociation from an anchoring protein, such as GRIP, to initiate endocytosis.

An exciting piece of evidence in AMPA receptor trafficking was obtained from *stargazer* mice. The main cerebellar defect is found in cerebellar granule neurons and is expressed as a complete lack of synaptic AMPA receptors. The protein encoded by the *stargazer* gene, *stargazin*, is able to bind both AMPA receptors and PSD-95. Importantly, the lack of surface and synaptic AMPA receptors could be restored by the recombinant expression of wild-type *stargazin* in cultured neurons from these mice. In contrast, when these neurons were transfected with a mutant *stargazin* lacking the PDZ-binding domain, both the PSD-95 interaction and synaptic clustering were not observed. However, this mutant could still interact with, and deliver, AMPA receptors to the cell surface. This observation led to the discovery of the dual role performed by *stargazin*; to deliver AMPA receptors to the cell surface (at extrasynaptic sites) and following their subsequent lateral diffusion into the synapse, to hand over its charge to PSD-95 (Schnell *et al.* 2002).

The possibility of AMPA receptor recruitment as being the site of expression of LTP was spurred on by the identification of silent synapses. At resting membrane potentials, when NMDA receptors are blocked by Mg^{2+}, only AMPA receptors respond to glutamate. In the absence of functional AMPA receptors, glutamatergic synapses are therefore silent. The majority of synapses are silent at early stages of development but silent synapses are uncommon in the adult brain. A morphological correlate of silent synapses, using the immunohistochemical localization of GluR1, suggested that silent synapses may lack AMPA receptors.

Conclusive evidence that postsynaptic AMPA receptor recruitment does occur during LTP was provided by an investigation into the trafficking of recombinant GluR1-GFP. GluR1-GFP was introduced into neurons in organotypic hippocampal slice cultures by a Sindbis virus expression system. The distribution of GluR1-GFP was determined directly by GFP fluorescence and surface receptors identified using GFP antibodies in non-permeabilized cells. As for endogenous GluR1, the GluR1-GFP is predominantly intracellular. However, upon brief tetanic stimulation (LTP-inducing), GluR1-GFP is recruited to both silent and active synapses.

In keeping with a role for CaMKII in regulated exocytosis and LTP, when a constitutively active form of CaMKII (tCaMKII-GFP) is expressed in neurons, enhanced synaptic transmission and GluR1-GFP recruitment occurs. GluR1-GFP recruitment required its PDZ-binding domain implying that GluR1 recruitment as well as clustering (Cattabeni, this volume Chapter 7) requires PDZ domain proteins.

The accumulation of recombinantly expressed GluR2 requires an association with NSF, ABP (AMPA receptor-binding protein), and/or GRIP (glutamate receptor interacting protein). In contrast to the findings for GluR1, GluR2-GFP was constitutively expressed at synapses by an activity-independent mechanism. A careful examination of the synaptic targeting of recombinantly expressed GluR1 versus GluR2 revealed that GluR1 is retained intracellularly until recruited to active or silent synapses upon activation, whereas GluR2 receptors appear to be excluded from silent synapses. However, endogenous AMPA receptors are predominantly composed of both GluR1 and GluR2 subunits. The recombinant expression of GluR1/2 receptors revealed a trafficking itinerary indistinguishable from recombinant GluR1 homomeric receptors. New GluR1 or GluR1/2 receptors appear transiently as micropuncta in non-synaptic areas prior to their accumulation at synapses. In contrast, GluR2 appear to be delivered directly, and rapidly, to synapses. The initially non-synaptic targeting of GluR1-containing receptors may be explained in one of two ways. Firstly, GluR1-containing receptors might diffuse laterally within the plasma membrane (Borgdorff and Choquet 2002) until 'captured' by synaptic proteins (Schnell *et al.* 2002). Alternatively, it is possible that synaptic localization may be achieved by random surface delivery, followed by receptor concentration into clathrin-coated vesicles, transport into a sorting endosome and subsequent targeting to synapses.

By performing these recombinant studies it has been possible to tease apart the relative subunit contributions to endogenous GluR1/2 receptor targeting. The most likely value of these differential targeting mechanisms is that GluR1 might be required for activity-dependent plasticity that recruits new receptors to silent or active synapses. Synaptic potentiation achieved in this way, may be maintained by the constitutive re-delivery of recycling receptors, by virtue of GluR2-dependent interactions. With the development of these molecular biological strategies to the study of LTP, has come the shift in popularity from the expression of LTP being exclusively presynaptic to predominantly postsynaptic. However, the pendulum of opinion may soon return to neutrality, with the discovery of an increased targeting of the presynaptic vesicle-associated protein, synaptophysin, to synapses during LTP (Antonova *et al.* 2001). Thus, the enhancement of synaptic strength during LTP may be achieved by either an increase in neurotransmitter release, or a concomitant increase in AMPA receptor numbers, or both. Of course, increasing either of these components would be futile if they are already present in excess. Variability between synapses of different brain regions may account for the discrepancies observed in the relative contributions of presynaptic and postsynaptic components to LTP.

4.8 Postsynaptic removal of receptors

A commonly used strategy for neurons to reduce responsiveness to a particular environmental cue (e.g. ligand) is to remove receptors from the cell surface. There is also evidence that endocytosis of certain G-protein-coupled receptors (GPCRs) may be a prerequisite for the activation of some signal transduction pathways (Koenig, this volume Chapter 9).

4.8.1 Mechanism and pathways available

Internalization from the plasma membrane typically occurs by the recruitment of adaptor (AP-2) proteins to specific endocytosis signals. These adaptor proteins then recruit clathrin, to instigate membrane invagination and endocytosis. The best characterized endocytosis motifs are the tyrosine-based (Yxxθ and NPxY) and dileucine signals, which recruit μ or β subunits of AP-2 respectively. Specialist adaptors also exist, such as arrestin and ubiquitin. Arrestin binding to GPCRs appears to facilitate receptor internalization by its ability to associate with clathrin and AP-2. The role of ubiquitin in targeting surface receptors to the endocytic pathway is less clear, but may involve the recruitment of AP-2 or clathrin, the release from receptor anchoring or the recruitment of receptors to sites of endocytosis. Neuronal endocytosis occurs at hot spots throughout the dendritic tree in young neurons. In mature dendrites, endocytosis appears to occur at fixed endocytic zones, including locations close to dendritic spines (excitatory synapses)(Blanpied *et al.* 2002).

Following internalization, multiple routes are available from the early endosome (Buckley *et al.* 2000; Kittler and Moss 2001). Receptors may be recycled rapidly to their site of origin in the plasma membrane (Fig. 4.3*A*) or delivered to a late endosome for sorting (Fig. 4.3*B*). From the late endosome, receptors may be delivered to the TGN and subsequently recycled to their original location (Fig. 4.3*C*) or a new location (Fig. 4.3*D*). The remaining receptors in the late endosome may be delivered to (or the compartment may mature into) lysosomes (Fig. 4.3*E*).

The endosomal system in hippocampal neurons has been studied by the recombinant expression of a syntaxin 13-GFP chimaera in conjunction with time-lapse microscopy. An extensive tubulovesicular endosomal network exists in both dendrites and axons, and comprises both mobile and stationary components. After

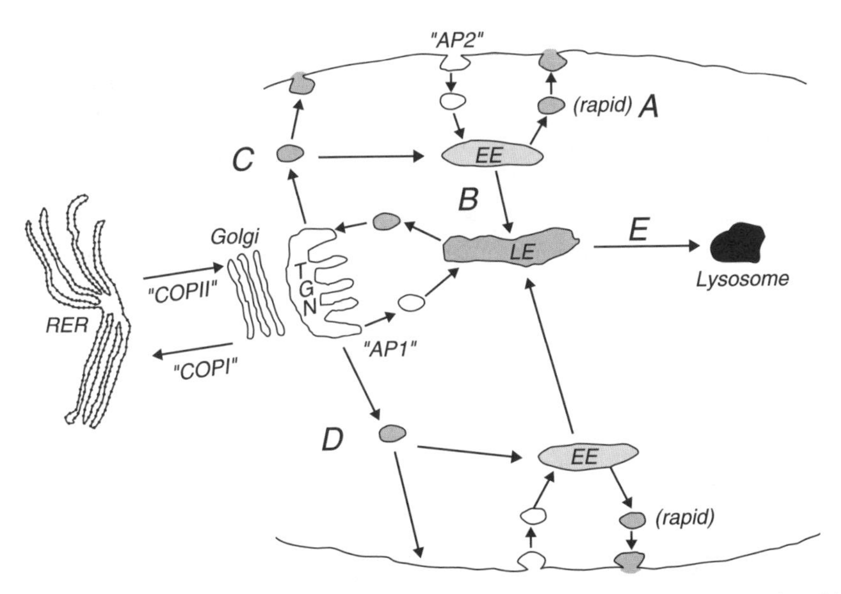

Fig. 4.3 Transport pathways. Newly synthesized integral membrane proteins are produced in the rough endoplasmic reticulum and transported forward (via COPII vesicles) through the Golgi to the *trans*-Golgi network (TGN). Quality control and regulatory steps occur to prevent the surface expression of proteins by recruitment into COPI vesicles and returned to the ER. Protein targeting to subdomains of the plasma membrane begins in the TGN, with proteins delivered either directly to the surface or via the late endosomes (LE) via AP1-dependent clathrin-coated vesicles. Surface receptors may be internalized by AP2-dependent clathrin-coated pits into early endosomes (EE). From early endosomes, proteins may be rapidly recycled to the cell surface (A), or diverted into late endosomes (B) prior to delivery back to the cell surface via the TGN (C), to a new location (D) or to the lysosome for degradation (E).

photobleaching of stationary compartments, new delivery of syntaxin 13-GFP was observed, supporting the dynamic intermixing of distinct endosomal elements. Furthermore, endosomal movement was microtubule-dependent, bidirectional, and occurred in a saltatory manner, consistent with the directional transport between stationary endosomes interconnected by microtubule tracks.

4.8.2 **Receptor endocytosis**

Agonist-dependent downregulation of $GABA_A$ receptors has been observed following prolonged treatment with GABA, benzodiazepines, barbiturates, and neurosteroids (Barnes 1996) and has been associated with LTP_{AMPA}, and a concomitant LTD_{GABA}. $GABA_A$ receptors constitutively recycle between the cell and the endosomes in both recombinant expression systems and neurons. $GABA_A$ receptors are recruited into endocytic vesicles via a direct interaction of their β and γ subunits with α and β adaptins of the AP2 complex (Kittler and Moss 2001). Following endocytosis, receptors containing the $\gamma 2$ subunit are diverted from early endosomes to late endosomes.

The identification of specific endocytosis motifs within the $GABA_A$ receptors has been hindered by the existence of multiple putative signals (Yxxθ and dileucine) present in both β and γ subunits. Molecular studies on the NMDA receptor NR2B subunit revealed a significant contribution of a tyrosine-based signal, but not a dileucine motif, to receptor endocytosis. However, mutant receptors were still capable of internalization at 50% of wild-type rates, suggesting the presence of other unidentified endocytosis motifs (Roche *et al.* 2001).

The modulation of synaptic strength by AMPA receptor recruitment (LTP) has been discussed as a molecular correlate to learning and memory. However, the traffic is not unidirectional, long-term depression (LTD) of synaptic strength can also occur by AMPA receptor endocytosis via clathrin-coated pits (Kittler and Moss 2001; Sheng and Lee 2001; Snyder *et al.* 2001). As we have seen for $GABA_A$ receptors, cell surface receptor number is a balance between two competing processes, delivery and removal of receptors. Like the $GABA_A$ receptors, AMPA receptors constitutively recycle between the cell surface and endosomes. Thus, LTD could arise by either an increase in AMPA receptor internalization or a decrease in recycling to the surface.

The observation that the perturbation of the NSF-GluR2 interaction causes a receptor rundown that occludes the expression of LTD, suggests that the same receptor pool may be involved in both forms of receptor downregulation. This hypothesis is supported by the observation that in both cases internalized AMPA receptors remain localized at synapses and are potentially available for re-insertion into the synaptic membrane should neuronal activity change. In contrast, AMPA receptors internalized in response to direct activation by AMPA, are not retained locally but transported primarily to the cell soma and degraded in lyosomes (reviewed in Malinow and Malenka 2002).

AMPA receptor downregulation can also be induced by insulin (in contrast to the upregulation of $GABA_A$ and NMDA receptors). Consistent with the conserved use of the endocytic pathways described above, insulin-dependent endocytosis is GluR2-dependent and mutually occlusive with LTD. A potential point of convergence of these common pathways is the retargeting of internalized receptor to the cell surface. Thus, AMPA receptors endocytose constitutively via the GluR2 subunit, whereupon recycling occurs constitutively by interactions with NSF, PICK1 (protein interacting with C kinase), ABP or GRIP (Hirbec *et al.* 2003). However, blocking any of these interactions with GluR2 (by PKC phosphorylation during LTD?) blocks receptor recycling in the face of continued endocytosis, leading to a rundown of surface receptors.

Given the bidirectional, activity-dependent modulation of synaptic strength and postsynaptic trafficking, an important question arises, are these mechanisms inversely regulated and controlling the same receptor pool (Fig. 4.4). An important study has correlated LTP, LTD, and the reversal of LTP by LTD (and vice versa) *in vivo*, with changes in AMPA receptor levels in subsequently purified synaptoneurosomes, confirming that these two inversely related processes can antagonize each other, but not that the same pool of receptors are involved. Intriguingly, the induction of AMPA receptor LTP in CA1 pyramidal cells is coincident with LTD of $GABA_A$ receptor-mediated inhibitory postsynaptic potentials. Although it is not known how this LTD_{GABA} is expressed (beyond receptor dephosphorylation by calcineurin)(Wang *et al.* 2003), it is tempting to speculate the existence of a parallel recycling pathway for $GABA_A$ receptors, operating in reverse, in response to the same signals controlling AMPA receptor trafficking.

4.8.3 Lateral diffusion

Although receptor anchoring at synaptic sites is now the central dogma for maintaining synaptic integrity, the interactions between receptors and the postsynaptic scaffold are recognized to be dynamic. The existence of a labile (frequently estimated to be 30–40% of total) population of receptors is observed for both $GABA_A$ and AMPA receptors, highlighting the presence of significant population of unrestrained receptors

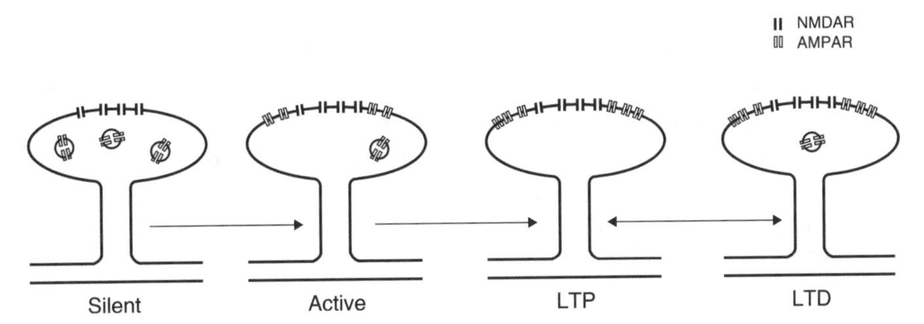

Fig. 4.4 The modulation of synaptic strength by alterations in postsynaptic AMPA receptors. Early in development, the majority of glutamatergic synapses are 'silent' at resting membrane potentials. This results from the presence of NMDA but not AMPA receptors in the postsynaptic membrane. Synapses become activated by an NMDA receptor-dependent process, leading to the recruitment of AMPA receptors. Synaptic strength may be increased further, in response to high-frequency activity (Long-term potentiation; LTP), by the further recruitment of AMPA receptors. This process may be reversed by low-frequency activity (Long-term depression; LTD), by the removal of AMPA receptors by endocytosis.

at synapses that may also be capable of lateral diffusion. Despite the potential for receptor diffusion to synaptic membranes as a mechanism for targeting and the potential significance of receptor migration between synapses, little research has focused on this issue.

A recent advance has been made by the use of single particle tracking to follow the lateral movement of glycine receptors expressed in the plasma membrane of transfected fibroblasts (Meier *et al.* 2001). The authors tracked the mobility of glycine receptors in cells coexpressing the glycine receptor (myc epitope-tagged) clustering molecule, gephyrin (GFP-tagged). The glycine receptors were tracked by labelling with a large (0.5 μm) bead coated with anti-myc antibodies. Receptors were found to alternate between periods of rapid lateral movement and relative immobility when associated with gephyrin-GFP clusters. This highlights the first important point, glycine receptors are not permanently restrained within gephyrin clusters. Secondly, the particles were found to be promiscuous, moving between gephyrin clusters. These findings are likely to represent an underestimate of receptor mobility due to the binding of the bead to multiple receptors, pulling in different directions, and the requirement for only a few receptor-gephyrin interactions to immobilize the bead. Despite these technical limitations, it will be important to perform similar studies on epitope-tagged receptors in neurons. Similar studies on the AMPA receptors have yielded the same pattern of diffusion, with receptors becoming immobilized at synaptic sites as neurons mature (Borgdorff and Choquet 2002), presumably due to their interaction with PSD-95 (Schnell *et al.* 2002).

4.10 **Future prospects**

It is clear that multiple transport pathways and sorting stations exist within neurons. Understanding how and where distinct sorting compartments are generated and regulated will be an important future challenge. Although this is a daunting prospect, the identity of many proteins that diverge into discrete routes are known. The use of multi-coloured (GFP derivatives) protein chimaeras and fluorescence recovery after photobleaching (FRAP) in real-time studies, combined with the study of particular mutants, will help identify such pathways. Combining these approaches with fluorescent energy transfer (FRET) between CFP and YFP chimaeric proteins will enable the role of specific protein interactions to be determined. Initially, it is expected that such studies will be performed by the transfection of neurons in culture and organotypic slices. Ultimately, these studies may be repeated *in vivo* by the generation of conditional knock-in transgenic mice.

References

Agno, F., Pin, J-P., Tu, J. C., Xiao, B., Worley, P. F., Bockaert, J., ***et al.*** (2000) Dendritic and axonal targeting of type 5 metabotropic glutamate receptor is regulated by homer 1 proteins and neuronal excitation. *Neuroscience*, **20**, 8710–16.

Antonova, I., Arancio, O., Trillet, A-C., Wang, H-G., Zablow, L., Udo, H., *et al.* (2001) Rapid increase in clusters of presynaptic proteins at onset of long-lasting potentiation. *Science*, **294**, 1547–50.

Banker, G. (2003) Pars, PI 3-kinase, and the establishment of neuronal polarity. *Cell*, **112**, 4–5.

Barnes Jr, E. M. (1996) Use-dependent regulation of $GABA_A$ receptors. *Int. Rev. Neurobiol.*, **39**, 53–76.

Blanpied, T. A., Scott, D. B., and Ehlers, M. D. (2002) Dynamics and regulation of clathrin coats at specialized endocytic zones of dendrites and spines. *Neuron*, **36**, 435–49.

Bollan K., King, D., Robertson, L. A., Brown, K., Taylor, P. M., Moss, S. J., *et al.* (2003) GABA (A) receptor composition is determined by distinct assembly signals within alpha and beta subunits. *J. Biol. Chem.*, **278**, 4747–55.

Borgdorff, A. J. and Choquet, D. (2002) Regulation of AMPA receptor lateral movements. *Nature*, **417**, 649–53.

Boyd, G. W., Doward, A. I., Kirkness, E. F., Millar, N. S., and Connolly, C. N. (2003) Cell surface expression of 5-hydroxytryptamine type 3 receptors is controlled by an endoplasmic reticulum retention signal. *J. Biol. Chem.*, **278**, 27681–7.

Boyd, G. W., Low, P., Dunlop, J. I., Robertson, L. A., Vardy, A., Lambert, J. J., *et al.* (2002) Assembly and cell surface expression of homomeric and heteromeric 5-HT3 receptors: the role of oligomerization and chaperone proteins. *Mol. Cell. Neurosci.*, **21**, 38–50.

Buckley, K. M., Melikian, H. E., Provoda, C. J., and Waring, M. T. (2000) Regulation of neuronal function by protein trafficking: a role for the endosomal pathway. *J. Physiol.*, **525.1**, 11–19.

Burack, M. A., Silverman, M. A., and Banker, G. (2000) The role of selective transport in neuronal protein sorting. *Neuron*, **26**, 465–72.

Connolly, C. N., Krishek, B. J., McDonald, B. J., Smart, T. G., and Moss, S. J. (1996) Assembly and cell surface expression of heteromeric and homomeric gamma-aminobutyric acid type A receptors. *J. Biol. Chem.*, **271**, 89–96.

El-Husseini, A.E.-D., Craven, S. E., Brock, S. C., and Bredt, D. S. (2001) Polarized targeting of peripheral membrane proteins in neurons. *J. Biol. Chem.*, **276**, 44984–92.

Fukata, Y., Kimura, T., and Kaibuchi, K. (2002) Axon specification in hippocampal neurons. *Neurosci. Res.*, **43**, 305–15.

Gardiol, A., Racca, C., and Triller, A. (1999) Dendritic and postsynaptic protein synthetic machinery. *J. Neurosci.*, **19**, 168–79.

Girault, J. A. and Peles, E. (2002) Development of nodes of Ranvier. *Curr. Opin. Neurobiol.*, **12**, 476–85.

Greger, I. H., Khatri, L., and Ziff, E. B. (2002) RNA editing at arg607 controls AMPA receptor exit from the endoplasmic reticulum. *Neuron*, **34**, 759–72.

Grosshans, D. R., Clayton, D. A., Coultrap, S. J., and Browning, M. D. (2001) LTP leads to rapid surface expression of NMDA but not AMPA receptors in adult rat CA1. *Nat. Neurosci.*, **5**, 27–33.

Hirbec, H., Francis, J. C., Lauri, S. E., Braithwaite, S. P., Coussen, F., Mulle, C., *et al.* (2003) Rapid and differential regulation of AMPA and kainate receptors at hippocampal mossy fibre synapses by PICK1 and GRIP. *Neuron*, **37**, 625–38.

Ikonen, E. (2001) Roles of lipid rafts in membrane transport. *Curr. Opin. Cell Biol.*, **13**, 470–77.

Imamura, F., Maeda, S., Doi, T., and Fujiyoshi, Y. (2002) Ligand binding of the second PDZ domain regulates clustering of PSD-95 with the Kv1.4 potassium channel. *J. Biol. Chem.*, **277**, 3640–6.

Jareb, M. and Banker, G. (1998) The polarized sorting of membrane proteins expressed in cultured hippocampal neurons using viral vectors. *Neuron*, **20**, 855–67.

Job, C. and Eberwine, J. (2001) Localization and translation of mRNA in dendrites and axons. *Nat. Rev. Neurosci.*, **2**, 889–98.

Kaether, C., Skehel, P., and Dotti, C. G. (2000) Axonal membrane proteins are transported in distinct carriers: a two-colour video microscopy study in cultured hippocampal neurons. *Mol. Biol. Cell*, **11**, 1213–24.

Kamiguchi, H. and Lemmon, V. (1998) A neuronal form of the cell adhesion molecule L1 contains a tyrosine-based signal required for sorting to the axonal growth cone. *J. Neurosci.*, **18**, 3749–56.

Kittler, J. T. and Moss, S. J. (2001) Neurotransmitter receptor trafficking and the regulation of synaptic strength. *Traffic*, **2**, 437–48.

Kneussel, M. (2002) Dynamic regulation of GAB(A) receptors at synaptic sites. *Brain Res. Brain Res. Dev.*, **39**, 74–83.

Ma, D., Zerangue, N., Lin, Y.-F., Collins, A., Yu, M., Jan, Y. N., *et al.* (2001) Role of ER export signals in controlling surface potassium channel numbers. *Science*, **291**, 316–19.

Malinow, R. and Malenka, R. C. (2002) AMPA receptor trafficking and synaptic plasticity. *Annu. Rev. Neurosci.*, **25**, 103–26.

Meddows, E., Le Bourdelles, B., Grimwood, S., Wafford, K., Sandhu, S., Whiting, P., *et al.* (2001) Identification of molecular determinants that are important in the assembly of *N*-methyl-D-aspartate receptors. *J. Biol. Chem.*, **276**, 18795–803.

Meier, J., Vannier, C., Serge, A., Triller, A., and Choquet, D. (2001) Fast and reversible trapping of surface glycine receptors by gephyrin. *Nat. Neurosci.*, **4**, 253–60.

Passafaro, M., Piech, V., and Sheng, M. (2001) Subunit-specific temporal and spatial patterns of AMPA receptor exocytosis in hippocampal neurons. *Nat. Neurosci.*, **4**, 917–26.

Prange, O. and Murphy, T. H. (2001) Modular transport of postsynaptic density-95 clusters and association with stable spine precursors during early development of cortical neurons. *J. Neurosci.*, **21**, 9325–33.

Rasband, M. N. and Trimmer, J. S. (2001) Developmental clustering of ion channels at and near the node of Ranvier. *Dev. Biol.*, **236**, 5–16.

Rivera, J. F., Ahmed, S., Quick, M. W., Liman, E. R., and Arnold, D. B. (2003) An evolutionarily conserved dileucine motif in Shal K(+) channels mediates dendritic targeting. *Nat. Neurosci.*, **6**, 243–50.

Roche, K. W., Standley, S., McCallum, J., Ly, C. D., Ehlers, M. D., and Wenthold, R. J. (2001) Molecular determinants of NMDA receptor internalization. *Nat. Neurosci.*, **4**, 794–802.

Sampo, B., Kaech, S., Kunz, S., and Banker, G. (2003) Two distinct mechanisms target membrane proteins to the axonal surface. *Neuron*, **37**, 611–24.

Schnell, E., Sizemore, M., Karimzadegan, S., Chen, L., Bredt, D. S., and Nicoll, R. A. (2002) Direct interactions between PSD-95 and stargazin control synaptic AMPA receptor number. *Proc. Natl. Acad. Sci., USA*, **99**, 13902–7.

Schwappach, B., Zerangue, N., Jan, Y. N., and Jan, L. Y. (2000) Molecular basis for K(ATP) assembly: transmembrane interactions mediate association of a K^+ channel with an ABC transporter. *Neuron*, **26**, 155–67.

Scott, D. B., Blanpied, T. A., Swanson, G. T., Zhang, C., and Ehlers. M. D. (2001) An NMDA receptor ER retention signal regulated by phosphorylation and alternative splicing. *J. Neurosci.*, **21**, 3063–72.

Sheng, M., Lee, S. H. (2001) AMPA receptor trafficking and the control of synaptic transmission. *Cell*, **105**, 825–8.

Silverman, M. A., Kaech, S., Jareb, M., Burack, M. A., Vogt, L., Sonderegger, P., *et al.* (2001) Sorting and directed transport of membrane proteins during development of hippocampal neurons in culture. *Proc. Natl. Acad. Sci. USA*, **98**, 7051–7.

Snyder, E. M., Philpot, B. D., Huber, K. M., Dong, X., Fallon, J. R., and Bear, M. F. (2001) Internalization of ionotropic glutamate receptors in response to mGluR activation. *Nat. Neurosci.*, **4**, 1079–85.

Stockklausner, C. and Klocker, N. (2003) Surface expression of inward rectifier potassium channels is controlled by selective Golgi export. *J. Biol. Chem.* [epub ahead of print].

Stowell, J. N. and Craig, A. M. (1999) Axon/dendrite targeting of metabotropic glutamate receptors by their cytoplasmic carboxy-terminal domains. *Neuron*, **22**, 525–36.

Takeda, T., Yamazaki, H., and Farquhar, M. G. (2003) Identification of an Apical Sorting Determinant in the Cytoplasmic Tail of Megalin. *Am. J. Physiol. Cell Physiol.* [epub ahead of print].

Tang, B. L. (2001) Protein trafficking mechanisms associated with neurite outgrowth and polarized sorting in neurons. *J. Neurochem.*, **79**, 923–30.

Tsui-Pierchala, B. A., Encinas, M., Milbrandt, J., and Johnson Jr, E. M. (2002) Lipid rafts in neuronal signaling and function. *Trends Neurosci.*, **25**, 412–7.

Votsmeier, C. and Gallwitz, D. (2001) An acidic sequence of a putative yeast Golgi membrane protein binds COPII and facilitates ER export. *EMBO J.*, **20**, 6742–50.

Wang, J., Liu, S., Haditsch, U., Tu, W., Cochrane, K., Ahmadian, G., *et al.* (2003) Interaction of calcineurin and type-A GABA receptor gamma 2 subunits produces long-term depression at CA1 inhibitory synapses. *J. Neurosci.*, **23**, 826–36.

Winckler, B., Forscher, P., and Mellman, I. (1999) A diffusion barrier maintains distribution of membrane proteins in polarized neurons. *Nature*, **397**, 698–701.

Zerangue, N., Malan, M. J., Fried, S. R., Dazin, P. F., Jan, Y. N., Jan, L. Y., *et al.* (2001) Analysis of endoplasmic reticulum trafficking signals by combinatorial screening in mammalian cells. *Proc. Natl. Acad. Sci. USA*, **98**, 2431–6.

Zhai, R., Vardinon-Friedman, H., Cases-Langhoff, C., and Becker, B. (2001) Assembling the presynaptic active zone: Characterization of an active zone precursor vesicle. *Neuron*, **29**, 131–43.

Ziv, N. E. and Garner, C. C. (2001) Principles of glutamatergic synapse formation: Seeing the forest for the trees. *Curr. Opin. Neurobiol.*, **11**, 536–43.

Chapter 5

Ion channels and electrical activity

Mauro Pessia

5.1 Introduction

Action potentials are the fundamental elements permitting the flow of electrical signals within neurons. The basic phenomena regulating action potential generation and the electrical properties of nerve cells were clarified by the pioneering work of a number of scientists, over several centuries.

Luigi Galvani, a physiologist at the University of Bologna, published the results of his famous experiments with the frog preparation between 1791 and 1797. By performing the first electro-physiological experiments he set the basis of this new science and proposed his 'animal electricity' theory. Stimulated by Galvani's findings, Alessandro Volta, a physicist at the University of Pavia, invented the electrical battery. The electrical and physiological effects stimulated by Volta in tongue, skin, ears, and eyes of human beings and frog preparations, by using his newly developed physical instrument, definitively inaugurated the modern era of neurophysiology (for review, see Piccolino 1997 and 2000).

A century later Julius Bernstein hypothesized the existence of bioelectric potentials in excitable membranes and proposed that the permeability of this structure changes upon stimulation. During the late 1930s the breakthrough experiments of Kenneth Cole and Howard Curtis at Woods Hole (Massachusetts, U.S.A.) confirmed Bernstein's theory. Indeed, by using squid giant axons they showed that the conductance of the membrane increases during an action potential (Cole and Curtis 1939). Alan Hodgkin and Andrew Huxley at Plymouth (U.K.) surprisingly observed that the membrane potential overshot zero and reversed sign during an impulse. A decade later, Hodgkin and Katz unequivocally demonstrated that the influx of sodium ions dictates the rising phase of an action potential, its overshoot to positive potentials, and its propagation (Hodgkin and Katz 1949). Soon thereafter, by using their newly devised *voltage-clamp* method, they successfully dissected the sodium and potassium ionic currents of axons and established their voltage-dependency, kinetic properties and role in the generation and propagation of impulses (see Fig. 5.1; Hodgkin and Huxley 1952).

Hodgkin and Huxley inferred the involvement of pore-forming proteins in the generation of electrical signals in neurons. Although, the existence of 'pores' was

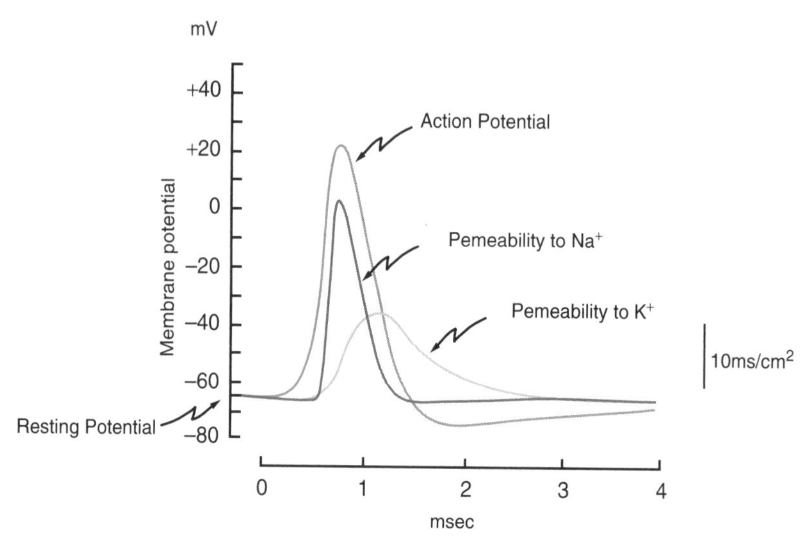

Fig. 5.1 Action potential recorded from a squid giant axon. In 1936 the English zoologist John Z. Young discovered that the squid *Loligo* possessed giant axons mistakenly believed to be blood vessels. The dimension of the axon (1 mm) permitted the electrophysiological recording of an action potential and of the underlying ionic conductances. The figure shows the rapid increase of sodium permeability, which determines the rising phase of the action potential. The Na^+ conductance also contributes to action potential overshoot to positive potentials. On the other hand, the potassium permeability shows a delayed onset (delayed-rectifying) and a much slower time course. This conductance determines the falling phase of the action potential and the after-hyperpolarization (adapted from Hodgkin A.L. and Huxley A.F. 1952).

originally proposed by Ernst Brücke in 1843, single-channel currents were not measured until the 1970s, when Erwin Neher and Bert Sakmann developed the *patch-clamp* technique (Neher and Sakmann 1976; Hamill *et al.* 1981). Thanks to these fascinating discoveries we now know that the concerted opening and closing of a number of voltage-dependent sodium, potassium, and calcium channels play key roles in neurotransmission. In particular, they generate and shape up action potentials, modulate the release of neurotransmitters, and control the excitability, electrical properties, and firing pattern of central and peripheral neurons (Hille 2001).

Most of these ion channels have been cloned during the past two decades. The wealth of information resulting from these studies inaugurated an intense period of structural and functional analysis of ion channels. In addition, the detailed elucidation of the relevant genes structure lead to the discovery of mutations that cause a number of inherited ion channel diseases. Recently, this successful era has been crowned by the original description of the crystal structure of the potassium channel KcsA, by X-ray crystallography (Doyle *et al.* 1998). The pore structure of the KcsA channel from *Streptomyces lividans* is homologous to voltage-gated channels, which allows important functional correlations to be drawn (MacKinnon *et al.* 1998).

This chapter aims to provide a concise description of the main structural determinants regulating the function of some voltage-gated Na^+, K^+, and Ca^{++} channels and their fundamental role in cell excitability. However, it does not presume to give a comprehensive review of the large literature related to these topics.

5.2 **The voltage sensor of ion channels**

Voltage-dependent Na^+, K^+, and Ca^{++} channels are generally closed at the resting membrane potential of nerve cells, which is approximately –60 to –70 mV. However, the transmembrane potential of neurons undergoes continuous changes that are caused by incoming stimuli. In particular, depolarizing inputs trigger the opening of voltage-gated channels, which therefore allow the flow of electrical signals throughout the nervous system. By contrast, membrane repolarization closes these channels and terminates the propagation of the impulse.

Almost 50 years ago, Hodgkin and Huxley hypothesized the presence of a voltage-sensing region within the ion channel protein that detected membrane potential changes. The molecular cloning of voltage-gated Na^+, K^+, and Ca^{++} channels revealed that the fourth transmembrane segment (S4) of these channels is composed of regularly spaced positively charged arginines and lysines (see Fig. 5.2). It appeared evident to investigators that this α-helical structure could be the voltage-sensor hypothesized for such a long time. A molecular model named the *'sliding helix'* was then proposed, whereby a rotation of the S4 segment upon membrane depolarization would open the channel (Catterall 1986). Subsequent structure–function experiments have confirmed that this segment comprises the main voltage-sensor of these channel types. However, the negatively charged residues located in the S2 and S3 segments of Kv channels also regulate this process (Papazian *et al.* 1995; Planelles-Cases *et al.* 1995).

How does the sensor really operate? The overall channel activity is determined by the probability of opening (P_0). The probability of finding a channel in the open conformation increases as the transmembrane potential is depolarized. The voltage sensor regulates this process directly. Thus, the P_0 of a given channel approaches zero at hyperpolarized potential, when the voltage-sensor is in the *off* position, and it is maximal ($P_0 = 0.8 - 1$) at depolarized potentials, when the voltage sensor is in the *on* position.

The outward translocation of the positive charges associated with the S4 segment or the reorientation of a dipole, generates detectable currents named *gating currents* that were first predicted by Hodgkin and Huxley (Hodgkin and Huxley 1952). During the early 1970s gating currents were successfully recorded from skeletal muscle and giant axons (Armstrong and Bezanilla 1973; Schneider and Chandler 1973). More recently these tiny and transient currents have been recorded and characterized by using high-resolution techniques such as the *patch-clamp* and the *cut-open* oocyte (Taglialatela *et al.* 1992). These studies also permitted an accurate estimate of the number of gating charges necessary to open a channel (~12–13 e_0; Schoppa *et al.* 1992; Hirschberg *et al.* 1995).

SCAM analysis (substituted cysteine accessibility method) provided further evidence on the primary role played by the S4 segment in the gating mechanism. This method

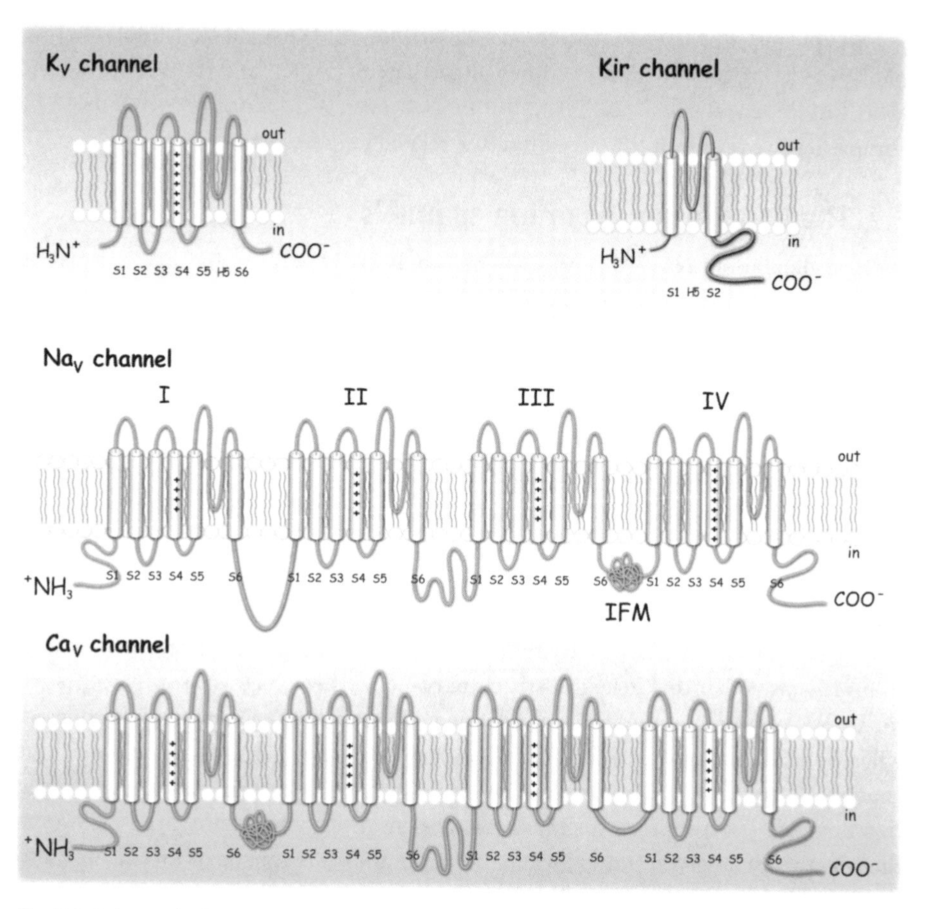

Fig. 5.2 Schematic diagram illustrating the predicted secondary structure of voltage-gated Na^+, K^+, and Ca^{++} channels. A Kv subunit is composed of a voltage-sensing region (S1 through S4) and of a pore region (S5-H5-S6). By contrast, a Kir subunit is composed of just a pore region (S1-H5-S2), which has a similar structure to the homologous region of a Kv subunit. Both the N- and C-termini reside inside the cell. Four subunits form a functional channel. Voltage-gated Na^+ and Ca^{++} channels are large molecules made up of four linked domains (I–IV). Nevertheless, the overall membrane topology of each domain is similar to a Kv subunit. The main voltage sensors of all these channel types are labelled S4 and the positive charges depicted point out the arginines and lysines composing this transmembrane domain. The critical residues involved in Na_V channel inactivation are labelled IFM (isoleucine–phenylalanine–methionine). The *hinged-lid* mechanism of Na_V channel inactivation appears to be adopted by Ca_V channels as well. However, the intracellular loop connecting the first and second domain of the latter channel type plays a role in the inactivation process. It has been shown that this loop forms the binding site for the G-protein $\beta\gamma$ subunits as well (DeWaard *et al.* 1997).

allows the detection of the state-dependent labelling of a cysteine residue by methanethiosulfonate (MTS) compounds (Akabas *et al.* 1992). By artificially introducing a cysteine at a specific location within the S4 segment, investigators were able to show that certain residues could be exposed and therefore labelled only upon depolarization. This result implies that the S4 segment undergoes a conformational rearrangement upon membrane depolarizations (Yang and Horn 1995; Larsson *et al.* 1996; Yang *et al.* 1996). More recently, evidence obtained by using fluorescence resonance energy transfer (FRET) and its modified version LRET (lanthanide-based) has shown that the S4 segment undergoes a rotation of approximately 180° upon depolarization (Cha *et al.* 1999; Glauner *et al.* 1999; Bezanilla 2000).

This helical twist movement is likely to be transferred, via the S4–S5 linker, to the gates that are pulled apart and open the channel. Consequently, more than ten million ions per second flow under their electrochemical gradients. These lower gates are physically located in the cytoplasmic side of the channel (Armstrong 1966). In particular, the carboxy-terminal portion of the S6 segment contributes to form the K^+ channel gates that may be opened as an iris diaphragm and give access to the central cavity (Liu *et al.* 1997; Yellen 1998; del Camino and Yellen 2001). This overall process is reversed by membrane repolarization that consequently stops the ion flux.

5.3 **Voltage-gated sodium channels**

The first biochemical identification and molecular cloning studies of sodium channels were reported in the early 1980s (Agnew *et al.* 1980; Beneski and Catterall 1980; Noda *et al.* 1984). These studies revealed that the α subunit of a Na^+ channel is a large polypeptide (1681–2016 amino acids) composed of four homologous domains (I through IV), each of which contains six transmembrane spanning regions (S1 through S6; Fig. 5.2).

The H5 loops form part of the ion-conducting pore. A specific glutamate residue located within these structures forms the binding site for tetrodotoxin (TTX) and saxitoxin (STX), which block the channel by physically plugging it (Noda *et al.* 1989). These negatively charged residues also impart ion selectivity. Indeed, it is possible to convert a sodium channel into a calcium channel by increasing the number of negatively charged residues in the pore (Heinemann *et al.* 1992). Several Na_V channels, however, show resistance to TTX block. This is caused by sequence differences identified in the region preceding the selectivity filter (Backx *et al.* 1992). Local anaesthetics, antiarrhythmic, and anticonvulsant drugs block Na_V channels as well. But, their binding sites are located on the cytoplasmic side of the pore.

Sodium channels inactivate within 1–2 milliseconds. This process is of pivotal physiological importance as it determines the action potential duration, the termination of the signal, and regulates the refractory period of nerve cells. By using antibodies and site-directed mutagenesis, investigators have demonstrated that the Ile-Phe-Met residues located in the cytoplasmic loop linking the third and fourth domain are the structural determinants of fast inactivation in sodium channels (Fig. 5.2; Vassilev *et al.* 1988; Stuhmer *et al.* 1989*a*; West *et al.* 1992). These residues seem to be part of a 'hinged lid' domain which occludes the channel after its opening by a swinging movement and by interacting with the carboxy-terminal portion of S6 in the fourth domain and with the S4–S5 linker of domains III and IV (McPhee *et al.* 1995; Lerche *et al.* 1997; Smith and Goldin 1997). This mechanism is reminiscent of the inactivation of some allosteric enzymes.

The brain sodium channel is a complex composed of a pore-forming α subunit and two auxiliary subunits: β1 and β2 (Fig. 5.3). The β1 subunit simply associates with the α subunit. By contrast, a disulfide bond anchors the β2 subunit to the α subunit. The β subunits are structurally composed of a large extracellular domain containing immunoglobulin-like folds, a transmembrane spanning segment and a short cytoplasmic carboxy terminus (Isom *et al.* 1995).

The β1 subunit tunes up the biophysical properties of the resulting channel. In particular, it accelerates the kinetics of fast inactivation of the α subunit, shifts the voltage dependence of inactivation to more hyperpolarized potentials and regulates its functional expression in *Xenopus* oocytes (Isom *et al.* 1992). However, the β1 and β3 isoforms may also regulate the sodium channel targeting to nodes of Ranvier (Ratcliffe *et al.* 2001).

Molecular cloning has identified nine different Na^+ channel types. They have been classified by using the numerical system $Na_vy.x$ whereby the prefix specifies both the permeating ion (Na) and the voltage-dependence of the channel (v). According to this nomenclature originally proposed for Kv channels by George Chandy, the nine members of the voltage-gated sodium channel have been assigned to a single family and have been classified $Na_v1.1$ through $Na_v1.9$ (Goldin *et al.* 2000). Other sodium-channel related proteins, not yet fully characterized, have been grouped and named Na_x. Alternative-splicing isoforms of Na_v channels have also been described ($Na_v1.3_a$, $Na_v1.3_b$, etc.) and up to the present four β subunits have been cloned and classified as follows: $Na_v\beta1.1$, $Na_v\beta1.1_a$, $Na_v\beta2.1$, and $Na_v\beta3.1$.

Skeletal muscles and heart express $Na_v1.4$ and $Na_v1.5$ channel types, respectively. The remaining members of the sodium channel family have been generally localized within the central and peripheral nervous system. In particular, the $Na_v1.6$ channel type is highly expressed at the node of Ranvier where it contributes to the saltatory propagation of signals. This localized expression involves interactions with anchoring and scaffolding molecules. Na^+ channels are held in place by ankyrin G and form a macromolecular complex with cell adhesion molecules as NrCAM, NF186 that may regulate channel localization. Extracellular matrix components, such as tenascins and phosphacan also interact with the extracellular sodium channel domains (Srinivasan *et al.* 1988, 1998; Zhou *et al.* 1998).

Recently, the three-dimensional structure of sodium channels has been elucidated at 19 Å resolution (Catterall 2001; Sato *et al.* 2001). The images, obtained by helium-cooled cryo-electron microscopy, show that the channel is bell-shaped and confirm that the four homologous domains are assembled to form a tetrameric structure.

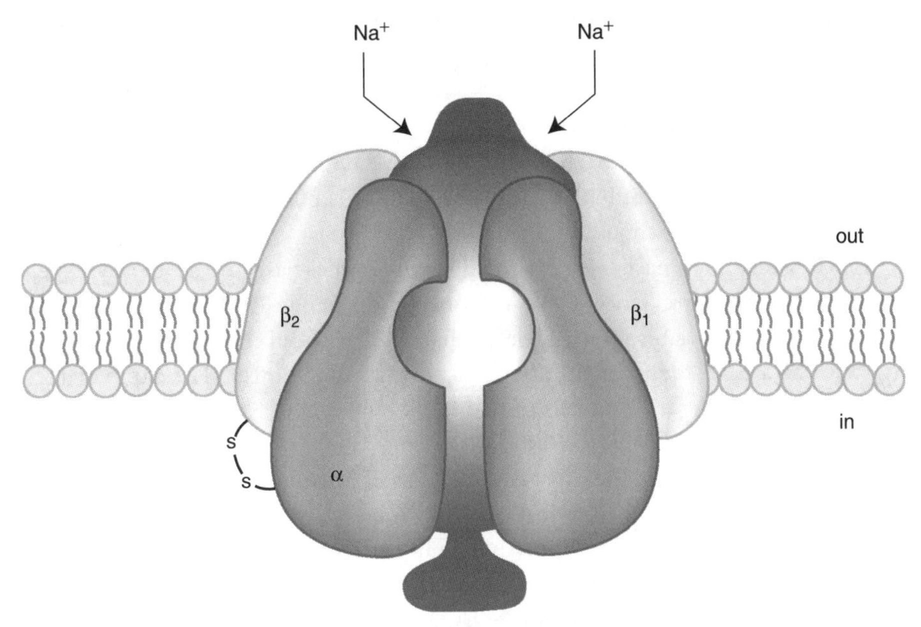

Fig. 5.3 Cartoon showing a voltage-gated sodium channel and the associated auxiliary subunits. Electron microscopy images of Nav channels from the eel Electrophorus electricus showed that the α-subunit possessed a bell-shaped structure with its wider side residing into the cytoplasm (Sato *et al*. 2001). Moreover, the channel contained several connected cavities, through which Na^+ ions may flow into the cytoplasm. The scheme also depicts the β1 and β2 subunits associated with the α subunit to form a brain sodium channel complex.

More interestingly and unexpectedly, the channel possesses several internal cavities that converge to form four extracellular orifices and several intracellular openings (Fig. 5.3). The voltage sensor S4 of each domain appear to reside in four cavities or *gating pores* that may co-ordinate its function. These observations also suggest that voltage-gated Na^+, K^+, and Ca^{++} channels may have a similar 3D-structure.

5.4 **Voltage-gated potassium channels**

Voltage-gated potassium channels (Kv) play key roles in neurotransmission and nerve cell physiology. In particular, these ion channels shorten action potential duration, modulate the release of neurotransmitters, and control the excitability, electrical properties, and fining pattern of central and peripheral neurons (Hille 1992).

5.4.1 **Molecular characterization of Kv channels**

The Jans and their colleagues at the University of California, San Francisco, first cloned Kv channels from the *Shaker* mutant of *Drosophila* in 1987 (Tempel *et al.* 1987). These fruit flies are characterized by a shaking phenotype induced by ether as the relevant K^+

channel gene is mutated (Kaplan and Trout 1969; Jan *et al.* 1977; Tanouye and Ferrus 1985). Since then, several other Kv channel genes have been identified from *Drosophila melanogaster* and vertebrates.

The predicted topology and structural features of a K^+ channel subunit are similar to a voltage-gated sodium channel domain (Fig. 5.2). Four subunits are assembled, probably in the endoplasmic reticulum and targeted to the plasma membrane to form a functional tetrameric channel.

Cloned Kv channels have been classified in subfamilies based on sequence relatedness by using the abbreviation Kv*y.x* (Gutman and Chandy 1993). According to this standardized nomenclature *Shaker*-related channels have been classified in the subfamily Kv1.x and each member numbered Kv1.1 through Kv1.9. The same criteria have been used to classify channels related to the subfamilies *Shab* (Kv2.1 and Kv2.2), *Shaw* (Kv3.1 to 3.4) and *Shal* (Kv4.1 to Kv4.3).

Potassium channels are the most diverse class of ion channels. They may exist as homomers, whenever four identical α-subunits are assembled. However, different types of α-subunits may heteropolymerize to form channels with properties that are different from the parental homomeric channels (Isacoff *et al.* 1990; Ruppersberg *et al.* 1990). This phenomenon greatly enhances potassium channel diversity. Notable examples are the heteromeric channels Kv1.1/Kv1.2 or Kv1.2/Kv1.4 that have been localized in several structures within the nervous system (Wang *et al.* 1993, 1994; Sheng *et al.* 1994; D'Adamo *et al.* 1999). Heteropolymerization has been reported to occur only among members of the same subfamily of Kv channels (Covarrubias *et al.* 1991). However, some exceptions have been observed amongst inwardly rectifying potassium channels (Pessia *et al.* 1996).

It is generally assumed that heteromeric channels are composed of two subunits of each type placed in a tandem order (e.g. A-B-A-B). However, the relative order of subunits within the channel may be different (e.g. A-A-B-B). These types of arrangements give rise to channels with different functional properties, at least in some inwardly rectifying K^+ channels (Pessia *et al.* 1996). Thus, subunit position within the tetramer may confer additional functional diversity. Finally, alternative-splicing also increases Kv channel diversity by generating a series of related proteins with distinct properties. The need for such a large number of potassium channels resulting from all these regulatory mechanisms remains unclear.

5.4.2 *Xenopus* oocytes as a heterologous expression system for studying cloned ion channels

Xenopus laevis is a fully aquatic amphibian species native to Africa and belonging to the family of *Pipidae*, which are mistakenly called 'frogs'. Thousands of oocytes might be collected from the ovaries of a female *Xenopus*. These cells may be as large as 1 mm in diameter and display morphological polarity, consisting of a yellowish vegetal pole and an animal pole, which appears dark due to the presence of pigment vesicles. Such polarity is very evident in mature oocytes (stage IV–VI).

John Gurdon originally showed that mature oocytes are able to faithfully and efficiently translate foreign genetic information. At the beginning of the 1980s, Eric Barnard and colleagues showed that *Xenopus* oocytes express ion channels on their plasma membrane after a microinjection of exogenous mRNAs into their cytoplasm (Sumikawa *et al.* 1981). Therefore, they became a useful tool for molecular neurobiology research and for the heterologous expression of ion channels.

Xenopus oocytes have been largely used for the structural and functional analysis of delayed-rectifying potassium channels. For instance, the injection of mRNA encoding for Kv1.1 into these cells normally results in functional channels after several hours of incubation. The potassium currents flowing through these channels can be recorded in *two-electrode voltage-clamp* mode (Fig. 5.4). On the other hand, single channel activity can be recorded by using the *patch-clamp* technique (Fig. 5.5). The biophysical properties of the channel can be determined by properly analysing these recordings.

5.4.3 Potassium selectivity of Kv channels

Molecular cloning facilitated detailed structure–function studies aimed at understanding the role played by each domain within a K^+ channel subunit.

Yellen and MacKinnon provided compelling evidence showing that the H5 loop contributes to the ion-conducting pore (Yellen *et al.* 1991). Mutagenesis directed to the

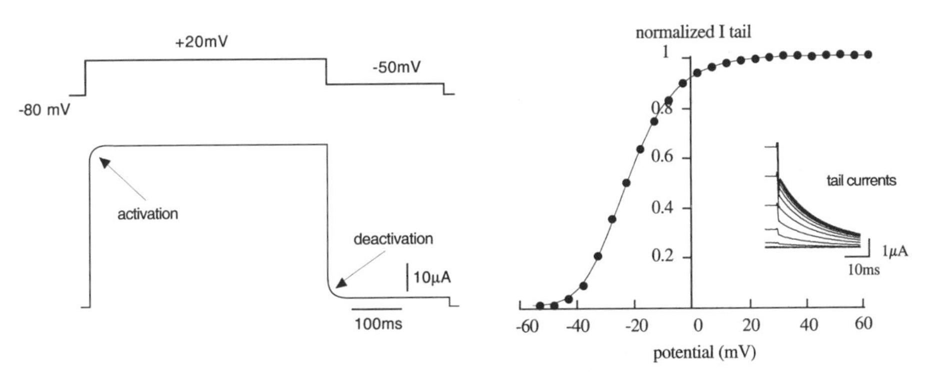

Fig. 5.4 Electrophysiological characterization of Kv1.1 channels heterologously expressed in *Xenopus* oocytes. Representative whole-oocyte current recorded in two-electrode *voltage-clamp* configuration showing the time course of activation and deactivation of Kv1.1 channels (left panel). On the top, the voltage protocol used to elicit the delayed-rectifier potassium current is reported. The panel on the right hand side shows the current–voltage relationship of Kv1.1 currents. It is generally obtained by plotting the normalized peak tail currents, recorded at −50 mV, as a function of the pre-pulse potentials. The solid line represents the fit with the Boltzmann function: $I = 1/[1+\exp^{-(V-V1/2)/k}]$ from which the half-maximal activation voltage of the channel ($V_{1/2}$) and the steepness of its voltage-dependence (slope factor k) are calculated. The inset shows the Kv1.1 tail currents recorded after 500 ms pre-pulses to different potentials.

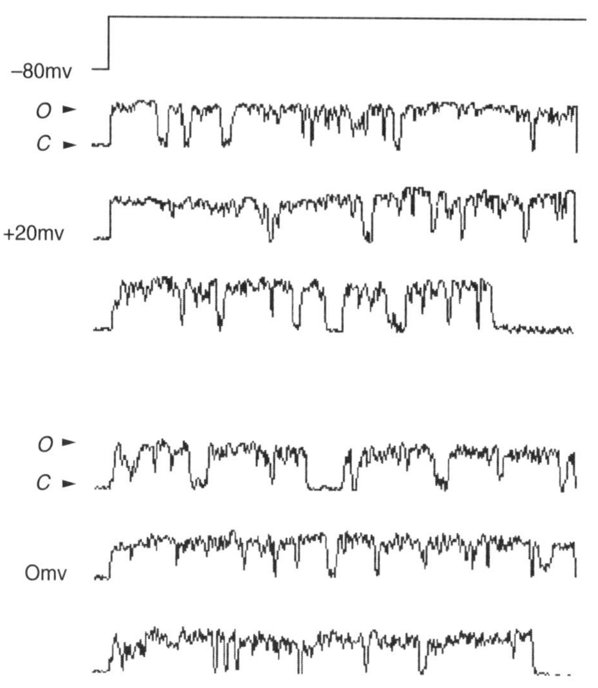

Fig. 5.5 Typical inside-out patch recording from an oocyte expressing the heteromeric channel Kv1.2–1.1. Depolarizing pulses to the indicated voltages evoke single-channel openings (holding potential of –80 mV). The traces are leak subtracted by using averaged sweeps with no opening. On the top, an electrode with an excised inside-out patch containing a single channel and a schematic representation of the Kv1.2–1.1 concatemer is shown. Ten glutamine residues (green rectangles) have been used to link both subunits tandemly (adapted from D'Adamo *et al.* 1999).

residues 'GYG' in the H5 loop, recognized early on as a potassium channel signature sequence was found to alter the potassium selectivity of the channel, allowing the permeation of other cations (Heginbotham *et al.* 1992).

The crystal structure of KcsA channels confirmed that these residues act as a potassium selectivity filter, and finally demonstrated the biophysical principle regulating potassium selectivity and permeation (Fig. 5.6). Briefly, the hydrated K^+ ion located in the inner cavity is dehydrated upon entering the narrow, 12 Å long selectivity filter. At this level a K^+ ion is coordinated by eight carbonyl oxygen atoms, which stabilize the ion simulating its hydrated state. The overall process is therefore energetically very favourable. Normally, two potassium ions separated by a water molecule reside within the selectivity filter. The entering of a third K^+ ion within this region expels the first one, allowing ion flux (Doyle *et al.* 1998; Morais-Cabral *et al.* 2001; Zhou *et al.* 2001). The permeation of sodium through a K^+ channel is seldom permitted, although it has a smaller ionic radius than potassium. The binding of the strongly hydrated Na^+ ion

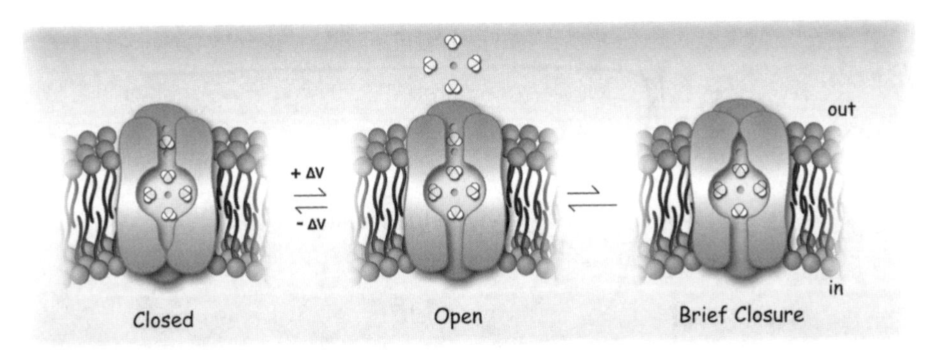

Fig. 5.6 Cartoon recapitulating the main structural changes linked to K^+ ion permeation. The scheme on the left shows a potassium channel in the closed state with the intracellular gates shut. Normally, eight water molecules hydrate the K^+ ion present in the central cavity. However, only four water molecules are shown for clarity. The selectivity filter can accommodate four K^+ ions. Nevertheless, only two K^+ are present within this region, separated by a water molecule. The membrane depolarization opens the intracellular gates and the K^+ residing in the cavity sheds its hydration water upon entering the selectivity filter (central scheme). The K^+ ion at the front is then forced to leave the channel and gets hydrated by extracellular water. This mechanism of conduction is very efficient and energetically favourable as the ions residing within the selectivity filter move concertedly between the positions 1,3 and 2,4. In addition, the selectivity filter undergoes conformational changes that are regulated by the K^+ concentration. The scheme on the right shows the non-conducting structure of the selectivity filter. The fluctuation between the non-conducting and the conducting structure (central scheme) may be regarded as the *selectivity filter gating* which dictates the brief closures observed in single-channel recordings (adapted from Zhou *et al.* 2001).

within this structure is predicted to be less well coordinated and therefore its dehydration is energetically unfavourable.

5.4.4 **Inactivation mechanisms of Kv channels**

The inactivation of delayed-rectifier potassium channels is a physiologically relevant process as it controls the firing properties of neurons and their response to input stimuli (Aldrich *et al.* 1979).

Kv channels show two principal types of inactivation, namely the N- and C-type. Hodgkin and Huxley first observed the fast *N-type* or *A-type* inactivation that is caused by a 'ball-and-chain' mechanism of pore occlusion. The studies of Richard Aldrich and his colleagues provided the first molecular confirmation of this model that was originally proposed by Clay Armstrong and Francisco Bezanilla in 1977. They were able to show that the occluding ball was indeed formed by the first 20 amino acids located in the N-terminus of *Shaker* channels (Hoshi *et al.* 1990; Zagotta *et al.* 1990). Four inactivating particles have been found in these channels even though only one is sufficient to occlude it. The availability of four particles renders the mechanism of block very effective. Indeed, these channels show a faster time course of inactivation than those that had only one particle.

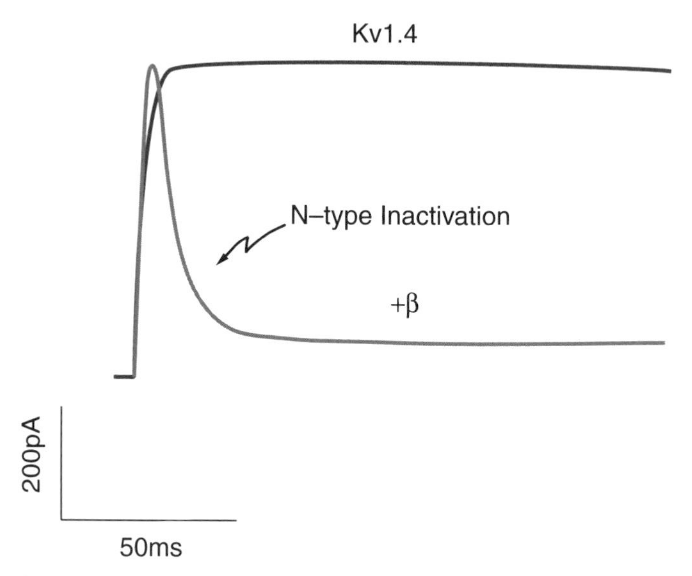

Fig. 5.7 Fast inactivation of voltage-gated K^+ channels. Two-electrode voltage-clamp recording of a *Xenopus* oocyte expressing Kv1.4 channels from which the inactivation particle has been removed (red trace). A fast inactivating current can be recorded from cells co-expressing Kv1.4 channels and the β subunits, which provide the inactivation particles (blue trace; adapted from Zhou *et al*. 2001).

Fast inactivation, however, may be conferred to non-inactivating channels by auxiliary subunits as Kvβ1.1 and Kvβ1.2 (see Fig. 5.7; Rettig *et al.* 1994). The initial 20 amino acids of these subunits have the same chemical characteristics of the *Shaker* inactivation ball, with the first 10 amino acids being hydrophobic and the following 10 hydrophilic. Four β subunits participate in the ion channel complex. They are bound to a structure called the T1 domain that is formed by approximately 100 amino acid residues located in the N-terminus of the Kv channel α subunit. The crystal structures of both the *Shaker* channel T1 domain and of the β subunit have also been determined and Fig. 5.8 shows a schematic cartoon of the T1-β subunit complex (Kreusch *et al.* 1998; Gulbis *et al.* 1999, 2000).

Recently, MacKinnon and co-workers provided a complete elucidation of the fast N-type inactivation mechanism by using crystallographic data (Zhou *et al.* 2001). The new model shows that the inactivation particle works its way through one of the four lateral vestibules that provide the access to the central pore. Then, the flexible ball domain reaches its final binding site by sneaking into the central cavity located below the selectivity filter. The blocking process is now complete and the potassium flux then terminates (Fig. 5.8).

Interestingly, sequence analysis showed that the Kvβ subunits have an aldo-keto reductase related core region (McCormack and McCormack 1994). Indeed, X-ray

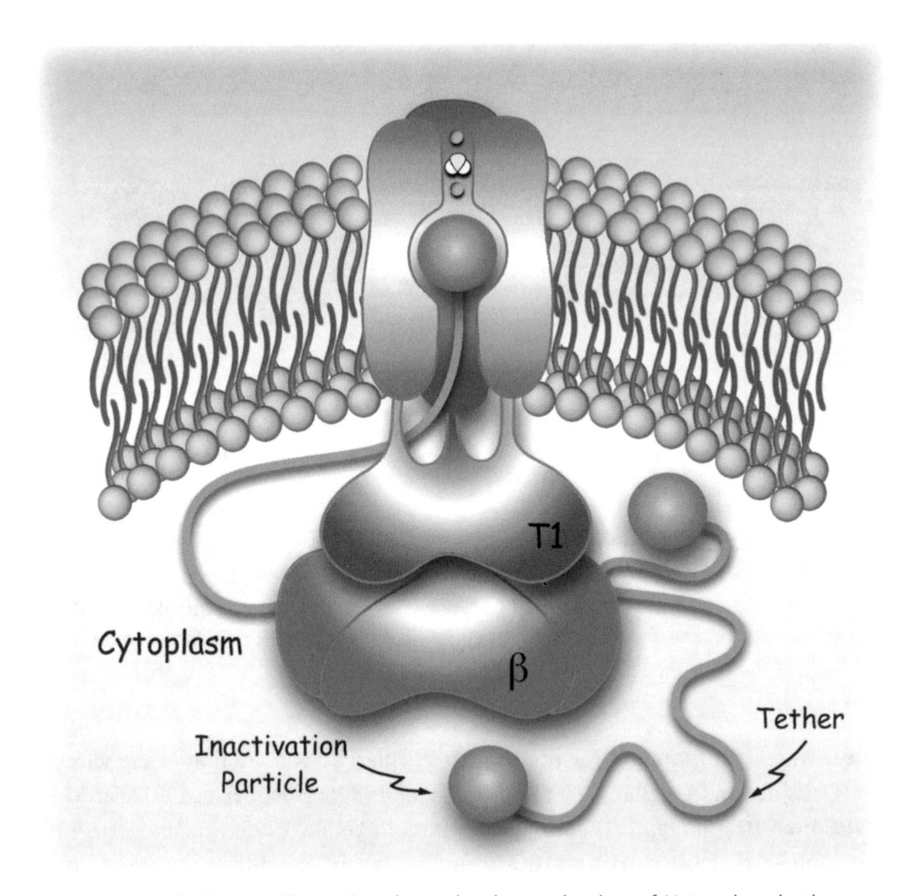

Fig. 5.8 Schematic diagram illustrating the molecular mechanism of N-type inactivation. The intracellular gates open upon membrane depolarization and the positively charged inactivation particle blocks the channel by entering the central cavity through one of the four windows formed by the T1 domains and the T1-S1 linkers. Four 'balls and chain' are provided by the corresponding auxiliary subunits that are anchored to the T1 domains. However, only three of them are visible in the figure, as both the Kv subunit and the β subunit at front have been removed for clarity (adapted from Zhou *et al*. 2001).

crystallography has revealed that Kvβ subunits bind nicotinamide adenine dinucleotide phosphate ($NADP^+$). The relevant biochemical role of these subunits complexed to the channel is currently under investigation.

Some Kv channels are characterized by a slower process of inactivation (*e.g.* Kv1.1, Kv1.2), which has been named *C-type* and *P-type*, depending on the structural determinants at the C-terminus or pore region associated with the process (Fig. 5.9). During intense neuronal activity the C-type inactivation of Kv channels can accumulate, modifying both the firing rate and the shape of the action potential (Aldrich *et al.* 1979). The molecular mechanism of this process involves conformational modification

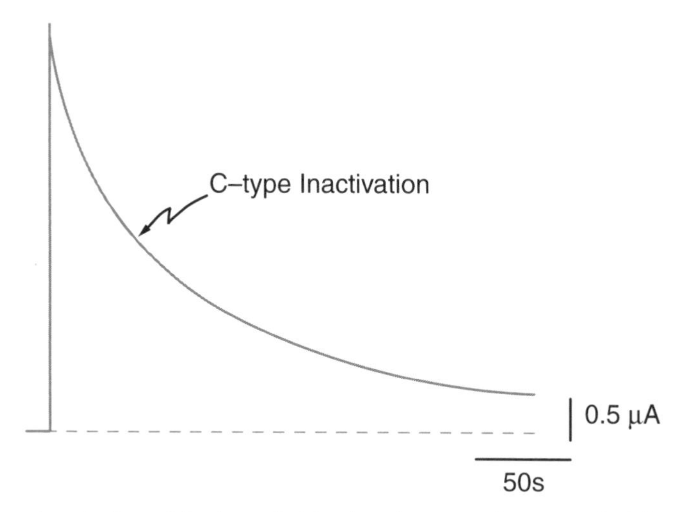

Fig. 5.9 C-type inactivation of Kv channels. The decaying trace shows a typical time course of slow C-type inactivation. The current has been elicited by a depolarizing pulse to +20 mV from a holding potential of –80 mV in an oocyte expressing the heteromeric hKv1.2–1.1 channels. The dashed line represents the zero current level (adapted from D'Adamo *et al.* 1999).

of the extracellular mouth of the pore and, particularly, a constriction of the selectivity filter (Grissmer and Cahalan 1989; Stühmer *et al.* 1989*b*; Hoshi *et al.* 1991; Pardo *et al.* 1992; López-Barneo *et al.* 1993; Baukrowitz and Yellen 1995; Molina *et al.* 1997). Surprisingly, disease-causing mutations located in cytoplasmic regions of the pore accelerate the slow inactivation of Kv channels (Adelman *et al.* 1995; D'Adamo *et al.* 1998).

5.4.5 Functional role and modulation of *Shaker*-like potassium channels

Action potentials propagate rapidly in myelinated axons by saltatory conduction. The velocity of signal propagation is dictated by axon insulation with a myelin sheath resulting from Schwann cells in the PNS and oligodendrocytes in the CNS. The structural organization of a myelinated axon is subdivided into several specialized domains. They include, in order, the node of Ranvier, paranode, juxtaparanode, and internode. A macromolecular membrane complex has been identified in the juxtaparanodal domain that is composed of Kv1.1, Kv1.2 heteromeric channels, their accessory subunit Kvβ1.2 and Caspr2, a contactin-associated protein (paranodin; Wang *et al.* 1993; Poliak *et al.* 1999).

Caspr2 is a new member of the neurexin superfamily that associates with the K^+ channel through a common protein containing a PDZ motif. Indeed, *Shaker*-like K^+ channels bind PSD-95, a protein purified from postsynaptic densities that belongs to the family of membrane-associated guanilate kinase (Kim *et al.* 1995). This protein contains three PDZ, an SH3 and a guanylyl kinase domain. It has been shown that the PDZ-1 and PDZ-2 domains of PSD-95 bind the C-terminus of the Kv channel subunit (Kim *et al.* 1995). This protein–protein interaction results in Kv channels clustering and determine their subcellular distribution in axons, neuromuscular junction etc. (Kim *et al.* 1995; Tejedor *et al.* 1997; Zito *et al.* 1997). Therefore, Caspr2 and PDZ-containing proteins may promote Kv channels subcellular targeting and clustering at the juxtaparanode where they regulate membrane repolarization, resting potential, and the efficient conduction of impulses (Zhou *et al.* 1998). Indeed, it has been shown in a heterologous expression system that heteromeric channels composed of hKv1.1 and hKv1.2 subunits may contribute to setting the resting potential of membranes in which this conductance predominates (D'Adamo *et al.* 1999).

Heteromeric Kv1.1/Kv1.2 channels are expressed at the cerebellar Pinceau, a structure composed of several basket cell terminals, which embrace the Purkinje axon hillock and proximal axon segment (McNamara *et al.* 1993; Wang *et al.* 1993, 1994; Laube *et al.* 1996). Purkinje axons represent the only output system of the cerebellar cortex and the Pinceau may regulate the generation of action potentials at the axon hillock. Patch-clamp recordings from Purkinje cells have revealed that α-DTX selectively blocks the Kv1.1 and Kv1.2 potassium channels from basket cell presynaptic terminals, and increases both the amplitude and frequency of spontaneous IPSCs mediated by $GABA_A$ receptor activation (Southan and Robertson 1998). These findings suggest that Kv1.1 and Kv1.2 heteromeric channels may contribute to the excitability and the rapid repolarization phase of action potentials in myelinated axons and basket cell terminals, where they modulate the release of the neurotransmitter γ-aminobutyric acid (GABA) onto Purkinje cells.

The electrical activity of neurons is also tuned up by mechanisms of ion channel modulation. Voltage-gated potassium channels can be dynamically regulated by several events, including neurotransmitter stimulated biochemical cascades mediated by G protein-coupled receptors. A number of receptors have been involved in this process, such as the muscarinic acetylcholine receptor M1, the serotonergic receptor 5-HT_{2C}, the epidermal growth factor receptor (EGF) or uncharacterized receptors responding to insulin treatment (Hoger *et al.* 1991; Lev *et al.* 1995; Holmes *et al.* 1996; Bowlby *et al.* 1997; Fadool *et al.* 1997; Cachero *et al.* 1998; Felsch *et al.* 1998; Imbrici *et al.* 2000). In particular, the activation of M1 and 5-HT_{2C} receptors causes a suppression of Kv1.1 and Kv1.2 channel activity (Tsai *et al.* 1999; Imbrici *et al.* 2000). It has been shown that downstream effectors, such as protein kinase C (PKC), the GTP-binding protein RhoA, the protein tyrosine kinases PYK2 and Src, play an important role in suppressing Kv1.x currents (Peretz *et al.* 1996; Boland and Jackson 1999; Moran *et al.* 1991; Cachero *et al.* 1998; Lev *et al.* 1995; Holmes *et al.* 1996). In contrast, RPTPα, a receptor protein tyrosine phosphatase highly expressed in the central nervous system, stimulates the activity of Kv1.2 channels. RPTPα physically associates with the intracellular domains of Kv1.2 reducing the tyrosine phosphorylation of the channel (Sap *et al.* 1990; Tsai *et al.* 1999).

The phosphorylation of multiple tyrosine residues has been involved in Kv channel modulation. Recently, it has been proposed that the 5-HT$_{2C}$ receptor suppresses hKv1.1 current by phosphorylating the tyrosine 120 and 149 and that this process is coordinated by RPTPα (Imbrici *et al.* 2000). Interestingly, these residues are located in a region linking the T1 domain to the first transmembrane segment (T1-S1 linker), which therefore may play a role in this phenomenon.

Several gating properties of the Kv channels may undergo modulation. ATP treatment provokes a positive shift of the voltage-dependence of activation of unidentified delayed-rectifier K^+ channels from squid giant axons (Augustine and Bezanilla 1990; Perozo *et al.* 1991*a*, 1991*b*). It has been suggested that the incorporated phosphoryl groups may interact electrostatically with the voltage sensor of the channel which therefore requires stronger depolarizations for activation (Perozo and Bezanilla 1990).

The fast N-type inactivation of *Shaker*-like potassium channels can also be modulated by protein phosphorylation. Several serine residues located within the inactivation particle of Kv3.4 channels may be phosphorylated by protein kinase C. The consequent formation of negatively charged phosphoserines results in structural changes that inhibit inactivation (Covarrubias *et al.* 1994; Beck *et al.* 1998; Antz *et al.* 1999). On the other hand, specific serine residues located in the C-terminus of *Shaker*-channels need to be phosphorylated by protein kinase A to insure fast N-type inactivation (Drain *et al.* 1994).

Cysteine oxidation also plays a role in this process. Indeed, the inactivating particle of Kv1.4 channels has a cysteine residue (C13) that may exist in the oxidized or reduced form. Normal time course of N-type inactivation is observed when C13 is in the reduced state. By contrast cysteine oxidation markedly slows the inactivation by promoting the formation of disulfide bridges. Antioxidants as dithiothreitol or glutathione readily revert this effect (Ruppersberg *et al.* 1991).

All these observations imply that phosphorylation mechanisms or the red-ox state of a neuron may change its electrical and input integration properties by modifying the activity and the inactivation kinetics of certain Kv channels.

5.5 **Inwardly rectifying potassium channels and cell excitability**

Inwardly rectifying potassium (Kir) channels are found in almost every cell type where they play key roles in controlling membrane resting potential, cell excitability and K^+ homeostasis (Nichols and Lopatin 1997; Reimann and Ashcroft 1999).

Kir subunits possess two transmembrane domains separated by the signature K^+ pore sequence and can assemble as both homotetramers and heterotetramers (Fig. 5.2). Since the initial isolation of Kir1.1 by Steven Hebert and his collaborators (Ho *et al.* 1993), a large number of clones have been identified which can be divided into seven major subfamilies, Kir1.x–Kir7.x (Doupnik *et al.* 1995).

Generally, a Kir channel acts as a diode. Therefore, the inward current through these channels is greater at potentials which are more negative than the E_K, but at more positive potentials the outward flow is inhibited and the membrane potential is free to change. The rectifying nature of the conductance is due to a voltage-dependent block of the intracellular side of the pore by cytoplasmic polyamines and Mg^{++} ions (Matsuda *et al.* 1987; Lopatin *et al.* 1994; Lu and MacKinnon 1994; Stanfield *et al.* 1994).

The gram-positive bacterium *Streptomyces lividans* express KcsA, a Kir channel type highly homologous to *Shaker* channels (Schrempf *et al.* 1995). By using these bacteria Roderick MacKinnon and colleagues collected enough protein for X-ray crystallographic studies. In 1998 these investigators successfully described the first crystal structure of a potassium channel at 3.2 Å resolution (Doyle *et al.* 1998). The channel showed a tetrameric structure and a conic shape (see Fig. 5.10). The S1 and S2 segments were α-helices that delimited a central cavity with a diameter of 10 Å. The 12 Å long selectivity filter was formed by the amino acids GYG, confirming previous findings, and was lined by carbonyl oxygen atoms. This narrow pore structure contained two K^+ ions 7.5 Å apart and possessed an architecture designed for rapid conduction

Fig. 5.10 Crystal structure of KcsA channels. The cone-shaped channel possessed a wide central cavity and a narrow external pore that comprise the selectivity filter. The architecture of these channels can serve as a model for the core region of voltage-gated sodium, potassium, and calcium channels. The structural representation shows two side-subunits and was generated by using Protein Data Bank entry 1BL8 and the WebLab™ ViewerLite 3.2 software (from Doyle *et al.* 1998).

and high selectivity. These results represented a milestone in the field of ion channel biophysics.

Neurotransmitters such as dopamine, opioids, somatostatin, acetylcholine, serotonin, adenosine, GABA exert their inhibitory actions by activating G-protein coupled inward rectifiers. These channels belong to the Kir3.x family and regulate the excitability of the brain and heart. The heart rate slowing upon vagal nerve stimulation represents a notable example. This physiological response is due to vagal release of acetylcholine that by stimulating the atrial muscarinic receptors activate the intracellular heterotrimeric G proteins $\alpha\beta\gamma$. The dissociated $\beta\gamma$ subunits then stimulate heteromeric Kir3.1/Kir3.4 channel activity by physically interacting with their intracellular termini (Krapivinsky *et al.* 1995; Slesinger *et al.* 1995). Finally, the activation of these channels results in an outward flux of K^+ ions that causes membrane hyperpolarization and cell inhibition.

Receptor-operated Kir3.2 channels also modulate the electrical activity of neurons in the CNS. Indeed, the activation of Kir3.2 channels by D_2 and $GABA_B$ receptors hyperpolarize midbrain dopaminergic neurons and inhibit their firing rate (Guatteo *et al.* 2000). Interestingly, *weaver* mice dopaminergic neurons are excited by these neurotransmitters (Guatteo *et al.* 2000). This contrasting effect is due to the *weaver* mutation G156S located in the pore of Kir3.2 channels (Patil *et al.* 1995). This point mutation disrupts the potassium selectivity filter of the channel allowing the permeation of sodium and calcium ions (Slesinger *et al.* 1996; Tucker *et al.* 1996). The activation of this mixed conductance by D_2 and $GABA_B$ receptors, likely accounts for the membrane depolarization, increased firing activity, and degeneration of *weaver* dopaminergic neurons.

Another important group of inward rectifier potassium channels are the K_{ATP} channels. The secretion of insulin from pancreatic β-cells is mediated by the closure of these channels caused by increased levels of cytoplasmic ATP. This class of channels has been cloned and assigned to the Kir6.x subfamily (Inagaki *et al.* 1995). However, a functional K_{ATP} channel is formed of four Kir6.2 subunits and four sulfonylurea receptors, which confer sensitivity to drugs, and to the MgADP stimulatory effects of MgADP (Aguilar-Bryan *et al.* 1995; Nichols *et al.* 1996). By contrast, ATP inhibits the channel opening by interacting mainly with the Kir6.2 subunits (Tucker *et al.* 1997). K_{ATP} channels are also expressed in the heart and brain where they couple the metabolic state of the cell to electrical activity.

5.6 **Voltage-gated calcium channels**

The activity of voltage-gated calcium channels is normally initiated by depolarizing stimuli of different degrees. The subsequent opening of these channels triggers an influx of Ca^{++} that is driven by a steep electrochemical gradient. As a result, several neurophysiological mechanisms and cellular responses related to Ca^{++}-signalling are modulated by this ion, which is an important *second messenger*. Calcium current may by itself generate and shape-up action potentials, or it may regulate the release of neurotransmitters, excitation–contraction coupling, the secretion of hormones, and gene expression.

In 1984 Curtis and Catterall first reported the purification of the calcium channel protein and showed that it was associated with some auxiliary subunits (Curtis and Catterall 1984, 1985). A few years later, Numa and his colleagues reported the cloning of the calcium channel $\alpha 1$ subunit (Tanabe *et al.* 1987). The $\alpha 1$ subunit is a large polypeptide that possesses a secondary structure similar to the Na^+ channel α subunit and has both a voltage-sensing and pore-forming region (Fig. 5.2).

Voltage-gated calcium channels can be activated by weak or strong depolarizing stimuli. Therefore, they have been subdivided into two main groups named low-threshold and high-threshold Ca^{++} channels, respectively. More detailed electrophysiological studies lead to the characterization of several calcium channel types named: L-, P-, Q-, N-, R-, and T-type, as listed in Table 5.1. However, molecular cloning has identified 10 $\alpha 1$ subunits to date and a new nomenclature was needed to classify them. Recently, an abbreviated form has been proposed for Ca^{++} channels as well ($Ca_v y.x$), which is based on the general criterion of sequence similarity and numerical system adopted for Kv and Na_v channels (Table 5.1; Chandy 1991; Ertel *et al.* 2000).

The first subfamily ($Ca_v 1.x$) identifies L-type Ca^{++} channels that are characterized by a high-threshold of activation, *long-lasting* openings and slow inactivation. These channel types are also distinguished by susceptibility to blocking by dihydropyridines, and by protein phosphorylation as their main regulatory mechanism.

The calcium channels that are activated by large depolarizations are normally heterooligomers including auxiliary subunits. This macromolecular complex includes the $\alpha 1$ subunit, a transmembrane spanning γ subunit, an α_2 that is bound to a δ by means of a disulfide bridge and an intracellularly associated β subunit (Fig. 5.11).

$Ca_v 1.1$ channels are mainly expressed in skeletal muscles in which they play a role as *voltage-sensors* in the mechanism of excitation–contraction coupling. The cytoplasmic

Table 5.1 Voltage-gated calcium channel types

Channel Type	Former Nomenclature		Voltage-dependence
Cav 1.1	α_{1S}		high-threshold (>−10 mV
Cav 1.2	α_{1C}	L-type (25pS)	
Cav 1.3	α_{1D}		
Cav 1.4	α_{1F}		
Cav 2.1	α_{1A}	P/Q-type	high-threshold (>−20 mV)
Cav 2.2	α_{1B}	N-type (13pS)	
Cav 2.3	α_{1E}	R-type	
Cav 3.1	α_{1G}		Low-threshold (>−70 mV)
Cav 3.2	α_{1H}	T-type (8pS)	
Cav 3.3	α_{1I}		

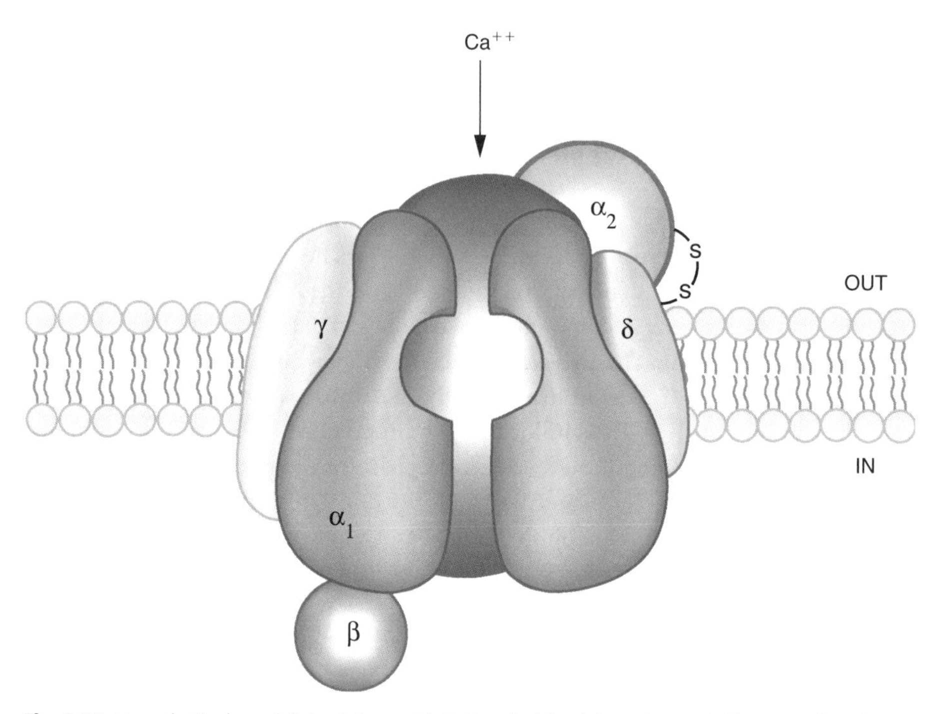

Fig. 5.11 Hypothetical model depicting a high-threshold calcium channel. The pore-forming α1 subunit of these channel types forms a complex with a $Ca_v\beta$, a $Ca_v\gamma$ and the $Ca_{vs}\alpha 2\delta$ subunits. A single gene encodes for the latter subunits that are bridged together via a disulide bond. Three $Ca_v\alpha 2\delta_{1-3}$ subunits and five $Ca_v\gamma_{1-5}$ subunits have been identified to date.

loop between domains II and III of Ca_v1.1 represents the structural region that functionally couple the α1 subunit at the transverse tubule with the ryanodine-sensitive Ca^{++} release channel at the sarcoplasmic reticulum (Tanabe *et al.* 1990).

In the heart, β-adrenergic receptors stimulate the activity of Ca_v1.2 channels, which in turn enhance cardiac contractility and excitability. However, the entry of Ca^{++} ions rapidly inactivates these channels. This calcium-dependent inactivation appears due to the formation of a calcium–calmodulin complex, which triggers a conformational change in the Ca_v1.2 channel by interacting with its C-terminal domain (Zühlke *et al.* 1999).

Ca_v1.2 and Ca_v1.3 channel types are widely expressed in the brain. The excitability and plasticity of hippocampal neurons appear to be particularly regulated by these channels. Indeed, the activation of NMDA receptors in these neurons may cause a proteolytic cleavage of the Ca_v1.2 C-terminus resulting in channels with an increased probability of opening (Wei *et al.* 1994; Hell *et al.* 1996). Ca_v1.3 channels also modulate the heart pacemaker activity and have been involved in the proper functioning of the inner hair cells of the organ of Corti (Platzer *et al.* 2000).

The Ca_v2.x subfamily includes P/Q- N- and R-type calcium channels that are mainly modulated by G proteins. Ca_v2.1 channels (P/Q-type) are specifically blocked by ω-Agatoxin IVA from the spider *Agelenopsis aperta.* They are abundantly and broadly expressed throughout the central nervous system where they regulate fast synaptic transmission, nerve cell survival, excitability, gene expression, and plasticity. In particular, they have been localized at the neuromuscular junction, in the brainstem, in the cerebellar Purkinje and granule cells and in most of the presynaptic terminals of the cerebellum (Westenbroek *et al.* 1995; Ludwig *et al.* 1997). At the presynaptic terminal, these channels play a pivotal role in the mechanisms of neurotransmitter release, as they are located within the release site and close to the docked vesicle (Wu *et al.* 1999). Indeed, the calcium-dependent direct interaction of $Ca_v2.1$ and $Ca_v2.2$ channels with the SNARE protein complex modulates the exocytosis of synaptic vesicles (Bezprozvanny *et al.* 1995). Recently, it has been shown that the P-type and Q-type channels are generated by alternative splicing of the $\alpha1_A$ gene (Bourinet *et al.*, 1999). This conclusion is also supported by the absence of P/Q-type currents from Purkinje and cerebellar granule neurons of mice lacking the $\alpha1_A$ gene (Jun *et al.* 1999). These animals are characterized by neurological deficits as ataxia, dystonia, and Purkinje cell death.

$Ca_v2.2$ channels (N-type) are expressed in the sympathetic nervous system where they play a role in the neurotransmission of signals (Ino *et al.* 2001). The voltage-dependence of these channels is modulated by neurotransmitters via G-protein coupled receptors. Interestingly, the activation of $G\beta\gamma$ inhibits $Ca_v2.2$ calcium channels (N-type) by reducing the mobility of its voltage-sensor (Jones *et al.* 1997). As a result the currents show slower time courses of activation and reduced voltage-dependence. This channel type is blocked by ω-Conotoxin GVIA or MVIIA from the shells *Conus geographus* and *Conus magus*, respectively.

The third subfamily ($Ca_v3.x$) identifies the low-threshold T-type Ca^{++} channels, which are characterized by *transient kinetics*, small single channel conductance and fast inactivation. Low threshold-activated channels are normally formed by the $\alpha1$ subunit alone.

Several auxiliary elements have been identified and assigned to three major categories: $Ca_v\beta$, $Ca_v\alpha2\delta$, and $Ca_v\gamma$ subunits. They play a role in Ca_v channel trafficking and modify several biophysical properties of the $\alpha1$ subunit.

Four $Ca_v\beta$ subunits have been cloned ($Ca_v\beta_{1-4}$) and some of which give rise to several splice variants. These proteins interact with the linker connecting the first and second domains of the $\alpha1$ subunit (De Waard *et al.* 1994; Pragnell *et al.* 1994). $Ca_v\beta$ and $Ca_v\alpha2\delta$ association increases the expression level of the resulting channel and shift its voltage-dependence to more negative potentials. In addition, $Ca_v\beta$ plays a major role in calcium channel inactivation (Lacerda *et al.* 1991). In particular, $Ca_v\beta_{1b}$ and $Ca_v\beta_3$ confer rapid inactivation properties to calcium channels. By contrast, $Ca_v\beta_{2a}$ markedly slower this process. Lutz Birnbaumer and colleagues showed that the structural differences located in the N-terminal domains of these subunits determine the different inactivation rates of the channel (Olcese *et al.* 1994).

The $\alpha1$ subunit possesses intrinsic voltage-dependent inactivation properties. Recently, Gerald Zamponi and colleagues proposed a model whereby voltage-gated calcium and sodium channels possess a similar molecular mechanism of inactivation (Stotz *et al.* 2000). According to this model the I-II linker may function as a 'hinged-lid' that occludes the pore by a swinging movement (Fig. 5.2).

5.7 Channelopathies

The involvement of defective ion channels in the pathophysiology of several neurological disorders had been postulated for years. Genetic linkage studies and mutation analysis have now demonstrated that a variety of inherited diseases are indeed ion channel diseases, named also 'channelopathies' (see for review Ashcroft 1999; Lehmann-Horn and Jurkat-Rott 1999).

5.7.1 Episodic ataxia type-1, a *Shaker*-like K^+ channel disease

Episodic ataxia type-1 (EA-1) is an autosomal dominant disorder affecting both the central and peripheral nervous systems (VanDyke *et al.* 1975; Brunt and van Weerden 1990). The hallmark of the disease is continuous myokymia and episodic attacks of spastic contractions of skeletal muscles, which often results in loss of balance. Attacks of ataxia are characterized by incoordination, impaired speech, and by jerking movements of the head, arms, and legs. They may be brought on by fever, startle, vestibulogenic stimulation, emotional stress, exercise, or fatigue. EA-1 affected patients also show constant muscle rippling (myokymia), associated with continuous muscle unit activity.

Genetic linkage studies of several EA-1 affected families led in 1994 to the localization of the disease gene on chromosome 12p13, in a region encoding a voltage-dependent potassium channel (Litt *et al.* 1994). The subsequent screen of the *KCNA1* gene for mutations resulted in the discovery of a number of point mutations in the *Shaker*-related potassium channel hKv 1.1 (Browne *et al.* 1994). To date, 18 point mutations have been identified in several affected families (see Fig. 5.12).

The molecular pathophysiology of episodic ataxia type-1 syndrome has been investigated by determining the biophysical properties of wild-type and several mutant channels in *Xenopus* oocytes or in mammalian cell lines. The EA-1 mutations altered the gating and expression characteristics of the channels, which showed impaired delayed-rectifying function (Adelman *et al.* 1995; D'Adamo *et al.* 1998; Zerr *et al.* 1998*a*, 1998*b*; Bretschneider *et al.* 1999; Eunson *et al.* 2000; Manganas *et al.* 2001; Rea *et al.* 2002). Thus, the overall effect of these genetic mutations is a reduction of the outward flow of potassium ions that may become insufficient to repolarize the membrane potential. Therefore, the neurons expressing mutated hKv1.1 channels might be hyperexcitable and have prolonged action potentials.

EA-1 mutations also alter the function of heteromeric channels composed of Kv1.1 and Kv1.2 subunits, which are expressed in presynaptic basket cell membranes (D'Adamo *et al.* 1999). Therefore, it has been proposed that an increased excitability of these terminals, which might be in a facilitated state, causes an abnormal release of γ-aminobutyric acid from such terminals onto Purkinje cells, altering the output of the entire cerebellar cortex to the rest of the brain of EA-1 patients (D'Adamo *et al.* 1999). These possible pathogenic mechanisms are consistent with behavioural and electrophysiological studies carried out by using Kv1.1-null mice. Indeed, these animals display overt ataxia, which begins during the third to fourth postnatal week, spontaneous seizure, altered axonal action potential conduction in sciatic nerve, and an increased GABAergic inhibition of Purkinje cells (Smart *et al.* 1998; Zhang *et al.* 1999).

5.7.2 Episodic ataxia type-2, a voltage-gated Ca^{++} channel disease

Episodic ataxia type-2 (EA-2) is an autosomal dominant neurological disorder characterized by attacks of generalized ataxia accompanied by nausea and vertigo. The attacks, which may last for hours or days, may be brought on by emotional stress and exercise but not by startle. Anxiety, alcohol, and coffee may increase the probability of undergoing an attack. The onset of the disease occurs in childhood, as in EA-1. However, the symptoms that clearly distinguish EA-2 from EA-1 are migraines, interictal nystagmus, and progressive atrophy of the anterior vermis, which causes permanent motor disability. The attacks of EA-2 may be effectively prevented by treatment with acetazolamide.

The locus for EA-2 was mapped to chromosome 19p13 by linkage studies (Kramer *et al.* 1995; Teh *et al.* 1995; Von Brederlow *et al.* 1995). Successively, mutation analysis of EA-2 families revealed two mutations in the *CACNL1A4* gene, which encodes for the α_{1A}-subunit of P/Q-type Ca^{++} channels ($Ca_v2.1$; Ophoff *et al.* 1996): a basepair deletion that resulted in a frame-shift and in the formation of a premature stop codon, and a splice-site mutation (Fig. 5.12). Therefore, it is expected that the calcium channel formed would be non-functional (Ophoff *et al.* 1996, 1998; Fig. 5.13). Although the evidence may predict an altered cerebellar neurotransmission in EA-2 syndrome, the detailed pathophysiological mechanisms of this neurologic disease remain poorly understood (Jen 1999).

Two other neurologic diseases result from mutations in the same *CACNL1A4* gene. Spinocerebellar ataxia type-6 (SCA6) is an autosomal dominant disorder characterized

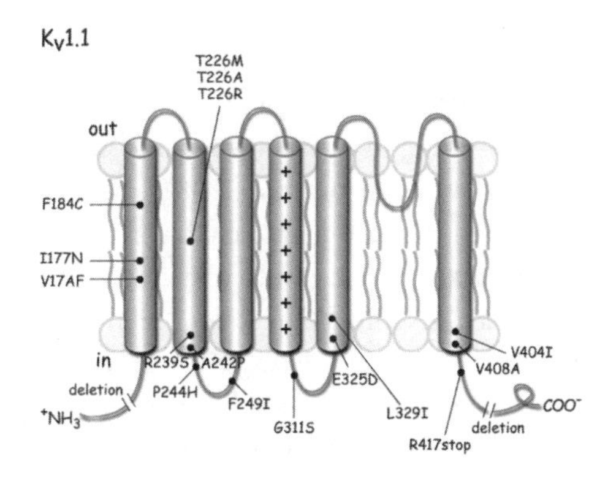

Fig. 5.12 Genetic mutations identified in patients affected by Episodic Ataxia Type-1. Predicted membrane topology of a hKv1.1 subunit indicating the position of the missense point mutations identified in EA-1 patients.

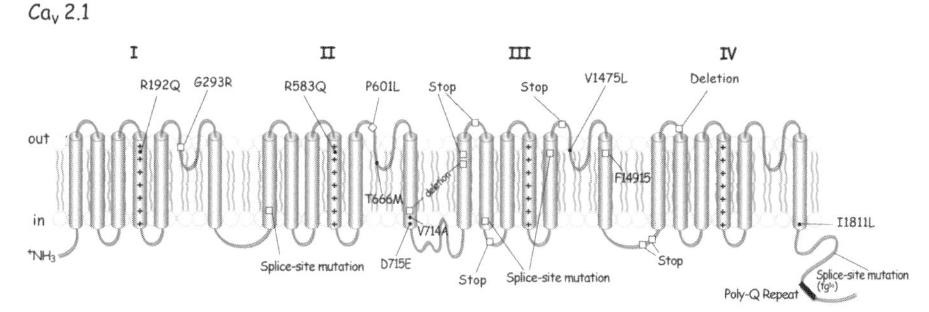

Fig. 5.13 Genetic mutations identified in the $Ca_V2.1$ channel. The diagram shows a $Ca_V2.1$ channel and the position of the mutations identified in patients affected by Episodic Ataxia Type-2 (open squares), Spinocerebellar Ataxia Type 6 (SCA6, Poly-Q repeat) and Familial Hemiplegic Migraine (FHM, closed circles). The tottering mouse (diamond) and leaner mouse tg^{la} (*) mutations are also shown.

by ataxia, nystagmus, and cerebellar degeneration. Affected individuals are eventually confined to a wheelchair. The genotypic analysis of a number of SCA6 patients revealed an expansion of CAG repeats in the *CACNL1A4* gene. The resulting Poly-Q repeat is located within the C-terminal region of the $Ca_V2.1$ channel (Fig. 5.12; Zhuchenko *et al.* 1997).

Familial hemiplegic migraine (FHM) is also a rare autosomal dominant syndrome of childhood onset. It is characterized by migraines with aura, ictal hemiparesis lasting for hours or days, and by progressive cerebellar atrophy. Some patients report cognitive impairment particularly before attacks. Mutation analysis of FHM patients resulted in the identification of several mutations in the *CACNL1A4* gene (Fig. 5.12; Ophoff *et al.* 1996; Battistini *et al.* 1999; Carrera *et al.* 1999; Ducros *et al.* 1999). Functional studies have shown that point mutations associated with FHM may cause both a gain and loss of $Ca_V2.1$ channel function, which make genotype–phenotype correlations hard to draw (Hans *et al.* 1999).

The leaner mouse (Cacna1a $^{tg-la}$) phenotype is characterized by severe ataxia, stiffness, retarded motor activity, neurodegeneration of the cerebellum (loss of Golgi, Purkinje, and granule neurons), and typical intermittent seizures that resemble absence epilepsy. Recently, a splice site mutation has been found in the *CACNL1A* gene of leaner mice resulting in a great decrease of $Ca_V2.1$ current density (Fletcher *et al.* 1996; Dove *et al.* 1998; Lorenzon *et al.* 1998). The reduction in calcium currents that results from this genetic mutation appears to be a key determinant of cerebellar neurodegeneration.

A number of neuromuscular channelopathies involving mutated sodium channels have been described. The long QT3 syndrome and febrile seizures have been associated with mutations in the $Na_V1.5$ channel and $Ca_V\beta_1$ subunit, respectively (Bennett *et al.* 1995; Wang *et al.* 1995; Wallace *et al.* 1998). Moreover, mutations in the skeletal Na_V channels result in hyperkalaemic periodic paralysis, paramyotonia congenita, and potassium-aggravated myotonia (see for review Cannon 1996; Ashcroft 1999; Lehmann-Horn and Jurkat-Rott 1999).

5.8 Concluding remarks

The work of biophysicists is shedding a new light on the molecular machinery of channel gating by using higher resolution technologies. The description of the entire structure of a voltage-gated channel, by X-ray crystallography, will be regarded as a milestone in the history of ion channels. However, the elucidation of the atomic structure of the channel in its open and closed conformation will finally provide a fascinating view of this protein in movement. Unfortunately, due to the obvious technical difficulties these results will not be achieved for quite some time.

On the other hand, physiologists and biochemists are refining their tools as well to determine the functional role played by ion channels, their modulation and protein–protein interactions. Mouse strains with inducible genes have been developed by using genetic engineering methodologies. The possibility to *turn on* and *turn off* a specific gene at any time during animal development overcomes many of the problems associated with traditional gene knockout experiments. This animal model will be particularly useful for studying the physiological role played by ion channels and the many gene products of unknown function revealed by the disclosure of the human genome. Jockusch and Schmitt-John (this volume, Chapter 2) present an overview of the use of mouse genetics in studying nervous system function.

The detailed genetic, behavioural, and electrophysiological analysis of mice carrying naturally or artificially occurring mutations in their ion channel genes is currently providing important information on the causes of several human diseases. Many new genes encoding for ion channels have been recently identified; however, their role in channelopathies awaits discovery.

While this chapter was in press a number of landmark papers appeared in the literature related to voltage-gated ion channels. Some of them include the description of the entire structure of the bacterial voltage-gated and inwardly rectifying potassium channels, by X-ray crystallography (Jiang *et al.* 2003*a*; Kuo *et al.* 2003). MacKinnon and his collaborators also proposed a new model by which the voltage-sensors of the potassium channel KvAP operate. Although controversial, it appears that the S4 segment does not rotate as a sliding helix, but it moves back and forth as a bird's wing (Jiang *et al.* 2003*a*,*b*).

Acknowledgments

I thank Armando Lagrutta for critically reading the manuscript. The financial support of Telethon-Italy (Grant no. 1083 and GGP030159), of the MIUR-COFIN 2003 and of COMPAGNIA di San Paolo (Turin) is gratefully acknowledged. I thank Antonella Cusimano, Maria Casamassima, and Maria Cristina D'Adamo for their help in preparing this manuscript and Lucia Simigliani for the art work.

References

Adelman, J. P., Bond, C. T., Pessia, M., and Maylie, J. (1995) Episodic ataxia results from voltage-dependent potassium channels with altered functions. *Neuron,* **15**, 1449–54.

Agnew, W. S., Moore, A. C., Levinson, S. R., and Raftery, M. A. (1980) Identification of a large molecular weight peptide associated with a tetrodotoxin binding protein from the electroplax of Electrophorus electricus. *Biochem. Biophys. Res. Commun.*, **92**, 860–6.

Aguilar-Bryan, L., Nichols, C. G., Wechsler, S. W., Clement 4th, J. P., Boyd 3rd, A. E. , Gonzalez, G., *et al.* (1995) Cloning of the beta cell high-affinity sulfonylurea receptor: a regulator of insulin secretion. *Science*, **268**, 423–6.

Akabas, M. H., Stauffer, D. A., Xu, M., and Karlin, A. (1992) Acetylcholine receptor channel structure probed in cysteine-substitution mutants. *Science*, **258**, 307–10.

Aldrich Jr, R. W., Getting, P. A., and Thompson, S. H. (1979) Mechanism of frequency-dependent broadening of molluscan neurone soma spikes. *J. Physiol.*, **291**, 531–44.

Antz, C., Bauer, T., Kalbacher, H., Frank, R., Covarrubias, M., Kalbitzer, H. R., *et al.* (1999) Control of K^+ channel gating by protein phosphorylation: structural switches of the inactivation gate. *Nat. Struct. Biol.*, **6**, 146–50.

Armstrong, C. M. (1966) Time course of TEA(+)-induced anomalous rectification in squid giant axons. *J. Gen. Physiol.*, **50,** 491–503.

Armstrong, C. M. and Bezanilla, F. (1973) Currents related to movement of the gating particles of the sodium channels. *Nature*, **242**, 459–61.

Armstrong, C. M. and Bezanilla, F. (1977) Inactivation of the sodium channel. II. Gating current experiments. *J. Gen. Physiol.*, **70**, 567–90.

Ashcroft, F. M. (1999) *Ion Channels and Disease.* Academic Press, London.

Augustine, C. K. and Bezanilla, F. (1990) Phosphorylation modulates potassium conductance and gating current of perfused giant axons of squid. *J. Gen. Physiol.*, **95**, 245–71.

Backx, P. H., Yue, D. T., Lawrence, J. H., Marban, E., and Tomaselli, G. F. (1992) Molecular localization of an ion-binding site within the pore of mammalian sodium channels. *Science*, **257**, 248–51.

Battistini, S., Stenirri, S., Piatti, M., Gelfi, C., Righetti, P. G., Rocchi, R., *et al.* (1999) A new CACNA1A gene mutation in acetazolamide-responsive familial hemiplegic migraine and ataxia. *Neurology*, **53**, 38–43.

Baukrowitz, T. and Yellen, G. (1995) Modulation of K^+ current by frequency and external $[K^+]$: a tale of two inactivation mechanisms. *Neuron*, **15**, 951–60.

Beck, E. J., Sorensen, R. G., Slater, S. J., and Covarrubias, M. (1998) Interactions between multiple phosphorylation sites in the inactivation particle of a K^+ channel. Insights into the molecular mechanism of protein kinase C action. *J. Gen. Physiol.*, **112**, 71–84.

Beneski, D. A. and Catterall, W. A. (1980) Covalent labeling of protein components of the sodium channel with a photoactivable derivative of scorpion toxin. *Proc. Natl. Acad. Sci. USA*, **77**, 639–43.

Bennett, P. B., Yazawa, K., Makita, N., and George Jr, A. L. (1995) Molecular mechanism for an inherited cardiac arrhythmia. *Nature*, **376**, 683–85.

Bezanilla, F. (2000) The voltage sensor in voltage-dependent ion channels. *Physiol., Rev.*, **80**, 555–92.

Bezprozvanny, I., Scheller, R. H., and Tsien, R. W. (1995) Functional impact of syntaxin on gating of N-type and Q-type calcium channels. *Nature*, **378**, 623–6.

Boland, L. M. and Jackson, K. A. (1999) Protein kinase C inhibits Kv1.1 potassium channel function. *Am. J. Physiol.*, **277**, C100–C110.

Bourinet, E., Soong, T. W., Sutton, K., Slaymaker, S., Mathews, E., Monteil, A., *et al.* (1999) Splicing of alpha 1A subunit gene generates phenotypic variants of P- and Q-type calcium channels. *Nat. Neurosci.*, **2**, 407–15.

Bowlby, M. R., Fadool, D. A., Holmes, T. C., and Levitan, I. B. (1997) Modulation of the Kv1.3 potassium channel by receptor tyrosine kinases. *J. Gen. Physiol.*, **110**, 601–10.

Bretschneider, F., Wrisch, A., Lehmann-Horn, F., and Grissmer, S. (1999) Expression in mammalian cells and electrophysiological characterization of two mutant Kv1.1 channels causing episodic ataxia type 1 (EA-1). *Eur. J. Neurosci.*, **11**, 2403–12.

Browne, D. L., Gancher, S. T., Nutt, J. G., Brunt, E. R., Smith, E. A., Kramer, P., *et al.* (1994) Episodic ataxia/myokymia syndrome is associated with point mutations in the human potassium channel gene, KCNA1. *Nat. Genet.*, **8**, 136–40.

Brunt, E. R. P. and van Weerden, T. W. (1990) Familial paroxysmal kinesigenic ataxia and continuous myokymia. *Brain*, **113**, 1361–82.

Cachero, T. G., Morielli, A. D., and Peralta, E. G. (1998) The small GTP-binding protein RhoA regulates a delayed rectifier potassium channel. *Cell*, **93**, 1077–85.

Cannon, S. C. (1996) Sodium channel defects in myotonia and periodic paralysis. *Annu. Rev. Neurosci.*, **19**, 141–64.

Cannon, S. C. (1997) From mutation to myotonia in sodium channel disorders. *Neuromuscul. Disord.*, **7**, 241–9.

Carrera, P., Piatti, M., Stenirri, S., Grimaldi, L. M., Marchioni, E., Curcio, M., *et al.* (1999) Genetic heterogeneity in Italian families with familial hemiplegic migraine. *Neurology*, **53**, 26–33.

Catterall, W. A. (1986) Voltage-dependent gating of sodium channels: correlating structure and function. *Trends Neurosci.*, **9**, 7–10.

Catterall, W. A. (2001) A 3D view of sodium channels. *Nature*, **409**, 988–9.

Cha, A. and Bezanilla, F. (1997) Characterizing voltage-dependent conformational changes in the *Shaker* K^+ channel with fluorescence. *Neuron*, **19**, 1127–40.

Cha, A., Snyder, G. E., Selvin, P. R., and Bezanilla, F. (1999) Atomic scale movement of the voltage-sensing region in a potassium channel measured via spectroscopy. *Nature*, **402**, 809–13.

Chandy, K. G. (1991) Simplified gene nomenclature. *Nature*, **352**, 26.

Cole, K. S. and Curtis, H. J. (1939) Electrical impedance of the squid giant axon during activity. *J. Gen. Physiol.*, **22**, 649–70.

Covarrubias, M., Wei, A. A., and Salkoff, L. (1991) Shaker, Shal, Shab, and Shaw express independent K^+ current systems. *Neuron*, **7**, 763–73.

Covarrubias, M., Wei, A., Salkoff, L., and Vyas, T. B. (1994) Elimination of rapid potassium channel inactivation by phosphorylation of the inactivation gate. *Neuron*, **13**, 1403–12.

Curtis, B. M. and Catterall, W. A. (1984) Purification of the calcium antagonist receptor of the voltage-sensitive calcium channel from skeletal muscle transverse tubules. *Biochemistry*, **23**, 2113–8.

Curtis, B. M. and Catterall, W. A. (1985) Phosphorylation of the calcium antagonist receptor of the voltage-sensitive calcium channel by cAMP-dependent protein kinase. *Proc. Natl. Acad. Sci. USA*, **82**, 2528–32.

D'Adamo, M. C., Liu, Z., Adelman, J. P., Maylie, J., and Pessia, M. (1998) Episodic ataxia type-1 mutations in the hKv1.1 cytoplasmic pore region alter the gating properties of the channel. *EMBO J.*, **17**, 1200–7.

D'Adamo, M. C., Imbrici, P., Sponcichetti, F., and Pessia, M. (1999) Mutations in the *KCNA1* gene associated with episodic ataxia type-1 syndrome impair heteromeric voltage-gated K^+ channel function. *FASEB J.*, **13**, 1335–45.

De Waard, M., Liu, H., Walker, D., Scott, V. E. S., Gurnett, C. A., and Campbell, K. P. (1997) Direct binding of G-protein complex to voltage-dependent calcium channels. *Nature*, **385**, 446–50.

De Waard, M., Pragnell, M., and Campbell, K. P. (1994) Ca2+ channel regulation by a conserved beta subunit domain. *Neuron*, **13**, 495–503.

del Camino, D. and Yellen, G.(2001) Tight steric closure at the intracellular activation gate of a voltage-gated K(+) channel. *Neuron*, **32**, 649–56.

Doupnik, C. A., Davidson, N., and Lester, H. A. (1995) The inward rectifier potassium channel family. *Curr. Opin. Neurobiol.*, **5**, 268–77.

Dove, L. S., Abbott, L. C., and Griffith, W. H. (1998) Whole-cell and single-channel analysis of P-type calcium currents in cerebellar Purkinje cells of leaner mutant mice. *J. Neurosci.*, **18**, 7687–99.

Doyle, D. A., Morais Cabral, J., Pfuetzner, R. A., Kuo, A., Gulbis, J. M., Cohen, S. L., *et al.* (1998) The structure of the potassium channel, molecular basis of K^+ conduction and selectivity. *Science*, **280**, 69–77.

Drain, P., Dubin, A. E., and Aldrich, R. W. (1994) Regulation of Shaker K^+ channel inactivation gating by the cAMP-dependent protein kinase. *Neuron*, **12**, 1097–109.

Ducros, A., Denier, C., Joutel, A., Vahedi, K., Michel, A., Darcel, F., *et al.* (1999) Recurrence of the T666M calcium channel CACNA1A gene mutation in familial hemiplegic migraine with progressive cerebellar ataxia. *Am. J. Hum. Genet.*, **64**, 89–98.

Ertel, E. A., Campbell, K. P., Harpold, M. M., Hofmann, F., Mori, Y., Perez-Reyes, E., *et al.* (2000) Nomenclature of voltage-gated calcium channels. *Neuron*, **25**, 533–5.

Eunson, L. H., Rea, R., Zuberi, S. M., Youroukos, S., Panayiotopoulos, C. P., Liguori, R., *et al.* (2000) Clinical, genetic, and expression studies of mutations in the potassium channel gene KCNA1 reveal new phenotypic variability. *Ann. Neurol.*, **48**, 647–56.

Fadool, D. A., Holmes, T. C., Berman, K., Dagan, D., and Levitan, I. B. (1997) Tyrosine phosphorylation modulates current amplitude and kinetics of a neuronal voltage-gated potassium channel. *J. Neurophysiol.*, **78**, 1563–73.

Felsch, J. S., Cachero, T. G., and Peralta, E. G. (1998) Activation of protein tyrosine kinase PYK2 by the m1 muscarinic acetylcholine receptor. *Proc. Natl. Acad. Sci. USA*, **95**, 5051–6.

Fletcher, C. F., Lutz, C. M., O'Sullivan, T. N., Shaughnessy Jr, J. D., Hawkes, R., Frankel, W. N., *et al.* (1996) Absence epilepsy in tottering mutant mice is associated with calcium channel defects. *Cell*, **87**, 607–17.

Glauner, K. S., Mannuzzu, L. M., Gandhi, C. S., and Isacoff, E. (1999) Spectroscopic mapping of voltage sensor movement in the *Shaker* potassium channel. *Nature*, **402**, 813–17.

Goldin, A. L., Barchi, R. L., Caldwell, J. H., Hofmann, F., Howe, J. R., Hunter, J. C., *et al.* (2000) Nomenclature of voltage-gated sodium channels. *Neuron*, **28**, 365–8.

Grissmer, S. and Cahalan, M. (1989) TEA prevents inactivation while blocking open K^+ channels in human T lymphocytes. *Biophys. J.*, **55**, 203–6.

Guatteo, E., Fusco, F. R., Giacomini, P., Bernardi, G., and Mercuri, N. B. (2000) The weaver mutation reverses the function of dopamine and GABA in mouse dopaminergic neurons. *J. Neurosci.*, **20**, 6013–20.

Gulbis, J. M., Mann, S., and MacKinnon, R. (1999) Structure of a voltage-dependent K^+ channel beta subunit. *Cell*, **97**, 943–52.

Gulbis, J. M., Zhou, M., Mann, S., and MacKinnon, R. (2000) Structure of the cytoplasmic β subunit-T1 assembly of voltage-dependent K^+ channels. *Science*, **289**, 123–7.

Gutman, G. A. and Chandy, K. G. (1993) Nomenclature of mammalian voltage-dependent potassium channel genes. *Semin. Neurosci.*, **5**, 101–6.

Hamill, O. P., Marty A., Neher, E., Sakmann, B., and Sigworth, F. J. (1981) Improved patch-clamp techniques for high-resolution current recording from cells and cell-free membrane patches. *Pflugers Arch.*, **391**, 85–100.

Hans, M., Luvisetto, S., Williams, M. E., Spagnolo, M., Urrutia, A., Tottene, A., *et al.* (1999) Functional consequences of mutations in the human alpha1A calcium channel subunit linked to familial hemiplegic migraine. *J. Neurosci.*, **19**, 1610–19.

Heginbotham, L., Abramson, T., and MacKinnon, R. (1992) A functional connection between the pores of distantly related ion channels as revealed by mutant K^+ channels. *Science*, **258**, 1152–5.

Heinemann, S. H., Terlau, H., Stuhmer, W., Imoto, K., and Numa, S. (1992) Calcium channel characteristics conferred on the sodium channel by single mutations. *Nature*, **356**, 441–3.

Hell, J. W., Westenbroek, R. E., Breeze, L. J., Wang, K. K., Chavkin, C., and Catterall, W. A. (1996) *N*-methyl-D-aspartate receptor-induced proteolytic conversion of postsynaptic class C L-type calcium channels in hippocampal neurons. *Proc. Natl. Acad. Sci. USA*, **93**, 3362–7.

Hille, B. (2001) *Ionic Channels of Excitable Membranes, Third Edition.* Sinauer, Sunderland, MA.

Hirschberg, B., Rovner, A., Lieberman, M., and Patlak, J. (1995) Transfer of twelve charges is needed to open skeletal muscle Na+ channels. *J. Gen. Physiol.*, **106**, 1053–68.

Ho, K., Nichols, C. G., Lederer, W. J., Lytton, J., Vassilev, P. M., Kanazirska, M. V., *et al.* (1993) Cloning and expression of an inwardly rectifying ATP-regulated potassium channel. *Nature*, **362**, 31–8.

Hodgkin, A. L. and Huxley, A. F. (1952) A quantitative description of membrane current and its application to conduction and excitation in nerve. *J. Physiol.*, **117**, 500–44.

Hodgkin, A. L. and Katz, B. (1949) The effect of sodium ions on the electrical activity of the giant axon of the squid. *J. Physiol. (Lond.)* **108**, 37–77.

Hoger, J. H., Walter, A. E., Vance, D., Yu L., Lester, H. A., and Davidson, N. (1991) Modulation of a cloned mouse brain potassium channel. *Neuron*, **6**, 227–36.

Holmes, T. C., Fadool, D. A., Ren, R., and Levitan, I. B. (1996) Association of Src tyrosine kinase with a human potassium channel mediated by SH3 domain. *Science*, **274**, 2089–91.

Hoshi, T., Zagotta, W. N., and Aldrich, R. W. (1990) Biophysical and molecular mechanisms of Shaker potassium channel inactivation. *Science*, **250**, 533–8.

Hoshi, T., Zagotta, W. N., and Aldrich, R. W. (1991) Two types of inactivation in Shaker K^+ channels: effects of alterations in the carboxy-terminal region. *Neuron*, **7**, 547–56.

Imbrici, P., Tucker, S. J., D'Adamo, M. C., and Pessia, M. (2000) Role of receptor protein tyrosine phosphatase alpha (RPTPalpha) and tyrosine phosphorylation in the serotonergic inhibition of voltage-dependent potassium channels. *Pflugers Arch.*, **441**, 257–62.

Inagaki, N., Gonoi, T., Clement 4th, J. P., Namba, N., Inazawa, J., Gonzalez, G., *et al.* (1995) Reconstitution of IKATP: an inward rectifier subunit plus the sulfonylurea receptor. *Science*, **270**, 1166–70.

Ino, M., Yoshinaga, T., Wakamori, M., Miyamoto, N., Takahashi, E., Sonoda, J., *et al.* (2001) Functional disorders of the sympathetic nervous system in mice lacking the alpha 1B subunit (Cav 2.2) of N-type calcium channels. *Proc. Natl. Acad. Sci. USA*, **98**, 5323–8.

Isacoff, E. Y., Jan, N. Y., and Jan, L. Y. (1990) Evidence for the formation of heteromultimeric potassium channels in *Xenopus* oocytes. *Nature*, **345**, 530–4.

Isom, L. L., De Jongh, K. S., Patton, D. E., Reber, B. F., Offord, J., Charbonneau, H., *et al.* (1992) Primary structure and functional expression of the beta 1 subunit of the rat brain sodium channel. *Science*, **256**, 839–42.

Isom, L. L., Ragsdale, D. S., De Jongh, K. S., Westenbroek, R. E., Reber, B. F., Scheuer, T., *et al.* (1995) Structure and function of the beta 2 subunit of brain sodium channels, a transmembrane glycoprotein with a CAM motif. *Cell*, **83**, 433–42.

Jan, Y. N., Jan, L. Y., and Dennis, M. J. (1977) Two mutations of synaptic transmission in Drosophila. *Proc. R. Soc. Lond. B Biol. Sci.*, **198**, 87–108.

Jen, J. (1999) Calcium channelopathies in the central nervous system. *Curr. Opin. Neurobiol.*, **9**, 274–80.

Jiang, Y. , Lee, A., Chen, J., Ruta, V., Cadene, M., Chait, B. T., *et al.* (2003*a*) X-ray structure of a voltage-dependent K^+ channel. *Nature*, **423**, 33–41.

Jiang, Y., Ruta, V., Chen, J., Lee, A., and Mackinnon, R. (2003*b*) The principle of gating charge movement in a voltage-dependent K^+ channel. *Nature*, **423**, 42–8.

Jones, L. P., Patil, P. G., Snutch, T. P., and Yue, D. T. (1997) G-protein modulation of N-type calcium channel gating current in human embryonic kidney cells (HEK 293) *J. Physiol.*, **498**, 601–10.

Jun, K., Piedras-Renteria, E. S., Smith, S. M., Wheeler, D. B., Lee, S. B., Lee, T. G., *et al.* (1999) Ablation of P/Q-type Ca(2+) channel currents, altered synaptic transmission, and progressive ataxia in mice lacking the alpha(1A)-subunit. *Proc. Natl. Acad. Sci. USA*, **96**, 15245–50.

Kaplan, W. D. and Trout, W. E. (1969) The behaviour of four neurological mutants of *Drosophila*. *Genetics*, **61**, 399–409.

Kim, E., Niethammer, M., Rothschild, A., Jan, Y. N., and Sheng, M. (1995) Clustering of Shaker-type K^+ channels by interaction with a family of membrane-associated guanylate kinases. *Nature*, **378**, 85–8.

Kramer, P. L., Yue, Q., Gancher, S. T., Nutt, J. G., Baloh, R., Smith, E., *et al.* (1995) A locus for the nystagmus-associated form of episodic ataxia maps to an 11cM region on chromosome 19p. *Am. J. Hum. Genet.*, **57**, 182–5.

Krapivinsky, G., Gordon, E. A., Wickman, K., Velimirovic, B., Krapivinsky, L., and Clapham, D. E. (1995) The G-protein-gated atrial K^+ channel IKACh is a heteromultimer of two inwardly rectifying K(+)-channel proteins. *Nature*, **374**, 135–41.

Kreusch, A., Pfaffinger, P. J., Stevens, C. F., and Choe, S. (1998) Crystal structure of the tetramerization domain of the Shaker potassium channel. *Nature*, **392**, 945–8.

Kuo, A., Gulbis, J. M., Antcliff, J. F., Rahman, T., Lowe, E. D., Zimmer, J., *et al.* (2003) Crystal structure of the potassium channel KirBac 1.1 in the closed state. *Science*, **300**, 1922–6.

Lacerda, A. E., Kim, H. S., Ruth, P., Perez-Reyes, E., Flockerzi, V., Hofmann, F., *et al.* (1991) Normalization of current kinetics by interaction between the alpha 1 and beta subunits of the skeletal muscle dihydropyridine-sensitive Ca2+ channel. *Nature*, **352**, 527–30.

Larsson, H. P., Baker, O. S., Dhillon, D. S., and Isacoff, E. Y. (1996) Transmembrane movement of the *Shaker* K^+ channel S4. *Neuron*, **16**, 387–97.

Laube, G., Roper, J., Pitt, J. C., Sewing, S., Kistner, U., Garner, C. C., *et al.* (1996) Ultrastructural localization of *Shaker*-related potassium channel subunits and synapse-associated protein 90 to separate like junctions in rat cerebellar Pinceaux. *Brain Res. Mol. Brain Res.*, **42**, 51–61.

Lehmann-Horn, F. and Jurkat-Rott, K. (1999) Voltage-gated ion channels and hereditary disease. *Physiol. Rev.*, **79**, 1317–72.

Lerche, H., Peter, W., Fleischhauer, R., Pika-Hartlaub, U., Malina, T., Mitrovic, N., *et al.* (1997) Role in fast inactivation of the IV/S4-S5 loop of the human muscle Na+ channel probed by cysteine mutagenesis. *J. Physiol.*, **505**, 345–52.

Lev, S., Moreno, H., Martinez, R., Canoll, P., Peles, E., Musacchio, J. M., *et al.* (1995) Protein tyrosine kinase PYK2 involved in Ca(2+)-induced regulation of ion channel and MAP kinase functions. *Nature*, **376**, 737–45.

Litt, M., Kramer, P., Browne, D., Gancher, S., Brunt, E. R., Root, D., *et al.* (1994) A gene for episodic ataxia/myokymia maps to chromosome 12p13. *Am. J. Hum. Genet.*, **55**, 702–9.

Liu, Y., Holmgren, M., Jurman, M. E., and Yellen, G. (1997) Gated access to the pore of a voltage-dependent K^+ channel. *Neuron*, **19**, 175–84.

Lopatin, A. N., Makhina, E. N., and Nichols, C. G. (1994) Potassium channel block by cytoplasmic polyamines as the mechanism of intrinsic rectification. *Nature*, **372**, 366–9.

López-Barneo, J., Hoshi, T., Heinemann, S. H., and Aldrich, R. W. (1993) Effects of external cations and mutations in the pore region on C-type inactivation of Shaker potassium channels. *Receptors Channels*, **1**, 61–71.

Lorenzon, N. M., Lutz, C. M., Frankel, W. N., and Beam, K. G. (1998) Altered calcium channel currents in Purkinje cells of the neurological mutant mouse leaner. *J. Neurosci.*, **18**, 4482–9.

Lu, Z. and Mackinnon, R. (1994) Electrostatic tuning of Mg2+ affinity in an inward-rectifier K^+ channel. *Nature*, **371**, 243–6.

Ludwig, A., Flockerzi, V., and Hofmann, F. (1997) Regional expression and cellular localization of the alpha1 and beta subunit of high voltage-activated calcium channels in rat brain. *J. Neurosci.*, **15**, 1339–49.

MacKinnon, R., Cohen, S. L., Kuo, A., Lee, A., and Chait, B. T. (1998) Structural conservation in prokaryotic and eukaryotic potassium channels. *Science*, **280**, 106–9.

Manganas, L. N., Akhtar, S., Antonucci, D. E., Campomanes, C. R., Dolly, J. O., and Trimmer, J. S. (2001) Episodic ataxia type-1 mutations in the Kv1.1 potassium channel display distinct folding and intracellular trafficking properties. *J. Biol. Chem.*, **276**, 49427–34.

Matsuda, H., Saigusa, A., and Irisawa, H. (1987) Ohmic conductance through the inwardly rectifying K channel and blocking by internal Mg2+. *Nature*, **325**, 156–9.

McCormack, T. and McCormack, K. (1994) Shaker K^+ channel beta subunits belong to an NAD(P)H-dependent oxidoreductase superfamily. *Cell*, **79**, 1133–5.

McNamara, N. M., Muniz, Z. M., Wilkin, G. P., and Dolly, J. O. (1993) Prominent location of a K^+ channel containing the alpha subunit Kv1.2 in the basket cell nerve terminals of rat cerebellum. *Neuroscience*, **57**, 1039–45.

McPhee, J. C., Ragsdale, D. S., Scheuer, T., and Catterall, W. A. (1995) A critical role for transmembrane segment IVS6 of the sodium channel alpha subunit in fast inactivation. *J. Biol. Chem.*, **270**, 12025–34.

Molina, A., Castellano, A. G., and Lopez Barneo, J. (1997) Pore mutations in Shaker K^+ channels distinguish between the sites of tetraethylammonium blockade and C-type inactivation. *J. Physiol. (Lond.)*, **499**, 361–7.

Morais-Cabral, J. H., Zhou, Y., and MacKinnon, R. (2001) Energetic optimization of ion conduction rate by the K^+ selectivity filter. *Nature*, **414**, 37–42.

Moran, O., Dascal, N., and Lotan, I. (1991) Modulation of a Shaker potassium A-channel by protein kinase C activation. *FEBS Lett.*, **279**, 256–60.

Neher, E. and Sakmann, B. (1976) Single-channel currents recorded from membrane of denervated frog muscle fibres. *Nature*, **260**, 779–802.

Nichols, C. G. and Lopatin, A. N. (1997) Inwardly rectifying potassium channels. *Annu. Rev. Physiol.*, **59**, 171–91.

Nichols, C. G., Shyng, S. L., Nestorowicz, A., Glaser, B., Clement 4th, J. P., Gonzalez, G., *et al.* (1996) Adenosine diphosphate as an intracellular regulator of insulin secretion. *Science*, **272**, 1785–87.

Noda, M., Shimizu, S., Tanabe, T., Takai, T., Kayano, T., Ikeda, T., *et al.* (1984) Primary structure of Electrophorus electricus sodium channel deduced from cDNA sequence. *Nature*, **312**, 121–7.

Noda, M., Suzuki, H., Numa, S., and Stuhmer, W. (1989) A single point mutation confers tetrodotoxin and saxitoxin insensitivity on the sodium channel II. *FEBS Lett.*, **259**, 213–6.

Olcese, R., Qin, N., Schneider, T., Neely, A., Wei, X., Stefani, E., *et al.* (1994) The amino terminus of a calcium channel beta subunit sets rates of channel inactivation independently of the subunit's effect on activation. *Neuron*, **13**, 1433–8.

Ophoff, R. A., Terwindt, G. M., Vergouwe, M. N., van Eijk, R., Oefner, P. J., Hoffman, S. M. G., *et al.* (1996) Familial hemiplegic migraine and episodic ataxia type-2 are caused by mutations in the Ca^{2+} channel gene CACNL1A4. *Cell*, **87**, 543–52.

Ophoff, R. A., Terwindt, G. M., Frants, R. R., and Ferrari, M. D. (1998) P/Q-type Ca^{2+} channel defects in migraine, ataxia and epilepsy. *Trends Pharmacol. Sci.*, **19**, 121–7.

Papazian, D. M., Shao, X. M., Seoh, S. A., Mock, A. F., Huang, Y., and Wainstock, D. H. (1995) Electrostatic interactions of S4 voltage sensor in Shaker K^+ channel. *Neuron*, **14**, 1293–301.

Pardo, L. A., Heinemann, S. H., Terlau, H., Ludewig, U., Lorra, C., Pongs, O., *et al.* (1992) Extracellular K^+ specifically modulates a rat brain K^+ channel. *Proc. Natl. Acad. Sci. USA*, **89**, 2466–70.

Patil, N., Cox, D. R., Bhat, D., Faham, M., Myers, R. M., and Peterson, A. S. (1995) A potassium channel mutation in weaver mice implicates membrane excitability in granule cell differentiation. *Nat. Genet.*, **11**, 126–9.

Peretz, T., Levin, G., Moran, O., Thornhill, W. B., Chikvashvili, D., and Lotan, I. (1996) Modulation by protein kinase C activation of rat brain delayed rectifier K^+ channel expressed in Xenopus oocytes. *FEBS Lett.*, **381**, 71–6.

Perozo, E. and Bezanilla, F. (1990) Phosphorylation affects voltage gating of the delayed rectifier K^+ channel by electrostatic interactions. *Neuron*, **5**, 685–90.

Perozo, E., Jong, D. S., and Bezanilla, F. (1991*a*) Single channel studies of the phosphorylation of K^+ channels in the squid giant axon. II. Nonstationary conditions. *J. Gen. Physiol.*, **98**, 19–34.

Perozo, E., Vandenberg, C. A., Jong, D. S., and Bezanilla, F. (1991*b*) Single channel studies of the phosphorylation of K^+ channels in the squid giant axon. I. Steady-state conditions. *J. Gen. Physiol.*, **98**, 1–17.

Pessia, M., Tucker, S. J., Lee, K., Bond, C. T., and Adelman, J. P. (1996) Subunit positional effects revealed by novel heteromeric inwardly rectifying K^+ channels. *EMBO J.*, **15**, 2980–7.

Piccolino, M. (1997) Luigi Galvani and animal electricity: two centuries after the foundation of electrophysiology. *Trends Neurosci.*, **20**, 443–8.

Piccolino, M. (2000) The bicentennial of the Voltaic battery (1800–2000): the artificial electric organ. *Trends Neurosci.*, **23**, 147–51.

Planells-Cases, R., Ferrer-Montiel, A. V., Patten, C. D., and Montal, M. (1995) Mutation of conserved negatively charged residues in the S2 and S3 transmembrane segments of a mammalian K^+ channel selectively modulates channel gating. *Proc. Natl. Acad. Sci. USA*, **92**, 9422–6.

Platzer, J., Engel, J., Schrott-Fischer, A., Stephan, K., Bova, S., Chen, H., *et al.* (2000) Congenital deafness and sinoatrial node dysfunction in mice lacking class D L-type Ca^{2+} channels. *Cell*, **102**, 89–97.

Poliak, S., Gollan, L., Martinez, R., Custer, A., Einheber, S., Salzer, J. L., *et al.* (1999) Caspr2, a new member of the neurexin superfamily, is localized at the juxtaparanodes of myelinated axons and associates with K^+ channels. *Neuron*, **24**, 1037–47.

Pragnell, M., De Waard, M., Mori, Y., Tanabe, T., Snutch, T. P., Campbell, K.P. (1994) Calcium channel beta-subunit binds to a conserved motif in the I-II cytoplasmic linker of the alpha 1-subunit. *Nature*, **368**, 67–70.

Ratcliffe, C. F., Westenbroek, R. E., Curtis, R., Catterall, W. A. (2001) Sodium channel beta1 and beta3 subunits associate with neurofascin through their extracellular immunoglobulin-like domain. *J. Cell Biol.*, **154,** 427–34.

Rea, R., Spauschus, A., Eunson, L. H., Hanna, M. G., and Kullmann, D. M. (2002) Variable K(+) channel subunit dysfunction in inherited mutations of KCNA1. *J. Physiol.*, **538**, 5–23.

Reimann, F. and Ashcroft, F. M. (1999) Inwardly rectifying potassium channels. *Curr. Opin. Cell Biol.*, **11**, 503–8.

Rettig, J., Heinemann, S. H., Wunder, F., Lorra, C., Parcej, D. N., Dolly, J. O., *et al.* (1994) Inactivation properties of voltage-gated K^+ channels altered by presence of beta-subunit. *Nature*, **369**, 289–94.

Ruppersberg, J. P., Schroter, K. H., Sakmann, B., Stocker, M., Sewing, S., and Pongs, O. (1990) Heteromultimeric channels formed by rat brain potassium-channel proteins. *Nature*, **345**, 535–7.

Ruppersberg, J. P., Stocker, M., Pongs, O., Heinemann, S. H., Frank, R., and Koenen, M. (1991) Regulation of fast inactivation of cloned mammalian IK(A) channels by cysteine oxidation. *Nature*, **352**, 711–14.

Sap, J., D'Eustachio, P., Givol, D., and Schlessinger, J. (1990) Cloning and expression of a widely expressed receptor tyrosine phosphatase. *Proc. Natl. Acad. Sci. USA*, **87**, 6112–16.

Sato, C., Ueno, Y., Asai, K., Takahashi, K., Sato, M., Engel, A., *et al.* (2001) The voltage-sensitive sodium channel is a bell-shaped molecule with several cavities. *Nature*, **409**, 1047–51.

Schneider, M. F. and Chandler, W. K. (1973) Voltage dependent charge movement of skeletal muscle: a possible step in excitation-contraction coupling. *Nature*, **242**, 244–6.

Schoppa, N. E., McCormack, K., Tanouye, M. A., and Sigworth, F. J. (1992) The size of gating charge in wild-type and mutant Shaker potassium channels. *Science*, **255**, 1712–15.

Schrempf, H., Schmidt, O., Kummerlen, R., Hinnah, S., Muller, D., Betzler, M., *et al.* (1995) A prokaryotic potassium ion channel with two predicted transmembrane segments from Streptomyces lividans. *EMBO J.*, **14**, 5170–8.

Sheng, M., Tsaur, M. L., Jan, Y. N., and Jan, L. Y. (1994) Contrasting subcellular localization of the Kv1.2 K^+ channel subunit in different neurons of rat brain. *J. Neurosci.*, **14**, 2408–17.

Slesinger, P. A., Reuveny, E., Jan, Y. N., and Jan, L. Y. (1995) Identification of structural elements involved in G protein gating of the GIRK1 potassium channel. *Neuron*, **15**, 1145–56.

Slesinger, P. A., Patil, N., Liao, Y. J., Jan, Y. N., Jan, L. Y., and Cox, D. R. (1996) Functional effects of the mouse weaver mutation on G protein-gated inwardly rectifying K^+ channels. *Neuron*, **16**, 321–31.

Smart, S. L., Lopantsev, V., Zhang, C. L., Robbins, C. A., Wang, H., Chiu, S. Y., *et al.* (1998) Deletion of the Kv1.1 potassium channel causes epilepsy in mice. *Neuron*, **20**, 809–19.

Smith, M. R. and Goldin, A. L. (1997) Interaction between the sodium channel inactivation linker and domain III S4-S5. *Biophys. J.*, **73**, 1885–95.

Southan, A.P. and Robertson, B. (1998) Patch-clamp recordings form cerebellar basket cell bodies and their presynaptic terminals reveal an asymmetric distribution of voltage-gated potassium channels. *J. Neurosci.*, **18**, 948–55.

Srinivasan, Y., Elmer, L., Davis, J., Bennett, V., and Angelides, K. (1988) Ankyrin and spectrin associate with voltage-dependent sodium channels in brain. *Nature*, **333**, 177–80.

Srinivasan, J., Schachner, M., and Catterall, W. A. (1998) Interaction of voltage-gated sodium channels with the extracellular matrix molecules tenascin-C and tenascin-R. *Proc. Natl. Acad. Sci. USA*, **95**, 15753–7.

Stanfield, P. R., Davies, N. W., Shelton, P. A., Khan, I. A., Brammar, W. J., Standen, N. B., *et al.* (1994) The intrinsic gating of inward rectifier K^+ channels expressed from the murine IRK1 gene depends on voltage, K^+ and $Mg2^+$. *J. Physiol.*, **475**, 1–7.

Stotz, S. C., Hamid, J., Spaetgens, R. L., Jarvis, S. E., and Zamponi, G. W. (2000) Fast inactivation of voltage-dependent calcium channels. A hinged-lid mechanism? *J. Biol. Chem.*, **275**, 24575–82.

Stuhmer, W., Conti, F., Suzuki, H., Wang, X. D., Noda, M., Yahagi, N., *et al.* (1989*a*) Structural parts involved in activation and inactivation of the sodium channel. *Nature*, **339**, 597–603.

Stuhmer, W., Ruppersberg, J. P., Schroter, K. H., Sakmann, B., Stocker, M., Giese, K. P., *et al.* (1989*b*) Molecular basis of functional diversity of voltage-gated potassium channels in mammalian brain. *EMBO J.*, **8**, 3235–44.

Sumikawa, K., Houghton, M., Emtage, J. S., Richards, B. M., and Barnard, E. A. (1981) Active multi-subunit ACh receptor assembled by translation of heterologous mRNA in Xenopus oocytes. *Nature*, **292**, 862–4.

Taglialatela, M., Toro, L., and Stefani, E. (1992) Novel voltage clamp to record small, fast currents from ion channels expressed in xenopus oocytes. *Biophys. J.*, **61**, 78–82.

Tanabe, T., Beam, K. G., Adams, B. A., Niidome, T., and Numa, S. (1990) Regions of the skeletal muscle dihydropyridine receptor critical for excitation-contraction coupling. *Nature*, **346**, 567–9.

Tanabe, T., Takeshima, H., Mikami, A., Flockerzi, V., Takahashi, H., Kangawa, K., *et al.* (1987) Primary structure of the receptor for calcium channel blockers from skeletal muscle. *Nature*, **328**, 313–18.

Tanouye, M. A. and Ferrus, A. (1985) Action potentials in normal and Shaker mutant drosophila. *J. Neurogenet.*, **2**, 253–71.

Teh, B. T., Silburn, P., Lindblad, K., Betz, R., Boyle, R., Schalling, M., *et al.* (1995) Familial periodic cerebellar ataxia without myokymia maps to a 19-cM region on 19p13. *Am. J. Hum. Genet.*, **56**, 1443–9.

Tejedor, F. J., Bokhari, A., Rogero, O., Gorczyca, M., Zhang, J., Kim, E., *et al.* (1997) Essential role for dlg in synaptic clustering of Shaker K^+ channels in vivo. *J. Neurosci.*, **17**, 152–9.

Tempel, B. L., Papazian, D. M., Schwarz, T. L., Jan, Y. N., and Jan, L.Y. (1987) Sequence of a probable potassium channel component encoded at Shaker locus of Drosophila. *Science*, **237**, 770–5.

Tsai, W., Morielli, A. D., Cachero, T. G., and Peralta, E. G. (1999) Receptor protein tyrosine phosphatase alpha participates in the m1 muscarinic acetylcholine receptor-dependent regulation of Kv1.2 channel activity. *EMBO J.*, **18**, 109–18.

Tucker, S. J., Pessia, M., Moorhouse, A. J., Gribble, F., Ashcroft, F. M., Maylie, J., *et al.* (1996) Heteromeric channel formation and Ca(2+)-free media reduce the toxic effect of the weaver Kir 3.2 allele. *FEBS Lett.*, **390**, 253–57.

Tucker, S. J., Gribble, F. M., Zhao, C., Trapp, S., and Ashcroft, F. M. (1997) Truncation of Kir6.2 produces ATP-sensitive K^+ channels in the absence of the sulphonylurea receptor. *Nature*, **387**, 179–83.

VanDyke, D. H., Griggs, R. C., Murphy, M. J., and Goldstein, M. N. (1975) Hereditary myokymia and periodic ataxia. *J. Neurol. Sci.*, **25**, 109–18.

Vassilev, P. M., Scheuer, T., and Catterall, W. A. (1988) Identification of an intracellular peptide segment involved in sodium channel inactivation. *Science*, **241**, 1658–61.

Von Brederlow, B., Hahn, A. F., Koopman, W. J., Ebers, G. C., and Bulman, D. E. (1995) Mapping the gene for acetazolamide responsive hereditary paryoxysmal cerebellar ataxia to chromosome 19p. *Hum. Mol. Genet.*, **4**, 279–84.

Wallace, R. H., Wang, D. W., Singh, R., Scheffer, I. E., George Jr, A. L., Phillips, H. A., *et al.* (1998) Febrile seizures and generalized epilepsy associated with a mutation in the Na+-channel beta1 subunit gene SCN1B. *Nat. Genet.*, **19**, 366–70.

Wang, H., Kunkel, D. D., Martin, T. M., Schwartzkroin, P. A., and Tempel, B. L. (1993) Heteromultimeric K^+ channels in terminal and juxtaparanodal regions of neurons. *Nature*, **365**, 75–9.

Wang, H., Kunkel, D. D., Schwartzkroin, P. A., and Tempel, B. L. (1994) Localization of Kv1.1 and Kv1.2, two K^+ channel proteins, to synaptic terminals, somata, and dendrites in the mouse brain. *J. Neurosci.*, **14**, 4588–99.

Wang, Q., Shen, J., Splawski, I., Atkinson, D., Li, Z., Robinson, J. L., *et al.* (1995) SCN5A mutations associated with an inherited cardiac arrhythmia, long QT syndrome. *Cell*, **80**, 805–11.

Wei, Y., Waltz, D. A., Rao, N., Drummond, R. J., Rosenberg, S., and Chapman, H. A. (1994) Identification of the urokinase receptor as an adhesion receptor for vitronectin. *J. Biol. Chem.*, **269**, 32380–8.

West, J. W., Patton, D. E., Scheuer, T., Wang, Y., Goldin, A. L., and Catterall, W. A. (1992) A cluster of hydrophobic amino acid residues required for fast Na(+)-channel inactivation. *Proc. Natl. Acad. Sci. USA*, **89**, 10910–14.

Westenbroek, R. E., Sakurai, T., Elliott, E. M., Hell, J. W., Starr, T. V., Snutch, T. P., *et al.* (1995) Immunochemical identification and subcellular distribution of the alpha 1A subunits of brain calcium channels. *J. Neurosci.*, **15**, 6403–18.

Wu, L. G., Westenbroek, R. E., Borst, J. G., Catterall, W. A., and Sakmann, B. (1999) Calcium channel types with distinct presynaptic localization couple differentially to transmitter release in single calyx-type synapses. *J. Neurosci.*, **15**, 726–36.

Yang, N. and Horn, R. (1995) Evidence for voltage-dependent S4 movement in sodium channels. *Neuron*, **15**, 213–18.

Yang, N., George Jr, A. L., and Horn, R. (1996) Molecular basis of charge movement in voltage-gated sodium channels. *Neuron*, **16**, 113–22.

Yellen, G. (1998) The moving parts of voltage-gated ion channels. *Q. Rev. Biophys.*, **31**, 239–95.

Yellen, G., Jurman, M. E., Abramson, T., and MacKinnon, R. (1991) Mutations affecting internal TEA blockade identify the probable pore-forming region of a K^+ channel. *Science*, **251**, 939–42.

Zagotta, W. N., Hoshi, T., and Aldrich, R. W. (1990) Restoration of inactivation in mutants of *Shaker* potassium channels by a peptide derived from ShB. *Science*, **250**, 568–71.

Zerr, P., Adelman, J. P., and Maylie, J. (1998*a*) Episodic ataxia mutations in Kv1.1 alter potassium channel function by dominant negative effects or Haploinsufficiency. *J. Neurosci.*, **18**, 2842–8.

Zerr, P., Adelman, J. P., and Maylie, J. (1998*b*) Characterization of three episodic ataxia mutations in the human Kv1.1 potassium channel. *FEBS Lett.*, **431**, 461–4.

Zhang, C. L., Messing, A., and Chiu, S. Y. (1999) Specific alteration of spontaneous GABAergic inhibition in cerebellar purkinje cells in mice lacking the potassium channel Kv1. 1. *J. Neurosci.*, **19**, 2852–64.

Zhou, D., Lambert, S., Malen, P. L., Carpenter, S., Boland, L. M., and Bennett, V. (1998) AnkyrinG is required for clustering of voltage-gated Na channels at axon initial segments and for normal action potential firing. *J. Cell Biol.*, **143**, 1295–304.

Zhou, Y., Morais-Cabral, J. H., Kaufman, A., and MacKinnon, R. (2001) Chemistry of ion coordination and hydration revealed by a K^+ channel-Fab complex at 2.0 A resolution. *Nature*, **414**, 43–8.

Zhuchenko, O., Bailey, J., Bonnen, P., Ashizawa, T., Stockton, D. W., Amos, C., *et al.* (1997) Autosomal dominant cerebellar ataxia (SCA6) associated with small polyglutamine expansions in the alpha 1A-voltage-dependent calcium channel. *Nat. Genet.*, **15**, 62–9.

Zito, K., Fetter, R. D., Goodman, C. S., and Isacoff, E. Y. (1997) Synaptic clustering of Fascilin II and Shaker: essential targeting sequences and role of Dlg. *Neuron*, **19**, 1007–16.

Zühlke, R. D., Pitt, G. S., Deisseroth, K., Tsien, R. W., and Reuter, H. (1999) Calmodulin supports both inactivation and facilitation of L-type calcium channels. *Nature*, **399**, 159–62.

Chapter 6

Molecular biology of neurotransmitter release

Julie Staple and Stefan Catsicas

6.1 Introduction

Neurons function to receive, integrate, and transmit signals to other neurons or to nonneuronal target cells. While the process of neurotransmission appears relatively simple — neurotransmitter released from one cell binds to a receptor on another cell which results in depolarization or other effects in the target — there are many potentially regulatable components involved in each step. The characteristics of neurotransmission and the opportunities for modification of this process are the subject of intense study since it is these properties and the potential for change with experience that is believed to underlie complex cognitive functions as well as simple forms of adaptation.

6.1.1 Historical perspective on molecular biology at synapses

The neuron. The idea of the synapse necessarily implies a relation between two neurons. Although this statement may seem simplistic, it is interesting to note that it has only been 100 years since neurons were defined as individual cells. Indeed, less than 50 years have passed since synapses were first seen and the 'neuron theory' supported by use of the electron microscope (Palade and Palay 1954; and see Robertson 1987). As an introduction to the topic of molecular biology of synapses, a brief outline of the history of experiment and observation which led to the definition of neurons and synapses in terms of histology, physiology, and biochemistry is presented in this chapter.

The concept of the synapse was essentially dependent on the definition of the neuron in histological terms. A number of debates were waged concerning the relation of axons to cell bodies (it was not clear that these were parts of the same cells) and the relation of one neuron to another (some workers saw all neurons as interconnected, others saw them as individual cells). Steps forward in these arguments were dependent on advances in techniques of fixation and staining (especially on Golgi's silver staining technique) as well as on increased resolution due to better microscopes. More than

150 years later, Robert Remak described for the first time the continuity of a neuronal cell body with an axon in a preparation of sympathetic ganglion.

The relationship of the axon to its own cell body having been established, the controversy which was to interest neurohistologists for 75 years came to the forefront when Joseph von Gerlach published observations of an extensive axonal and dendritic network of neuronal processes. The substance of this argument concerned the direct interconnection of one nerve cell with another ('reticular theory') as opposed to their existence as individual cells ('nerve cell theory'). Although many others supported this position, Camillo Golgi, became known as the leading proponent of the 'reticular theory'. While Golgi denied the existence of the dendritic net proposed by Gerlach, he affirmed an axonal reticular formation based on his observations using his newly developed staining technique. A major contributor to the 'nerve cell theory' was August Forel who believed that nerve cells were separate entities whose individuality was often hard to discern because of the close contacts which they made with each other. Forel based this hypothesis on the pattern of degeneration induced by axonal lesions. He saw that a specific population of cells degenerated in response to axonal injury and that this degeneration was limited in scope, which he believed would not be the case if all cells were connected directly. Santiago Ramon y Cajal, regarded as the main proponent of the 'nerve cell theory' used a modification of the Golgi stain to demonstrate instances of axonal terminations. Wilhelm von Waldeyer introduced the term 'neuron' in 1891 to describe individual nerve cells which were not directly connected to each other. High resolution pictures made possible by the advent of electron microscopy in the 1950s allowed visualization of synapses and showed the membranes that separate the presynaptic cell from the postsynaptic cell (Palade and Palay 1954; and see Robertson 1987).

The synapse. It was in the context of this neuroanatomical debate that Charles Sherrington first used the term 'synapse'.

'... we are led to think that the tip of a twig of the arborescence is not contiguous with but merely in contact with the substance of the dendrite or cell body on which it impinges. Such a special connection of one nerve cell with another might be called a synapse' (Foster 1897, p. 57).

The individual nature of neurons led to questions about how a current propagated along an axon might be transmitted to another cell. Two main theories were proposed in this context, the first being direct transmission of an electrical current and the second mediation between a neuron and its target by a molecule released from the presynaptic terminal. Evidence for the possibility that a chemical signal could 'transmit' information was derived especially from work on acetylcholine (ACh) and its actions both at sympathetic ganglia and at the neuromuscular junction. Otto Loewi showed that ACh was the substance which mediated the vagus nerve-induced slowing of cardiac muscle contraction. The 'acetylcholine hypothesis' of neurotransmission was based on a number of observations (detailed in Eccles 1937) summarized here:

(i) ACh was shown to be released after direct nerve stimulation both from preparations of sympathetic ganglia and from neuromuscular preparations.

(ii) Addition of exogenous ACh or agonists to these preparations mimics the effects on the target of direct nerve stimulation.

(iii) When a nerve is stimulated to exhaustion or when preganglionic fibers degenerate no more ACh is released.

(iv) Direct stimulation of the target itself in the absence of a motoneuron produces no ACh.

(v) Compounds which block the action of ACh, for example curare, have a similar chemical structure, providing further basis for the idea that it is ACh which is involved in neurotransmission.

The development of intracellular recording techniques allowed detailed analysis of the effects of ACh at neuromuscular junctions and showed that these were inconsistent with direct electrical transmission (Fatt and Katz 1951).

6.2 Modern concepts in neurotransmitter release

6.2.1 The synaptic vesicle

Intracellular recording techniques also identified discrete units of neurotransmission: small, stereotyped membrane potential changes known as 'quanta'. The smallest events of neurotransmission involve membrane potential changes of single quantum size and larger events always occur as multiples of quanta. Thus, neurotransmission may involve 5 or 200 quanta, but not 2.5 or 154.2 quanta. What is the physical basis for quantal neurotransmission? Soon after the development of intracellular recording, technical advances in electron microscopy made it possible to obtain high resolution images of synapses for the first time. These images showed not only clearly identifiable presynaptic and postsynaptic elements separated by a synaptic cleft, but also a newly discovered synaptic characteristic: vesicles, which were immediately recognized as a possible basis for the release of 'quanta' (Fig. 6.1). Two main types of vesicles are found at synapses; small, clear vesicles of about 50 nm (SVs) and larger (~100 nm) vesicles characterized by electron-dense cores (LVs) (Fig. 6.2). These two vesicle types differ functionally as well as morphologically. SVs contain conventional transmitters such as acetylcholine, glutamate, caminobutyric acid (GABA), and glycine whereas LVs contain catecholamines and soluble peptides.

6.2.2 Aspects of neurotransmission

When intracellular vesicles fuse with the plasmalemma, the vesicle contents are secreted into the extracellular space and the components of the vesicle membrane become part of the plasmalemma. Thus, vesicle-mediated secretion plays two fundamental roles: intercellular communication and maintenance or renewal of membrane components. There are four main pathways for secretion in all eukaryotic cells. Secretory vesicles can originate from the Golgi apparatus or from endoplasmic reticulum-derived endosomes. Distinct vesicles from both origins contribute to

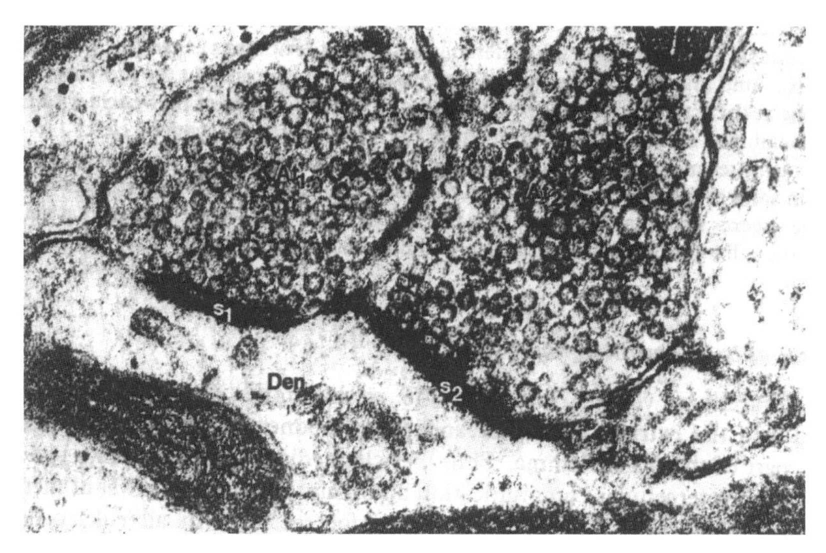

Fig. 6.1 Electron micrograph of two synapses (s_1 and s_2) between two axon terminals (Atl and At2) and a dendrite (den). The axon terminals contain many synaptic vesicles and the active zone of the synapses is demarcated by an intercellular cleft and a prominent coating of dense material. Reproduced from *The Fine Structure of the Nervous System: Neurons and Their Supporting Cells,* 3rd Edn, by Alan Peters, Sanford L. Palay and Henry deF. Webster. Copyright © 1990 by Alan Peters. Reprinted by permission of Oxford University Press.

regulated or constitutive secretion. Constitutive fusion is, by definition, independent of extracellular signals and mediates transport and insertion of membrane components. It is thought to play a key role in membrane turn-over as well as membrane insertion underlying changes in cell shape and motility (Catsicas *et al.* 1994). Regulated secretion occurs upon transient and often fast stimulatory events that induce a cellular response. At the synapse, regulated secretion involves SVs derived from endosomes and LVs derived from the Golgi (see Kelly 1993). Release of neurotransmitters by SVs requires sequential steps of vesicle trafficking. In the terminal, vesicles can be stored in the so-called reserve pool consisting of vesicles bound to actin filaments (Figs. 6.2 and 6.3). Vesicles in the reserve pool must be mobilized to progress towards the plasma membrane (Greengard *et al.* 1993). At the membrane, vesicles are docked to microdomains within the presynaptic active zones and form the 'releasable pool' (Greengard *et al.* 1993). Docking is an additional storage step as docked vesicles must be primed to become ready to fuse. Membrane fusion and transmitter release are then induced when fast and localized Ca^{2+} gradients are generated by incoming action potentials (see Hu *et al.* 1993). In contrast, LV progression to the membrane and subsequent fusion seems to involve a single, Ca^{2+}-dependent, cytoskeletal destabilization step (Trifaró and Vitale 1993). LVs are located at a distance from the presynaptic membrane and their progression is induced by slow and diffuse Ca^{2+} gradients through channels that are distributed more homogeneously in the membrane of the

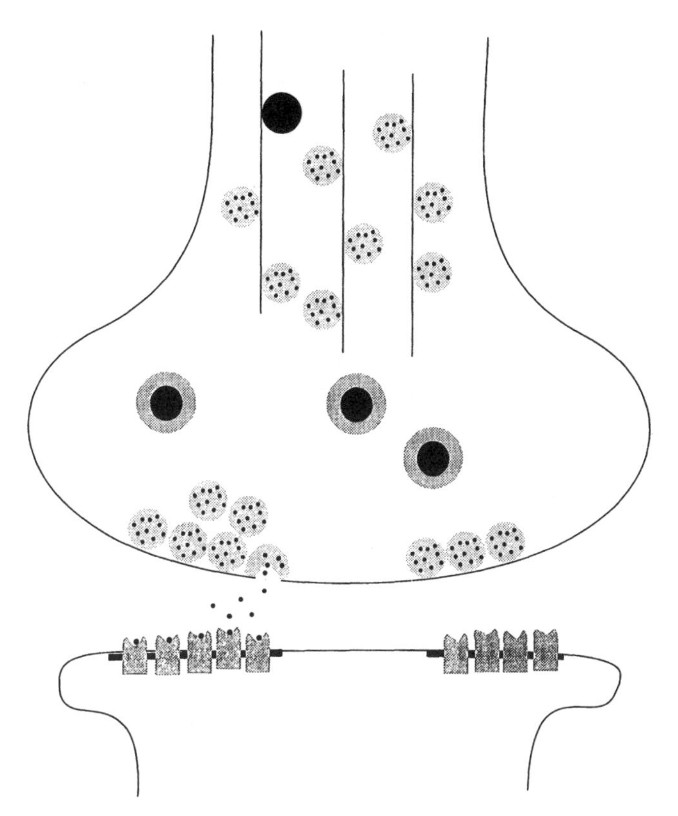

Fig. 6.2 Schematic representation of some basic aspects of neurotransmission. Synaptic vesicles (pale orange spheres) filled with neurotransmitter (small black circles) are present in nerve terminals as two functionally different populations. The 'reserve pool' of synaptic vesicles is bound to actin filaments and the releasable pool is present at the presynaptic membrane at the active zone. Following docking of vesicles, fusion of the vesicle with the plasma membrane occurs and neurotransmitter is released into the synaptic cleft. Neurotransmitter molecules then bind to specific receptors present on the postsynaptic cell (shown in solid orange) resulting in signal transmission. Large, dense-core vesicles (pictured as orange spheres with black centers) may be present in the same nerve terminals as small vesicles, though biogenesis and regulation of their fusion is distinct (see text).

nerve terminal (see Hu *et al.* 1993). Following fusion, vesicles are retrieved by specific endocytosis mechanisms. SVs remain in the terminal to be filled with transmitters and used again, whereas LVs return to the *trans*-Golgi network to be packaged with newly synthesized peptides (see Hu *et al.* 1993). Each of these steps and the main proteins known to be involved are described in the following sections.

6.2.3 **Vesicle storage and mobilization**

Synaptic vesicles which contain neurotransmitter are sometimes known as 'mature' synaptic vesicles, a term which suggests their ability to participate in exocytosis. However, since neurotransmission is highly regulated, there are mechanisms which

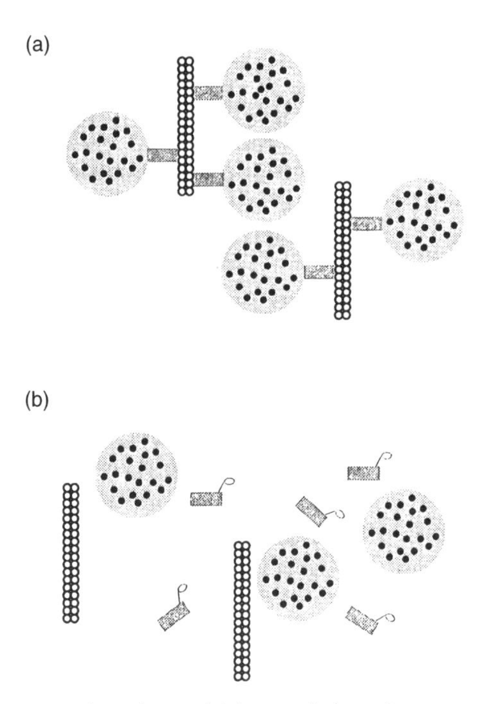

Fig. 6.3 Schematic representation of a model for regulation of synaptic vesicle mobilization by synapsin. (a) Neurotransmitter-filled synaptic vesicles (pale orange spheres with small black circles) are shown bound via dephosphorylated synapsin (orange bars) to actin filaments (open circles). (b) Depolarization causes Ca^{2+}-dependent phosphorylation of synapsin I (orange-tagged bars) and results in dissociation of synaptic vesicles from actin filaments. The vesicles are now free to move to the active zone to join the releasable pool.

restrain the process of vesicle fusion, and further steps necessary to 'prime' vesicles for exocytosis. A mechanism proposed for restraint of vesicle fusion is the tethering of vesicles to the cytoskeleton in the synaptic terminal. This attachment would prevent vesicles from diffusing to the active zone, but keeping them close to the synapse for recruitment from a 'reserve pool' when needed (Figs. 6.2 and 6.3).

6.2.4 The synapsin family of phosphoproteins

The synapsins, a family of four phosphoproteins which are associated with synaptic vesicles, have been implicated in reversible binding of vesicles to actin filaments. These abundant proteins, which make up approximately 0.6% of total brain protein and 9% of synaptic vesicle protein (De Camilli *et al.* 1990; Valtorta *et al.* 1992), were first characterized as neuronal substrates for cAMP-dependent phosphorylation by Greengard and colleagues (see Greengard *et al.* 1993). The synapsin family consists of two homologous genes, *synapsin I* and *synapsin II* (reviewed in Südhof *et al.* 1989). *Synapsin I* exists in two isoforms, Ia (apparent molecular weight of 86 kDa, 704 a.a. in rat) and

Ib (80 kDa, 668 a.a. in rat), which are splice products of a single gene. The *synapsin II* gene also encodes two alternatively spliced products, synapsin IIa and IIb, proteins of approximately 74 kDa (586 a.a. in rat) and 55 kDa (479 a.a in rat). Synapsin I isoforms have 70% average sequence identity with synapsin II isoforms over the first 420 amino acids which contain a conserved phosphorylation site and proposed actin binding domains. Comparison of the C-terminal regions of synapsins I and II shows that synapsin II isoforms are missing two Ca^{2+}/calmodulin-dependent kinase phosphorylation sites found in the synapsin I isoforms.

Experiments by Greengard and collaborators suggest that dephosphorylated synapsin I mediates the attachment of SVs to actin filaments, keeping SVs in the 'reserve pool' until needed. Electron microscopic and biochemical techniques have shown that synapsins are peripheral membrane proteins associated with synaptic vesicles (DeCamilli *et al.* 1983; Valtorta *et al.* 1988; Benfenati *et al.* 1989; Thiel *et al.* 1990) and, additionally, that synapsin I binds actin via specific domains (Bähler and Greengard 1987; Petrucci and Morrow 1987; Ceccaldi *et al.* 1995). The binding of phosphorylated synapsin I both to synaptic vesicles and to actin is reduced compared to the dephosphorylated protein (Huttner *et al.* 1983; Schiebler *et al.* 1986; Sihra *et al.* 1989). Since synapsin I has been shown to be phosphorylated by Ca^{2+}/calmodulin-dependent kinase II in response to various kinds of synaptic stimulation (Forn and Greengard 1978; Nestler and Greengard 1982) it has been suggested that activity-dependent phosphorylation regulates the association of SVs with actin, and allows translocation of the vesicles from the stored pool to the releasable pool, near the active zone (Fig. 6.4). Consistent with this, dephosphorylated synapsin I had an inhibitory effect on neurotransmission at intact synapses (Llinás *et al.* 1985, 1991; Hackett *et al.* 1990; Lin *et al.* 1990) and in synaptosomes (Nichols *et al.* 1992), whereas phosphorylated synapsin I had no effect on neurotransmission in these studies.

Apparently, neither synapsin I nor synapsin II is essential for transmitter release. Recent generation of mice lacking synapsin I, synapsin II or both, results in changes in short-term plasticity but no change has been found in basic neurotransmission or in the ability to induce long-term changes such as those which occur in learning and memory (see Section 6.3; Chin *et al.* 1995; Li *et al.* 1995; Rosahl *et al.* 1995; Spillane *et al.* 1995). However, altered expression of synapsin I or synapsin II has effects on both the developmental process of synaptogenesis and functional characteristics of mature synapses (Han *et al.* 1991; Lu *et al.* 1992; Ferreira *et al.* 1994, 1995; Chin *et al.* 1995; Rosahl *et al.* 1995). These observations are consistent with a role for synapsins in regulation of synaptic vesicle mobilization and further suggest that mobilization is a check point involved in short-term, fast actions on transmitter release, and long-term structural changes.

6.2.5 Vesicle docking

A number of proteins have been recently identified with suspected prominent roles in the docking process. Their functional organization is depicted schematically in

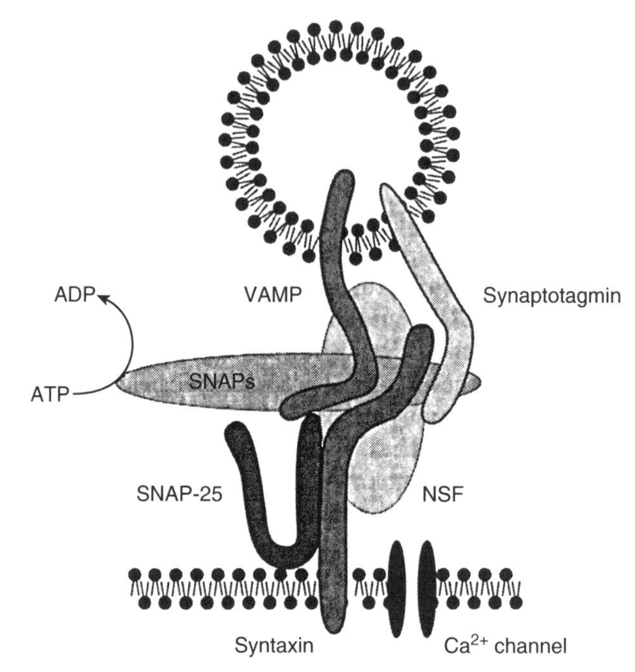

Fig. 6.4 An outline of the SNARE hypothesis depicts interactions between vesicles and release sites at the membrane which are mediated by specific proteins. An intrinsic vesicle protein (VAMP, a v-SNARE) binds to plasma membrane proteins (SNAP-25 and syntaxin, both t-SNAREs). The association of these three proteins allows binding of $\alpha/\beta/\gamma$ SNAPs which in turn act as receptors for NSF. The hydrolysis of ATP by NSF results in the dissociation of the complex of proteins and is followed by vesicle fusion. Modified from *Trends Neurosci.*, vol. 17, pp. 368–373, with permission from Elsevier Trends Journals.

Fig. 6.5. Bassoon and Piccolo (aka aczonin) are extremely large proteins (420 and 530 kDa) localized to active zones of most, but not all, synapses (tom Dieck *et al.* 1998). They contain various protein interaction domains, and are likely to act as scaffolding molecules promoting the assembly of the exocytotic multiprotein complex.

Rab3 is a small GTPase that is bound to the synaptic vesicles in the active zone, but which dissociates from the vesicles during exocytosis. Rab3 binds a protein called rabphilin, which is thought to mediate an interaction with actin. The eponymous Rab3-interacting molecules RIM1 and RIM2 (aka oboe) bind to the active (GTP bound) form of Rab3, and are suggested to be components of the scaffold that retains SVs at the active zone. Apart from Rab3, RIM1 can also interact with synaptotagmins, SNAP-25, RIM-binding proteins (which link RIMs to Ca^{2+} channels), and a class of proteins involved in active zone organization known as α-liprins. Deletion of RIM1 in mice has no gross effects on synaptic transmission, but, as observed with Rab3 deletion, impairs long-term potentiation at hippocampal mossy fiber synapses (Castillo *et al.* 2002).

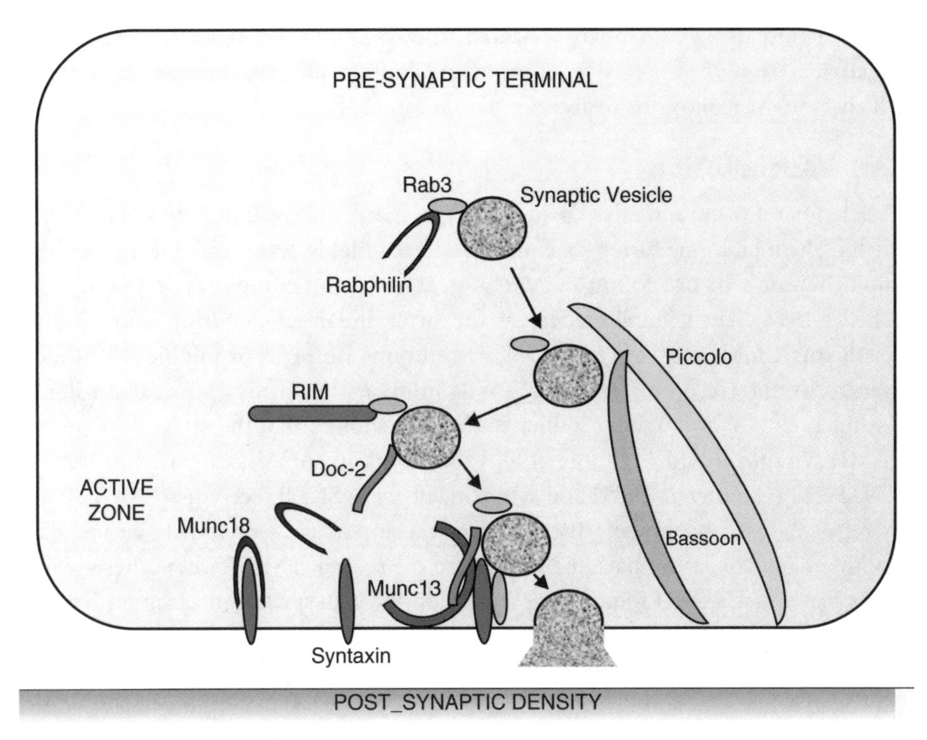

Fig. 6.5 Schematic diagram of the molecular interactions regulating SV docking and priming.

Munc-13 is present at the majority of central synapses, and binds diacylglycerol and Ca^{2+}. The diacylglycerol binding activity raised the possibility that the well-known effects of phorbol esters on neurotransmitter release might be mediated via Munc-13 rather than PKC, and this has proved to be the case (Rhee *et al.* 2002). Diacylglycerol binding promotes the association of Munc-13 with SVs via an interaction with Doc-2, and overexpression of Munc-13 increases neurotransmitter release (Betz *et al.* 1998; Duncan *et al.* 1999). Absence of Munc-13 abolishes evoked or spontaneous transmitter release (Tokumaru and Augustine 1999), but leaves the number of vesicles in the active zone unaltered, suggesting that Munc-13 functions after the docking stage. Munc-13 binds to syntaxin, an integral protein at the active zone (Bennett *et al.* 1992) with a central role in SV fusion (see below), and the Munc-13-1 isoform (but not the Munc-13-2 isoform) also binds to RIM1 (Betz *et al.* 2001). Hence the Munc-13-1/RIM1 interaction may be critical to prime SVs ready for fusion.

As with Munc-13, absence of Munc-18 (aka nsec1, rbsec1, and msec1) results in the loss of spontaneous and evoked transmitter release. Munc-18 binds syntaxin, and is thought to prevent syntaxin functional complexes with the SVs, by binding to a region of syntaxin that overlaps with the Munc-13 binding site (see Martin 2002), hence regulating the formation of the vesicle fusion complex. Munc-18 could also retain SVs in the vicinity of Ca^{2+} channels. A protein known as tomosyn reportedly encourages the dissociation of Munc-18 from syntaxin (Fujita *et al.* 1998; Masuda *et al.* 1998), which could then allow the fusion complex to associate.

6.2.6 **Vesicle fusion**

Vesicle fusion occurs in all cells and between many different membrane compartments. The membrane fusion machinery has been highly conserved during evolution and different cells use common general protein fusion components (Pevsner and Scheller 1994). The general machinery for intracellular vesicle fusion comprises the *N*-ethylmaleimide sensitive factor (NSF) and three isoforms of soluble NSF attachment proteins (α, β and γ SNAPs) (Wilson *et al.* 1992). Biochemical studies by Rothman and his colleagues (Söllner *et al.* 1993) showed that the integral SV protein VAMP/synaptobrevin (Trimble *et al.* 1988), synaptosomal-associated protein 25 (SNAP-25) (Oyler *et al.* 1989) and syntaxin, all act as SNAP receptors (SNARES) and thereby can form a complex with the general fusion machinery. Based on these data, Rothman and colleagues have then proposed the so-called SNARE hypothesis, which states that vesicles select their targets for fusion through specific interactions between vesicle and target SNARES (Fig. 6.4).

The SNARE hypothesis (Fig. 6.4)implies that pairing between proteins specifically expressed on vesicles and their target membranes (v- or t-SNARES) mediates selective vesicle docking and fusion (Söllner *et al.* 1993; Rothman and Warren 1994). Recognition between v- and t-SNARES would mediate docking while binding of SNARES to proteins involved in the membrane fusion machinery would mediate vesicle fusion. Thus, VAMP was postulated to be a v-SNARE and SNAP-25 and syntaxin t-SNARES, involved in SV docking and fusion at the presynaptic membrane.

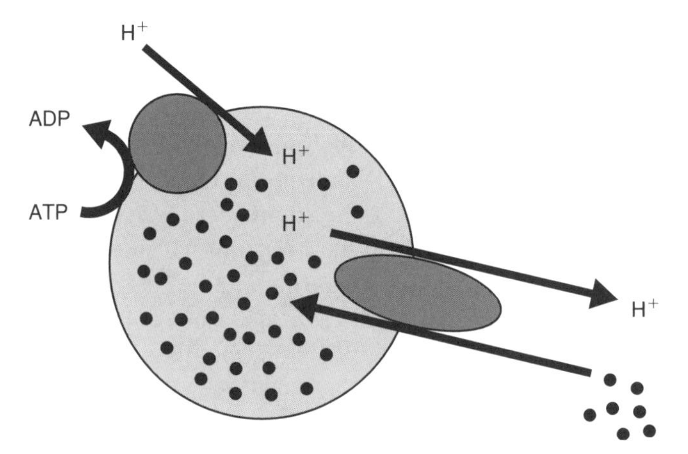

Fig. 6.6 Neurotransmitter transport into synaptic vesicles is dependent on two proteins. An H^+-ATPase (orange circle) is essential for creating a H^+ gradient across the vesicle membrane. The neurotransmitter transporter (orange oval) is an H^+ transmitter antiporter.

Consistent with this, under appropriate conditions, the three proteins form a stable heterotrimeric complex *in vitro* (Chapman *et al.* 1994; Hayashi *et al.* 1994). Functional evidence that VAMP, SNAP-25, and syntaxin are involved in transmitter release came from research on the targets of the tetanus and botulinum bacterial neurotoxins (TeNT and BoNT). TeNT and BoNT are metalloproteases that block transmitter release and induce paralysis of skeletal muscles followed by rapid death. Studies (Schiavo *et al.* 1992; Blasi *et al.* 1993), showed that blockade of transmitter release by the toxins is due to the proteolytic cleavage of each of the three SNARE proteins identified by Rothman and colleagues. TeNT and BoNT type B, D, F, and G cleave VAMP, while BoNT/A and /E cleave SNAP-25 and BoNT/C cleaves syntaxin and SNAP-25 (for review see Ferro-Novick and Jahn 1994; Montecucco and Schiavo 1994; see also Osen-Sand *et al.* 1996, for the role of BoNT/C). These data are consistent with the hypothesis that SNARES are involved in vesicle fusion, and with evidence that docking can occur in the absence of the SNARES (Schweizer *et al.* 1995; and see Südhof 1995; Pellegrini *et al.* 1995; Osen-Sand *et al.* 1996). Hence docking and fusion require separate molecular interactions.

It is also worth noting that, in addition to their role in release, SNAP-25 and syntaxin may have a structural role in the entire axonal compartment. In differentiated neurons *in vitro*, VAMP or SNAP-25 cleavage with clostridial toxins causes a strong inhibition of spontaneous and evoked release, and does not result in major morphological changes. In contrast, BoNT/C, which cleaves SNAP-25 and syntaxin, induces massive disruption of the axons and neuronal death (Osen-Sand *et al.* 1996). The fact that cleavage of SNAP-25 by BoNT/A did not induce similar morphological effects suggests that syntaxin plays a major role in axonal maintenance and cell viability. Consistent with this, although knock-out experiments in *Drosophila* indicate that syntaxin 1A is not essential for cell survival at early stages (Schuize *et al.* 1995), lack of the protein eventually results in massive neurite disruption and degeneration of the nervous system (Schulze and Bellen, personal communication).

The fast toxic effects induced by simultaneous cleavage of SNAP-25 and syntaxin may reflect co-operativity of the two proteins for essential membrane exchanges within the axonal compartment. Consistent with this, syntaxin and SNAP-25 form a heterodimer *in vitro* and this association is believed to exist *in vivo* on the cytosolic face of the axonal membrane (Chapman *et al.* 1994; Hayashi *et al.* 1994). In fact, very recent immunochemical studies indicate an extensive co-localization of SNAP-25 and syntaxin on the plasmalemma (Garcia *et al.* 1995) and a widespread distribution of both proteins throughout the axon (Duc and Catsicas 1995; Garcia *et al.* 1995).

As noted above, SV docking does not necessarily result in fusion (see Kelly 1993). Electrophysiological evidence from time-resolved capacitance measurements in goldfish bipolar cells suggests that only a fraction of the releasable vesicles fuse in response to the rise of cytosolic Ca^{2+} generated by the action potential (von Gersdorff and Matthews 1994). These and other data suggest that only a fraction of docked vesicles are ready (primed) to fuse. An additional problem that neurons must face is that secretion (fusion) must be highly regulated in time. The rapid arrest of release following the drop of cytosolic Ca^{2+} at the end of the action potential implies the presence of a low affinity Ca^{2+} sensor within the microdomains where fusion occurs (Almers 1994). Several independent approaches indicate that synaptotagmin, another integral synaptic vesicle protein (Matthew *et al.* 1981), is a possible Ca^{2+} sensor. Ca^{2+} binds to synaptotagmin and its binding regulates the capacity of the protein to bind phospholipids (Brose *et al.* 1992) and RIMs. This could in turn affect the possible role of synaptotagmin in fusion. In addition, and most importantly, synaptotagmin can bind to the SNARE complex formed by VAMP, SNAP-25, and syntaxin, and competes with a-SNAP for a common site on the complex (Söllner *et al.* 1993). a-SNAP can displace synaptotagmin which allows binding of the NSF to the complex and vesicle fusion (Söllner *et al.* 1993). One possible interpretation of these data is that synaptotagmin acts as a negative regulator, preventing fusion in the absence of appropriate signals. Consistent with this hypothesis, analysis of synaptotagmin mutants of *Caenorhabditis elegans* and *Drosophila* shows that although the protein is not essential for release, it has modulatory functions (DiAntonio *et al.* 1993; Nonet *et al.* 1993). There is also additional indirect evidence that Ca^{2+} may regulate protein–protein interactions within the fusion complex, since syntaxin can interact with N-type Ca^{2+} channels within the active zone (Bennet *et al.* 1992). Taken together, these studies suggest that synaptotagmin is a Ca^{2+}-dependent regulator of release efficiency, acting downstream of SV docking.

6.2.7 Membrane retrieval and neurotransmitter loading

While the cellular mechanisms involved in vesicle docking and fusion are the focus of many studies of neurotransmission, they represent only the initial steps of the exocytotic/endocytotic cycle. Following neurotransmitter release, SVs which fuse with the synaptic membrane do not generally become a permanent part of the plasmalemma (but see Section 6.3). Instead, SV membrane is retrieved to re-form vesicles which are then re-filled with neurotransmitters and take part in further rounds of synaptic activity. The molecules involved in these phases of synapse function have begun to be characterized: a selection of them are described in the following sections.

Membrane retrieval following fusion. The specific mechanisms of vesicle retrieval remain a matter of debate. The two main hypotheses involve the formation of clathrin coats and subsequent endocytosis, or the rapid opening and closure of a transient exocytotic fusion pore (reviewed by DeCamilli and Takei 1996). A role for clathrin-coated vesicles in vesicle retrieval seems likely since the number of clathrin-coated vesicles increases after stimulation and these vesicles contain synaptic vesicle proteins (Heuser and Reese 1973; Maycox *et al.* 1992). Recently, some of the molecules involved in formation of clathrin-coated pits have been characterized (see DeCamilli and Takei 1996).

Clathrin coats are formed from heavy (180 kDa) and light (35–40 kDa) clathrin chains, and assembly particles (340 kDa, composed of two copies of three proteins) which form characteristically shaped hetero-oligomers called triskelia. While the components of clathrin coats are found at the plasmalemma as well as in the *trans*-Golgi network, there are proteins particular to each compartment. A heterotetrameric protein complex, AP2 [comprised of two 100–115 kDa subunits (a and b) and two subunits of 50 and 17 kDa, respectively], functions at the plasmalemma, possibly as an adaptor between clathrin triskelia and membrane proteins (see DeCamilli and Takei 1996). In addition, synaptotagmin I has been shown to have a role in endocytosis (Fukuda *et al.* 1995; Jorgensen *et al.* 1995) and to bind AP2 (Li *et al.* 1995) suggesting that there are neuron-specific membrane proteins involved in clathrin-mediated vesicle retrieval at synapses.

As indicated above, the initial steps of vesicle endocytosis involve recruitment of clathrin coat components to the synaptic plasma membrane and formation of invaginations. A subsequent step has been identified in which these invaginations break off from the membrane to complete the endocytotic event. A neuron-specific molecule, dynamin I, has been implicated in this process at synapses. Dynamin was first identified as a neuronal microtubule-binding protein which has GTPase activity (Shpetner and Vallee 1989; Obar *et al.* 1990). The protein is phosphorylated by protein kinase C (PKC) and it is dephosphorylated in response to neuronal stimulation (Robinson *et al.* 1993). A number of functional studies, as well as its concentration in nerve terminals (Takei *et al.* 1995), suggest that dynamin has a major role in endocytosis (for recent review see DeCamilli and Takei 1996). Dynamin is a mammalian homolog of shibire, a *Drosophila* protein. Mutations of shibire cause arrest of endocytosis after formation of coated pits and subsequent paralysis (Koenig and Ikeda 1989). Further evidence for the involvement of dynamin in mammalian endocytosis comes from studies of the GTPase activity of this protein. When GTPase-defective mutants of dynamin are transfected into mammalian cells, endocytosis is blocked (Vallee and Okamoto 1995). Ultrastructural studies of shibire mutants show numerous invaginations with electron dense ring-like structures around a narrow neck (Koenig and Ikeda 1989). Dynamin has been shown to self-assemble into rings *in vitro* (Hinshaw and Schmid 1995), and similar structures which are immunopositive for dynamin form when synaptosomes are treated with GTPγS (Takei *et al.* 1995). Finally, it appears that separation of an endosomal pit from the plasma membrane is due to a concerted conformational change in the molecules forming the dynamin 'collar', (Sweitzer and Hinshaw 1999).

Various dynamin-interacting proteins containing SH3 domains, such as amphiphysins, endophilins, and syndapins, have also been implicated in endocytosis of calthrin-coated pits. Their roles include stimulation of GTPase activity of dynamin via SH3 domain binding, scaffolding structures, and targeting of dynamin to clathrin. In particular, interference with the normal function of endophilins impairs endocytosis and depletes SVs (Ringstad *et al.* 1999; Schmidt *et al.* 1999), while syndapins inhibit endocytosis through their SH3 domain interaction with dynamin, and may also link endocytosed vesicles with the actin substructure (Qualmann and Kelly 2000).

Vesicle loading by neurotransmitter transporters. Following exocytosis and membrane retrieval, newly formed vesicles must be reloaded with neurotransmitter. Two classes of neurotransmitter transport activities have been characterized at synapses. These are a family of plasma membrane transporters which move neurotransmitters from the synaptic cleft to the synaptic cytoplasm, and a distinct family of transporters which take up neurotransmitters from the cytoplasm and concentrate them in synaptic vesicles. The plasma membrane transporters were extensively characterized earlier than vesicular transporters (for review see Borowsky and Hoffman 1995) due to the difficulties involved in obtaining pure synaptic vesicles. Following improvements in isolation techniques, vesicular transport has been demonstrated in synaptic vesicles from various sources for many transmitters, including ACh (Parsons and Koenigsberger 1980; Toll and Howard 1980; Anderson *et al.* 1982), monoamines (Johnson *et al.* 1978), glutamate (Shioi *et al.* 1989; Tabb *et al.* 1992), GABA (Fykse and Fonnum 1988; Hell *et al.* 1988), and glycine (Kish *et al.* 1989). Vesicular transport is dependent on ATP and a proton gradient across the vesicle membrane (Naito and Ueda 1983; Maycox *et al.* 1988; Tabb *et al.* 1992). This dependence highlights the fact that two independent proteins are involved in vesicular neurotransmitter transport (Fig. 6.6).

The first protein needed is a proton pump which generates an electrochemical gradient across the vesicle membrane. The vesicular H^+-ATPase has been characterized and shown to be similar to the vacuolar H^+-ATPase (see Gluck 1993). The second protein necessary for vesicular neurotransmitter transport is a neurotransmitter-H^+ antiporter (see Schuldiner *et al.* 1995; Usdin *et al.* 1995). The transport activities of the vesicular monoamine transporter, and the vesicular GABA and glutamate transporters reside in single polypeptides (Maycox *et al.* 1988, 1990; Carlson *et al.* 1989; Hell *et al.* 1990).

The sequences of a number of vesicle transporters have been isolated. These proteins share a generally similar structure with a large number of transport proteins, including bacterial drug-resistance genes, sugar transporters, plasma membrane neurotransmitter transporters, and the vesicular membrane protein SV2, although sequence homology between these widely differing transporter types is limited. Two vesicular monoamine transporters (VMATs) have been identified (Erickson *et al.* 1992; Liu *et al.* 1992), which have overall identity of 62% (Schuldiner *et al.* 1995) and are encoded by distinct genes. Sequence and hydropathy analyses predict 521-amino acid proteins with 12 putative transmembrane segments which differ mostly between the first and second transmembrane domains and at the amino and carboxyl terminals. Functional studies of the two isoforms were performed in stable transfectants of Chinese hamster ovary cells and showed that VMAT2 has a higher affinity for all the monoamines (Peter *et al.* 1994). VMAT clones from species other than rat are more similar to VMAT2 than VMAT1 based on sequence comparisons (Vandenburg *et al.* 1992; Erickson and Eiden 1993; Peter *et al.* 1993; Howell *et al.* 1994).

A vesicular ACh transporter (AChT) cDNA has been isolated from *Torpedo marmorata, C. elegans,* rat, and human (Alfonso *et al.* 1993; Erickson *et al.* 1994). The clone predicts a protein of 532 amino acids, including 12 transmembrane domains, with 40 and 38% identity to VMAT1 and VMAT2, respectively (Usdin *et al.* 1995). Interestingly, the vesicular AChT is encoded within the same gene as an enzyme involved in ACh synthesis, choline acetyltransferase, suggesting that the regulation of transmitter synthesis and transport may be closely related (Usdin *et al.* 1995). It is not known whether similar mechanisms may be involved for other transmitters.

More recently, the sequences of a GABA transporter (vGAT) (McIntire *et al.* 1997), and two glutamate transporters (vGlut1 and vGlut2)(Fremeau *et al.* 2001) have also been identified. They possess 10 (vGAT) or 8 (vGlut1 and vGlut2) potential transmembrane domains. The closer sequence homology between the VMAT and VAChT vesicle transporters defines these proteins as a family, more distantly related to vGluts and vGAT, and to other transporters, including those neurotransmitter transporters found at the plasma membrane.

6.3 Learning and synapses

6.3.1 The neuronal software

The amount of transmitter released at a given synapse may change according to previous 'experience' of the synapse. This parameter is commonly defined as synaptic strength. It is clearly very tempting to relate this adaptive property of the synapse to high order functions of the nervous system. Indeed, learning has now been associated in a variety of systems and animal models with changes in the strength of connections between neurons (Hawkins *et al.* 1993). Such changes are often referred to as 'synaptic plasticity'. To date, two main types of synaptic plasticity have been identified. Although both types of plasticity can occur at the same synapses, they involve distinct cellular and molecular mechanisms. Short-term changes in synaptic efficiency involve strengthening of existing connections, possibly through modification of the structure and function of pre-existing proteins, such as ion channels, protein kinases, and receptors (Hawkins *et al.* 1993). The resulting effects include activation of postsynaptic receptors and modulation of transmitter release itself (see Chapter 14). These adaptations can occur rapidly and usually last for hours but sometimes for days. Long-term changes involve structural modifications of existing synapses or formation of new ones. This type of adaptation has been suggested as one of the possible cellular mechanisms that contributes to the storage of memory. Long-term changes seem to rely on the activation of gene expression and new protein synthesis (see Chapter 14).

Changes in the efficiency, size, and number of synapses imply that the adult nervous system is a dynamic network where new terminals can grow and where synapses can be activated, added, or removed in response to behavioral experience (Hawkins *et al.* 1993). Are there common effectors for short- and long-term synaptic plasticity? Are some of the mechanisms involved in synaptic plasticity similar to those that regulate axonal growth and synaptogenesis during development? Both questions are still unanswered, but transmitter release and growth of new connections have at least one mechanism in common: membrane fusion. Indeed, membrane fusion is necessary for vesicle secretion and it is also necessary in order to add new patches of membrane (a prerequisite for growth). The possible relevance of the mechanisms that control

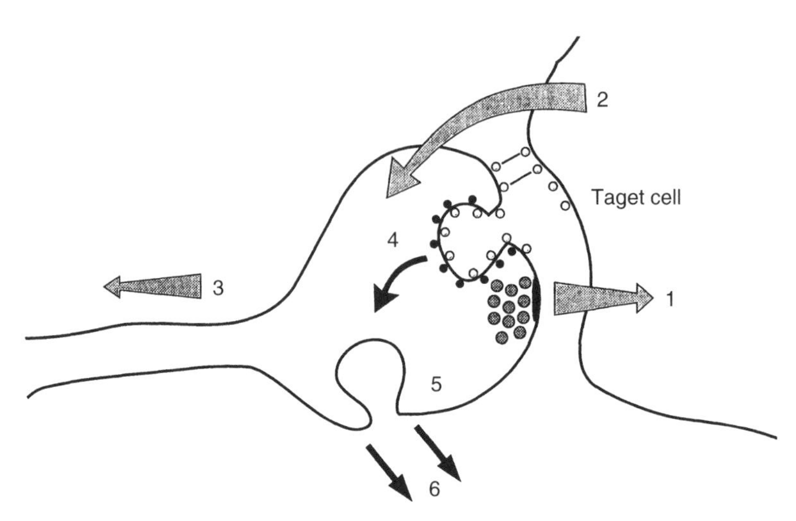

Fig. 6.7 Diagram of a 'remodeling' synapse illustrating the possible participation of membrane fusion events during structural changes associated with learning (based on Bailey *et al.* 1992; Hu *et al.* 1993). Stimulation of the presynaptic neuron increases the fusion-dependent release of transmitter from SVs (gray circles, 1). When the stimulus is sustained, a target-derived factor (2) induces changes in gene expression in the presynaptic cell (3). In turn, this activates the endosomal pathway (4) and leads to fusion-dependent redistribution of membrane components at the sites of new growth (5,6). Black circles: clathrin. Orange circles: adhesion molecules. Modified from *Trends Neurosci.*, vol. 17, pp. 368–373, with permission from Elsevier Trends Journals.

membrane fusion events in synaptic plasticity has been clearly demonstrated by the work of Kandel and his collaborators. Figure 6.7 summarizes the different stages that lead to a plastic response of a synapse and each stage that can involve membrane fusion is schematically illustrated. According to this model, modulation of transmitter release is the first step of a cascade leading to changes in gene expression and redistribution of membrane components that result in new synapse formation (Bailey *et al.* 1992; Hu *et al.* 1993). Each step of vesicle trafficking and fusion described in the previous chapters could contribute to regulated changes of transmitter release.

6.3.2 Functional synaptic plasticity

Altered efficacy of existing synapses is often referred to as functional synaptic plasticity. One of the most impressive example is long-term potentiation (LTP). Bliss and Lomo, discovered that brief high frequency trains of action potentials produced an increase of synaptic strength in synapses of the hippocampus that use glutamate as a transmitter (Bliss and Lomo 1973). This was the first demonstration that synaptic strength could change as a function of previous experience. LTP can last for hours or even days and seems to depend on both postsynaptic and presynaptic events. At the level of the postsynaptic cell, high levels of presynaptic stimulation result in the activation of glutamate receptors that are functionally blocked under normal conditions. At the level of the presynaptic cell, LTP involves an increase in transmitter release. This increase is likely to be at least partially the result of a retrograde signal originating from the postsynaptic cell. However, the possibility that covalent modifications of the proteins involved in vesicle trafficking and fusion could play a role is also under extensive study. For example, as mentioned in Section 6.2.4, studies on the synapsins indicate that vesicle mobilization may contribute to the number of vesicles available for release. Regulating synapsin function may therefore have an indirect effect on synaptic strength. Also, the function of proteins such as Rab3 and RIM1 is clearly important for some forms of LTP (see Section 6.2.5). Data by Staple and colleagues (Staple *et al.* 1995) suggest that the relative ratio of proteins involved at different stages of transmitter release can vary in different synaptic boutons of the same neuron and that this variability correlates with synaptic strength.

6.3.3 Morphological synaptic plasticity

Long-term memory has been clearly associated with changes in synaptic structure in a variety of experimental systems (Wallace *et al.* 1991). Quantitative ultrastructural analysis indicates that four basic morphological parameters of the synapse are strongly correlated. These are total vesicle number, active zone size, presynaptic bouton volume, and postsynaptic spine volume (Pierce and Mendell 1993; Pierce and Lewin 1994). Interestingly, co-ordinated increases in size of pre- and postsynaptic elements have been reported after LTP in the dentate gyrus. These experiments have shown expansion of the average active zone and apposed surface areas, as well as synaptic bouton volume (see Wallace *et al.* 1991). These increases imply the fusion and incorporation of new membrane into the synaptic structure. What are the mechanisms involved? There are at least two possibilities. Net changes in presynaptic surface could arise from any imbalance between the amount of membrane that is fused and then retrieved during transmitter release and vesicle recycling. Consistent with this, ultrastructural analysis of the synapses of the shibire mutant (the *Drosophila* homolog of dynamin, which has a defect in vesicle endocytosis, see Section 6.2.7) shows enlarged synapses. However, the observation that membrane retrieval is stimulus dependent (von Gersdorff and Matthews 1994) suggests that it may be regulated in response to previous signals received by the neuron. This would provide a mechanism of activity-dependent (or use-dependent or experience-dependent) control of membrane surface. The data showing that dynamin (see Section 6.2.7) is phosphorylated by PKC and dephosphorylated by electrical activity (van der Bliek and Mcycrowitz 1991; Robinson *et al.* 1993), is also consistent with this hypothesis.

An alternative way to increase synaptic surface would be the fusion of membrane vesicles other than SVs. Pfenninger and colleagues have shown that plasmalemmal expansion involves the fusion of large clear vesicles that were identified as plasmalemmal precursors (PPV) (Pfenninger *et al.* 1991). These vesicles accumulate in the growth cones of developing axons (Pfenninger and Friedman 1993). Interestingly, data obtained with a cell-free growth cone-expansion assay suggest that PPV fusion with the plasma membrane is controlled by Ca^{2+} influx (Lockerbie *et al.* 1991), thereby providing an activity-dependent mechanism of membrane expansion. If the same mechanism was maintained in adult neurons, growth of new connections, or morphological changes of existing ones, could be activated by stimulating the membrane-expansion pathway. If this hypothesis is correct, the cell must be capable of regulating vesicle fusion for release and vesicle fusion for expansion through distinct mechanisms. Of particular relevance to these observations, is the finding that SNAP-25 is involved in both transmitter release and in axonal growth. Inhibition of SNAP-25 expression with antisense oligonucleotides prevented neurite outgrowth in PC12 cells and cortical neurons *in vitro,* and in chick retina neurons *in vivo* (Osen-Sand *et al.* 1993). These effects could result from inhibition of transmitter release, but it is also possible that SNAP-25 is involved in the fusion of PPV. Very recent studies support the latter interpretation. Using clostridial toxins on primary neurons *in vitro,* Osen-Sand *et al.* (1996) showed that the fusion machineries for transmitter release and axonal growth involve common but also distinct SNARES. The v-SNARE VAMP was found to be exclusively involved in transmitter release, whereas the t-SNARES SNAP-25 and syntaxin have a role in release and growth. These data suggest that transmitter release (and SV fusion) is not necessary for axonal growth and that additional v-SNARES must be involved in vesicle fusion for membrane expansion.

In addition to changes in synaptic size, the structural rearrangements that occur during morphological synaptic plasticity involve the formation of new synapses and are likely to involve the co-ordinated action of genes that regulate membrane expansion as well as vesicle storage and fusion. Consistent with this, candidate plasticity genes encoding products involved in the endo- and exocytotic pathways have been isolated following stimulation of *N*-methyl-D-aspartate receptors in rat hippocampus (Nedivi *et al.* 1993), during long-term facilitation *in Aplysia* (Hu *et al.* 1993) and during synapse formation in the developing chick retina (see Osen-Sand *et al.* 1993, and in preparation). These findings demonstrate the importance of the fusion machinery during nerve terminal remodeling in the adult and suggest that the mechanisms involved are also at work during development. Studies of SNAP-25 isoform expression support this hypothesis. SNAP-25 exists in at least two alternatively spliced variants, differentially regulated during development (Bark 1993; Bark and Wilson 1994*a*; Bark *et al.* 1995; Boschert *et al.* 1996). SNAP25a mRNA is highly and transiently expressed during axonal growth, while SNAP25b is induced during synapse formation and its levels are maintained throughout adulthood (Bark 1993; Bark and Wilson 1994*b*; Bark *et al.* 1995; Boschert *et al.* 1996). Further to these observations, Boschert and colleagues (1996) have shown that SNAP-25a (but not b) expression is induced in the adult hippocampus following lesions that are known to induce reactive sprouting. Although indirect, these observations suggest that the two SNAP-25 isoforms may have different roles in transmitter release and membrane expansion. The regulated expression of the two SNAP-25 isoforms, and of additional SNARES, may provide a molecular framework for differential use of the fusion machinery during specific stages of maturation of the terminal. However, Roberts *et al.* (1998) have demonstrated increased SNAP-25 gene expression in hippocampal granule cells following the induction of LTP in the afferent synapses, with both isoforms being affected.

6.4 **Concluding remarks**

Fusion and re-uptake of specialized membrane is involved in a number of processes which affect neuronal morphology and function including, intracellular vesicle trafficking, neurotransmitter release, and membrane expansion. We are only beginning to understand the molecular mechanisms of exocytosis/endocytosis which occur at synapses and the role of regulation of these steps on complex cellular behaviors. Our increasingly better understanding of these phenomena will have considerable implications for the molecular basis of synaptic adaptation.

References

Alfonso, K., Grundahl, K., Duerr, J. S., Han, H. P., and Rand, J. B. (1993) The *Caenorhabditis elegans* unc-17 gene — *a* putative vesicular acetylcholine transporter. *Science*, **261**, 617–19.

Almers, W. (1994) Synapses. How fast can you get? *Nature*, **367**, 682–3.

Anderson, D. C., King, S. C., and Parsons, S. M. (1982) Proton gradient linkage to active uptake of [^{3}H]acetylcholine by *Torpedo* electric organ synaptic vesicles. *Biochemistry*, **21**, 3037–43.

Bähler, M. and Greengardm P. (1987) Synapsin I bundles F-actin in *a* phosphorylation-dependent manner. *Nature*, **326**, 704–7.

Bailey, C. H., Chen, M., Keller, F., and Kandel, E. R. (1992) Serotonin-mediated endocytosis of apCAM, an early step of learning-related synaptic growth in *Aplysia*. *Science*, **256**, 645–9.

Bark, I. C. (1993) Structure of the chicken gene for SNAP-25 reveals duplicated exon encoding distinct isoforms of the protein. *J. Mol. Biol.*, **233**, 67–76.

Bark, I. C. and Wilson, M. C. (1994*a*) Human cDNA clones encoding two different isoforms of the nerve terminal protein SNAP-25. *Gene*, **139**, 291–2.

Bark, I. C. and Wilson, M. C. (1994*b*) Regulated vesicular fusion in neurons: snapping together the details. *Proc. Natl Acad. Sci. USA*, **91**, 4621–4.

Bark, I. C., Hahn, K. M., Ryabinin, A. E., and Wilson, M. C. (1995) Differential expression of SNAP-25 protein isoforms during divergent vesicle fusion events of neural development. *Proc. Natl Acad. Sci. USA*, **92**, 1510–14.

Benfenati, F., Bähler, M., Jahn, R., and Greengard, P. (1989) Interaction of synapsin I with small synaptic vesicles: distinct sites in synapsin I bind to vesicle phospholipids and vesicle proteins. *J. Cell Biol.*, **108**, 1863–72.

Bennett, M. K., Calakos, N., and Scheller, R. H. (1992) Syntaxin: a synaptic protein implicated in docking of synaptic vesicles at presynaptic active zones. *Science*, **257**, 255–9.

Betz, A., Ashery, U., Rickmann, M., Augustin, I., Neher, E., Sudhof, T. C., *et al.* (1998) Munc-13 is a presynaptic phorbol ester receptor that enhances neurotransmitter release. *Neuron*, **21**, 123–36.

Betz, A., Thakur, P., Junge, H., Ashery, U., Rhee, J. S., Scheuss, V., *et al.* (2001) Functional interaction of the active zone proteins Munc13-1 and RIM1 in synaptic vesicle priming. *Neuron*, **30**, 183–96.

Blasi, J., Chapman, E. R., Link, E., Binz, T., Yamasaki, S., De Camilli, P., *et al.* (1993) Botulinum neurotoxin A selectively cleaves the synaptic protein SNAP-25. *Nature*, **365**, 160–3.

Bliss, T. V. and Lomo, T. (1973) Long-lasting potentiation of synaptic transmission in the dentate area of the anaesthetized rabbit following stimulation of the perforant path. *J. Physiol (Land.)*, **232**, 331–56.

Borowsky, B. and Hoffman, B. J. (1995) Neurotransmitter transporters: molecular biology, function, and regulation. *Int. Rev. Neurobiol.*, **38**, 139–99.

Boschert, U., O'Shaughnessy, C., Dickenson, R., Tessari, M., Bendotti, C., Catsicas, S., *et al.* (1996) Developmental and plasticity-related differential expression of two SNAP-25 isoforms in the rat brain. *J. Conp. Neurol.*, **367**, 177–93.

Brose, N., Petrenko, A. G., Südhof, T. C., and Jahn, R. (1992) Synaptotagmin: a calcium sensor on the synaptic vesicle surface. *Science*, **256**, 1021–5.

Carlson, M. D., Kish, P. E., and Ueda, T. (1989) Characterization of the solubilized and reconstituted ATP-dependent vesicular glutamate uptake system. *J. Biol. Chem.*, **264**, 7369–76.

Castillo, P. E., Schoch, S., Schmitz, F., Sudhof, T. C., and Malenka, R. C. (2002) RIM1 alpha is required for presynaptic long-term potentiation. *Nature*, **415**, 327–30.

Catsicas, S., Grenningloh, G., and Pich, E. M. (1994) Nerve-terminal proteins; to fuse to learn. *Trends Neurosci.*, **17**, 368–73.

Ceccaldi, P., Grohovaz, F., Benfenati, F., Chieregattim E., Greengard, P., and Valtorta, F. (1995) Dephosphorylated synapsin I anchors synaptic vesicles to actin cytoskeleton: an analysis by videomicroscopy. *J. Cell Biol.*, **128**, 905–12.

Chapman, E. R., An, S., Barton, N., and Jahn, R. (1994) SNAP-25, a t-SNARE which binds to both syntaxin and synaptobrevin via domains that may form coiled coils. *J. Biol. Chem.*, **269**, 27427–32.

Chin, L. S., Li, L., Ferreira, A., Kosik, K. S., and Greengard, P. (1995) Impairment of axonal development and of synaptogenesis in hippocampal neurons of synapsin I-deficient mice. *Proc. Natl. Acad. Sci. USA*, **92**, 9230–4.

De Camilli, P. and Takei, K. (1996) Molecular mechanisms in synaptic vesicle endocytosis and recycling. *Neuron*, **16**, 481–6.

De Camilli, P., Harris, S. M., Huttner, W. B., and Greengard, P. (1983) Synapsin I (protein I), a nerve terminal-specific phosphoprotein. II. Its specific association with synaptic vesicles demonstrated by immunocytochemistry in agarose-embedded synaptosomes. *J. Cell Biol.*, **96**, 1209–11.

De Camilli, P., Benfenati, F., Valtorta, F., and Greengard, P. (1990) The synapsins. *Annu. Rev. Cell Biol.*, **6**, 433–60.

DiAntonio, A., Parfitt, K. D., and Schwarz, T. L. (1993) Synaptic transmission persists in synaptotagmin mutants of *Drosophila*. *Cell*, **73**, 1281–90.

Duc, C. and Catsicas, S. (1995) Ultrastructural localization of SNAP-25 within the rat spinal cord and peripheral nervous system. *J. Comp. Neurol.*, **357**, 1–12.

Duncan, R. R., Betz, A., Shipston, M. J., Brose, N., and Chow, R. H. (1999) Transient, phorbol ester-induced DOC2-Munc13 interactions in vivo *J. Biol. Chem.*, **274**, 27347–50 (see also ibid **275**, 2246).

Eccles, J. C. (1937) Synaptic and neuro-muscular transmission. *Physiol. Rev.*, **17**, 538–55.

Erickson, J., Eiden, L., and Hoffman, B. (1992) Expression cloning of a reserpine-sensitive vesicular monoamine transporter. *Proc. Natl. Acad. Sci. USA*, **89**, 10993–7.

Erickson, J. and Eiden, L. (1993) Functional identification and molecular cloning of a human brain vesicle monoamine transporter. *J. Neurochem.*, **61**, 2314–17.

Erickson, J. D., Varoqui, H., Schafer, M. K. H., Modi, W., Diebler, M. F., Weihe, E., *et al.* (1994) Functional identification of a vesicular acetylcholine transporter and its expression from a 'cholinergic' gene locus. *J. Biol. Chem.*, **269**, 21929–32.

Fatt, P. and Katz, B. (1951) An analysis of the end-plate potential recorded with an intracellular electrode. *J. Physiol. (Land.)*, **115**, 320–70.

Ferreira, A., Kosik, K. S., Greengard, P., and Han, H-Q. (1994) Aberrant neurites and synaptic vesicle protein deficiency in synapsin II-depleted neurons. *Science*, **264**, 977–9.

Ferreira, A., Han, H-Q., Greengard, P., and Kosik, K. S. (1995) Suppression of synapsin II inhibits the formation and maintenance of synapses in hippocampal culture. *Proc. Natl. Acad. Sci. USA*, **92**, 9225–9.

Ferro-Novick, S. and Jahn, R. (1994) Vesicle fusion from yeast to man. *Nature*, **370**, 191–3.

Forn, J. and Greengard, P. (1978) Depolarizing agents and cyclic nucleotides regulate the phosphorylation of specific neuronal proteins in rat cerebral cortex slices. *Proc. Natl. Acad. Sci. USA*, **75**, 5195–9.

Foster, M. (1897) A *Text-book of Physiology*. Macmillan, London.

Fremeau, R. T., Troyer, M. D., Pahner, I., Nygaard, G. O., Tran, C. H., Reimer, R. J., *et al.* (2001) The expression of vesicular glutamate transporters defines two classes of excitatory synapse. *Neuron*, **31**, 247–60

Fujita, Y., Shirataki, H., Sakisaka, T., Asakura, T., Ohya, T., Kotani, H., *et al.* (1998) Tomosyn: a syntaxin-ninding protein that forms a novel complex in the neurotransmitter release process. *Neuron*, **25**, 905–15.

Fukuda, M., Moreira, J. E., Lewis, F. M. T., Sugimori, M., Niinobe, M., Mikoshiba, K., *et al.* (1995) Role of the C2B domain of synaptotagmin in vesicular release and recycling as determined by specific antibody injection into the squid giant synapse preterminal. *Proc. Natl. Acad. Sci. USA*, **92**, 10708–12.

Fykse, E. M. and Fonnum, F. (1988) Uptake of gamma aminobutyric acid by a synaptic vesicle fraction isolated from rat brain. *J. Neurochem.*, **50**, 1237–42.

Garcia, E. P., McPherson, P. S., Chilcote, T. J., Takei, K., and DeCamilli, P. (1995) rbSeclA and B colocalize with syntaxin 1 and SNAP-25 throughout the axon, but are not in a stable complex with syntaxin. *J. Cell Biol.*, **129**, 105–20.

Gluck, S. L. (1993) Thc vacuolar H(+) ATPascs: vcrsatile proton pumps participating in constitutive and specialized functions of eukaryotic cells. *Int. Rev. Cytol.*, **137**, 105–37.

Greengard, P., Valtorta, F., Czernik, A. J., and Benfenati, F. (1993) Synaptic vesicle phosphoproteins and regulation of synaptic function. *Science*, **259**, 780–5.

Hackett, J. T., Cochran, S. L., Greenfield Jr, L. J., Brosius, D. C., and Ueda, T. (1990) Synapsin I injected presynaptically into goldfish mauthner axons reduces quantal synaptic transmission. *J. Neurophysiol.*, **63**, 701–6.

Han, H-Q., Nichols, R. A., Rubin, M. R., Bähler, M., and Greengard, P. (1991) Induction of formation of presynaptic terminals in neuroblastoma cells by synapsin IIb. *Nature*, **349**, 697–700.

Hawkins, R. D., Kandel, E. R., and Siegelbaum, S. A. (1993) Learning to modulate transmitter release: themes and variations in synaptic plasticity. *Annu. Rev. Neurosci.*, **16**, 625–65.

Hayashi, T., McMahon, H., Yamasaki, S., Binz, T., Hata, Y., Südhof, T. C., *et al.* (1994) Synaptic vesicles membrane fusion complex: action of clostridial neurotoxins on assembly. *EMBO J.*, **13**, 5051–61.

Hell, J. W., Maycox, P. R., Stadler, H., and Jahn, R. (1988) Uptake of GABA by rat brain synaptic vesicles isolated by a new procedure. *EMBO J.*, **7**, 3023–9.

HeII, J. W., Maycox, P. R., and Jahn, R. (1990) Energy dependence and functional reconstitution of the gamma-aminobutyric acid carrier from synaptic vesicles. *J. Biol. Chem.*, **265**, 2111–17.

Heuser, J. E. and Reese, T. S. (1973) Evidence for recycling of synaptic vesicle membrane during transmitter release at the frog neuromuscular junction. *J. Cell Biol.*, **57**, 315–44.

Hinshaw, J. E. and Schmidt, S. L. (1995) Dynamin self-assembles into rings suggesting a mechanism for coated vesicle budding. *Nature*, **374**, 190–2.

Hu, Y., Barzulai, A., Chen, M., Bailey, C. H., and Kandel, E. R. (1993) 5-HT and cAMP induce the formation of coated pits and vesicles and increase the expression of clathrin light chain in sensory neurons of aplysia. *Neuron*, **10**, 921–9.

Huttner, W. B., Schiebler, W., Greengard, P., and De Camilli, P. (1983) Synapsin I (Protein I), a nerve terminal-specific phosphoprotein. III. Its association with synaptic vesicles studied in a highly purified synaptic vesicle preparation. *J. Cell Biol.*, **96**, 1374–88.

Johnson, R. G., Carlson, N. J., and Scarpa, A. (1978) Delta pH and catecholamine distribution in isolated chromaffin granules. *J. Biol. Chem.*, **253**, 1512–21.

Jorgensen, E. M., Hartwieg, E., Schuske, K., Nonet, M. L., Jin, Y., and Horwitz, H. R. (1995) Defective recycling of synaptic vesicles in synaptotagmin mutants of *Caenorhabditis elegans. Nature*, **378**, 196–9.

Kelly, R. B. (1993) Storage and release of neurotransmitters. *Cell*, **10**, 43–53.

Kish, P. E., Fischer-Bovenkerk, C., and Ueda, T. (1989) Active transport of gamma aminobutyric acid and glycine into synaptic vesicles. *Proc. Natl. Acad. Sci. USA*, **86**, 3877–81.

Koenig, J. H. and Ikeda, K. (1989) Disappearance and reformation of synaptic vesicle membrane upon transmitter release observed under reversible blockage of membrane retrieval. *J. Neurosci.*, **9**, 3844–60.

Li, C., Ullrich, B., Zhang, J. Z., Anderson, R. G., Brose, N., and Südhof, T. C. (1995) Ca^{++}-dependent and independent activities of neural and nonneural synaptotagmins. *Nature*, **375**, 594–9.

Lin, J. W., Sugimori, M., Llinás, R., McGuinness, T. L., and Greengard, P. (1990) Effects of synapsin I and calcium/calmodulin-dependent protein kinase II on spontaneous neurotransmitter release in the squid giant synapse. *Proc. Natl. Acad. Sci. USA*, **87**, 8257–61.

Liu, Y., Peter, D., Roghani, S., Schuldiner, G., Prive, D., Eisenberg, N., *et al.* (1992) A cDNA that suppresses MPP^+ toxicity encodes a vesicular amine transporter. *Cell*, **70**, 539–51.

Llinás, R., McGuiness, T. L., Leonard, C. S., Sugimori, M., and Greengard, P. (1985) Intraterminal injection of synapsin I or calcium/calmodulin-dependent kinase II alters neurotransmitter release at the squid giant synapse. *Proc. Natl. Acad. Sci. USA*, **82**, 3035–9.

Llinás, R., Gruner, J. A., Sugimori, M., McGuinness, T., and Greengard, P. (1991) Regulation of synapsin I and Ca^{++} calmodulin-dependent protein kinase II of the transmitter release in squid giant synapse. *J. Physiol. (London)*, **436**, 257–82.

Lockerbie, R. O., Miller, V. E., and Pfenninger, K. H. (1991) Regulated plasmalemmal expansion in nerve growth cones. *J. Cell Biol.*, **112**, 1215–27.

Lu, B., Greengard, P., and Poo, M-M. (1992) Exogenous synapsin I promotes functional maturation of developing neuromuscular synapses. *Neuron*, **8**, 521–9.

Martin, T. F. J. (2002) Prime movers of synaptic vesicle exocytosis, *Neuron*, **34**, 9–12.

Masada, E. S., Huang, B. C., Fisher, J. M., Luo, Y., and Scheller, R. H. (1998) Tomosyn binds t-SNARE proteins via a VAMP-like coiled coil. *Neuron*, **21**, 479–90.

Matthew W. D., Tsavaler, L., and Reichardt, L. F. (1981) Identification of a synaptic vesicle-specific membrane protein with a wide distribution in neuronal and neurosecretory tissue. *J. Cell Biol.*, **91**, 257–69.

Maycox, P. R., Deckwerth, T., Hell, J. W., and Jahn, R. (1988) Glutamate uptake by brain synaptic vesicles. Energy dependence of transport and functional reconstitution in proteoliposomes. *J. Biol. Chem.*, **263**, 15423–8.

Maycox, P. R., Deckwerth, T., and Jahn, R. (1990) Bacteriorhodopsin drives the glutamate transporter of synaptic vesicles after co-reconstitution. *EMBO J.*, **9**, 1465–9.

Maycox, P. R., Link, E., Reetz, A., Morris, S. A., and Jahn, R. (1992) Clathrin-coated vesicles in nervous tissues are involved primarily in synaptic vesicle recycling. *J. Cell Biol.*, **118**, 1379–88.

McIntire, S. L., Reimer, R. J., Schuske, K., Edwards, R. H., and Jorgensen, E. M. (1997) Identification and characterization of the vesicular GABA transporter. *Nature*, **389**, 870–6.

Montecucco, C. and Schiavo, G. (1994) Mechanism of action of tetanus and botulinum neurotoxins. *Mol. Microbiol.*, **13**, 1–8.

Naito, S. and Ueda, T. (1983) Adenosine triphosphate-dependent uptake of glutamate into protein I-associated synaptic vesicles. *J. Biol. Chem.*, **258**, 696–9.

Nedivi, E., Hevroni, D., Naot, D., Israeli, D., and Citri, Y. (1993) Numerous candidate plasticity-related genes revealed by differential cDNA cloning. *Nature*, **363**, 718–22.

Nestler, E. and Greengard, P. (1982) Distribution of protein I and regulation of its state of phosphorylation in the rabbit superior cervical ganglion. *J. Neurosci.*, **2**, 1011–23.

Nichols, R. A., Chilcote, T. J., Czernik, A. J., and Greengard, P. (1992) Synapsin I regulates glutamate release from rat brain synaptosomes. *J. Neurochem.*, **58**, 783–5.

Nonet, M. L., Grundahl K., Meyer, B. J., and Rand, J. B. (1993) Synaptic function is impaired but not eliminated in *C. elegans* mutants lacking synaptotagmin. *Cell*, **73**, 1291–305.

Obar, R. A., Collins, C. A., Hammarback, J. A., Shpetner, H. S., and Vallee, R.B. (1990) Molecular cloning of the microtubule-associated mechanochemical enzyme dynamin reveals homology with a new family of GTP-binding proteins. *Nature*, **347**, 256–61.

Osen-Sand, A., Catsicas, M., Staple, J. K., Jones, K. A., Ayala, G., Knowles, J., *et al.* (1993) Inhibition of axonal growth by SNAP-25 antisense oligonucleotides *in vitro* and *in vivo*. *Nature*, **364**, 445–8.

Osen-Sand, A., Staple, J. K., Naldi, E., Schiavo, G., Rossetto, O., Petitpierre, S., *et al.* (1996) Common and distinct fusion proteins in axonal growth and transmitter release. *J. Comp. Neurol.*, **367**, 222–34.

Oyler, G. A., Higgins, G. A., Hart, R. A., Battenberg, E., Billingsley, M., Bloom, F. E., *et al.* (1989) The identification of a novel synaptosomal-associated protein, SNAP-25, differentially expressed by neuronal subpopulations. *J. Cell Biol.*, **109**, 3039–52.

Palade, G. E. and Palay, S. L. (1954) Electron microscope observations of interneuronal and neuromuscular synapses. *Anat. Rec.*, **118**, 335–6.

Parsons, S. M. and Koenigsberger, R. (1980) Specific stimulated uptake of acetylcholine by *Torpedo* electric organ synaptic vesicles. *Proc. Natl. Acad. Sci. USA*, **77**, 6234–8.

Peter, D., Jimenez, J., Liu, Y., Kim, J., and Edwards, R. H. (1994) The chromaffin granule and synaptic vesicle amine transporters differ in substrate recognition and sensitivity to inhibitors. *J. Biol. Chem.*, **269**, 7231–7.

Peter, D., Liu, Y., Sternini, C., de Giorgio, R., Brecha, N., and Edwards, R. H. (1995) Differential expression of two vesicular monoamine transporters. *J. Neurosci.*, **15**, 6179–88.

Petrucci, T. C. and Morrow, J. S. (1987) Synapsin I: an actin-bundling protein under phosphorylation control. *J. Cell Biol.*, **105**, 1355–63.

Pellegrini, L. L, O'Connor, V., Lottspeich, F., and Betz, H. (1995) Clostridial neurotoxins compromise the stability of a low energy SNARE complex mediating NSF activation of synaptic vesicle fusion. *EMBO J.*, **14**, 4705–13.

Pevsner, J. and Scheller, R. H. (1994) Mechanisms of vesicle docking and fusion: insights from the nervous system. *Curr. Opin. Cell Biol.*, **6**, 555–60.

Pfenninger, K. H. and Friedman, L. B. (1993) Sites of plasmalemmal expansion in growth cones. *Dev. Brain Res.*, **71**, 181–92.

Pfenninger, K. B., de la Houssaye, B. A., Frame, L., Helmke, S., Lockerbie, R. O., Lohse, K., *et al.* (1991) In: *The Growth Cone* (ed. P.C. Letourneau, S. B. Kater, E. R. Macagno). Raven Press, New York, pp. 111–23.

Pierce, J. P. and Lewin, G. R. (1994) An ultrastructural size principle. *Neuroscience*, **58**, 441–6.

Pierce, J. P. and Mendell, L. M. (1993) Quantitative ultrastructure of Ia boutons in the ventral horn: scaling and positional relationships. *J. Neurosci.*, **13**, 4748–63.

Qualmann, B. and Kelly, R. B. (2000) Syndapin isoforms participate in receptor-mediated endocytosis and actin organisation. *J. Cell Biol.*, **148**, 1047–61.

Rhee, J.-S., Betz, A., Pyott, S., Varoqueaux, F., Augustin, I., Hesse, D., *et al.* (2002) Beta phorbol ester- and diacylgylcerol-induced augmentation of transmitter release is mediated by Munc 13s and not by PKCs. *Cell*, **108**, 121–33.

Ringstad, N., Gad, H., Low, P., Dipaolo, G., Brodin, L., Shupliakov, O., *et al.* (1999) Endophilin/SH3p4 is required for the transition from early to late stages in clathrin-mediated synaptic vesicle endocytosis. *Neuron*, **24**, 143–54.

Roberts, L. A., Morris, B. J., and O'Shaughnessy, C. T. (1998) Involvement of two isoforms of SNAP-25 in the expression of long-term potentiation. *Neuroreport*, **9**, 33–6.

Robertson, J. D. (1987) The early days of electron microscopy of nerve tissue and membranes. *Int. Rev. Cytol.*, **100**, 129–201.

Robinson, P. J., Sontag, J. M., Liu, J. P., Fykse, E. M., Slaughter, C., McMahon, H., *et al.* (1993) Dynamin GTPase regulated by protein kinase C phosphorylation in nerve terminals. *Nature*, **365**, 163–6.

Rosahl, T. W., Spillane, D., Missler, M., Herz, J., Selig, D., Wolff, J. R., *et al.* (1995) Essential function of synapsins I and II in synaptic vesicle regulation. *Nature*, **375**, 488–93.

Rothman, J. E. and Warren, G. (1994) Implications of the SNARE hypothesis for intracellular membrane topology and dynamics. *Curr. Biol.*, **4**, 220–33.

Schiavo, G., Benfenati, F., Poulain, B., Rossetto, O., Polverino de Laureto, P., Das Gupta, B. R., *et al.* (1992) Tetanus and botulinum-B neurotoxins block neurotransmitter release by proteolytic cleavage of synaptobrevin. *Nature*, **359**, 832–5.

Schmidt, A., Wolde, M., Thiele, C., Fest, W., Kratzin, H., Podtelejnikov, A. V., *et al.* (1999) Endophilin 1 mediates synaptic vesicle formation by transfer of arachidonate to lysophosphatidic acid. *Nature*, **41**, 133–41

Schuldiner, S., Shirvan, A., and Linial, M. (1995) Vesicular neurotransmitter transporters: from bacteria to humans. *Physiol. Rev.*, **75**, 369–92.

Schulze K. L., Broadie K., Perin M. S., and Bellen H. J. (1995) Genetic and electrophysiological studies of *Drosophila* syntaxin 1A demonstrate its role in nonneuronal secretion and neurotransmission. *Cell*, **80**, 311–20.

Schiebler, W., Jahn, R., Doucet, J.-P., Rothlein, J., and Greengard, P. (1986) Characterization of synapsin I binding to small synaptic vesicles. *J. Biol. Chem.*, **261**, 8383–90.

Schweizer, F. E., Betz, H., and Augustine, G. J. (1995) From vesicle docking to endocytosis: intermediate reactions of exocytosis. *Neuron*, **14**, 689–96.

Shpetner, H. S. and Vallee, R. B. (1989) Identification of dynamin, a novel mechanochemical enzyme that mediates interactions between microtubules. *Cell*, **59**, 421–32.

Shioi, J., Naito, S., and Ueda, T. (1989) Glutamate uptake into synaptic vesicles of bovine cerebral cortex and electrochemical potential difference of protons across the membrane. *Biochem. J.*, **258**, 499–504.

Sihra, T. S., Wang, J. K. T., Gorelick, F. S., and Greengard, P. (1989) Translocation of synapsin I in response to depolarization of isolated nerve terminals. *Proc. Natl. Acad. Sci. USA*, **86**, 8108–12.

Söllner, T., Bennett, M. K., Whiteheart, S. W., Scheller, R. H., and Rothman, J. E. (1993) A protein assembly–diassembly pathway *in vitro* that may correspond to sequential steps of synaptic vesicle docking, activation, and fusion. *Cell*, **75**, 409–18.

Staple, J. K., Osen-Sand, A., Benfenati, F., Merlo Pich, E., and Catsicas, S. (1995) Molecular and functional diversity of individual nerve terminals of isolated cortical neurons. *Sci. Neurosci. Abstr.*, 21, 331.

Südhof, T. C. (1995) The synaptic vesicle cycle: a cascade of protein–protein interactions. *Nature*, **375**, 645–53.

Südhof, T. C., Czernik, A. J., Kao, H.-T., Takei, K., Johnston, P. A., Horiuchi, A., *et al.* (1989) Synapsins: mosaics of shared and individual domains in a family of synaptic vesicle phosphoproteins. *Science*, **245**, 1474–80.

Sweitzer, S. M. and Hinshaw, J. E. (1998) Dynamin undergoes a GTP-dependent conformational change causing vesiculation. *Cell*, **93**, 1021–9.

Tabb, J., Kish, P., Vandyke, R., and Ueda, T. (1992) Glutamate transport into synaptic vesicles — roles of membrane potential, pH gradient, and intravesicular pH. *J. Biol. Chem.*, **267**, 15 412–18.

Takei, K., McPherson, P. S., Schmid, S. L., and DeCamilli, P. (1995) Tubular membrane invaginations coated by dynamin rings are induced by GTP-gS in nerve terminals. *Nature*, **374**, 186–90.

Thiel, G., Südhof, T. C., and Greengard, P. (1990) Synapsin II: mapping of a domain in the NH_2-terminal region which binds to small synaptic vesicles. *J. Biol. Chem.*, **265**, 16 527–33.

Tokumaru, H. and Augustine, G. J. (1999) Unc-13 and neurotransmitter release. *Nat. Neurosci.*, **2**, 929–30.

Toll, L. and Howard, B. D. (1980) Evidence that an ATPase and a protonmotive force function in the transport of acetylcholine into storage vesicles. *J. Biol. Chem.*, **255**, 1787–89.

tomDieck, S., Sanmarti-Vila, L., Langnaese, K., Richter, K., Kindler, S., Soyke, A., *et al.* (1998) Bassoon, a novel zinc-finger CAG/glutamine repeat protein selectively localised at the active zone of presynaptic nerve terminals. *J. Cell Biol.*, **142**, 499–509.

Trifaró, J. M. and Vitale, M. L. (1993) Cytoskeleton dynamics during neurotransmitter release. *Trends Neurosci.*, **16**, 466–72.

Trimble, W. S., Cowan D. M., and Scheller R. H. (1988) *Proc. Natl. Acad. Sci. USA*, **85**, 4538–42.

Usdin, T. B., Eiden, L. E., Bonner, T. I., and Erickson, J. D. (1995) Molecular biology of the vesicular ACh transporter. *Trends Neurosci.*, **18**, 218–24.

Vallee, R. B. and Okamoto, P. M. (1995) The regulation of endocytosis: identifying dynamin's binding partners. *Trends Cell Biol.*, **5**, 43–7.

Valtorta F., Villa A., Jahn R., De Camilli P., Greengard P., and Ceccarelli, B. (1988) Localization of synapsin I at the frog neuromuscular junction. *Neuroscience*, **24**, 593–603.

Valtorta, F., Greengard, P., Fesce, R., Chieregatti, E., and Benfenati, F. (1992) Effects of the neuronal phosphoprotein synapsin I on actin polymerization. I. Evidence for a phosphorylation-dependent nucleating effect. *J. Biol. Chem.*, **267**, 11 281–8.

Vandenbergh, D., Persico, A., and Uhl, G. (1992) A human dopamine transporter cDNA predicts reduced glycosylation, displays a novel repetitive element and provides racially-dimorphic Taqi RFLPs. *Mol. Brain Res.*, **15**, 161–6.

Van der Bliek, A. M. and Meyerowitz, E. M. (1991) Dynamin-like protein encoded by the *Drosophila shibire* gene associated with vesicular traffic. *Nature*, **351**, 411–4.

von Gersdorff, H. and Matthews, G. (1994) Dynamics of synaptic vesicle fusion and membrane retrieval in synaptic terminals. *Nature*, **367**, 735–9.

Wallace, C. S., Hawrylak, N., and Greenough, W. T. (1991) In: *Long-term Potentiation* (ed. M. Baudry, J.L. Davis). MIT Press, Cambridge, MA, pp. 189–32.

Wilson, D. W., Whiteheart, S. W., Wiedmann, M., Brunner, M., and Rothman, J. E. (1992) A multisubunit particle implicated in membrane fusion. *J. Cell Biol.*, **117**, 531–8.

Chapter 7

Molecular biology of postsynaptic structures

Flaminio Cattabeni, Fabrizio Gardoni, and Monica Di Luca

Synapses are specialized sites of communication between neurons in the brain that are vital for interneuronal signaling and necessary for the processing and integration of information. Efficient and plastic signal transduction at synapses is critical for the correct functioning of the synapse and information processing in the nervous system. The strength of individual central nervous system (CNS) synapses is thought to be controlled by signaling machinery that regulates the number and activity of postsynaptic receptor ion channels.

The neurotransmitter glutamate mediates the majority of excitatory synaptic transmission in the brain. Excitatory glutamatergic synapses feature a prominent thickening at the cytoplasmic surface of the postsynaptic membrane at sites of close opposition to the presynaptic terminal for which the term Post Synaptic Density (PSD) was coined. Electron-microscopy studies have identified in the 1950s the PSD as an electron-dense structure beneath the postsynaptic membrane in register with the active zone of the presynaptic compartment (Palay 1956). The thickness and density of PSD is variable and falls into two categories: type I, where PSD is electron dense and its size exceeds that of nerve terminals (excitatory glutamatergic synapse); it can be described as a kind of web adhering to the postsynaptic membrane; type II, where PSD is less electron-dense and its size is similar to the presynaptic thickening (GABAergic synapse).

7.1 Structural features and components of the excitatory PSD

Type I isolated PSD (and from now on the term PSD will be referred only to type I PSD) appears as semicircular bands about 400 nm long and 40 nm wide. In the 1970s two groups of cell biologists, one led by P. Siekevitz (Carlin *et al.* 1980), and the other by Cotman *et al.* (1974), developed methods for the isolation of subcellular fractions that were enriched in structures that appeared to be PSDs by morphological criteria.

Preparation of the PSD fraction is now a relatively simple procedure that involves first the purification of synaptosomes or synaptic membranes, followed by extraction of the membranes with detergents to dissolve lipids. The extraction leaves behind disk-shaped protein structures with the apparent shape, size, and morphology of the PSD.

PSDs isolated by Cotman *et al.* (1974) were purified from synaptosomal membranes washed with 3%-*N*-lauroyl sarcosinate, and contain approximately 10–15 protein bands visible on Coomassie blue stained polyacrylamide gels. The PSD fraction isolated by Siekevitz's group was purified from synaptosomes washed with 0.5% Triton X-100, a milder detergent than sarcosinate, and has a more complex protein composition, consisting of about 25–30 protein bands.

Biochemical fractionation showed that the PSD structure includes four major classes of components (see Table 7.1): i) *plasma membrane proteins* such as ionotropic and

Table 7.1

Protein	Binding partners
i) plasma membrane proteins	
AMPA	SAP-97, GRIP
mGluR	Homer
Neuroligin	SAP-97
NMDA	CaM, α-CaMKII, PSD-95
ii) cytoskeletal proteins	
Actin	α-actinin
α-actinin	NMDA, actin, CaMKII
Spectrin	
Tubulin	
iii) signaling proteins	
Calmodulin	CaMKII, NMDA
CaMKII	NMDA
Fyn-Src	PSD-95
IP3	Homer
nNOS	PSD-95
Ras	SynGap
SynGAP	PSD-95, Ras
iv) linker proteins	
GKAP	PSD-95, Shank
GRIP	AMPA
Homer	mGluR, IP3
PSD-95	NMDA, nNOS, GKAP, SynGAP
SAP-97	AMPA, Neuroligin
Shank	GKAP, Homer
Yotiao	NMDA, PP1, PKA

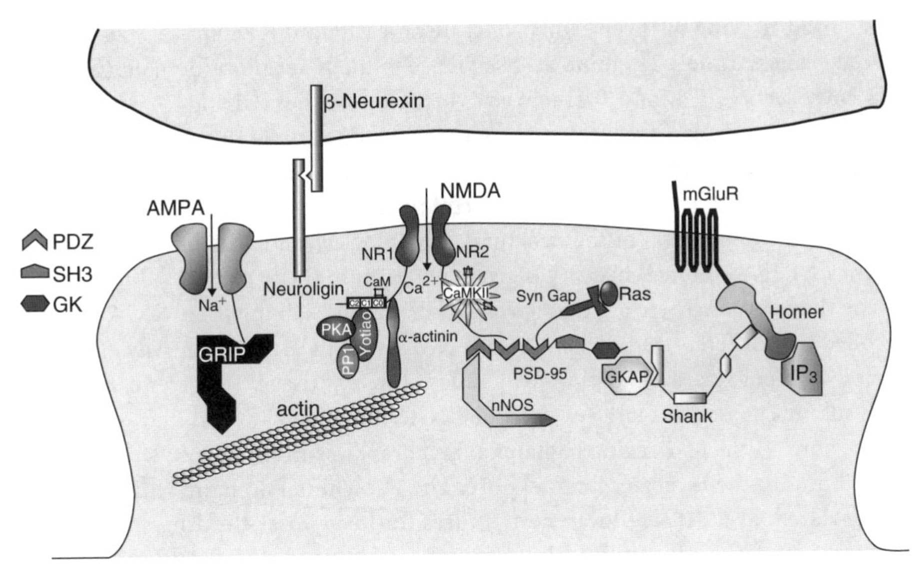

Fig. 7.1

metabotropic glutamate receptor subunits (Kennedy 2000), and neuroligins (Kennedy 1997), ii) *signalling proteins* such as Ca^{2+}/Calmodulin-dependent kinase II (CaMKII; Kennedy *et al.* 1983), tyrosine kinases, i.e. *Fyn* and *Src* (Elliss *et al.* 1988) and SynGap (Chen *et al.* 1998; Kim *et al.* 1998), iii) *cytoskeletal proteins* such as actin, spectrin, tubulin (Kelly and Cotman 1978; Carlin *et al.* 1980), and iv) *linker proteins* such as members of PSD-95/SAP family (Gomperts 1996; Sheng and Kim 1996). Hence the PSD contains the receptors with associated signalling and scaffolding proteins that organize signal transduction pathways near the postsynaptic membrane (Fig. 7.1).

7.1.1 Plasma membrane proteins

The ionotropic glutamate *N*-Methyl-D-Aspartate (NMDA) and 1-α-amino-3-hydroxy-5-methyl-4-isoxazole propionic acid (AMPA) type receptors are concentrated in the PSD of excitatory synapses (Kennedy 2000). More controversial is the presence in the PSD structure of mGluR1α and mGluR5 subunits of metabotropic glutamate receptors.

7.1.1.1 NMDA receptors

NMDA receptors are oligomeric complexes formed by the coassembly of members of 3 receptor subunit families: NR1, NR2 subfamily (NR2A–D; Hollman and Heinemann 1994), and NR3A (Das *et al.* 1998). One of these, the NR1 subunit, is a ubiquitous and necessary component of functional NMDA receptor channels. Diverse molecular forms of the NR1 subunit are present and generated by alternative RNA splicing;

differential splicing of three exons generate at least eight NR1 splice variants; the spliced exons encode a 21 amino acid sequence in the N-terminus domain (N1) and adjacent sequences of 37 and 38 amino acids in the C-terminus (C1 and C2 respectively). Splicing out the exon segment that encodes for the C2 cassette removes the first stop codon resulting in a new open reading frame that encodes an unrelated sequence of 22 amino acids (C2′) before a second stop codon is reached. These alternative splice processes cause alteration of the structural, physiological, and pharmacological properties of NR1. Additional diversity is given by the multiplicity of NR2 subunits composing the receptor. There are four species of the second subunit type, NR2A–D, each encoded by a different gene. The different NR2 subunits are differentially expressed during different stages of development and in different tissues (Monyer *et al.* 1994). Each subunit possesses a large extracellular N-terminal domain and four membrane (M) regions — the M-2 region contains a membrane-reentrant hairpin structure that contributes to the receptor channel pore. The N-terminal domain, which is large, glycosylated, and extracellular, contributes to the agonist-binding site. NMDA receptors bind two agonist ligands: glutamate and the coagonist, glycine. The NR1 subunit contains the binding site for glycine and the NR2 subunit, the binding site for glutamate.

NMDA receptors respond to agonists more slowly than AMPA receptors, and require greater than 2 ms to open. However, they have a higher affinity for glutamate, and their currents persist longer than AMPA receptor currents. The NMDA receptor admits both Na^+ and Ca^{2+} ions. At the resting potential of the cell, a Mg^{2+} ion blocks the NMDA receptor pore, but the Mg^{2+} ion is released from the pore upon cell depolarization. Therefore, opening of the channel requires binding of ligand and simultaneous depolarization of the cell. The channel thus operates as a coincidence detector that admits current only when agonist binding and cell depolarization take place simultaneously.

NMDA receptors display activity-dependent current decreases of several types, two of which are ligand dependent. One of these, glycine-dependent desensitization, occurs following receptor stimulation by glutamate when glycine concentrations are subsaturating, in the nanomolar range. The second, glycine-independent desensitization takes place in the presence of saturating glycine, at concentrations in the micromolar range. When glutamate binds to the receptor in the presence of low concentrations of glycine, the receptor rapidly enters a desensitized (low-conductance) state. This desensitizing transition is blocked by glycine. NMDA receptor desensitization may limit receptor currents during persistent stimulation by glutamate. Domains of NR2 that influence receptor desensitization characteristics have been defined. A different activity-dependent NMDA receptor current decrease is Ca^{2+}-dependent inactivation. Ca^{2+}-dependent inactivation may be induced by increases in intracellular $[Ca^{2+}]$ that follow activity-dependent fluxes of Ca^{2+} through the receptor. The increase in intracellular $[Ca^{2+}]$ triggers biochemical modifications of the receptor that decrease receptor mean opening time. It has recently been shown that Ca^{2+}-dependent inactivation results from binding of Ca^{2+}-calmodulin to the membrane proximal region of the C-terminal domain of the NR1 subunit.

7.1.1.2 AMPA receptors

AMPA receptors are complexes of four subunit types, GluR1–4, which may be homomeric or heteromeric. AMPA receptors account for the great majority of fast excitatory CNS synaptic transmission. AMPA receptors have a lower affinity for glutamate than NMDA receptors, and their currents are typically rapid, rising within less than 1 ms. AMPA receptor channels that contain GluR1, GluR3, or GluR4 subunits, or subunit GluR2 that is encoded by an unmodified GluR2 mRNA, can admit both Ca^{2+} ions and Na^{+} ions. RNA editing of GluR2 mRNA changes the structure of the GluR2 subunit by replacing glutamine with arginine at the "Q/R site" in the pore filter region, at the apex of the M2 hairpin. AMPA receptors containing GluR2 subunits encoded by edited mRNA are impermeable to Ca^{2+} ions. The effect of an edited subunit is dominant, such that inclusion of a single edited GluR2 subunit in an AMPA channel prevents Ca^{2+} entry through a receptor otherwise composed of GluR1, -3, -4, or unedited GluR2 subunits.

7.1.1.3 Metabotropic glutamate receptors

A third class of glutamate receptors present at many excitatory synapses is the metabotropic or heterotrimeric GTP-binding protein-linked glutamate receptors (mGluRs). Subtypes mGluR1 and mGluR5 are concentrated around the outer rim of glutamatergic PSDs as well as in decreasing concentration in the spine membrane as a function of distance from the PSD. A lattice of scaffold proteins may link the cytoplasmic face of mGluRs to IP3 receptors in the spine apparatus. The lattice may also be connected to PSD-95 and thus to the NMDA receptor complex.

In this aspect, the Shank protein family consists of proteins — core components of PSD structure — sharing a domain organization consisting of ankyrin repeats near the N-terminus followed by SH3 domain, PDZ domain, and a SAM domain at the C-terminus (Naisbitt *et al.* 1999). Consistent with a scaffolding function, Shank mediates multiple protein interactions: the PDZ domain binds the GKAP family of PSD-95 binding proteins, thereby linking Shank to NMDA receptor complexes, while two other distinct PDZ domains of Shank form the binding sites for Homer, which in turn binds mGluR1 and IP3 receptors. From these interactions Shank is acting as a glue between ionotropic and metabotropic receptors in the postsynaptic compartment.

7.1.2 **Signaling proteins**

PSDs contain different classes of enzymatic systems, most of which are responsible for regulating the phosphorylation state of several PSD substrates: CaMKII represents the most abundant signaling protein in the PSD fraction. There, the enzyme is ideally positioned to play a major role in synaptic plasticity events (Kennedy *et al.* 1983).

CaMKII is a target for transient Ca^{2+} entry through the NMDA channel and is necessary for normal synaptic plasticity in pyramidal neurons (Silva *et al.* 1992). CaMKII is a multisubunit protein having 8–12 subunits assembled in stochastic combinations from two closely related catalytic subunits, alpha and beta (Hanson and Schulman 1992). In the forebrain, the alpha subunit is about three times as abundant as the beta subunit.

A large body of evidence suggests that αCaMKII is a critical player in Long Term Potentiation (LTP), and it has special properties that make it an attractive candidate for exhibiting persistent changes and serving as a memory molecule (Lisman 1994). A simple and direct role for CaMKII in triggering and perhaps maintaining LTP is supported by studies in which CaMKII activity was acutely increased either with viral transfection (Pettit *et al.* 1994), or injection of calcium and calmodulin. In these cases synaptic transmission is enhanced and LTP is occluded. An important property of CaMKII is that when autophosphorylated on Thr286, its activity is no longer dependent on Ca^{2+}-calmodulin (CaM). This allows its activity to continue long after the Ca^{2+} signal has returned to baseline (see also Chapter 12). Biochemical studies have demonstrated that this autophosphorylation does in fact occur after triggering LTP (Liu *et al.* 1999). That CaMKII autophosphorylation is required for LTP was convincingly demonstrated by an elegant use of molecular genetic techniques in which replacement of endogenous CaMKII with a form of CaMKII containing a Thr286 point mutation was capable of shifting LTP towards LTD (Mayford *et al.* 1995). A final important piece of evidence implicating CaMKII in LTP is that it can directly phosphorylate the AMPA receptor subunit, GluR1, in situ, and this has been shown to occur following the generation of LTP.

The cytosolic tails of the NR2A/B subunits of the NMDA receptor bind to CaMKII and thus can serve as docking sites for it in the PSD (Gardoni *et al.* 1998; Strack and Colbran 1998). In addition, both the NR2A and NR2B subunits are phosphorylated by CaMKII (Omkumar *et al.* 1996; Gardoni *et al.* 2001). Docking of CaMKII to the tail of the NMDA receptor would position its catalytic domains near the receptor mouth, ideally located for activation by Ca^{2+} flowing through the channel.

In the PSD fraction, phosphotyrosine-mediated pathways interact both physically and biochemically with NMDA receptor (Tezuka *et al.* 1999). The NMDA receptor NR2B subunit is the major synaptic phosphotyrosine peptide of the excitatory synapse, although the precise effect of tyrosine phosphorylation on NR2B function at the synapse is not yet known.

However, the receptor associates via PSD-95 with Syn-GAP, a PSD protein with a GTPase-activating domain that induces hydrolysis of GTP in complexes with the G protein *Ras* (Chen *et al.* 1998). *Ras*, in its active, GTP-bound state transduces signals from a large number of tyrosine kinases. One of these, *Src*, is a nonreceptor tyrosine kinase that is abundant in the brain. A related kinase, *Fyn*, is also found in the PSD.

SynGAP is specifically expressed in neurons and is highly concentrated at synaptic sites in hippocampal neurons, where it is tightly colocalized with PSD-95 (Kim *et al.* 1998). SynGAP is almost as abundant in the PSD fraction as PSD-95 itself, suggesting that many synaptic PSD-95 molecules are bound to at least one copy of SynGAP.

The function of RasGAPs is to accelerate the intrinsic guanosine triphosphatase (GTPase) activity of *Ras*, thus accelerating the rate of inactivation of the GTP-bound form of *Ras*. Because the most common down-stream effect of GTP-Ras is activation of the MAP kinase (ERK 1 and ERK 2) cascade, RasGAPs can be thought of as brakes on the MAP kinase pathway. How might the function of SynGAP be linked to the NMDA receptor? The RasGAP activity is strongly inhibited by phosphorylation of SynGAP by CaMKII, an early target of calcium flowing through the NMDA receptor. Hence, activation of the NMDA receptor may lead directly to inhibition of SynGAP and release of the brake on the MAP kinase pathway. An important missing link in this scheme is the nature of signaling pathways at glutamatergic synapses that can activate Ras. Possible candidates include *Src* or *Fyn*, which can activate *Ras* through the N-Shc adaptor protein, or the BDNF and Ephrin/EPH pathways (see Chapter 11). Postulated dendritic targets for regulation by MAP kinase include A-type K^+ channels that modulate the sizes of EPSPs and of back-propagating action potentials and MAP2, which may mediate cellular remodeling.

Neuronal NOS, a Ca^{2+}-activated form of NOS, can bind to PSD-95 through a class III PDZ domain interaction in which its own amino-terminal PDZ domain binds to a PDZ domain of PSD-95 (Sattler *et al.* 1999). Neuronal NOS is not abundant in the PSD fraction and is not expressed at high levels in pyramidal neurons. However, it is highly expressed in certain [gamma]-aminobutyric acid-containing neurons, which also express members of the PSD-95 family. Therefore, PSD-95 may concentrate nNOS near the NMDA receptor at postsynaptic sites in these neurons.

Biochemical and pharmacological evidence indicates that a number of other signaling complexes are located in spines within or near the PSD. For example, AMPA receptors may be bound to their own unique set of signaling complexes. In addition, the cyclic adenosine monophosphate (cAMP) signaling pathway is implicated in the regulation of glutamatergic transmission. cAMP potentiates induction of LTP in the Schaeffer collateral pathway. The favored mechanism involves a regulatory cycle first postulated in liver and muscle and now well documented in dopaminergic transmission. Activated cAMP-dependent protein kinase phosphorylates a protein called Inhibitor-1 (DARPP-32 in the dopaminergic pathway). Upon phosphorylation, Inhibitor-1 becomes an inhibitor of protein phosphatase-1. This inhibition potentiates phosphorylation of proteins that can be dephosphorylated by protein phosphatase-1. In Schaeffer collateral synapses, the cAMP pathway 'gates' autophosphorylation of CaMKII and subsequent induction of LTP.

Phosphatase-1 and the cAMP-dependent protein kinase can be complexed with the NR1 subunit of the NMDA receptor by the scaffold protein Yotiao, a splice variant of a family of AKAP (A-kinase-associated protein) proteins that target the cAMP-dependent protein kinase to subcellular compartments (Westphal *et al.* 1999). Yotiao and a second AKAP, AKAP75/150, which targets protein kinase A (PKA), protein kinase C (PKC), and the Ca^{2+}-dependent protein phosphatase calcineurin to dendritic microtubules, can be detected by immunoblot in the PSD fraction and in immunoprecipitates of the NMDA receptor. Their relatively low abundance in the PSD fraction suggests that they may be present in a subset of PSDs in the brain. Differential distribution of AKAPs could alter the forms of synaptic plasticity displayed by different synapses.

Pharmacological evidence also indicates that PKC and the MAP kinase pathway participate in postsynaptic regulation of synaptic plasticity at glutamatergic synapses. The structural basis for their localization in spines is not yet firmly established.

Just as for protein kinases, appropriate location and regulation of protein phosphatases are crucial for proper metabolic control. The calcium-dependent protein phosphatase calcineurin is localized in dendritic spines, perhaps by AKAP75/150. Both Yotiao and the neurabin/spinophilin family of proteins could target protein phosphatase-1 (PP1) to dendritic spines where it can dephosphorylate a variety of substrates. The ubiquitous protein phosphatase 2A (PP2A) is regulated and targeted by tissue-specific subunits; its brain-specific regulatory subunits have just begun to be studied.

7.1.3 Cytoskeletal proteins

The major cytoskeletal components in PSD are actin, α-actinin, spectrin, tubulin, and an homologue of the neurofilament NF-L subunit. A close link between cytoskeleton and glutamate receptors is present in PSD. In fact, the rundown of NMDA channel activity can be prevented by the presence of the microfilament-stabilizing drug, phalloidin. In addition, NMDA currents are increased in hippocampal neurons lacking gelsolin, an actin-severing protein. A direct molecular binding exists between α-actinin and C-terminal domains of NMDA receptor subunits (Wyszynski *et al.* 1999). Inside the PSD, also brain spectrin, or fodrin, are able to interact directly with NMDA receptor complex. Actin filaments bind directly to the PSD cytoplasmic face and associate with PSD components. The integrity of microfilament web secures the synaptic localization of ionotropic glutamate receptors; dissolution of microfilaments within the dendritic spines shifts a portion of NMDA clusters from a synaptic to a non-synaptic localization over a 24 h period.

7.1.4 Linker proteins

Linker proteins belong to a family of synaptic proteins homologous to the product of the Drosophila gene *disc large* and comprises four closely related proteins called PSD-95 protein family (sometimes also MAGUK proteins), each of which contains five protein-binding domains (Cho *et al.* 1992). Three amino-terminal PDZ (PSD-95, Discs-large, ZO-1) domains are followed by an SH3 domain and a GK domain homologous to yeast guanylate kinase but lacking enzymatic activity. The first and second PDZ domains bind tightly to the tails of the NR2 subunits of the NMDA receptor. The three PDZ domains each have slightly different binding specificities and can interact with a variety of different neuronal membrane proteins. The tight colocalization of NMDA receptors and PSD-95 at synapses and the abundance of both proteins in the PSD fraction suggest that in the forebrain, many synaptic NMDA receptors are attached to the PDZ domains of PSD-95 or one of its family members.

Neuroligin is an adhesion molecule that is present throughout the soma and dendrites of many neurons and has been localized to the synaptic cleft and the postsynaptic density of some neurons. It has not been detected in substantial amounts in the PSD fraction; thus, its association with PSD-95 may be transient, or more easily disrupted, than that of other proteins by extraction with detergent during purification of the PSD fraction (Irie *et al.* 1997). Alternatively, it may associate with PSD-95 in a relatively small proportion of synapses. The recent finding that expression of neuroligin in heterologous cells can induce clustering of presynaptic vesicles in contacting axons suggests that neuroligin may help to induce synapse formation at potential postsynaptic sites that contain NMDA receptor-associated signaling complexes.

7.2 Functional interactions between PSD components

The vast array of cytoskeletal and regulatory proteins found in the PSD interact to constitute the glutamatergic postsynaptic signal transduction machinery, coordinating activity-dependent changes in postsynaptic structures, including LTP and LTD, the cellular bases for learning and memory. The picture emerging from studying these interactions is that these molecular components are essential for (i) clustering glutamate receptors just opposite to the presynaptic active zone, (ii) modulating glutamate receptor sensitivity, and (iii) inducing long-lasting modifications in preactivated synapses.

In PSD, NR2A and B subunits directly interact with PSD-95 (Kornau *et al.* 1995), and other members of the MAGUK family (Kim *et al.* 1996; Lau *et al.* 1996) through their intracellular extended COOH sequence. In particular, NR2A C-terminal motif tSDV is mandatory for efficient binding to PSD-95 PDZ domains (Bassand *et al.* 1999). The interaction with the PSD-95 protein family induces the clustering of the channel proteins, thus playing an important role in the molecular organization of NMDA receptors although more recent findings demonstrate that postsynaptic NMDA receptor clustering does not solely depend on PSD-95 family (Migaud *et al.* 1998). In addition, PSD-95 appears to be important in coupling NMDA receptor to biochemical intracellular pathways controlling bi-directional synaptic plasticity (Tezuka *et al.* 1999). Nevertheless, although it has been reported that the molecular interactions involving PSD-95 and the NMDA receptor are modified by pathological insults such as an ischemic challenge, the physiological conditions influencing association/dissociation of specific proteins to and from NMDA receptor are not yet fully understood. NR2 subunits are not solely associated to PSD-95: indeed the NMDA receptor complex has been also shown to bind αCaMKII (Strack and Colbran 1998; Leonard *et al.* 1999). Furthermore, the αCaMKII association to NR2A can be affected by activation of the NMDA receptor *in vitro* either by pharmacological tools or by induction of LTP. The increased αCaMKII binding to the receptor entails the detachment of PSD-95 from tSDV-NR2A both *in vitro* and *ex vivo* (Gardoni *et al.* 2001). NR1, in turn, provides a direct link to cytoskeletal proteins through the interaction with α-actinin-2 and other two-filamentous proteins.

AMPA receptors have been found able to bind PDZ 4 and 5 domains of GRIP probably through GluR2/3 subunits; PDZ 1 and 7 domains of GRIP seem to be anchored to cytoskeletal elements (O'Brien *et al.* 1998). On the other hand, GluR1 associates specifically with another member of the PSD-95 family, SAP-97.

The PSD-95 protein family has another important role: they bind other PSD components such as enzymes and modulatory molecules. Indeed, PSD-95 binds neuronal NO synthase via a PDZ–PDZ interaction holding NOS in the appropriate position to sense the calcium influx through NMDA receptor channel opening. The third PDZ domain of PSD-95 binds SynGAP (Chen *et al.* 1998) providing a link between the NMDA receptor and the mitogen-activated protein kinase pathway.

7.3 **Proposed functions of the PSD**

Proposed functions for PSD proteins include regulation of adhesion between presynaptic and postsynaptic membranes, control of postsynaptic receptor clustering and function (Sheng and Kim 1996; Ziff 1997; Meyer and Shen 2000), and signal transduction in response to receptor activation (Kennedy 2000).

Study of the proteins of the PSD fraction provides a unique and useful view of the glutamatergic synapse. It has already contributed to the characterization of at least one major mechanism of assembly of postsynaptic machinery and promises to reveal others. The PSD is formed very early during synaptogenesis of the CNS and little is known about its precise role in synaptogenesis or in the medically important process of regeneration. The identification of its molecular constituents will permit studies of these issues.

Interestingly, it has been observed that synaptic activity affects PSD morphology, by inducing its perforation and duplication, indicating that long-lasting biochemical events triggered by glutamate receptors activation entail mechanisms influencing not only signal transmission but also structural remodeling of synaptic components, such as the PSD.

7.4 **The PSD in synaptic plasticity**

The AMPA and NMDA receptors display different topologic distributions in the postsynaptic membrane. Electron microscopy of immunogold-labeled synapses has shown that NMDA receptors tend to be clustered near the center of the synapse, while AMPA receptors are distributed more uniformly across the postsynaptic membrane. This difference may reflect association of the two receptor types with different subsynaptic structures. NMDA receptors and AMPA receptors show other differences at synapses. They are transported to synapses at different times during development (as measured for cultured primary neurons); and once installed at the synapse, they differ in their ease of extraction by detergents, with the NMDA receptor being more firmly attached. Furthermore, a significant proportion of excitatory synapses lacks the AMPA receptor. This finding is important in view of the influences that one receptor type can exert on the other receptor's function. Opening of the NMDA receptor channel requires cell depolarization, which may arise when currents flow through the AMPA receptor. Thus, NMDA receptors may be inactive at synapses that lack functional AMPA receptors, unless depolarization is provided by some other means, such as by depolarization of a neighboring synapse that does contain functional AMPA receptors.

Recently synapses have been described that lack functional AMPA receptors, but for which AMPA receptors may be activated by tetanic stimulation. This finding has led to the formulation of the 'silent synapse' hypothesis. In this hypothesis, synaptic strength is increased by recruitment of AMPA receptors to synapses that lack functional forms of the AMPA receptor, a process sometimes called 'AMPAfication.' The hypothesis has provoked much interest in the mechanisms that govern the abundance of functional AMPA receptor at the synapse. Silent synapses are converted to 'talking' synapses by tetanic stimulation of the sort that induces LTP. Ca^{2+} fluxes through NMDA receptors following tetanic stimulation of silent synapses may induce biochemical pathways that trigger synaptic plastic changes. This may lead in turn to AMPA receptor functional activation. The mechanisms that govern AMPA receptor abundance at the synapse are not known. Recently, it has been shown that the *N*-ethylmaleimide-sensitive fusion (NSF) protein makes a specific complex with the GluR2 C-terminal domain. NSF is a chaperonin, which modulates protein–protein interactions. Electrophysiologic studies suggest that NSF function may be required to maintain AMPA receptor currents. NSF plays an important role in dissociating SNARE protein complexes following vesicle fusion with target membranes and may mediate protein interaction with GluR2, although its precise role with the AMPA receptors is not yet known.

The rapid increase in our understanding of molecular and cell biologic features of ionotropic glutamate receptors holds great promise for the field. We may anticipate a corresponding increase in our grasp of receptor function in all respects, from the molecular to the system levels.

7.5 **PSD structure and pathophysiology of the CNS**

From the studies described above, it has become increasingly evident that the PSD structure in general, and ionotropic glutamate receptor complexes in PSD in particular, are intimately involved in the regulation of synaptic plasticity events. On the other hand, recent findings showed a role for the macromolecular complex of the PSD in the dynamic regulation of different insults in the brain. In fact, because many of the initial events (i.e. activation of Glu receptors, Calcium influx, NOS activation) occur at/or in the vicinity of the PSD this appears to be a likely site for the pathogenic biochemical cascade that leads to neuronal cell death in numerous neurodegenerative disorders.

On this basis, several groups have described alterations of PSD structural organization in some models of CNS disorders.

7.5.1 Cerebral ischemia

In recent years, it has been shown that loss of calcium homeostasis may be an important mechanism of ischemic brain damage. Most of the studies focused on ischemia-induced changes in the PSD structure because of its temporal and spatial proximity to initial events which occur after an ischemic challenge. Changes in the composition and morphology of forebrain PSD have been reported to occur after an ischemic challenge. Ischemia has been shown to decrease CaMKII activity most likely due to selective post-translational modification of the enzyme leading to its inhibition (Aronowsky and Grotta 1996). In addition, molecular interactions involving PSD-95 are modified by an ischemic challenge, in the PSD of the most vulnerable CA1 region of the hippocampus. Ischemia also resulted in a decrease in the size of protein complexes containing PSD-95, but had only a small effect on the size distribution of complexes containing the NMDA receptor, indicating that molecular interactions involving PSD-95 and the NMDA receptor are modified by an ischemic challenge (Takagi *et al.* 2000). Furthermore, ischemia differentially affects the association of different tyrosine kinase with the PSD (Cheung *et al.* 2000): the level of PSD-associated PYK2 and trkB increased during the ischemic episode whether FAK levels decreased. *Src* and *Fyn* levels appear increased with a short delay after reperfusion.

7.5.2 Experimental models of impaired long-term potentiation and learning tasks

It has been mentioned previously that knock outs for CaMKII and protein tyrosine kinases abolish or impair certain forms of LTP and LTD, that induction of LTP promotes CaMKII activation (Barria *et al.* 1997) and constitutive activation of CaMKII changes the frequencies inducing LTP and LTD (Mayford *et al.* 1995). But there is also evidence that pathological changes of the CNS involving impairments of cognitive tasks and LTP have a counterpart in changes of CaMKII activity and NMDA receptor phosphorylation in the postsynaptic compartment.

7.5.3 Streptozotocin-diabetic rats

Moderate disturbances of learning and memory have been reported as a complication of diabetes mellitus in patients. In animal models of the diabetic pathology, such as the streptozotocin-diabetic rat, spatial learning impairments associated with changes in hippocampal synaptic plasticity have been reported, including an impaired expression of LTP and enhanced expression of LTD (Gispen and Biessels 2000). In situ hybridization histochemistry revealed that mRNA levels of NR2B in hippocampus of diabetic rats were reduced when compared to control rats (Di Luca *et al.* 1999). In addition, PSD-associated CaMKII activity as well as CaMKII-dependent phosphorylation of both NR2A and NR2B subunits of NMDA receptor were reduced in hippocampal PSDs of streptozotocin-diabetic rats as compared with controls.

7.5.4 Prenatal ablation of hippocampal neurons

To examine the implication of PSD proteins in synaptic plasticity an animal model characterized by developmentally induced targeted neuronal ablation within the cortex and the hippocampus and lack of long-term potentiation was used (Methylazoxymethanol treated rats; Cattabeni and Di Luca 1997). In this model the lack of long-term potentiation is due to a partial prenatal ablation of CA1 neurons, obtained by exposing embryos to the antiproliferative agent methylazoxymethanol acetate through their mothers at gestational day 15. The offsprings show a marked cellular ablation in the intermediate layers of the cortex and in the CA region of the hippocampus, in agreement with the neurogenetic gradient in rodents. Methylazoxy-methanol (MAM) treated rats develop normally, but in adulthood they show impairment of learning and memory. Interestingly, these neuroanatomical, behavioral, and electrophysiological abnormalities have their counterpart, at the molecular level, in a consistent increase of PKC localized in the membrane compartment of hippocampal synaptosomes (Di Luca *et al.* 1995; Caputi *et al.* 1996). As a consequence of the alteration in basal PKC translocation, a parallel increase in calcium-dependent glutamate release in MAM treated rats was observed (Di Luca *et al.* 1997). Moreover, long-term potentiation could be restored by D-serine (Ramakers *et al.* 1993), an agonist at the glycine site of NMDA receptors.

At the postsynaptic level, a marked decrease in CaMKII-activity and CaMKII-dependent phosphorylation of both NR2A and NR2B subunits of NMDA receptor are present when compared to controls, although the concentration of the enzyme in PSD is not altered (Caputi *et al.* 1999). The activity of CaMKII was reconstituted by incubating hippocampal slices with D-serine, a treatment previously shown to rescue long-term potentiation in MAM treated rats (Ramakers *et al.* 1993). D-serine acts as an agonist at the glycine site of NMDA receptors and, as mentioned previously, is able to prevent NMDA receptor desensitization in mouse hippocampal cultures (Vyklicky *et al.* 1990). These data taken together point to an important role of postsynaptic CaMKII activity in modulating synaptic plasticity through phosphorylation of NMDA receptors. Its physical association with NR2 subunits, as discussed above, places it in the core of the action responsible for the long-lasting changes associated with synaptic plasticity.

7.6 Conclusions

Progress in understanding the biochemical and structural basis of synaptic regulation has been rapid and exciting over the past few years. It has been fueled by the recognition among cell biologists that signaling specificity results from the formation of protein complexes that respond locally and discretely to signals from the membrane surface. A few synaptic signaling 'machines' have now been identified, but many more remain to be characterized before we unravel the intricacies of signal processing in the brain. We face two major challenges. The first is to understand to what extent the presence and organization of signaling machinery varies among different synaptic types. This information is crucial because the complement of signaling complexes at a synapse determines the rules by which it integrates and encodes information. The second challenge is to understand how the different signaling pathways interact with and feed back on each other to maintain homeostasis while processing, integrating, and storing rapidly changing information. Efforts to meet this challenge will be aided by promising new strategies for creating and testing spatially accurate computer simulations of complex biochemical signaling machinery. We have come a long way toward understanding how synapses work, but we still have far to go.

References

Aronowski, J. and Grotta, J. C. (1996) Ca^{2+}/calmodulin-dependent protein kinase II in postsynaptic densities after reversible cerebral ischemia in rats. *Brain Res.*, **709**, 103–10.

Barria, A., Muller, D., Derkach, V., Griffith, L. C., and Soderling, T. R. (1997) Regulatory phosphorylation of AMPA-type glutamate receptors by CaMKII during long-term potentiation. *Science*, **276**, 2042–5.

Bassand, P., Bernard, A., Rafiki, A., Gayet, D., and Khrestchatisky, M. (1999) Differential interaction of the tSXV motifs of the NR1 and NR2A NMDA receptor subunits with PSD-95 and SAP-97. *Eur. J. Neurosci.*, **11**, 2031–43.

Caputi, A., Rurale, S., Pastorino, L., Cattabeni, F., and Di Luca, M. (1996) Differential translocation of Protein Kinase C isozymes in rats characterized by a chronic lack of LTP induction and cognitive impairment. *FEBS Lett.*, **393**, 121–3.

Caputi, A., Gardoni, F., Cimino, M., Pastorino, L., Cattabeni, F., and Di Luca, M. (1999) CaMKII-dependent phosphorylation of NR2A and NR2B is decreased in animals characterized by hippocampal damage and impaired LTP. *Eur. J. Neurosci.*, **11**, 141–8.

Carlin, R. K., Grab, D. J., Cohen, R. S., and Siekevitz, P. (1980) Isolation and characterization of Post Synaptic Densities from various brain regions: enrichment of different types of Post Synaptic Densities. *J. Cell Biol.*, **86**, 831–43.

Cattabeni, F. and Di Luca, M. (1997) Developmental models of brain dysfunctions induced by targeted cellular ablations with methylazoxymethanol. *Physiol. Rev.*, **77**, 199–215.

Chen, H. J., Rojas-Soto, M., Oguni, A., and Kennedy, M. B. (1998) A synaptic Ras-GTPase activating protein (p135 SynGAP) inhibited by CaM kinase II. *Neuron*, **20**, 895–904.

Cheung, H. H., Takagi, N., Teves, L., Logan, R., Wallace, M. C., and Gurd, J. W. (2000) Altered association of protein tyrosine kinases with postsynaptic densities after transient cerebral ischemia in the rat brain. *J. Cereb. Blood Flow Metab.*, **20**, 505–12.

Cho, K. O., Hunt, C. A., and Kennedy, M. B. (1992) The rat brain postsynaptic density fraction contains a homolog of the Drosophila discs-large tumor suppressor protein. *Neuron*, **9**, 929–42.

Cotman, C. W., Banker, B., Churchill, L., and Taylor, D. (1974) Isolation of postsynaptic densities from rat brain. *J. Cell Biol.*, **63**, 441–5.

Das, S., Sasaki, Y. F., Rothe, T., Premkumar, L. S., Takasu, M., Crandall, J. E., *et al.*(1998) Increased NMDA current and spine density in mice lacking the NMDA receptor subunit NR3A. *Nature*, **393**, 377–81.

Di Luca, M., Caputi, A., Cinquanta, M., Cimino, M., Marini, P., Princivalle, A., *et al.* (1995) Changes in Protein Kinase C and its presynaptic substrate after intrauterine exposure to methylazoxymethanol, a treatment inducing cortical and hippocampal damage and cognitive impairments. *Eur. J. Neurosci.*, **7**, 899–906.

Di Luca, M., Caputi, A., Cattabeni, F., DeGraan, P. N. E., Gispen, W. H., Raiteri, M., *et al.* (1997) Increased presynaptic protein kinase C activity and glutamate release in rats with a prenatally induced hippocampal lesion. *Eur. J. Neurosci.*, **9**, 472–9.

Di Luca, M., Ruts, L., Gardoni, F., Cattabeni, F., Biessels, G. J., and Gispen, W. H. (1999) NMDA receptor subunits are modified trancriptionally and post-translationally in the brain of streptozotocin-diabetic rats. *Diabetologia*, **42**, 693–701.

Elliss, P. D., Bissoon, N., and Gurd, J. W. (1988) Synaptic protein tyrosine kinase: partial characterization and identification of endogenous substrates. *J. Neurochem.*, **51**, 611–20.

Gardoni, F., Caputi, A., Cimino, M., Pastorino, L., Cattabeni, F., and Di Luca, M. (1998) CaMKII is associated to NR2A/B subunits of NMDA receptor in Post Synaptic Densities. *J. Neurochem.*, **71**, 1733–41.

Gardoni, F., Schrama, L. H., Kamal, A., Gispen, W. H., Cattabeni, F., and Di Luca, M. (2001) Hippocampal synaptic plasticity involves competition between αCaMKII and PSD-95 for binding to the NR2A subunit of the NMDA receptor. *J. Neurosci.*, **21**, 1501–9.

Gispen, W. H. and Biessels, G. J. (2000) Cognition and synaptic plasticity in diabetes mellitus. *Trends Neurosci.*, **23**(11), 542–9.

Gomperts, S. N. (1996) Clustering membrane proteins: It's all coming together with the PSD-95/SAP90 protein family. *Cell*, **84**, 659–62.

Hanson, P. I. and Schulman, H. (1992) Inhibitory autophosphorylation of multifunctional Ca^{2+}/Calmodulin- dependent protein kinase analyzed by site-directed mutagenesis. *J. Biol. Chem.*, **267**, 17216–24.

Hollmann, M. and Heinemann, S. (1994) Cloned glutamate receptors. *Annu. Rev. Neurosci.*, **17**, 31–108.

Irie, M., Hata, Y., Takeuchi, M., Ichtchenko, K., Toyoda, A., Hirao, K., *et al.* (1997) Binding of neuroligins to PSD-95. *Science*, **277**, 1511–15.

Kelly, P. T. and Cotman, C. W. (1978) Synaptic proteins. Characterization of tubulin and actin and identification of a distinct postsynaptic density polypeptide. *J. Cell Biol.*, **79**, 173–83.

Kennedy, M. B., Bennett, M. K., and Erondu, N. E. (1983) Biochemical and immunochemical evidence that the "major postsynaptic density protein" is a subunit of a calmodulin-dependent protein kinase. *Proc. Natl. Acad. Sci. USA*, **80**, 7357–61.

Kennedy, M. B. (1997) The postsynaptic densities at glutamatergic synapses. *Trends Neurosci.*, **20**, 264–8.

Kennedy, M. B. (2000) Signal-processing machines at the postsynaptic density. *Science*, **290**, 750–4.

Kim, E., Cho, K. O., Rothschild, A., and Sheng, M. (1996) Heteromultimerization and NMDA receptor-clustering activity of Chapsyn-110, a member of the PSD-95 family of proteins. *Neuron*, **17**, 103–13.

Kim, J. H., Liao, D., Lau, L. F., and Huganir, R. L. (1998) SynGAP: a synaptic RasGAP that associates with the PSD-95/SAP90 protein family. *Neuron*, **20**, 683–91.

Kornau, H. C., Schenker, L. T., Kennedy, M. B., and Seeburg, P. H. (1995) Domain interaction between NMDA receptor subunits and the Post Synaptic Density protein PSD-95. *Science*, **269**, 1737–40.

Lau, L. F., Mammen, A., Ehlers, M. D., Kindler, S., Chung, W. J., Garner, C. C., *et al.* (1996) Interaction of the *N*-methyl-D-aspartate receptor complex with a novel synapse-associated protein, SAP102. *J. Biol. Chem.*, **271**, 21622–8.

Leonard, S., Lim, I. A., Hemsworth, D. E., Horne, M. C., and Hell, D. (1999) Calcium/Calmodulin-dependent protein kinase II is associated with the *N*-Methyl-D-Aspartate receptor *Proc. Nat. Acad. Sci. USA*, **96**, 3239–44.

Lisman, J. (1994) The CaM kinase II hypothesis for the storage of synaptic memory. *Trends Neurosci.*, **17**, 406–12.

Liu, J., Fukunaga, K., Yamamoto, H., Nishi, K., and Miyamoto, E. (1999) Differential roles of Ca^{2+}/calmodulin-dependent protein kinase II and mitogen-activated protein kinase activation in hippocampal long-term potentiation. *J. Neurosci.*, **19**, 8292–9.

Mayford, M., Wang, J., Kandel, E. R., and O'Dell, T. J. (1995) CaMKII regulates the frequency-response function of hippocampal synapses for the production of both LTD and LTP. *Cell*, **81**, 891–904.

Meyer, T. and Shen, K. (2000) In and out of the postsynaptic region: signalling proteins on the move. *Trends Cell Biol.*, **10**, 238–44.

Migaud, M., Charlesworth, P., Dempster, M., Webster, L. C., Watabe, A. M., Makhinson, M., *et al.* (1998) Enhanced long-term potentiation and impaired learning in mice with mutant postsynaptic density-95 protein. *Nature*, **396**, 433–9.

Monyer, H., Burnashev, N., Laurie, D. J., Sakmann, B., and Seeburg, P. H. (1994) Developmental and regional expression in the rat brain and functional properties of four NMDA receptors. *Neuron*, **12**, 529–40.

Naisbitt, S., Kim, E., Tu, J. C., Xiao, B., Sala, C., Valtschanoff, J., *et al.* (1999) Shank, a novel family of postsynaptic density proteins that binds to the NMDA receptor/PSD-95/GKAP complex and cortactin. *Neuron*, **23**, 569–82.

O'Brien, R. J., Lau, L. F., and Huganir, R. L. (1998) Molecular mechanisms of glutamate receptor clustering at excitatory synapses. *Curr. Opin. Neurobiol.*, **8**, 364–9.

Omkumar, R. V., Kiely, M. J., Rosenstein, A. J., Min, K. T., and Kennedy, M. B. (1996) Identification of a phosphorylation site for calcium/calmodulin dependent protein kinase II in the NR2B subunit of the *N*-methyl-D-aspartate receptor. *J. Biol. Chem.*, **271**, 31670–8.

Palay, S. L. (1956) Synapses in the central nervous system. *J. Biophys. Biochem. Cytol.*, **2**, 193–202.

Pettit, D. L., Perlman, S., and Malinow, R. (1994) Potentiated transmission and prevention of further LTP by increased CaMKII activity in postsynaptic hippocampal slice neurons. *Science*, **266**, 1881–5.

Ramakers, G. M., Urban, I. J., De Graan, P. N. E., Di Luca, M., Cattabeni, F., and Gispen, W. H. (1993) The impaired long-term potentiation in the CA1 field of the hippocampus of cognitive deficient microencephalic rats is restored by D-serine. *Neuroscience*, **54**, 49–60.

Sattler, R., Xiang, Z. G., Lu, W. Y., Hafner, M., MacDonald. J. F., and Tymianski, M. (1999) Specific coupling of NMDA receptor activation to nitric oxide neurotoxicity by PSD-95 protein. *Science*, **284**, 1845–8.

Schulman, H. and Hanson, I. P. (1993) Multifunctional Ca^{2+}/CaM dependent protein kinase. *Neurochem. Res.*, **18**, 65–77.

Sheng, M. and Kim, E. (1996) Ion channel associated proteins. *Curr. Opin.. Neurobiol.*, **6**, 602–8.

Silva, A. J., Stevens, C. F., Tonegawa, S., and Wang, Y. (1992) Deficient hippocampal long-term potentiation in α-calcium-calmodulin kinase II mutant mice. *Science*, **257**, 201–6.

Strack, S. and Colbran, R. J. (1998) Autophosphorylation-dependent targeting of Calcium Calmodulin kinase II by the NR2B subunit of the *N*-Methyl-D-Aspartate receptor. *J. Biol. Chem.*, **273**, 20689–92.

Takagi, N., Logan, R., Teves, L., Wallace, M. C., and Gurd, J. W. (2000) Altered interaction between PSD-95 and the NMDA receptor following transient global ischemia. *J. Neurochem.*, **74**, 169–78.

Tezuka, T., Umemori, H., Akiyama, T., Nakanishi, S., and Yamamoto, T. (1999) PSD-95 promotes Fyn-mediated tyrosine phosphorylation of the *N*-methyl-D-aspartate receptor subunit NR2A. *Proc. Natl. Acad. Sci. USA*, **96**, 435–40.

Vyklicky, L., Benveniste, M., and Mayer, M. L. (1990) Modulation of N-methyl-D-aspartate receptor desensitization by glycine in mouse hippocampal neurones. *J. Physiol.*, **428**, 313–31.

Westphal, R. S., Tavalin, S. J., Lin, J. W., Alto, N. M., Fraser, I. D., Langeberg, L. K., *et al.* (1999) Regulation of NMDA receptors by an associated phosphatase-kinase signalling complex. *Science*, **285**, 93–6.

Wyszynski, M., Valtschanoff, J. G., Naisbitt, S., Dunah, A. W., kim, E., Standaert, D. G., *et al.* (1998) Association of AMPA receptors with a subset of glutamate receptor-interacting protein in vivo. *J. Neurosci.*, **18**, 1383–92.

Ziff, E. B. (1997) Enlightening the postsynaptic density. *Neuron*, **19**, 1163–74.

Abbreviations

AKAP	A-kinase-associated protein
cAMP	cyclic adenosine monophosphate
AMPA	1-alpha-amino-3-hydroxy-5-methyl-4-isoxazole propionic acid
CA	Cornu Ammonis
CAM	Calmodulin
CaMKII	Ca2+/Calmodulin-dependent kinase II
CNS	Central Nervous System
GKAP	Guanylyl kinase associated protein
GRIP	Glutamate Receptor Interacting Protein
LTD	Long Term Depression
LTP	Long Term Potentiation
MAGUK	Membrane-associated guanylyl kinase
MAM	Methylazoxymethanol acetate
NMDA	*N*-methyl-D-aspartic acid
NOS	Nitric oxide synthase
NSF	N-ethylmaleimide sensitive fusion protein
PDBu	phorbol 12, 13-dibutyrate
PDZ	PSD-95 Discs-large ZO-1
PKA	protein kinase A
PKC	protein kinase C
PP1	Protein phosphatase 1
PP2A	Protein phosphatase 2A
PSD	Post Synaptic Density
SAM	Sterile alpha motif
SAP	Synapse Associated Protein
SDS-PAGE	Sodium dodecylsulfate-polyacrylamide gel electrophoresis
SynGAP	Synaptic Ras-GTPase-activating protein

Chapter 8

Signal reception: Ligand-gated ion channel receptors

R. Wayne Davies and Thora A. Glencorse

8.1 Introduction

Fast synaptic transmission is crucial for real-time functioning of the brain. All the receptor molecules that mediate fast transmission events are also ligand-gated ion channels, i.e. they are ion channels that undergo allosteric structural changes on binding a particular neurotransmitter molecule, resulting in the opening of the channel, the entry of selected ions into the neuron and subsequent signalling events. Their primary function is to receive signal input at postsynaptic membranes, where some also play central roles in synaptic plasticity. However, they are also found in postsynaptic membranes outside synapses, and in presynaptic terminals (MacDermott *et al.* 1999), where they are involved in the control of transmitter release.

The neurotransmitters that mediate fast synaptic transmission through these receptors are: glutamate, γ-amino butyric acid (GABA), glycine, serotonin (5HT, 5-hydroxytryptamine), acetylcholine and ATP. Since most of these neurotransmitters also act on G-protein linked receptors, the terms ionotropic and metabotropic receptor are used, respectively, for ligand-gated ion channel receptors and G-protein linked receptors that are activated by a particular neurotransmitter. Fast excitatory transmission is mediated primarily by glutamate receptors throughout the CNS, while fast inhibitory transmission is mediated primarily by GABA receptors of type A (termed GABA(A) receptors; GABA(B) receptors are metabotropic, and GABA(C) receptors are ionotropic but have a very limited distribution) in the CNS and by glycine receptors in the spinal cord. The other types of receptors have more restricted distributions in the CNS, and are likely to act as modulators rather than as primary receivers of fast transmission information (Role and Berg 1996). The postsynaptic membranes of excitatory synapses involving glutamate receptors are characterized by an electron-dense 30–40 nm disc-like thickening termed the postsynaptic density, which is a subcellular microdomain that contains all the proteins involved in synaptic plasticity as well as simple information transfer (review: Sheng 2001). Postsynaptic membranes of inhibitory synapses do not show such a microdomain, although the receptors are also known to be part of a protein complex. It appears that the excitatory and inhibitory protein complexes involving ligand-gated ion channel receptors are very, perhaps completely, different, since none of the proteins known to be involved in inhibitory protein complexes were detected among the proteomic analysis of the protein complex associated with the NMDA-type glutamate receptor (Husi *et al.* 2000).

All these receptors fall into three distinct molecular groups. All the proteins in each group are clearly evolutionarily related and each group probably derives from a single common ancestral gene and protein by gene duplication, genetic drift, and selection (Schofield *et al.* 1987). The three groups are: the glutamate receptors; the GABA(A) receptor group, traditionally exemplified for historical reasons by nicotinic acetylcholine receptors (nACh receptors; muscarinic acetylcholine receptors are G-protein linked) and P2X ATP receptors. The glutamate receptor group has three subdivisions, based historically on pharmacological distinctions but clearly reflecting both distinct molecular and functional categories. Because of the pharmacological classification, these subgroups of glutamate receptors have been named AMPA (affinity for α-amino-3-hydroxy-5-methylisoxazole-4-proprionate), KAIN (affinity for kainate), and NMDA (affinity for *N*-methyl-D-aspartate). The GABA(A)R or nAChR group consists of all types of GABA(A) receptors, GABA(C) receptors, all glycine receptors (GlyR), all types of neural and muscle nicotinic acetylcholine receptors, and the 5HT3 class of serotonin receptor. This group is sometimes called the Cys-loop superfamily, because they all have two conserved cysteine residues in the N-terminal extracellular domain, which form a disulfide bridge and define a subdomain, the Cys-loop. As an example of the degree of evolutionary relationship found, all GABA(A) subunit isoforms (e.g. α1, α2 etc.) exhibit >70 % primary sequence identity, while different GABA(A)R subunits (e.g. α1, β3, γ2 etc.) show between 35 % and 50 % identity. The sequence identity to subunits of different receptors in the same family is, for example, 34% with the GlyR α1 subunit (Grenningloh *et al.* 1987) and 20 % with the nAChR α1 subunit (Noda *et al.* 1983). Arbas *et al.* (1991) proposed that the acetylcholine-gated ion channel was first to diverge from the ancestral form, followed by the GABA/glycine receptor divergence. Subsequently, the primordial channel subunit gene of each class duplicated and diverged to give the current set of genes encoding subunit classes and subunit isoforms. Recent genomic sequence data shows that this divergence occurred a long time ago, since the fish *Fugu* has the same set of genes as man and rodents.

Molecular biological studies have established that each class of ion channel receptor is actually a complex set of multimeric molecules. While the final ion channels are tetrameric or pentameric (the number of subunits in P2X channels are unclear, but is at least three; Khakh 2001), there are more than four or five genes encoding subunits for each type of channel. There are, for example, seventeen genes in seven gene families encoding subunits of glutamate receptors (Dingledine *et al.* 1999), and sixteen genes for GABA(A) receptor subunits: six α, three β, three γ, and one each for δ, π, θ, and ε. The distinct subunit classes and subunit isoforms of each ion channel class have been identified, and the most important of these are shown in Table 8.1 (see also Boess and Martin 1995;

Table 8.1

Receptor family	Number of genes	Selectivity and ions gated	Natural ligand	Pharmacology: agonists	Pharmacology: antagonists	Associated diseases and therapeutics
nAChR (neuronal)	α2-9 β2-4	Nonselective cation channel Na^+ K^+ Ca^{++}	Acetylcholine	Nicotine Cysteine	α-bungarotoxin Atropine (+)tubocurarine (channel block)	Targets for local anaesthetics, sedatives, and hallucinogenic drugs. Nicotine enhances attention, diminishes anxiety, improves acquisition and retention of short-term memory
GABA(A)R	α1-6 β1-3(4) γ1-3 (4) δ ε θ π	Anion channel Cl^- HCO_3^-	γ-aminobutyric acid (GABA)	Muscimol Flunitrazepam Diazepam Clonazepam Zolpidem Alfaxalone Pentobarbital Zn^{++} sensitive	Bicuculline Flumenazil Ro15-4513 (partial inverse agonist) TBPS and picrotoxin (channel block)	Targets for anaesthetics and sedatives Anxiety Epilepsy Cleft palate Prader–Willi
GABA(C)R	ρ1-2					
GlyR	α1-4 β	Anion channel Cl^- HCO_3^-	Glycine	B-alanine Taurine Proline	Strychnine	Deficiency on GlyR leads to spasticity and loss of motor control. Mutations cause hyperekplexia, spastic paraplegia
5HT3R	5HT3	Nonselective cation channel Na^+ K^+ Ca^{++}	Serotonin (5HT, 5-hydroxy tryptamine)	2-methyl-5-hydroxytryptamine 1-phenyl-biguanidine mCPBG	Zacopride Ondansetron Granisetron Tropisetron (+)tubocurarine (channel block)	Emesis (general) Anxiety Anti-nociceptive Anti-psychotic

(continued)

Table 8.1 (continued)

Receptor family	Number of genes	Selectivity and ions gated	Natural ligand	Pharmacology: agonists	Pharmacology: antagonists	Associated diseases and therapeutics
ATPR	P2X1-7	Nonselective cation channel Na^+ K^+ Ca^{++}	ATP	2-methyl-thioATP αβ-methylene ATP Zn^{++} sensitive	Suramin Stilbene isothiocyanate (+)tubocurarine	Nociception Primary afferent neurotransmission
GluR: NMDAR	NMDAR1 NMDAR2A-2D	Cation channel Na^+ K^+ Ca^{++} Mg^{++} voltage dependent channel block	Glutamate Glycine = co-agonist	NMDA Aspartate	D-AP5 MK801 (dizocilpine)	LTP/LTD Learning and memory Neuronal plasticity Kindling (epilepsy) Anxiogenesis Ischaemia damage (local)
GluR: AMPAR	GluR1-4	Cation channel Na^+ K^+ (Ca^{++})	Glutamate	AMPA Domoate Quisqualate Kainate	NBQX CNQX Potentiators: Cyclothiazide Concanavalin A	LTP/LTD Learning and memory Neuronal plasticity Kindling (epilepsy) Anxiogenesis Ischaemia damage (global)
GluR: KAINR	GluR5-7 KA1-2	Cation channel Na+ K+ (Ca++)	Glutamate	Kainate Domoate Quisqualate	NBQX CNQX NS-102 Potentiator: Concanavalin A	LTP/LTD Learning and memory Neuronal plasticity Kindling (epilepsy) Anxiogenesis Ischaemia damage (global)

CNQX, 6-cyano-7-nitroquinoxaline-2,3-dione; D-AP5, D-2-amino-5-phosphopentanoic acid; mCPBG, *meta*-chloro-phenylbiguanide; NBQX, 6-nitro-7-sulfamobenzo-*f*-quinoxaline-2,3-dione; Ro15-4513, ethyl-8-azido-5,6-dihydro-5-methyl-6-oxo-4H-imidazo[1,5-a] [1,4] benzodiazepine-3-carboxylate; TBPS, *t*-butylbicyclophosphorothionate.

Fredholm 1995; Lindstrom *et al.* 1995; McKernan and Whiting 1996 and Dingledine *et al.* 1999). The task of defining the functional ion channel receptor proteins that are found in particular neurons and at particular synapses is massive and complex, and is still at an early stage for technical reasons. However, great strides forward have been made towards understanding how these molecules are assembled, transported, and maintained at the synapse and other membrane sites.

The complexity of these molecules and their transmembrane location has made the study of structure–function relationships very difficult and slow. However, significant steps forward have been made recently with the determination of atomic structures of extracellular domains of both nAChR and GluR by X-ray crystallography, and the determination of overall nAChR structure to 4.6 Å by electron crystallography. Together with a myriad of site-specific mutagenesis, pharmacological and functional studies, these data provide a basis for new insights into the molecular basis of the key functions of these molecules — ligand recognition, ion permeation, gating, and desensitization.

We present an overview of current knowledge of the molecular biology of these receptors. Since we intend to illuminate principles rather than cover all receptors in detail, we will use as examples whichever receptor has yielded the most appropriate data in each area of molecular characterization.

8.2 **The receptor molecules**

Ligand-gated ion channel receptors are multimeric transmembrane proteins, composed of more than one type of subunit, i.e. they are heteromeric. A few native receptors are homomeric, such as the GABA(C) receptor which is homomeric for the ρ subunit, the foetal spinal cord GlyR, which is a homomeric Glyα2 molecule (Takahashi *et al.* 1992), and the P2X7 channel. Some subunits can be assembled into homomeric receptors in cell culture or Xenopus oocytes (e.g rat nAChR α9, GlyR α1, AMPA and kainate receptor subunits, all P2X subunits except P2X6), but this seems to occur rarely *in vivo*. NMDA glutamate receptor subunits do not form homomeric receptors, even artificially. Even in cell culture, subunits of a protein family will usually only assemble with a subunit of the same family.

Receptors of the nAChR/GABA(A) group are pentameric, with the five subunits in a quasi-symmetrical arrangement, based on data from *Torpedo* nAChR (Unwin 1989, 1993, 1995). Glutamate receptors were also thought to be pentameric for some time, but recent evidence (Schorge and Colquhoun 2003) has confirmed other studies (e.g. Laube *et al.* 1998) indicating that they are tetrameric. Electrophysiological studies indicate that P2X receptors are at least trimeric (Khakh *et al.* 2001). The molecules are built on a modular basis, with each subunit having a similar overall structure, together surrounding a central channel, providing extracellular binding sites for their neurotransmitters and other ligands, and intracellular structures for protein interaction and regulation.

The subunit composition of native receptors is difficult to determine precisely. In some cases, such as muscle nAChR, only one combination of subunits is found (α, β, δ, α, γ), but generally receptors of any given type occur with different combinations of subunits and subunit splice variants, even in the same cell. NMDA glutamate receptors always have one or more NR1 subunit splice variant combined with a variety of subunits of the NR2 class, usually at least two different ones. Schorge and Colquhoun (2003) showed that the subunit arrangement of NMDA receptors was 1,1,2,2 rather than alternating. Among AMPA receptors there are preferential associations of subunits, such as GluR2/GluR1 and GluR2/GluR3 in the hippocampus (Wenthold *et al.* 1996). GABA(A) receptors have the potential for immense diversity of subunit combinations (Whiting *et al.*1995), but *in vivo* the diversity is significantly restricted, although more than twenty receptor subtypes (distinct subunit combinations) are known. The predominant GABA(A) stoichiometry is 2α, 1β, 2γ, with mixtures of different γ or α subunits occurring. The δ, π, ε and θ subunits appear to replace γ subunits. Predominant subunits in adults are α1, β2, and γ2. GABA(A) receptors targeted to synapses always contain a γ subunit, usually γ2, (or a γ-replacement subunit such as ε), but pure αβ receptors are found extrasynaptically.

Each subunit of the nAChR/GABA(A)R group has a large N-terminal extracellular domain, a series of four transmembrane segments called M1–4 with an extracellular loop between M2 and M3, and a short extracellular C-terminal sequence: intracellular parts of the receptor are the large loop between M3 and M4, where all protein–protein interaction and phosphorylation occurs, and the short M1–M2 loop (Fig. 8.1). In contrast, the glutamate receptor subunits only have three true transmembrane segments M1, M3, and M4, and a cytoplasm-facing re-entrant loop M2 which lines the channel (Fig. 8.1: Kuner *et al.* 1996, 2001). Thus glutamate receptor subunits have two major extracellular regions, the N-terminal segment and the M3–M4 loop. The primary intracellular sequence of these receptors for interaction with other proteins and intracellular regulation is the C-terminal segment, and there are also small intracellular M1–M2 and M2–M3 loops. The topology of ATP P2XR is the same as that of sodium channels, with two transmembrane segments, cytoplasmic amino and carboxy-terminal end domains, and an extracellular-facing re-entrant loop (Fig. 8.1). Correct topology is determined not only by the transmembrane domains, but also by charged residues flanking the transmembrane domains. For example, in the GlyR α1 subunit, a cluster of six charged amino acids (RFRRKRR) in the M3–M4 loop is important for topology since neutralization of one or more of these charged residues (but not of charged residues elsewhere in the M3–M4 loop) leads to aberrant translocation of the M3–M4 loop into the ER lumen (Sadtler *et al.* 2003).

8.3 **Molecular diversity and its control**

Each class of ligand-gated ion channel receptor is found in many neurons, and in order to fulfil the precise functional requirements in each cell type and circuit, great variety in the detailed action of the receptor is necessary. This is achieved by the generation of molecular diversity at several levels, and its control by gene regulation, assembly and delivery requirements and post-translational controls.

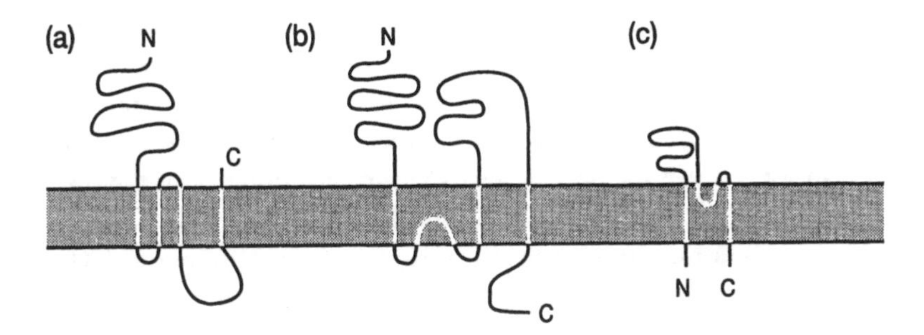

Fig. 8.1 Transmembrane topologies of ligand-gated ion channels. (a) Transmembrane topology of nAChR, GABA(A)R, GABA(C)R, GlyR, and 5HT3R. (b) Transmembrane topology of GluR. (c) Transmembrane topology of ATP P2XR. Transmembrane domains are shown in linear order in the primary protein sequence, not in their real spatial arrangement. They are named M1, M2, M3, and M4 from left to right in the nAChR class, and M1, M2 re-entrant loop, M3 and M4 from left to right in GluR. M2 provides the amino acids lining the pore of the nAChR class, and the M2 re-entrant loop provides the amino acids lining the cytoplasmic side of the pore of GluR. M2 also provides pore-lining residues in P2X receptors. In the nAChR class of receptors, it is known that M1–M4 of each subunit are arranged in a diamond-shape with a transmembrane domain at each corner oriented approximately normal to the membrane, and that putatively M2 is closest to the pore, M4 furthest away (Unwin *et al.* 2002).

8.3.1 Genes and gene expression

All the ion channel receptor genes in man and the mouse are now known, given the completion of the DNA sequences of these species, and may be identified using the online resources at the National Centre for Biotechnology Information website (www.ncbi.nlm.nih.gov). Many were identified earlier by cDNA clone isolation, sequencing and cross-hybridization. Glencorse and Davies (1997) provide a list of the genes encoding common subunits, with their map locations, and many details are provided by Conley and Brammar (1999). The sets of gene families encoding an ion channel receptor correspond to the sets of protein subunits of the same functional class, while other subunits occur in classes with single members and single genes (e.g. *GRIN1* encoding the NMDA glutamate receptor NR1 subunit). Thus there are gene families for the major α, β, and γ subunit groups for GABA(A) receptors, and for the GluR1–4 (AMPA), GluR5–7 (Kainate), KA-1 and KA-2 (Kainate), and NR2A–D (NMDA) glutamate receptor subunits. In general the genes for each receptor class are scattered over many chromosomes, with occasional clusters, e.g. the human GABA(A) receptor $\alpha 1$, $\alpha 6$, $\beta 2$, and $\gamma 2$ genes are all close together at q31–35 on chromosome 5.

Ion channel receptor genes are standard eukaryotic genes containing introns. Gene structure can be conserved between genes encoding different subunits of the same family, and even between families and species as for the mouse GABA(A) receptor δ-subunit gene (Sommer *et al.* 1990*a*) and the human β1-subunit gene (Kirkness *et al.* 1991). Thus the genes encoding glycine receptor subunits have remarkably similar genomic structures to GABA receptor genes, differing only in that they lack the intron that in GABA(A) receptor genes splits the exon that encodes the M2 transmembrane domain (Matzenbach *et al.* 1994). However, there is no conservation of the lengths of introns at any site, so that the size of a gene can vary significantly; e.g. the gene encoding the mouse GABA(A) receptor δ subunit is about 14 kb long, but the β1 subunit gene, which has the same intron–exon structure, is 65 kb long. In contrast to GABA(A) and glycine receptor genes, there is considerable variation in intron–exon structure among nACh receptor encoding genes. While the α2–α5 and β1–β3 subunit genes do all have the same pattern of six exons, the α7 encoding gene has ten exons, four of which are conserved with respect to other subunit genes, and the α9 subunit gene has five exons in a quite distinct pattern. The structure of the gene encoding the serotonin 5HT3 receptor most closely resembles the nACh receptor α7 gene, suggesting an evolutionary relationship.

Transcriptional control is the most fundamental of the processes ensuring that these molecules are present in neurons in the correct spatiotemporal pattern. The aspect of transcriptional control that has received most attention to date is its involvement in system establishment, such as the timing of gene expression during development, the mechanism of restriction of gene expression to neurons and types of neurons, and the regulation of receptor gene expression during synapse formation. However, little is known about the mechanism by which receptor gene expression responds to environmental signals impinging on an individual neuron. In addition to their role in maintaining information transfer, these mechanisms are of great importance for learning and memory, and for responses to exposure to artificial agonists or potentiator compounds (such as alcohol and benzodiazepines for GABA(A) receptors and nicotine for nACh receptors) leading to addictive behaviours.

As for all protein-encoding eukaryotic genes, ion channel receptor gene expression requires the assembly of a multiprotein core transcription complex composed of RNA polymerase II and auxiliary factors (Gill 1994). In many genes this complex binds at a sequence called the TATA box, which lies about 30 bp upstream of a transcription start, but many other genes do not have TATA boxes and use other initiator elements. The formation of the core complex on the TATA box is usually insufficient *in vivo* to reach physiological levels of transcription, and additional sequence-specific interactions of various transcription factors with *cis*-acting enhancer and silencer elements are required. There is a lot of evidence for a role for silencing in neurons, acting at two levels. First, there is global silencing of neuronal genes in non-neuronal cell types, and second there is silencing at a fine tuning level to restrict expression of neuronal genes to a subset of neurons. One sequence element that occurs in many neuron-specific genes and is involved in the inhibition of expression in non-neuronal cells is the neural restrictive silencer element (NSRE or NSR), also known as restrictive element-1 (RE-1)

(Grant and Wisden 1997). This sequence is recognized by a zinc finger transcription factor with homology to the GLI-Krüppel family (Schoenherr and Anderson 1995), and known as NSRF or REST. This protein is expressed in a wide range of non-neuronal tissues but is absent from fully differentiated adult neurons. There is evidence (Naruse *et al.* 1999) that NSRF acts in a complex with mSin3 to bind and activate histone deacetylase 1, indicating that NSRF-mediated silencing involves local deacetylation of histones. Neuronal cell-type specific expression is directed by a combination of multiple positive and negative regulatory *cis*-acting elements and their corresponding transcription factors. Many of these elements are found 5′ to the transcription start, but they can reside within introns as for the nestin gene (Zimmerman *et al.* 1994) or in the 3′ untranslated region. The details of these and their functional roles need to be determined separately for each gene.

Preliminary characterization of the 5′ flanking regions of a number of ion channel receptor subunit genes has been carried out, with the greatest effort on glutamate receptor genes. The NR1 (Bai and Kusiak 1995; Bai *et al.* 1998), NR2A (Desai *et al.* 2002), NR2B (Sasner and Buonanno 1996; Klein *et al.* 1998), NR2C (Suchanek *et al.* 1995, 1997), GluR1 (Borges and Dingledine 2001), GluR2 (Myers *et al.* 1998), and KA2 (Huang and Gallo 1997) glutamate receptor subunit genes have been studied. The promoters of these genes share several characteristics. They do not have TATA or CAAT sequences, they are GC-rich and have multiple transcription start sites within a CpG island. In most cases the promoters contain overlapping Sp1 and GSG recognition sites near the major transcription start sites, and an NSR silencer. Other regulatory sites are found, both up and downstream of the principal transcription start, in introns (Huang and Gallo 1997) and in exons (Tintrup *et al.* 2001). For several of these promoters, sequences from 800 bp down to 150 bp in length have been shown to confer neuron-specific expression in transgenic mice, but the mechanism of this specificity of expression is not understood. The NSR sequence in the NR1 and GluR2 genes has only a small modulatory effect with respect to neuronal specificity of expression (Bai *et al.* 1998; Myers *et al.* 1998). Indeed, the GluR2 NSR sequence is a site of mediation of the stimulatory effects on gene expression of the signalling pathways initiated by neurotrophic factors GDNF and BDNF (Brene *et al.* 2000), indicating a quite different role for this element in this gene at least. Sp1 elements have also been shown to play a role in neuronal specificity of expression of the glycine receptor β subunit gene (Tintrup *et al.* 2001).

Lüscher and coworkers (1997) used transgenic mouse lines carrying a lacZ gene driven by GABA(A)R δ subunit gene sequences to analyse promoter activity, showing that a 6.4 kb fragment determined correct adult but not developmental gene expression, and that *in vitro* a 267 bp minimal neuronal promoter could be defined. Mu and Burt (1999) studied promoter regions of the mouse and human GABA(A)R γ2 subunit genes. They showed that, like glutamate receptor genes, this gene had multiple transcription start sites (two of which are major), no TATA or CAAT box and a NRSE sequence, which lies within the first intron. In cell lines the NRSE played a role in suppressing expression in non-neuronal cells, and a 24 bp portion of the 5′ UTR preferentially promoted expression in neuronal cell lines. Fuchs and Celepirone (2002) showed that the GABA(A) receptor α2 subunit gene has three promoter regions, each 5′ proximal to a different first non-coding (5′ UTR) exon. Two of these three promoters lack TATA or CCAAT boxes, and all have multiple transcription initiation sites.

There are several well-established switches of ligand-gated ion channel subunit gene expression during development. Examples are: the replacement of the nAChR foetal γ subunits by functionally homologous ε subunits in muscle cells; the switch from foetal expression of only GlyR α2 subunits to adult expression of only GlyR α1 and α3 subunits in the spinal cord, which underlies the differential strychnine sensitivity of neonates and adults because of the different strychnine affinities of the respective GlyR subtypes; and the switch from NR2B to NR2C glutamate NMDA receptor subunits in cerebellar granule cells about two weeks after birth. The molecular basis of these switches remains unknown.

A particularly important stage in nervous system development is the formation of functional synapses, and the induction of ligand-gated ion channel gene expression in response to growth cone arrival is an important component of synapse development. A clear example of this is the dependence of nAChR gene expression on agrin and other factors such as neuregulin. Both neuronal and muscle nAChR genes are not expressed in mice lacking a functional agrin gene (Gautam *et al.* 1996*a*), while rapsyn is essential for receptor clustering (Gautam *et al.* 1996*b*). Swope (2002) reviews the extensive knowledge of muscle AchR gene expression and assembly at the neuromuscular junction.

8.3.2 Alternative splicing

Alternative splicing is not very widespread among the RNA transcripts of ligand-gated ion channel receptors, but does occur, giving rise to further structural and functional diversity. There are several examples of differential localization of alternatively spliced mRNAs and of splice variant derived proteins, and of differential developmental timing of splice variants, indicating specific functional roles for some splice variants (e.g. Glencorse *et al.* 1992; Paupard *et al.* 1997; Weiss *et al.* 1998). However, in most cases the functional differences remain to be elucidated.

In some cases, alternative splicing controls the presence or absence of a short amino acid sequence that contains a phosphorylation site, and may thus confer different responsiveness to the regulatory and cellular localization effects of phosphorylation. For example, the splice variant α1ins of the GlyR α1 subunit has an 8 amino acid insertion containing a potential phosphorylation site, and the important γ2 subunit of GABA(A) receptors occurs in two splice variants that differ only in the presence (γ2L) or absence (γ2S) of a short 8 amino acid exon encoding a protein kinase C phosphorylation site.

The GABA(A) receptor α2 gene produces six alternatively spliced transcripts, differing only in the 5′UTR region (Fuchs and Celepirone 2002), with the differential inclusion of all or part of three alternative non-coding exons, which have both internal and external splice donor sites. The multiple promoters and splice variants of the transcripts from this gene may enable differential activation and subunit delivery in various cell types during development, when this gene is primarily expressed.

Each of the four AMPA glutamate receptor subunits occurs in two alternatively spliced versions, called flip and flop. These correspond to the alternative inclusion of either of two adjacent exons (exons 14 and 15 in the GluR2 gene), which lie just N-terminal to the M4 transmembrane domain (Sommer *et al.* 1990*b*; Monyer *et al.* 1991). In this case there is good evidence for a functional difference: the flip forms of most subunits desensitize more slowly and to a lesser extent than the flop forms, and this process is affected differentially by various pharmacological agents (Dingledine *et al.* 1999). C-terminal splice variants are also found in AMPA genes GluR2 and GluR4 and in all the kainate receptor subunits. It is thought that alternative splicing at the C-terminus may influence intracellular targeting via altered interactions with targeting proteins. There is a PDZ protein recognition site (S/TXV/L) right at the C-terminus: in GluR1 this sequence is TGL. The intracellular C-terminal sequence is known to play an important part in determining cycling of AMPA receptors in and out of the synapse, as discussed below. The NR1 subunit of glutamate NMDA receptors has three alternatively spliced exons, resulting in eight splice variants. Variants involving different combinations of the two C-terminal exons have different receptor clustering properties (the C1 exon interacts with neurofilaments and the yotiao protein (Ehlers *et al.* 1998; Lin *et al.* 1998)), while only NR1 variants with the C2″ cassette interact with the postsynaptic density. Recombinant NR1 receptors lacking exon 5 show different Zn^{++} and proton sensitivity: when expressed without NR2 subunits Zn^{++} potentiates, but when expressed with NR2 subunits Zn^{++} blocks these receptors at a lower concentration compared with NR1exon5 receptors.

8.3.3 RNA editing

Nascent mRNAs transcribed from genes of members of the glutamate AMPA and kainate gene families (only) are subject to the process of RNA editing (Seeburg 1996), whereby selected adenosine residues are oxidatively deaminated to inosine. Inosine effectively base pairs like guanosine, thus changing the codon and resulting in the insertion of a different amino acid in the translated protein. RNA editing is carried out by either or both of two dsRNA adenosine deaminase enzymes ADAR1 and 2, previously known as dsRAD and RED1 (Rueter *et al.* 1995), and depends on the formation of double-stranded structures involving an intronic sequence (ECS: editing complementary site) which base pairs with the exonic sequences to be edited (Higuchi *et al.* 1993: Fig. 8.2). Other unknown protein factors have been implicated because some cells express the ADAR enzymes but cannot edit (Lai *et al.* 1997).

There are four sites of RNA editing. In the AMPA GluR2, GluR5, and GluR6 mRNAs, editing at the Q/R site (Fig. 8.2a) results in the replacement of a glutamine codon

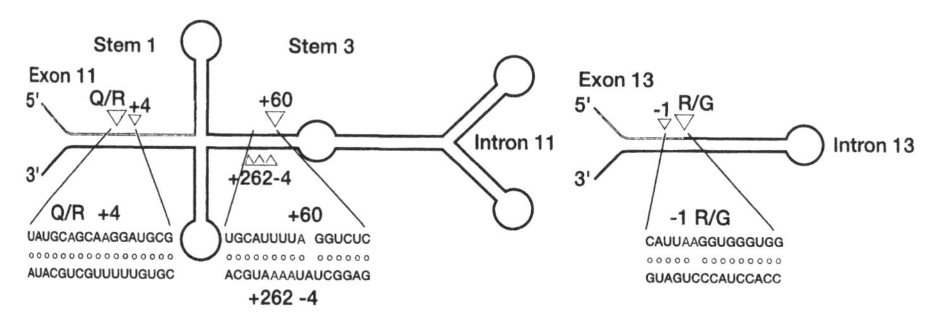

Fig. 8.2 RNA editing of the GluR2 pre-mRNA. (a) Q/R editing: The pre-mRNA is edited at the Q/R site and other sites (+4, +60 and +262 -4) in exon 11 and intron 11. (b) R/G editing at this and the −1 site in exon 13. Editing at all sites involves the oxidative deamination of adenosine to inosine. Editing frequencies at the site shown and other low-frequency sites vary with experimental conditions. Edited A's are indicated by triangles and by bold letters in the corresponding nucleotide sequence. Small open circles indicate Watson-Crick or G:U base pairing. Reprinted with permission from *Nature* vol, 380, pp. 391-392. © 1996 Macmillan magazines Ltd.

(CAG) by an arginine codon (CIG = CGG) and the insertion of Arg (R) into the M2 transmembrane domain. The AMPA GluR2, 3 and 4 mRNAs also undergo editing at the R/G site in exon 13, just N-terminal to the flip-flop region of alternative splicing (Fig. 2b).

The other two editing sites occur in mRNAs of the kainate receptor subunits GluR5 and GluR6. These sites are known as the I/V and Y/C sites, and both result in amino acid substitutions in the M1 transmembrane domain.

GluR2 with R at the Q/R site in M2 has low calcium permeability, low single channel conductance, and linear rectification properties (Hume *et al.* 1991; Verdoorn *et al.* 1991; Swanson *et al.* 1996). In KAIN receptors, Q/R site editing in GluR6 also controls anion permeability (Burnashev *et al.* 1996). R/G site editing in GluR5 and 6 has been shown to reduce desensitization and speed up recovery (Lomeli *et al.* 1994).

These functional properties are important — indeed the R insertion at the Q/R site is essential for life. Concomitantly, this is the most extensively edited site, with editing levels over 99 % in rat GluR2 AMPA receptor mRNA after day E14 of gestation. Removal of the ECS reduced the efficiency of Q/R site editing by about 25 %, and resulted in epilepsy and premature death (Brusa *et al.* 1995). It is not clear that the unedited Q form has any remaining role, since mice made permanently R were phenotypically normal (Kask *et al.* 1998). It is not clear why a mutation has not evolutionarily replaced the need for Q/R editing in GluR2. The introduction of an arginine residue has two important functional effects: it makes the AMPA receptors impermeable to calcium, which is critical because it limits excitotoxicity; and it determines a new endoplasmic reticulum retention signal, which must be masked by the assembly of GluR2 with a different subunit, thus driving the formation of heteromeric receptors (Greger *et al.* 2002). The extent of editing is lower at the other sites, resulting in a large number of functional variants. Editing at these sites increases gradually during development, reaching levels in the adult of between 50 % and 80 % depending on the subunit and species studied.

Q/R editing of kainate receptor subunits is not so extensive, and is subject to developmental regulation. For example, only about 50 % of GluR5 transcripts are edited in the adult rat brain (Sailer *et al.* 1999), while both GluR5 and GluR6 are unedited in the rat embryo with editing setting in before birth (Bernard and Khrestchatisky 1994). There is also regional regulation of the extent of kainate receptor editing in the brain, and significant changes in editing occur after seizures (Grigorenko *et al.* 1998; Bernard *et al.* 1999). Vissel *et al.* (2001) generated mice deficient in GluR6 Q/R site editing by deleting the intronic ECS site. In these mice, NMDA receptor-independent long-term potentiation (LTP) could be induced at the medial perforant path-dentate gyrus synapse. This does not occur in the wild-type, indicating that GluR6 editing suppresses this mechanism of synaptic plasticity, and that regulation of editing will result in differences in LTP. In addition, it was shown that mice deficient in editing were more susceptible to seizures, so that Q/R edited GluR6 is important for the suppression of seizure occurrence.

8.3.4 **Translational control**

In vitro experiments have shown that regulation of the NR2A and GluR2 subunits of glutamate receptors can occur at the level of translation. Many of the glutamate receptor RNAs possess quite long stretches of 5′ UTR upstream of at least some transcription start sites e.g. 772 bp upstream of the major start site in NR2C mRNA (Suchanek *et al.* 1995). For NR2A mRNA it has been shown that removal of a 15 bp sequence in the 5′ UTR that is involved in a putative stem-loop structure resulted in significant disinhibition of translation in rabbit reticulocytes and *Xenopus* oocytes (Wood *et al.* 1996). In NR2A and GluR2 mRNAs, removal of most of the upstream 5′ UTR resulted in about 100 fold and 30–60 fold disinhibition, respectively (Wood *et al.* 1996; Myers *et al.* 1998). Translational suppression of GluR2 mRNA was largely due to a broad region containing a repeat sequence near the 5′ end of the mRNA, which may affect various transcription start sites differentially. The alternative AUG start codons upstream of the major start site did not play a large role in either mRNA. Translational suppression is likely to involve specific mRNA secondary structure and proteins as yet unknown.

It is not known how widespread this phenomenon is among ligand-gated ion channel subunits. However, it is clear that levels of protein subunits should not necessarily be inferred from levels of mRNA. Most importantly, translational disinhibition is a potentially powerful way to deliver large amounts of a receptor subunit quickly to a synapse, since large amounts of mRNA can be held locally in the inactive form until required.

8.3.5 Post-translational modification

Ligand-gated ion channel subunits are extensively phosphorylated and glycosylated. These post-translational modifications play important roles in receptor assembly and delivery, and in the case of phosphorylation, a myriad of function modulation roles (reviewed by Dingledine *et al.* 1999; Swope *et al.* 1999). These receptors are phosphorylated by protein kinase A (PKA: cAMP-dependent protein kinase), protein kinase C (PKC) and calcium/calmodulin-dependent protein kinase II (CAMKII), while tyrosine phosphorylation also occurs, probably by a variety of tyrosine kinases such as Src (GABA(A)R $\gamma 2$ subunit: Brandon *et al.* 2001). Tyrosine phosphorylation has been implicated in the clustering of nACh receptors.

All phosphorylation sites on nAChR and GABA(A)R have been mapped to the large intracellular loop between transmembrane domains M3 and M4, in agreement with the standard topology model. For glutamate receptors, most of the phosphorylation events occur in the intracellular C-terminal domain. The situation with glutamate receptor phosphorylation was confused for some time because of reports that PKA, PKC, and CAMKII phosphorylation occurred in the loop between M3 and M4, which must now be regarded as incorrect. For many years, glutamate receptors were expected to have the same topology as nAChR and GABA(A)R, but subsequently it became clear that this was not so, and that the M3–M4 loop was extracellular and thus very unlikely to be phosphorylated. There had been early reports of non-phosphorylation of this loop by PKA (Moss *et al.* 1993; Tan *et al.* 1994) which are now seen to be correct, and it has now been shown that the putative M3–M4 CAMKII phosphorylation site at Ser627 is not actually phosphorylated (Yakel *et al.* 1995; Roche *et al.* 1996). In fact, Ser831 in the C-terminus of GluR1 is the site of CAMKII action (Roche *et al.* 1996; Mammen *et al.* 1997). Thus current evidence for the location of actual rather than potential phosphorylation sites is in agreement with the membrane topology models for these receptors.

Clear effects of phosphorylation state on channel function or delivery have been found in a number of cases. PKA phosphorylation potentiates native AMPA receptors in cultured neurons, apparently by increasing channel open time (Greengard *et al.* 1991) or the probability of the channel open state. Early evidence showed that phosphorylation of AMPA receptors by CAMKII and probably PKC is generated by electrical stimulation that induces LTP. Phosphorylation of AMPA receptors by CAMKII correlates temporally with increasing synaptic responses mediated by AMPA receptors (Barria *et al.* 1997). Thus phosphorylation by CAMKII (at least) is very likely to play a role in synaptic plasticity.

One role that has emerged is activity-dependent recruitment of glutamate receptor subunits to the synapse. AMPA receptors are heterooligomeric proteins formed by combinations of the subunits GluR1,2,3, and 4. These subunits have either long (GluR1 and GluR4) or short cytoplasmic tails. Receptors with only short cytoplasmic termini cycle continuously in and out of the synapse, while those with long cytoplasmic termini are driven into synapses by strong synaptic activity (Hayashi *et al.* 2000; Zhu *et al.* 2000; Passafaro *et al.* 2001; Shi *et al.* 2001). GluR1-containing receptors are driven into synapses by LTP or CAMKII activity (Shi *et al.* 1999; Hayashi *et al.* 2000). However, the CAMKII effect is mediated indirectly through phosphorylation of proteins other than GluR subunits (Hayashi *et al.* 2000), since mutations in the CAMKII site of the GluR1 subunit do not block. Recent work by Esteban *et al.* (2003) has shown that PKA phosphorylation of the AMPA receptor subunits GluR1 and GluR4 does directly control the incorporation of AMPA receptors containing these subunits into synapses. They showed that in organotypic slices of rat hippocampus, activity-dependent PKA phosphorylation of GluR4 was necessary and sufficient to relieve a retention interaction and drive receptors into synapses. If the GluR4 PKA site at Ser842 is deleted, the receptor is incorporated into synapses in an activity-independent manner. An explanation is that the hydroxyl group of Ser842 is necessary for retention of the receptor away from the synapse by association with unknown factors, and that phosphorylation by PKA breaks this interaction, liberating the receptor. The situation with GluR1 receptors is different. In this case the evidence suggests that PKA phosphorylation at Ser845 (the equivalent residue to Ser842 in GluR4) is necessary but not sufficient for receptor incorporation, and that CAMKII activity, working through an independent pathway not involving GluR1 phosphorylation, is required as well. There is evidence that PDZ-domain proteins are important for AMPA receptor delivery, and that the CAMKII effect works via these proteins. Thus, when GluR1 subunits with a mutation in the PDZ recognition sequence in the C-terminal tail were expressed in hippocampal slice neurons, the CAMKII effect on synaptic response amplitude and rectification were completely blocked (Hayashi *et al.* 2000).

It is now clear that ion-channel receptors are found in large molecular complexes, together with the signalling enzymes that regulate their activity. The efficiency of phosphorylation (and dephosphorylation) is dependent on the formation of these complexes, so that proteins that enable complex formation are often found to stimulate receptor phosphorylation and its functional consequences. For glutamate receptors the MAGUK protein PSD-95/SAP97 is important for clustering, while AKAPs (A-kinase anchoring protein) such as AKAP 79/150 anchor kinases and phosphatases to the complex. Phosphorylation of AMPA receptors is enhanced by a SAP97–AKAP79 complex that directs PKA to GluR1 (Colledge *et al.* 2000). AKAP 79/150 has also been shown to be critical for PKA-mediated phosphorylation of GABA(A)R subtypes containing a β1 or β3 subunit (Brandon *et al.* 2003). Similarly, PKC phosphorylation of GABA(A)R at Ser409 in β subunits is stimulated by the PKC targeting protein RACK-1, which binds to β subunits of the receptor at a different site from PKC (Brandon *et al.* 2002). This phosphorylation event is increased by activation of G-protein linked acetylcholine receptors, providing a molecular basis for functional receptor cross talk.

The extracellular regions of all ligand-gated ion-channel receptors are glycosylated, with both *N*- and *O*-glycosylation occurring (e.g. GABA(A)R α5, Sieghart *et al.* 1993). In glutamate receptors glycosylation occurs on the N-terminal extracellular domain and on the M3–M4 loop. AMPA receptors contain 4–6, kainate receptors 8–10, and NMDA receptors 6–12 glycosylation sites (Everts *et al.* 1997). Indeed, about 20 kD of the 120 kD NR1 and 180 kD NR2 proteins is carbohydrate, and 10 kD of the other NR2 subunits (Kawamoto *et al.* 1995*a*; Laurie *et al.* 1997), and glutamate receptors are major glycoproteins of the postsynaptic density (Clark *et al.* 1998). The carbohydrate side-chains on glutamate receptors are predominantly neutral oligosaccharides, 50 % of which are oligomannosidic glycans with from 5 to 9 mannoses and Man 5 being the major glycan, and much or the rest is complex oligosaccharides (Clark *et al.* 1998).

Glycosylation has been shown to increase the efficiency of receptor assembly and cell-surface delivery of GABA(A) receptors (Connolly *et al.* 1996), but it is not essential for these processes. It is strongly required for ER exit of assembled GlyR (Griffon *et al.* 1999) and nAChR (Gehle and Sumikawa 1991), and it appears that glycosylation plays a similar quality control role for all receptors. Apart from this, there does not seem to be a clear general role for glycosylation, with effects varying from protein to protein and dependent on the cellular system used to study the effects. In addition to protein trafficking, effects of glycosylation have been observed on ligand-binding affinity, channel function, and susceptibility to allosteric modulation (reviewed for glutamate receptors by Standley and Baudry 2000). Considerable attention has been given to the proposal that *N*-glycosylation is required for the formation of functional ligand-binding sites, because of the conservation of glycosylation sites (e.g. AMPA receptors have two conserved *N*-glycosylation sites in the ligand-binding domains) and other evidence (e.g. Kawamoto *et al.* 1995*b*). However, unglycosylated extracellular domains of GluR2 were crystallized successfully with kainate bound (Armstrong *et al.* 1998), and mutant GluR4 receptors with the conserved glycosylation sites at N407 and N414 eliminated showed no differences in ligand-binding properties, channel properties, or indeed surface delivery (Pasternack *et al.* 2003). However, lectins like ConA do have modulatory effects on channel properties that depend on binding to carbohydrate side chains (Everts *et al.* 1997), which must exert some influence on protein function.

There have also been reports of palmitoylation (Pickering *et al.* 1995), and calpain proteolysis (Bi *et al.* 1997, 1998), but the functional relevance of these modifications is unknown.

8.4 **Receptor assembly and trafficking**

Assembly of ligand-gated ion channels is a multistep process, involving folding reactions and post-translational modifications following polypeptide synthesis and insertion into the endoplasmic reticulum. This process displays strict selectivity in that only certain combinations of subunits are oligomerized and targeted to the plasma membrane via the Golgi. All ligand-gated ion channel polypeptides have N-terminal signal sequences that direct ER insertion. Protein folding occurs with the assistance of chaperone proteins, such as calnexin for nAChR subunits (Gelman *et al.* 1995) and BiP and calnexin for GABA(A) receptor subunits (Connolly *et al.* 1996). Association between N-terminal domains may be initiated before folding is complete in some cases, occurring as chaperone molecules like calnexin dissociate (Gelman *et al.* 1995). Assembly is driven by specific interactions between the ER-internal (extracellular) domains of different subunits, leading to correct stoichiometry and nearest-neighbour relationships. Each receptor assembles through a series of assembly intermediates. These have been reasonably well characterized for the muscle nAChR, which is a single molecular subtype with four subunits in the stoichiometry $\alpha,\alpha,\beta,\gamma,\delta$. The conformationally mature α subunit oligomerizes with either of the δ or γ subunits to form $\alpha\delta$ or $\alpha\gamma$ heterodimers. These then associate with one another and the β subunit to form $\alpha2\beta\gamma\delta$ complexes (Blount *et al.* 1990; Gu *et al.* 1991). Other evidence has suggested that $\alpha\beta\gamma$ trimers form and recruit first δ and then the final α subunit (Green and Claudio 1993). For Gly receptors (Griffon *et al.* 1999) the α subunits are able to form homomeric receptors when expressed by themselves, but the β subunit cannot. The presence of β subunits overrides the homomeric α assembly, and only receptors with the stoichiometry $\alpha3\beta2$ are assembled, via $2(\alpha\beta) + \alpha$. Particular N-terminal amino acid sequences, called assembly boxes, and subunit surface structures are important for these intersubunit interactions. Analysis of an assembly box in GABA(A) receptor $\alpha1$ subunits (Bollan *et al.* 2003) showed that amino acids 57–68 are required absolutely for assembly of $\alpha\beta$ receptors, and that the motif GKER in subunit $\beta3$ is able to direct assembly of $\beta\gamma$ receptors. Using site-directed mutagenesis and expression in *Xenopus* oocytes, Griffon *et al.* (1999) identified eight amino acid residues within the N-terminal region of the GlyR $\alpha1$ subunit that are required for the formation of homomeric channels. Some of these residues, and some of those known to be important for nAChR assembly (Kreienkamp *et al.* 1995) also occur in the intersubunit interface, but otherwise there is no correlation between the assembly boxes of different receptor families. These parts of the N-terminal domain have diverged to gain specificity, while the ligand-binding regions retain structural similarities (Devillers-Thiéry *et al.* 1993).

Another crucial level of sorting occurs between the ER and the plasma membrane, and only certain combinations of subunits are able to make this passage. This is driven by interactions between sequences on the receptor subunits and accessory proteins. Exit from the ER is also subject to a quality control step involving recognition of glycosylation of certain sites in the N-terminal domain. GlyR $\alpha1$ mutant subunits that are not glycosylated are retained in the ER (Griffon *et al.* 1999), and are rapidly degraded. This glycosylation-dependent exit applies to all receptors of this superfamily at least, although the stringency may vary. There are also ER retention signals in some subunits. The best example is the AMPA GluR2 subunit, where Q/R editing introduces an arginine at position 607, creating an ER retention signal. This has to be masked by assembly with other subunits to allow exit from the ER (Greger *et al.* 2002).

As they move from ER to Golgi and on to the plasma membrane, the receptors become associated with proteins that play key roles in targeting the receptors to the correct sites in the cell (reviewed by Connolly, this volume Chapter 4; Chapters on each ion-channel receptor family in Moss and Henley 2002; Kittler and Moss 2001;

Sheng and Lee 2001; Malinow and Malenka 2002; Kneussel 2002). An example of such a protein is GABARAP (GABA(A) receptor associated protein), which is found primarily at transport sites in the Golgi, interacts with NSF (*N*-ethylmaleimide-sensitive fusion protein), and plays a key role in synaptic vesicle fusion with target membranes. GABARAP associates specifically with γ subunit proteins of GABA(A) receptors, primarily γ2 which is the most common, and is also able to bind tubulin. Thus GABARAP probably selects vesicles with γ2-containing GABA(A) receptors and enables transport along microtubules to the synapse. GABA(A) receptors lacking a γ subunit are localized extrasynaptically, and must use a different molecular mechanism for transport. The AMPA receptor subunit GluR2 also interacts with NSF, this time directly. The *stargazin* protein plays a key role in AMPA receptor delivery, both to the plasma membrane in general, and via a PDZ-binding domain, in handing over receptors to the synapse complex. These transport processes are linked to receptor recycling and storage, and are regulated by signalling pathways acting through PKC and tyrosine kinases. It is to be expected that there is a specific component to each receptor family and mode of transport. Membrane delivery can be direct to the synapse, as for GABA(A) receptors containing γ subunits and for GluR2, or indirectly via random surface delivery followed by capture at the synapse or delivery to the synapse via clathrin vesicles and sorting endosomes, as for GluR1-containing glutamate receptors.

At the synapse, ligand-gated ion channel receptors are held in a dynamic relationship (30–40 % of synaptic GABA(A) and AMPA receptors are capable of lateral diffusion) with a protein complex by multiple interactions, with certain proteins playing key roles in receptor clustering and retention. The best characterized of these are gephyrin, the GlyR clustering protein, and rapsyn, the nAChR clustering protein. Rapsyn associates with the intracellular M3–M4 loop of nAChR (Unwin 1993), and mediates the action of the agrin-stimulated signalling pathway that drives nAChR into synapses (Gautam *et al.* 1996). GlyR and AMPA receptors have been shown in tracking experiments (Meier 2001; Borgdorff and Choquet 2002) to alternate between periods of movement in the membrane and of relative immobility at cluster sites together with the clustering protein.

Once at the synapse, the receptors become involved in extensive protein complexes with scaffolding proteins, signal transduction proteins, and cytoplasmic anchoring proteins linked to the cytoskeleton. This complex is particularly striking for excitatory synapses, where it is termed the post-synaptic density (review: Sheng 2001). NMDA receptors are relatively more fixed components of the synaptic complex than AMPA receptors, which move in and out all the time. These two families of glutamate receptor are involved in distinct protein sub-complexes. These differential protein interactions underlie the differential regulation of synaptic targeting of AMPA and NMDA receptors. PSD-95, a MAGUK family protein with multiple PDZ domains, binds the ESDV or ESEV motif in the intracellular C-terminal tails of NMDA receptor subunits, anchors the receptor to the cytoskeleton and couples the receptor to cytoplasmic signalling pathways such as calcium signalling acting via neuronal nitric oxide synthase, and also to adhesion molecules, receptor tyrosine kinases and ion channels. The NMDA synaptic complex has been isolated and studied by proteomic approaches (Husi *et al.* 2000).

Receptors are also removed from the cell surface and from synapses, and this is used to regulate responsiveness to environmental signals as well as for receptor husbandry. This typically occurs by the signalled recruitment of adaptor proteins, e.g. AP-2, arrestin, ubiquitin. These then recruit clathrin, which instigates membrane invagination and endocytosis, forming early endosomes. From there, receptors can be recycled rapidly to the plasma membrane or delivered to a late endosome for sorting: after sorting receptors may be recycled via the *trans*-Golgi network to the plasma membrane, or sent to lysosomes for degradation. These trafficking events all require protein–protein interactions, most of which have not been discovered. One possible example is the interaction of GABA(A) receptors with Plic-1, which only binds α and β subunits. Plic-1 is required for maintenance of cell surface receptor levels, but may achieve this indirectly by protecting receptors from degradation, either by blocking ubiquitinization or by channelling them into recycling. All of these trafficking events are subject to regulation. For example, αβγ-containing GABA(A) receptors recycle continuously between plasma membrane and internal compartments, and are routed through a late endosomal compartment, from which they can be recruited rapidly by insulin stimulation via tyrosine kinase activity, which can be blocked by PKC activation (Connolly, this volume Chapter 4).

8.5 **3D Structure and the molecular basis of receptor properties**

Knowledge of the 3D structure of ligand-gated ion channels is fundamental to understanding the properties of these important molecules, and how to modulate those properties with drugs. However, their complexity, the fact that they are membrane proteins, and some mistaken assumptions, notably that all such receptors would resemble the nAChRs, have made progress slow. Nevertheless, very significant progress has been made recently. In this brief discussion we will base our statements on physical data or 3D modelling data wherever possible, rather than on the immense body of site-directed mutagenesis and chimaeric receptor data. Effects of amino acid substitutions can be specific local effects of the amino acid exchange and/or effects on overall structure, and we have no way of deciding which pertains. It is helpful when considering the structure of nAChR and GluR to view each subunit as having a modular structure, with the extracellular module forming the agonist-binding site and directing subunit assembly, and a membrane/C-terminal module that forms the ion channel and links to intracellular structures and regulatory processes. Support for the modular concept is provided by the generation of functional chimeras between the GABA(C) ρ subunit extracellular domain and the glycine receptor membrane-spanning domain (Mihic *et al.* 1997), and between 5HT3 and nAChR α7 subunits (Eisele *et al.* 1993).

8.5.1 nACh receptor group structure

The nAChR is still assumed, probably correctly, to be a prototype for all receptors of the Cys loop, nAChR/GABA(A)R superfamily. Many years of electron crystallographic work by Unwin and coworkers on tubular crystals of *Torpedo* postsynaptic membranes, which are spectacularly enriched in a lower vertebrate nAChR, has given us an overall picture of the distribution of peptide backbone and side-chain densities throughout an entire receptor down to 4.6 Å resolution, and provided a framework for the interpretation of all other experiments (Unwin 1989, 1993, 1995, 1996; Miyazawa *et al.* 1999). More precise information about the structure of the extracellular domains of nAChR class receptors was obtained by the determination by X-ray crystallography of the 3D atomic structure of an acetylcholine-binding protein from glial cells of the snail *Lymnaea stagnalis* (AchBP; Brejc *et al.* 2001; post-structure review, Karlin 2002), and by the fitting of the electron data to the crystal structure (Unwin *et al.* 2002).

The nAChR is a large (~290 kDa) glycoprotein with five ~160 Å long rod-shaped subunits. The five subunits are arranged around a pseudo 5-fold axis, form a cation-selective pathway across the membrane when the channel is open, and form a robust barrier to ions when it is closed. The clockwise order of the subunits in the *Torpedo* receptor is γαβδα. The two α subunits are identical in sequence but slightly different in structure when assembled into the mature receptor. They are termed α(γ), which lies between γ and β, and α(δ) respectively. The cation-conducting pathway lies centrally along the axis of the receptor for most of its length. There are ~20 Å wide vestibules on either side of the membrane, and the pore is open and gets gradually wider towards the extracellular side, but is closed off at the cytoplasmic end both by the combined M3–M4 cytoplasmic loops and by the accessory protein rapsyn.

The primary feature of the cytoplasmic portion of each subunit is a 30 Å α-helical rod. The five rods come together beneath the receptor to form an inverted pentagonal cone. Thus cytoplasmic ions cannot access the cation channel directly, but must pass through narrow transverse openings, which are about 15 Å long and not more than 8 Å wide in order to access the channel: extracellular ions must take this route in reverse. The two major openings lie between the 30 Å α-helical rods of the α(δ) and δ subunits and between the α(γ) and β subunits. The amino acids lining these channels may play a role in cation selectivity. A conserved α-helical region (which is likely to be part of the 30 Å rod) in the M3–M4 loop near to M4 contains a set of charged amino acids. These amino acids are negatively charged in nAChR subunits but positively charged in GlyR subunits, and the overall net charge of the M3–M4 loop is negative for cation–conducting receptors and positive for the anion-conducting glycine and GABA(A) receptors. About 40 Å from the cytoplasmic membrane surface the pentagonal features cease to be discernable, and the remainder of the density is due to bound rapsyn.

The gate in the receptor is of particular interest. In the closed state, a narrow strip of density can be seen to close the pore. This is close to the middle of the membrane, almost 15 Å away from the cytoplasmic membrane surface. The width of the bridging density is only a fifth of the separation of the phospholipid headgroups (i.e. <6 Å), suggesting that it is made of no more than one or two rings of side chains. Cysteine-substitution mutagenesis experiments have indicated a more cytoplasmic location for the gate (Akabas *et al.* 1994; Wilson and Karlin 1998), and these experiments have not yet been reconciled with the physical data. The gate could be made of any non-polar amino acid, since in the absence of polarizable groups to provide electrostatic stabilization, the ion would retain its hydration shell but there would be restricted space for water, and the ion would be unable to pass through the central hole in the closed state. Unwin (1993) initially suggested that the set of Leu251 side chains in the M2 segments of the subunits could form the gate with a leucine-zipper effect, but more recently has proposed that a hydrophobic girdle of leucine and valine side chains constitutes the gate.

All the mutagenesis evidence (Karlin 2002) indicates that the M2 transmembrane segment is the primary pore-lining segment in all receptors of this class (e.g. 5HT3, Reeves *et al.* 2001; GABA(A), Xu and Akabas 1996), and that it is largely α-helical in structure. In agreement with the topology model, four rod-like densities perpendicular to the membrane are seen arranged in a diamond-shaped grouping with one corner density next to the pore (Unwin *et al.* 2002). These densities can be presumed to correspond to the four α-helices of the transmembrane domains M1 to M4. From other evidence (Karlin 2002), M2 is closest to the pore, and M4 furthest away. There is a kink of M2 into the pore at or close to the level of the gate. Although M2 residues line the pore, some residues in other transmembrane domains are water accessible; for example, some residues in M3 of the α1 subunit of GABA(A) receptors are accessible in the absence of GABA, others only when GABA is present (Williams and Akabas 1999, 2000).

The structure of the extracellular domain is defined by the atomic structure of AChBP (Brejc *et al.* 2001), which has 20–23 % sequence identity with the N-terminal 200 or so residues of nAChR subunits. The nAChR sequence can be fitted to this atomic structure, and the observed densities at 4.6 Å fit closely for the β, γ, and δ subunits and for α subunits when an agonist is bound (i.e. in the open state) (Unwin *et al.* 2002). The core of this structure is organized around two sets of β-strands, forming 'Greek key' motifs, which are linked together through the Cys-loop disulfide bond and folded into a curled β sandwich, consisting of inner and outer sheets. The sandwich is staggered, so that towards the synaptic cleft, only the 'inner' β-sheet is still present. Connecting loop structures are not conserved. In cross-section, the five subunits are seen to be relatively well separated and distinguishable close to the membrane, but their densities blend into a uniform ring further out where only the inner sheet is present.

The two ACh binding sites, one per β subunit, lie at the α(δ)–δ and α(γ)–γ interfaces on chemical labelling and mutagenesis evidence. A transverse channel in the extracellular domain that could accommodate an ACh molecule is found only in the α subunit, and has been suggested to be the ACh binding pocket (Miyazawa *et al.* 1999).

The α subunits are unique and play a crucial role. Their activated conformation is like that of the other subunits, and like AChBP. However, in the resting, closed channel state, they are distorted, presumably by intersubunit interactions in assembly. The binding of ACh results in the relaxation of these distortions, and the receptor takes up a more symmetrical and uniform structure. This is analogous to the conversion from T state to R state in allosteric proteins. How this occurs awaits higher resolution data, but it may be that the binding of ACh draws the C-loop inwards close to its position in AChBP, allowing the outer β-sheet to be reoriented and stabilized in its configuration prior to intersubunit interactions. This relaxation involves a 15–16° clockwise rotation of the inner β-sheet about an axis normal to the membrane, and 11° changes in tilt of the outer β-sheet. It is presumed that this structural change has transmitted effects on the transmembrane domains and the gate, leading to channel opening.

What is observed is that when ACh binds (i.e. comparing open to resting states) there is a rotation of the (putative) M2 transmembrane rod, so that the central kinks in the barrel of five M2 helices are rotated out of the pore, widening the aperture. The open pore has a tapered shape, narrowing towards the intracellular membrane surface, with a minimum diameter of ~10 Å. It now seems clear that the rotational relaxation of the α subunits is the key event, and that this conformational change is transmitted in turn to the M2 transmembrane domains. It is not known how the coupling of these two events is achieved, but the M2–M3 loop plays a crucial role in both muscle ACh receptor (Grosman *et al.* 2000) and GABA(A) receptor gating (Bera *et al.* 2002).

8.5.2 Glutamate receptor structure

Glutamate receptors appear to be built of a pore-forming transmembrane module with similarities to the structure of potassium channels in this region (Doyle *et al.* 1998; Kuner *et al.* 2003), and extracellular domains that in part resemble bacterial periplasmic amino acid binding proteins. Many useful models of extracellular structure were built by fitting to the bacterial protein structure, but a more definitive 3D structure was provided by Armstrong *et al.* (1998) who determined the atomic structure of the ligand-binding domain of GluR2 complexed with the agonist kainate. The N-terminal 400 amino acids play no role in ligand binding. The amino acid binding pocket is formed from amino acid side chains of two globular domains (S1 and S2) drawn from the sequence adjacent to the M1 domain and from the M3–M4 loop (which is extracellular in these receptors). Unlike nAChR, the binding site is completely contained within a single subunit. The two globular domains have been compared to the two parts of a clamshell. The ligand binds deep within the pocket between S1 and S2, and stabilizes the closed form of the clamshell. As for ACh receptors, the conformational change occurring on clamshell closure is presumably transmitted to the transmembrane domain, which then also undergoes a conformational change resulting in channel opening. The cytoplasmic vestibule, the inner half of the pore, and a central constriction are all provided by amino acids in the ascending or descending parts of the M2 segment. The M2 ascending limb is α-helical, while the descending limb has an extended structure (Kuner *et al.* 1996). This segment provides the functional gate of the channel, which is located at or just below the turn of the re-entrant loop with asparagine residues playing a key role. They also affect other key properties such as calcium permeation and voltage-dependent magnesium block of NMDA receptors. The outer half of the pore and extracellular vestibule are provided by the C-terminus of M3 and the N-terminus of M4, supported structurally by the sequence preceding M1 (Beck *et al.* 1999). Clearly, there is a more intimate relationship between the ligand-binding domain and the pore domain in these receptors.

8.5.3 P2X receptor structure

We do not have three dimensional structure information for P2X receptors. However, there is evidence that ATP binds to the region just extracellular to TM1 (Ennion *et al.* 2000; Jiang *et al.* 2000), and causes an unknown conformational change which in turn results in movement of TM1 (Haines *et al.* 2001; Jiang *et al.* 2001) with respect to TM2, thus opening the pore. The evidence that TM2 residues line all or most of the pore is that mutations in TM2 affect ion permeability (Khakh *et al.* 1999; Virginio *et al.* 1999) and that cysteine substitution mutants in TM2 are modified by chemical agents when the channel is open (Rassendren *et al.* 1997). The gate is likely to be near the conserved glycine G342 (Egan *et al.* 1998; Migita *et al.* 2001), while a conserved aspartate residue (D349) is internal to the gate. There is evidence for two open states I1 and I2, with the I2 pore being 60–75 Å greater in cross-sectional area than the I1 pore, and about 3 Å greater (14 Å) in diameter. These state changes of the pore are regulated by the disposition of side chains in the cytosolic C-terminal domain, and possibly by other protein partners (Kim *et al.* 2001), and lead to different permeability properties for the channel (Eickhorst *et al.* 2002).

8.6 Outlook

There is now some realistic hope that we can understand ligand recognition, ion permeation, gating and desensitization of ligand-gated ion channels. There is now a firm atomic structural foundation for the study of ligand binding, and the binding sites for the large number of natural modulators (such as glycine acting on the M3–M4 loop of NMDA receptors) and pharmacological agents can now be delineated in three dimensions. Further structural and mutagenesis work should throw light on how various types of potentiation and inhibition occur at the atomic level. Further attempts to understand ion-selectivity and permeation can be made; for example, in the nAChR class of molecules, the contributions of both pore-lining amino acids and the amino acid side chains lining the transverse pores described by Unwin can be studied using mutagenesis and functional analysis. Our understanding of gating and desensitization is basic and descriptive (Dingledine 1999), but progress can now be made using pharmacology and mutagenesis on a known 3D framework. Many investigators are studying the cell biology of these receptors, their roles in synaptic plasticity (Morris BJ, this volume Chapter 14) and in nervous system function and disease. Successful modulation of the functions of these receptors continues to depend on increasing our comprehension of how these amazing molecules work at the level of atomic structure and biophysics.

References

Akabas, M. H., Kaufmann, C., Archdeacon, P., and Karlin, A. (1994) Identification of acetylcholine receptor channel-lining residues in the entire M2 segment of the α subunit. *Neuron*, **13**, 919–27.

Arbas, E. A., Meinertzhagen, I. A., and Shaw, J. R. (1991) Evolution in nervous systems. *Ann. Ker. Neurosci.,* **14**, 9–38.

Armstrong, N., Sun, Y., Chen, G.-Q., and Goaux, E. (1998) Structure of a glutamate-receptor ligand-binding core in complex with kainate. *Nature*, **395**, 913–17.

Bai, G. and Kusiak, J. W. (1995) Functional analysis of the proximal 5′-flanking region of the *N*-methyl-D-aspartate receptor subunit gene, NMDAR1. *J. Biol. Chem.*, **270**, 7737–44.

Bai, G., Norton, D. D., Prenger, M. S., and Kusiak, J. W. (1998) Single-stranded DNA binding proteins and neuron-restrictive silencer factor participate in cell-specific transcriptional control of the NMDAR1 gene. *J. Biol. Chem.*, **273**, 1086–91.

Barria, A., Muller, D., Derkach, V., Griffith, L. C., and Soderling, T. R. (1997) Regulatory phosphorylation of AMPA-type glutamate receptors by CAM-KII during long-term potentiation. *Science*, **276**, 2042–5.

Beck, C., Wollmuth, L. P., Seeburg, P. H., Sakmann, B., and Kuner, T. (1999) NMDAR channel segments forming the extracellular vestibule inferred from the accessibility of substituted cysteines. *Neuron*, **22**, 559–70.

Bera, A. K., Chatav, M., and Akabas, M. H., (2002) GABA(A) receptor M2-M3 loop secondary structure and changes in accessibility during channel gating. *J. Biol. Chem.*, **277**, 43002–10.

Bernard, A. and Khrestchatisky, M. (1994) Assessing the extent of RNA editing in the TMII regions of GluR5 and GluR6 kainate receptors during rat brain development. *J. Neurochem.*, **62**, 2057–60.

Bernard, A., Ferhat, L., Dessi, F., Charton, G., Represa, A., Ben-Ari, Y., *et al.* (1999) Q/R editing of the rat GluR5 and GluR6 kainate receptors in vivo and in vitro: evidence for independent developmental, pathological and cellular regulation. *Eur. J. Neurosci.*, **11**, 604–16.

Bi, X., Chen, J., Dang, S., Wenthold, R. J., Tocco, G., and Baudry, M. (1997) Characterization of calpain-mediated proteolysis of GluR1 subunits of α-amino-3-hydroxy-5-methylisoxazole-4-proprionate receptors in rat brain. *J. Neurochem.*, **68**, 1484–94.

Bi, X., Rong, Y., Chen, J., Dang, S., Wang, Z., and Baudry, M. (1998) Calpain-mediated regulation of NMDA receptor structure and function. *Brain Res.*, **790**, 245–53.

Blount, P., Smith, M. M., and Merlie, J. P. (1990) Assembly intermediates of the mouse muscle nicotinic acetylcholine receptor in stably transfected fibroblasts. *J. Cell Biol.*, **111**, 2601–11.

Boess, F. G. and Martin, I. L. (1995) Molecular biology of 5-HT receptors. *Neuropharmacology*, **33**, 275–317.

Bollan, K., King, D., Robertson, L. A., Brown, K., Taylor, P. M., Moss, S. J., *et al.* (2003) GABA(A) receptor composition is determined by distinct assembly signals within alpha and beta subunits. *J. Biol. Chem.*, **278**, 4747–55.

Borgdorff, A. J. and Choquet, D. (2002) Regulation of AMPA receptor lateral movements. *Nature*, **417**, 649–53.

Borges, K. and Dingledine, R. (2001) Functional organization of the GluR1 glutamate receptor promoter. *J. Biol. Chem.*, **276**, 25929–38.

Brandon, N. J., Delmas, P., Hill, J., Smart, T. G., and Moss, S. J. (2001) Constitutive tyrosine phosphorylation of the GABA(A) receptor gamma 2 subunit in rat brain. *Neuropharmacology*, **41**, 745–52.

Brandon, N. J., Jovanovic, J. N., Smart, T. G., and Moss, S. J. (2002) Receptor for activated C kinase-1 facilitates protein kinase C-dependent phosphorylation and functional modulation of GABA(A) receptors with the activation of G-protein-coupled receptors. *J. Neurosci.*, **22**, 6353–61.

Brandon, N. J., Jovanovic, J. N., Colledge, M., Kittler, J. T., Brandon, J. M., Scott, J. D., *et al.* (2003) A-kinase anchoring protein 79/150 facilitates the phosphorylation of GABA(A) receptors by cAMP-dependent protein kinase via selective interaction with receptor beta subunits. *Mol. Cell Neurosci.*, **22**, 87–97.

Brejc, K., van Dijk, W. J., Klaassen, R. V., Schurmanns, M., van der Oost, J., Smit, A. B., *et al.* (2001) Crystal structure of an ACh-binding protein reveals the ligand-binding domain of nicotinic receptors. *Nature*, **411**, 269–76.

Brene, S., Messer, C., Okado, H., Hartley, M., Heinemann, S. F., and Nestler, E. J. (2000) Regulation of GluR2 promoter activity by neurotrophic factors via the neuron-restrictive silencer element. *Eur. J. Neurosci.*, **12**, 1525–33.

Brusa, R., Zimmermann, F., Koh, D.-S., Feldmeyer, D., Gass, P., Seeburg, P. H., *et al.* (1995) Early-onset epilepsy and postnatal lethality associated with an editing-deficient GluR-B allele in mice. *Science*, **270**, 1677–80.

Burnashev, N., Villa roel, A., and Sakmann, B. (1996) Dimensions and ion selectivity of recombinant AMPA and kainate receptor channels and their dependence on Q/R site residues. *J. Physiol.*, **496**, 165–73.

Clak, R. A., Gurd, J. W., Bissoon, N., Tricaud, N., Molnar, E., Zamze, S. E., *et al.* (1998) Identification of lectin-purified neural glycoproteins, GPs 180, 116 and 110, with NMDA and AMPA receptor subunits: conservation of glycosylation at the synapse. *J. Neurochem.*, **70**, 2594–605.

Colledge, M., Dean, R. A., Scott, G. K., Langeberg, L. K., Huganir, R. L., and Scott, J. D. (2000) Targeting of PKA to glutamate receptors through a MAGUK-AKAP complex. *Neuron*, **27**, 107–19.

Conley, E. C. and Brammar, W. J. (1999) *The Ion Channel facts book.* Third Edition, Academic Press, New York.

Connolly, C. N., Krishek, B. J., McDonald, B. J., Smart, T., and Moss, S. J. (1996) Assembly and cell surface expression of heteromeric and homomeric γ-aminobutyric acid type A receptors. *J. Biol. Chem.*, **271**, 89–96.

Desai, A., Turetsky, D., Vasudevan, K., and Buonanno, A. (2002) Analysis of transcriptional regulatory sequences of the *N*-methyl-D-aspartate receptor 2A subunit gene in cultured cortical neurons and transgenic mice. *J. Biol. Chem.*, **277**, 46374–84.

Devillers-Thiéry, A., Galzi, J. L., Eisele, J. L., Bertrand, S., Bertrand, D., and Changeux, J. P. (1993) Functional architecture of the nicotinic acetylcholine receptor: a prototype of ligand-gated ion channels. *J. Membr. Biol.*, **136**, 97–112.

Dingledine, R., Borges, K., Bowie, D., and Traynelis, S. F. (1999) The glutamate receptor ion channels. *Pharmacol. Rev.*, **51**, 7–61.

Doyle, D. A., Cabral, J. M., Pfuetzner, R. A., Kuo, A., Gulbis, S. L., Chait, B. T., *et al.* (1998) The structure of the potassium channel: molecular basis of K^+ conduction and selectivity. *Science*, **280**, 69–77.

Eickhorst, A., Berson, A., Cockayne, D., Lester, H. A., and Khakh, B. S. (2002) Control of P2X2 receptor permeability by the cytosolic domain. *J. Gen. Physiol.*, **120**, 119–31.

Eisele, J. l., Bertrand, S., Galzi, J. L., Devillers-Thiery, A., Changeux, J. P., and Bertrand, D. (1993) Chimaeric nicotinic-serotonergic receptor combines distinct ligand binding and channel specificities. *Nature*, **366**, 479–83.

Egan, T. M., Haines, W. R., and Voigt, M. M. (1998) A domain contributing to the ion channel of ATP-gated P2X2 receptors identified by the substituted cysteine accessibility method. *J. Neurosci.*, **18**, 2350–9.

Ehlers, M. D., Fung, E. T., O'Brien, R. J., and Huganir, R. L. (1998) Splice variant-specific interaction of the NMDA receptor subunit NR1 with neuronal intermediate filaments. *J. Neurosci.*, **18**, 720–30.

Ennion, S., Hagan, S., and Evans, R. J. (2000) The role of positively charged amino acids in ATP recognition by human P2X1 receptors. *J. Biol. Chem.*, **275**, 29361–67.

Esteban, J. A., Shi, S-H., Wilson, C., Nuriya, M., Huganir, R. L., and Malinow, R. (2003) PKA phosphorylation of AMPA receptor subunits controls synaptic trafficking underlying plasticity. *Nat. Neurosci.*, **6**, 136–43.

Everts, I., Villman, C., and Hollmann, M. (1997) N-glycosylation is not a prerequisite for glutamate receptor function but is essential for lectin modulation. *Mol. Pharmacol.*, **52**, 861–73.

Fredholm, B. B. (1995) Purinoreceptors in the nervous system. *Pharmacol. Toxicol.*, **76**, 228–39.

Fuchs, K. and Celepirone, N. (2002) The 5′- flanking region of the rat GABA(A) receptor alpha2-subunit gene (Gabra 2) *J. Neurochem.*, **82**, 1512–23.

Gautam, M., Noakes, P. G., Moscoso, L., Rupp, F., Scheller, R. H., Merlie, J. P. *et al.* (1996*a*) Defective neuromuscular synaptogenesis in agrin-deficient mutant mice. *Cell*, **85**, 525–35.

Gautam, M., Noakes, P. G., Mudd, J. M., Nichol, M., Chu, G. C., Sanes, J. R., *et al.* (1996*b*) Failure of postsynaptic specialization to develop at the neuromuscular junction of rapsyn-deficient mice. *Nature*, **377**, 232–6.

Gehle, V. M. and Sumikawa, K. (1991) Site-directed mutagenesis of the conserved N-glycosylation site on the nicotinic acetylcholine receptor subunits. *Brain Res. Mol. Brain Res.*, **11**, 17–25.

Gelman, M. S., Chang, W., Thomas, D. Y., Bergeron, J. J., and Prives, J. M. (1995) Role of the endoplasmic reticulum chaperone calnexin in subunit folding of nicotinic acetylcholine receptors. *J. Biol. Chem.*, **270**, 15085–92.

Gill, G. (1994) Taking the initiative. *Curr. Biol.*, **4**, 374–6.

Glencorse, T. A., Bateson, A. N., and Darlison, M. G. (1992) Differential localization of two alternatively-spliced GABA(A) receptor γ2-subunit mRNAs in the chick brain. *Eur. J. Neurosci.*, **4**, 271–7.

Glencorse, T. A. and Davies, R. W. (1997) Molecular biology of neurotransmitter-gated ion channel receptors. In *Molecular Biology of the Neuron* (ed. R. W. Davies, B. J. Morris), Oxford University Press, Oxford, U.K., pp. 205–40.

Grant, A. L. and Wisden, W. (1997) Neuron-specific gene expression. In *Molecular Biology of the Neuron* (ed. R. W. Davies, B. J. Morris), Oxford University Press, Oxford, U.K., pp. 67–93.

Green, W. N. and Claudio, T. (1993) Acetylcholine receptor assembly: subunit folding and oligomerization occur sequentially. *Cell*, **74**, 57–69.

Greengard, P., Jen, J., Nairn, A. C., and Stevens, C. F. (1991) Enhancement of the glutamate response by cAMP-dependent protein kinase in hippocampal neurons. *Science*, **253**, 1135–8.

Greger, I. H., Khatri, L., and Ziff, E. B. (2002) RNA editing at arg607 controls AMPA receptor exit from the endoplasmic reticulum. *Neuron*, **34**, 759–72.

Grenningloh, G., Rienitz, A., Schmitt, B., Methfessel, C., Zensen, M., Beyreuther, K., *et al.* (1987) The strychnine-binding subunit of the glycine receptor shows homology with nicotinic acetylcholine receptors. *Nature*, **328**, 215–20.

Griffon, N., Büttner, C., Nicke, A., Kuhse, J., Schmalzing, G., and Betz, H. (1999) Molecular determinants of glycine receptor assembly. *EMBO J.*, **18**, 4711–21.

Grigorenko, E. V., Bell, W. L., Glazier, S., Pons, T., and Deadwyler, S. (1998) Editing status at the Q/R site of the GluR2 and GluR6 glutamate receptor subunits in the surgically excised hippocampus of patients with refractory epilepsy. *Neuroreports*, **9**, 2210–24.

Grosman, C., Salamone, F. N., Sine, S. M., and Auerbach, A. (2000) The extracellular linker of muscle acetylcholine receptor channels is a gating control element. *J. Gen. Physiol.*, **116**, 327–40.

Gu, Y., Gamacho, P., Gardner, P., and Hall, Z. W. (1991) Assembly of the mammalian muscle acetylcholine receptor in transfected COS cells. *J. Cell. Biol.*, **114**, 799–807.

Haines, W. R., Migita, K., Cox, J. A., Egan, T. M., and Voigt, M. M. (2001) The first transmembrane domain of the P2X receptor subunit participates in the agonist-activated opening of the channel. *J. Biol. Chem.*, **276**, 32793–8.

Hayashi, Y., Shi, S.-H., Estaban, J. A., Piccini, A., Poncer, J.-C., and Malinow, R. (2000) Driving AMPA receptors into synapses by LTP and CAMKII: requirement for GluR1 and PDZ domain interaction. *Science*, **287**, 2262–7.

Higuchi, M., Single, F. N., Köhler, M., Sommer, B., Sprengel, R., and Seeburg, P. H. (1993) RNA editing of AMPA receptor subunit GluR-B: a base-paired intron-exon structure determines position and efficiency. *Cell*, **75**, 1361–70.

Huang, F. and Gallo, V. (1997) Gene structure of the rat kainate receptor subunit KA2 and characterization of an intronic negative regulatory region. *J. Biol. Chem.*, **272**, 8618–27.

Hume, R. I., Dingledine, R., and Heinemann, S. F. (1991) Identification of a site in glutamate receptor subunits that controls calcium permeability. *Science*, **253**, 1028–31.

Husi, H., Ward, M., Choudhary, J., Blackstock, W., and Grant, S. (2000) Proteomic analysis of NMDA receptor-adhesion protein signalling complexes. *Nat. Neurosci.*, **3**, 661–9.

Jiang, L. H., Rassendren, F., Surprenant, A., and North, R. A. (2000) Identification of amino acid residues contributing to the ATP binding site of a P2X receptor. *J. Biol. Chem.*, **275**, 34190–6.

Jiang, L. H., Rassendren, F., Spelta, V., Surprenant, A., and North, R. A. (2001) Amino acid residues involved in gating identified in the first transmembrane domain of the rat P2X2 receptor. *J. Biol. Chem.*, **276**, 14902–8.

Karlin, A. (2002) Emerging structure of the nicotinic acetylcholine receptors. *Nat. Rev. Neurosci.*, **3**, 102–14.

Kask, K., Zamanillo, D., Rozov, A., Burnashev, N., Sprengel, R., and Seeburg, P. H. (1998) The AMPA receptor subunit GluRB in its Q/R site-unedited form is not essential for brain development and function. *Proc. Natl. Acad. Sci. USA*, **95**, 13777–82.

Kawamoto, S., Uchino, S., Hattori, S., Hamajima, K., Mishina, M., Nakajima-Iijima S., *et al.* (1995*a*) Expression and characterization of the ζ1 subunit of the *N*-methyl-D-aspartate (NMDA) receptor channel in a baculovirus system. *Mol. Brain Res.*, **30**, 137–48.

Kawamoto, S., Hattori, S., Sakimura, K., Mishina, M., and Okuda, K. (1995*b*) N-linked glycosylation of the α-amino-3-hydroxy-5-methylisoxazole-4-proprionate (AMPA)-selective glutamate receptor channel α2 subunit is essential for the acquisition of ligand binding activity. *J. Neurochem.*, **64**, 1258–66.

Khakh, B.S. (2001) Molecular physiology of the P2X receptor and ATP signalling in synapses. *Nat. Rev. Neurosci.*, **2**, 165–74.

Khakh, B.S., Bao, X., Labarca, C., and Lester, H. A. (1999) Neuronal P2X receptor-transmitter-gated cation channels change their ion selectivity in seconds. *Nat. Neurosci.*, **2**, 322–30.

Kirkness, E. F., Kusiak, J. W., Fleming, T. J., Menninger, J., Gocayne, J. D., Ward, D. C., *et al.* (1991) Isolation, characterization and localization of human genomic DNA encoding the β1 subunit of the GABA(A) receptor (GABR1) *Genomics*, **10**, 985–95.

Kim, M., Jiang, L. H., Wilson, H. L., North, R. A., and Surprenant, A. (2001) Proteomic and functional evidence for a P2X7 receptor signalling complex. *EMBO J.*, **20**, 6347–58.

Kittler, J. T. and Moss, S. J. (2001) Neurotransmitter receptor trafficking and the regulation of synaptic strength. *Traffic*, **2**, 437–48.

Klein, M., Pieri, I., Uhlman, F., Pfizenmaier, K., and Eisel, U. (1998) Cloning and characterization of promoter and 5′-UTR of the NMDA receptor subunit ε2: evidence for alternative splicing of 5′-non-coding exon. *Gene*, **208**, 259–69.

Kneussel, M. (2002) Dynamic regulation of GABA(A) receptors at synaptic sites. *Brain Res. Dev.*, **39**, 74–83.

Kreienkamp, H. J., Maeda, R. K., Sine, S. M., and Taylor, P. (1995) Intersubunit contacts governing assembly of the mammalian nicotinic acetylcholine receptor. *Neuron*, **14**, 635–44.

Kuner, T., Wollmuth, L. P., Karlin, A., Seeburg, P. H., and Sakmann, B. (1996) Structure of the NMDA receptor channel M2 segment inferred from the accessibility of cysteines. *Neuron*, **17**, 343–52.

Kuner, T., Beck C., Sakmann, B., and Seeburg, P. H. (2001) Channel-lining residues of the AMPA receptor M2 segment: structural environment of the Q/R site and identification of the selectivity filter. *J. Neurosci.*, **21**, 4162–72.

Kuner, T., Seeburg, P. H., and Guy, H. (2003) A common architecture for K+ channels and ionotropic glutamate receptors. *Trends Neurosci.*, **26**, 27–32.

Lai, F., Chen, C.-X., Lee, V. M. Y., and Nishikura, K. (1997) Dramatic increase of the RNA editing for glutamate receptor subunits during terminal differentiation of clonal human neurons. *J. Neurochem.*, **69**, 43–52.

Laube, B., Kuhse, J., and Betz, H. (1998) Evidence for a tetrameric structure of recombinant NMDA receptors. *J. Neurosci.*, **18**, 2954–61.

Laurie, D. J., Bartke, I., Schoepfer, R., Naujoks, K., and Seeburg, P. H. (1997) Regional, developmental and interspecies expression of the four NMDAR2 subunits, examined using monoclonal antibodies. *Mol. Brain Res.*, **51**, 23–32.

Lin, J. W., Wyszynski, M., Madhavan, R., Sealock, R., Kim, J. U., and Sheng, M. (1998) Yotiao, a novel protein of neuromuscular junction and brain that interacts with specific splice variants of NMDA receptor subunit NR1. *J. Neurosci.*, **18**, 2017–27.

Lindstrom, J., Anand, R., Peng, X., Gerzanich, V., Wang, F., and Li, Y. (1995) Neuronal nicotinic receptor subtypes. *Ann. NY Acad. Sci.*, **757**, 100–16.

Lomeli, H., Mosbacher, J., Melcher, T., Höger, T., Geiger, J. R. P., Kuner, T., *et al.* (1994) Control of kinetic properties of AMPA receptor channels by nuclear RNA editing. *Science*, **266**, 1709–13.

Lüscher, B., Hauselmann, R., Leitgeb, S., Rulicke, T., and Fritschy, J. M. (1997) Neuronal subtype-specific expression directed by the GABA(A) receptor delta subunit gene promoter/upstream region in transgenic mice and in cultured cells. *Brain Res. Mol. Brain Res.*, **51**, 197–211.

MacDermott, A. B., Role, L. W., and Siegelbaum, S. A. (1999) Presynaptic ionotropic receptors and the control of transmitter release. *Annu. Rev. Neurosci.*, **22**, 443–85.

Malinow, R. and Malenka, R. C. (2002) AMPA receptor trafficking and synaptic plasticity. *Annu. Rev. Neurosci.*, **25**, 103–26.

Mammen, A., Kameyama, K., Roche, K. W., and Huganir, R. L. (1997) Phosphorylation of the α-amino-3-hydroxy-5-methylisoxazole-4-proprionic acid receptor GluR1 subunit by calcium/calmodulin-dependent kinase II. *J. Biol. Chem.*, **272**, 32528–33.

Matzenbach, B., Maulet, Y., Sefton, L., Courtier, B., Avner, P., Guenet, J. L., *et al.* (1994) Structural analysis of mouse glycine receptor alpha subunit genes. Identification and chromosomal localization of a novel variant. *J. Biol. Chem.*, **269**, 2607–12.

McKernan, R. M. and Whiting, P. J. (1996) Which GABA(A) receptor subtypes really occur in the brain. *Trends Neurosci.*, **19**, 139–43.

Meier, J., Vannier, C., Serge, A., Triller, A., and Choquet, D. (2001) Fast and irreversible trapping of surface glycine receptors by gephyrin. *Nat. Neurosci.*, **4**, 917–26.

Migita, K., Haines, W. R., Voigt, M. M., and Egan, T. M. (2001) Polar residues of the second transmembrane domain influence cation permeability of the ATP-gated P2X2 receptor. *J. Biol. Chem.*, **276**, 30934–41.

Mihic, S. J., Ye, Q., Wick, M. J., Koltchine, V. V., Krasowski, M. D., Finn, S. E., *et al.* (1997) Sites of alcohol and volatile anaesthetic action on GABA(A) and glycine receptors. *Nature*, **389**, 385–9.

Miyazawa, A., Fujiyoshi, Y., Stowell, M., and Unwin, N. (1999) Nicotinic acetylcholine receptor at 4.6 Å resolution: transverse tunnels in the channel wall. *J. Mol. Biol.*, **288**, 765–86.

Monyer, H., Seeberg, P. H., and Wisden, W., (1991) Glutamate-operated channels: developmentally early and mature forms arise by alternative splicing. *Neuron*, **6**, 779–810.

Moss, S. J., Blackstone, C. D., and Huganir, R. L. (1993) Phosphorylation of recombinant non-NMDA glutamate receptors on serine and tyrosine residues. *Neurochem. Res.*, **18**, 105–10.

Mu, W., and Burt, D. R. (1999) Transcriptional regulation of GABA(A) receptor gamma 2 subunit gene. *Brain Res. Mol. Brain Res.*, **67**, 137–47.

Myers, S. J., Peters, J., Huang, Y., Comer, M. B., Barthel, F., and Dingledine, R. (1998) Transcriptional regulation of the GluR2 gene: neural-specific expression, multiple promoters and regulatory elements. *J. Neurosci.*, **18**, 6723–39.

Naruse, Y., Aoki, T., Kojima, T., and Mori, N. (1999) Neural restrictive silencer factor recruits mSin3 and histone deacetylase complex to repress neuron-specific target genes. *Proc. Natl. Acad. Sci. USA*, **96**, 13691–6.

Noda, M., Takahashi, H., Tanabe, T., Toyosato, M., Kikyotani, S., Furutani, Y., *et al.* (1983) Structural homology of *Torpedo californica* acetylcholine receptor subunits. *Nature*, **302**, 528–32.

Passafaro, M., Piech V., and Sheng, M. (2001) Subunit-specific temporal and spatial patterns of AMPA receptor exocytosis in hippocampal neurons. *Nat. Neurosci.*, **4**, 917–26.

Pasternack, A., Coleman, S. K., Fethiere, J., Madden, D. R., LeCaer, J. P., Rossier, J., *et al.* (2003) Characterization of the functional role of the N-glycans in the AMPA receptor ligand-binding domain. *J. Neurochem.*, **84**, 1184–92.

Paupard, M-C., Friedman, L. K., and Zukin, R. S. (1997) Developmental regulation and cell-specific expression of N-methyl-D-aspartate receptor splice variants in rat hippocampus. *Neuroscience*, **79**, 399–409.

Pickering, D. S., Taverna, F. A., Salter, M. W., and Hampson, D. R. (1995) Palmitoylation of the GluR6 kainate recepor. *Proc. Natl. Acad. Sci. USA*, **92**, 12090–4.

Rassendren, F., Buell, G., Newbolt, A., North, R. A., and Suprenant, A. (1997) Identification of amino acid residues contributing to the pore of a p2x receptor. *EMBO J.*, **16**, 3446–54.

Reeves, D. C., Goren, E. N., Akabas, M. H., and Lummis, S. C. (2001) Structural and electrostatic properties of the 5-HT3 receptor pore revealed by substituted cysteine mutagenesis. *J. Biol. Chem.*, **276**, 42035–42.

Roche, K. W., O'Brien, R. J., Mammen, A. L., Bernhardt, J., and Huganir, R. L. (1996) Characterization of multiple phosphorylation sites on the AMPA receptor GluR1 subunit. *Neuron*, **16**, 1179–88.

Role, L. W. and Berg, D. K. (1996) Nicotinic receptors in the development and modulation of CNS synapses. *Neuron*, **16**, 1077–85.

Rueter, S. M., Burns, C. M., Coode, S. A., Mookherjee, P., and Emeson, R. B. (1995) Glutamate receptor RNA editing in vitro by enzymatic conversion of adenosine to inosine. *Science*, **267**, 1491–94.

Sailer, A., Swanson, G. T., Perex-Otano, I., O'Leary, L., Malkmus, S. A., Dyck, R. H., *et al.* (1999) Generation and analysis of GluR5(Q636R) kainate receptor mutant mice. *J. Neurosci.*, **19**, 8757–64.

Sadtler, S., Laube, B., Lasub, A., Nicke, A., Betz, H., and Schmalzing, G. (2003) A basic cluster determines topology of the cytoplasmic M3-M4 loop of the glycine receptor alpha 1 subunit. *J. Biol. Chem.*, in press.

Sasner, N., and Buonanno, A. (1996) Distinct *N*-methyl-D-aspartate receptor 2B subunit gene sequences confer neural and developmental specific expression. *J. Biol. Chem.*, **271**, 21316–22.

Schoenherr, C. J., and Anderson, D. J. (1995) The neuron-restrictive silencer factor (NRSF): a coordinate repressor of multiple neuron-specific genes. *Science*, **267**, 1360–3.

Schofield, P. R., Darlison, M. G., Fujita, N., Burt, D. R., Stephenson, F. A, Rodriguez, H., *et al.* (1987) Sequence and functional expression of the GABA(A) receptor shows a ligand-gated receptor super-family. *Nature*, **328**, 221–7.

Schorge, S., and Colquhoun, D. (2003) Studies of receptor function and stoichiometry with truncated and tandem subunits. *J. Neurosci.*, **23**, 1151–8.

Seeburg, P. H. (1996) The role of RNA editing in controlling glutamate receptor channel properties. *J. Neurochem.*, **66**, 1–5.

Sheng, M. (2001) Molecular organization of the postsynaptic specialization. *Proc. Natl. Acad. Sci. USA*, **98**, 7058–61.

Shi, S-H., Hayashi, Y., Petralia, R., Zaman, S., Wenthold, R., Svoboda, K., *et al.* (1999) Rapid spine delivery and redistribution of AMPA receptors after synaptic NMDA receptor activation. *Science*, **284**, 1811–16.

Shi, S-H., Hayashi, Y., Esteban, J. A., and Malinow, R. (2001) Subunit-specific rules governing AMPA receptor trafficking to synapses in hippocampal pyramidal neurons. *Cell*, **105**, 331–43.

Sieghart, W., Item, C., Buchstaller, A., Fuchs, K., Höger, H., and Adamiker, D. (1993) Evidence for the existence of differential O-glycosylated α5-subunits of the α-butyric acid A receptor in rat brain. *J. Neurochem.*, **60**, 93–8.

Sommer, B., Poustka, A., Spurr, N. K., and Seeburg, P. H. (1990*a*) The murine GABA(A) receptor δ-subunit gene: structure and assignment to human chromosome 1. *DNA Cell Biol.*, **9**, 561–8.

Sommer, B., Keinänen, K., Verdoorn, T. A., Wisden, W., Burnashev, N., Herb, A., *et al.* (1990*b*) Flip and flop: a cell-specific functional switch in glutamate-operated channels of the CNS. *Science*, **249**, 1580–5.

Standley, S. and Baudry, M. (2000) The role of glycosylation in ionotropic glutamate receptor ligand binding, function and trafficking. *Cell. Mol. Life Sci.*, **57**, 1508–16.

Suchanek, B., Seeburg, P. H., and Sprengel, R. (1995) Gene structure of the murine *N*-methyl-D-aspartate receptor subunit NR2C. *J. Biol. Chem.*, **270**, 41–4.

Suchanek, B., Seeburg, P. H., and Sprengel, R. (1997) Tissue specific control regions of the murine *N*-methyl-D-aspartate receptor subunit NR2C promoter. *J. Biol. Chem.*, **278**, 929–34.

Swanson, G. T., Feldmeyer, D., Kaneda, M., and Cull-Candy, S. G. (1996) Effect of RNA editing and subunit coassembly on single-channel properties of recombinant kainate receptors. *J. Physiol. (Lond.)*, **492**, 129–42.

Swope, S. L. (2002) Targeting of acetlycholine receptors to the postsynaptic endplate of the nerve-muscle synapse. In *Receptors and Ion-channel trafficking*, Oxford University Press, Oxford.

Swope, S. L., Moss S. J., Raymond, L. A., and Huganir R. L. (1999) Regulation of ligand-gated ion channels by protein phosphorylation. *Adv. Second Messenger Phosphoprotein Res.*, **33**, 49–78.

Takahashi, T., Momiyama, A., Hira, K., Hishinuma, B., and Akagi, H. (1992) Functional correlation of fetal and adult forms of glycine receptors with developmental changes in inhibitory synaptic receptor channels. *Neuron*, **9**, 1155–61.

Tan, S. E., Wenthold, R. J., and Soderling, T. R. (1994) Phosphorylation of AMPA-type glutamate receptors by calcium/calmodulin-dependent protein kinase II and protein kinase C in cultured hippocampal neurons. *J. Neurosci.*, **14**, 1123–9.

Tintrup, H., Fischer, M., Betz, H., and Kuhse, J. (2001) Exonic Sp1 sites are required for neuron-specific expression of the glycine receptor beta subunit gene. *Biochem. J.*, **355**, 179–87.

Unwin, N. (1989) The structure of ion channels in the membranes of excitable cells. *Neuron*, **3**, 565–75.

Unwin, N. (1993) Nicotinic acetylcholine receptor at 9 Å resolution. *J. Mol. Biol.*, **229**, 1101–24.

Unwin, N. (1995) Acetylcholine receptor channel imaged in the open state. *Nature*, **373**, 37–43.

Unwin, N. (1996) Projection structure of the nicotinic acetylcholine receptor: distinct conformations of the α subunits. *J. Mol. Biol.*, **257**, 586–96.

Unwin, N. (2000) Nicotinic acetylcholine receptor and the structural basis of fast synaptic transmission. *Phil. Trans. Roy. Soc. Lon. B*, **355**, 1813–29.

Unwin, N., Miyazawa, A., Li, J., and Fujiyoshi, Y. (2002) Activation of the nicotinic acetylcholine receptor involves a switch in conformation of the subunits. *J. Mol. Biol.*, **319**, 1165–76.

Verdoorn, T. A., Burnashev, N., Monyer, H., Seeburg, P. H., and Sakmann, B. (1991) Structural determinants of ion flow through recombinant glutamate receptor channels. *Science*, **252**, 1715–18.

Virginio, C., MacKenzie, A., Rassendren, F. A., North, R. A., and Surprenant, A. (1999) Pore dilation of neuronal P2X receptor channels. *Nat. Neurosci.*, **2**, 315–24.

Vissel, B., Royle, G. A., Christie, B. R., Schiffer, H. H., Ghetti, A., Tritto, T., *et al.* (2001) The role of RNA editing of kainate receptors in synaptic plasticity and seizures. *Neuron*, **29**, 217–27.

Weiss, S. W., Albers, D. S., Iadorola, M. J., Dawson, T. M., Dawson, V. L., and Standaert, D. G. (1998) NMDAR1 glutamate receptor subunit isoforms in neostiatal, neocortical and hippocampal nitric oxide synthase neurons. *J. Neurosci.*, **18**, 1725–34.

Wenthold, R. J., Petralia, R. S., Blahos, J. I., and Niedzielski, A. S. (1996) Evidence for multiple AMPA receptor complexes in hippocampal CA1/CA2 neurons. *J. Neurosci.*, **16**, 1982–9.

Whiting, P., McKernan, R. M., and Wafford, K. A. (1995) Structure and pharmacology of vertebrate GABA(A) receptor subtypes. *Int. Rev. Neurobiol.*, **38**, 95–137.

Williams, D. B. and Akabas, M. H. (1999) Gamma-amino butyric acid increases the water accessibility of M3 membrane-spanning segment residues in gamma-amino butyric acid type A receptors. *Biophys. J.*, **77**, 2563–74.

Williams, D. B. and Akabas, M. H. (2000) Benzodiazepines induce a conformational change in the region of the gamma-amino butyric acid type A receptor alpha(1)-subunit M3 membrane-spanning segment. *Mol. Pharmacol.*, **58**, 1129–36.

Wilson, G. G. and Karlin, A. (1998) The location of the gate in the acetylcholine receptor channel. *Neuron*, **20**, 1269–81.

Wood, M. W., VanDongen, H. M. A., and VanDongen, A. M. J. (1996) The 5′-untranslated region of the *N*-methyl-D-aspartate receptor NR2A subunit controls efficiency of translation. *J. Biol. Chem.*, **271**, 8115–20.

Xu, M. and Akabas, M. H. (1996) Identification of channel-lining residues in the M2 membrane-spanning segment of the GABA(A) receptor alpha1 subunit. *J. Gen. Physiol.*, **107**, 195–205.

Yakel, J. L., Vissavaijhala, P., Derkach, V. A., Brickey, D. A., and Soderling, T. R. (1995) Identification of a Ca^{2+}/calmodulin-dependent protein kinase II regulatory phosphorylation site in non-*N*-methyl-D-aspartate glutamate receptors. *Proc. Natl. Acad. Sci. USA*, **92**, 1376–80.

Zhu, J. J., Esteban, J. A., Hayashi, Y., and Malinow, R. (2000) Postnatal synaptic potentiation: delivery of GluR4-containing AMPA receptors by spontaneous activity. *Nat. Neurosci.*, **3**, 1098–106.

Zimmerman, L., Lendahl, U., Cunningham, M., McKay, R., Parr, B., Gavin, B., *et al.* (1994) Independent regulatory elements in the nestin gene direct transgene expression in neural stem cells or muscle precursors. *Neuron*, **12**, 11–24.

Chapter 9

Signal reception: G protein-coupled receptors

Jennifer A Koenig

9.1 Introduction

9.1.1 The G protein-coupled receptor (GPCR) superfamily

The G protein-coupled receptor (GPCR) family is the largest known receptor family comprising more than 1% of the human genome (Bockaert and Pin 1999) and contains receptors for molecules as diverse as small neurotransmitters, odorants, lipids, neuropeptides, and large glycoprotein hormones. Setting aside the odorant receptor family which contains hundreds of genes, there are nearly 300 mammalian GPCR; genes which can be separated into three main subfamilies (Table 9.1): (1) the rhodopsin-like group which includes the majority of the GPCRs; (2) the glucagon-like group; and (3) the metabotropic glutamate (mGlu) and $GABA_B$ receptor family (Bockaert and Pin 1999). The inclusion of a receptor in a subfamily requires the presence of an overall percentage amino acid identity (>20%) and not any discrete motif. The rhodopsin-like subfamily has been sub-classified further based on common biochemical and pharmacological properties (Flower 1999; Gether 2000). Databases for the classification of receptors into subfamilies, phylogenetic trees, chromosome localization, ligand binding constants, and receptor mutations can be found at www.gpcr.org/7tm.

Almost all of these receptors comprise a number of subtypes; for example, there are five subtypes of dopamine receptors, 13 subtypes of 5HT receptors, eight subtypes of mGlu receptors, and five subtypes of muscarinic acetylcholine receptor. These receptor subtypes have been defined by their pharmacological and functional characteristics rather than strict sequence homology since some receptors for the same ligand can show remarkably little homology (e.g. histamine H3 and H4 have the lowest recorded homology (~20%) to other histamine receptors, H1 and H2).

Each of the GPCR families contains a number of so-called orphan receptors which have been identified as members of the GPCR superfamily by homology cloning but whose activating ligand is unknown (Stadel *et al.* 1997; Marchese *et al.* 1999). Recent developments in the use of high-throughput screening technology have allowed great

Table 9.1 GPCR subtypes and their associated G protein pathways

Receptor	Family #	Subtype	*Predominant G protein partners		
			G_s	$G_{q/11}$	G_i/G_o
acetylcholine (muscarinic)	1/amine	M_1, M_3, M_5		✓	
		M_2, M_4			✓
adenosine	1/nucleotide-like	A_1, A_3			✓
		A_{2A}, A_{2B}	✓		
adrenergic - α_1	1/amine	α_{1A}, α_{1B}, α_{1D}		✓	
adrenergic - α_2		α_{2A}, α_{2B}, α_{2C}			✓
adrenergic - β		β_1, β_2, β_3	✓		
angiotensin	1/peptide	AT_1		✓	
		AT_2			✓
bombesin	1/peptide	BB_1, BB_2		✓	
bradykinin	1/peptide	B_1, B_2		✓	
cannabinoid	1/cannabis	CB_1, CB_2			✓
chemokine	1/peptide	CCR_1–CCR_{10}			✓
		$CXCR_1$–$CXCR_5$			✓
		CX_3C			✓
chemotactic peptide	1/peptide	C_{3a}, C_{5a}, fMLP			✓
cholecystokinin	1/peptide	CCK_1,	✓	✓	
		CCK_2		✓	
dopamine	1/amine	D_1, D_5	✓		
		D_2, D_3, D_4			✓
endothelin	1/peptide	ET_A	✓	✓	
		ET_B		✓	✓
galanin	1/peptide	GAL1, GAL3			✓
		GAL2		✓	✓
gonadotropin releasing hormone	1/GnRH	GnRH		✓	
glycoprotein hormone	1/protein hormone	FSH (follicle stim. hormone)	✓		
		LSH (lutropin)	✓	✓	✓
		TSH (thyroid stim. hormone)	✓	✓	✓
histamine	1/amine	H_1		✓	
		H_2	✓		
		H_3, H_4			✓
5-HT (serotonin)	1/amine	$5\text{-}HT_{1A}$, $5\text{-}HT_{1B}$, $5\text{-}HT_{1D}$			✓
		$5\text{-}HT_{2A}$, $5\text{-}HT_{2B}$, $5\text{-}HT_{2C}$		✓	
		$5\text{-}HT_4$, $5\text{-}HT_6$, $5\text{-}HT_7$	✓		
leukotriene	1/leukotriene	BLT		✓	✓
		$CysLT_1$, $CysLT_2$		✓	

Table 9.1 (continued) GPCR subtypes and their associated G protein pathways

Receptor	Family#	Subtype	*Predominant G protein partners		
			G_s	$G_{q/11}$	G_i/G_o
lysophospholipid	1/lyso-phospholipid	edg1, edg2, edg6, edg8			✓
		edg3, edg4, edg7		✓	✓
		edg5		✓	
melanocortin	1/peptide	MC1–MC5	✓		
melatonin	1/melatonin	MT1, MT2			✓
neurokinin tachykinin	1/peptide	NK1, NK2, NK3		✓	
neuropeptide Y	1/peptide	Y1–Y5			✓
neurotensin	1/peptide	NTS1		✓	
opioid	1/peptide	μOR, δOR, κOR, N/OFQ			✓
platelet activating factor	1/PAF	PAF		✓	✓
prostanoid	1/prostanoid	DP, IP	✓		
		FP, TP		✓	
		EP_1		✓	
		EP_2	✓		
		EP_3	✓	✓	✓
		EP_4	✓		
protease activated	1/peptide	PAR1–PAR4		✓	✓
purinergic	1/nucleotide-like	P2Y1, P2Y2, P2Y4, P2Y6		✓	
		P2Y11	✓	✓	
		P2Y12			✓
somatostatin	1/peptide	SST1–SST5			✓
thyrotropin releasing hormone	1/protein hormone	TRH		✓	
urotensin II	1/peptide			✓	
vasopressin	1/peptide	V1A, V1B		✓	
		V2	✓		
		OT		✓	✓
calcitonin	2		✓	✓	
calcitonin gene related peptide (CGRP)	2		✓	✓	
corticotropin-releasing hormone	2	CRF1, CRF2	✓		
growth hormone releasing hormone	2		✓	✓	

(continued)

Table 9.1 (continued) GPCR subtypes and their associated G protein pathways

Receptor	Family #	Subtype	*Predominant G protein partners		
			G_s	$G_{q/11}$	G_i/G_o
glucagon	2		✓	✓	
secretin	2		✓		
vasoactive intestinal peptide	2	$VPAC_1$, $VPAC_2$, PAC_1	✓		
$GABA_B$	3	$GABA_B$			✓
glutamate	3	$mglu_1$, $mglu_5$		✓	
		$mglu_2$, $mglu_3$, $mglu_4$, $mglu_6$, $mglu_7$, $mglu_8$			✓

* The table lists receptors for which the G protein partners are well-established as indicated in the following sources.

The IUPHAR Compendium of Receptor Characterization and Classification 1998.

Trends in Pharmacological Sciences Receptor Nomenclature Supplement 2001.

G V Segre and S R Goldring (1993). Receptors for secretin, calcitonin, parathyroid hormone (PTH)/PTH-related peptide, vasoactive intestinal peptide, glucagon-like peptide 1, growth hormone-releasing hormone, and glucagon belong to a newly discovered G protein-linked receptor family. *Trends Endocrinol. Metab.*, **4**, 309–314.

Other GPCR sequences also exist and are listed in the website www.gpcr.org

#The three main families are 1: rhodopsin-like, 2: secretin-like, 3: metabotropic glutamate-like. Classification into the three main families (and subfamilies where indicated) is based on the website www.gpcr.org

advances in the speed at which the activating ligand can be identified (Howard *et al.* 2001; Kostenis 2001).

9.1.2 The receptor–G protein cycle

G protein-coupled receptors (GPCR) reside in the plasma membrane of neuronal cells and transduce the binding of extracellular neurotransmitters or hormones into changes in the activity of intracellular effector proteins. The G protein is the intermediary in this process. After binding of the ligand, the receptor catalyses the exchange of GTP for GDP on the G protein causing the receptor/G protein complex to dissociate (Fig. 9.1). Both the activated α subunit bearing GTP and the $\beta\gamma$ dimer can act upon one or more of a variety of effector molecules which may result in changes in the intracellular levels of second messengers such as cyclic AMP, cyclic GMP, inositol phosphates, calcium, and arachidonic acid. These second messengers can generate changes in neurotransmitter release, hormone secretion, protein phosphorylation, cytoskeletal structure, and gene transcription. G proteins also regulate the function of ion channels including voltage-gated calcium channels, potassium channels, and ligand-gated ion channels such as the nicotinic acetylcholine receptor. GPCRs have been implicated in the regulation of growth, synaptogenesis, and differentiation. The activity of the G protein α subunit is terminated by its intrinsic GTPase activity which converts

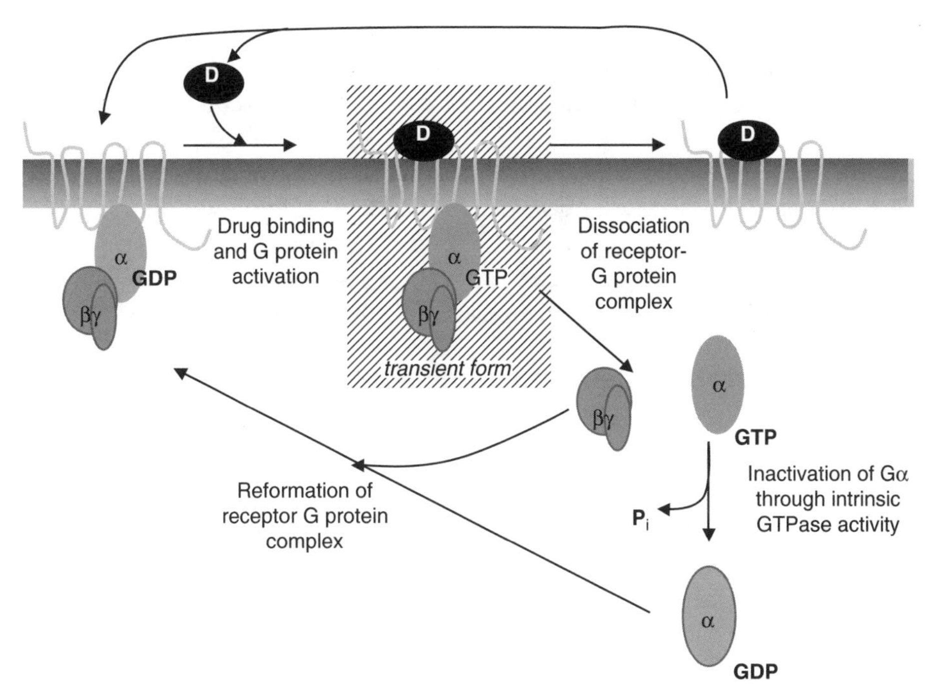

Fig. 9.1 The G protein cycle. Drug (D) binding to the receptor (shown in yellow) causes release of GDP from the α subunit of the heterotrimeric G protein. The resulting receptor-GTP-liganded G protein complex is transient and dissociates rapidly. The released α-GTP and βγ subunits are then able to interact with effectors. The intrinsic GTPase activity of the α subunit converts the GTP to GDP and the α-GDP is unable to interact with effectors. The α-GDP and βγ subunits reform the heterotrimeric G protein which is then able to rebind the receptor.

the GTP back to GDP. Reassociation of the α-GDP and βγ subunits with the unliganded receptor allows the cycle to begin again.

9.1.3 **Types of G proteins and their second messenger pathways**

G protein-coupled receptors can couple to one or more different types of G proteins which are classified by their α subunits, including G_s, G_i, G_o, G_q, G_{11}, G_{12}, G_{13}, G_t, and G_{olf} (reviewed in Morris and Malbon 1999). G_t and G_{olf} are found predominantly in visual and olfactory systems respectively. Generally, activation of G_s or G_{olf} causes stimulation of adenylate cyclase and regulation of calcium channels via α_s. Cholera toxin is often used as a diagnostic experimental tool for the identification of G_s pathways since it catalyses ADP-ribosylation of the α_s subunit. This causes the GTPase activity of α_s to be turned off and therefore results in continuous activity. Activation of G_q or G_{11} causes stimulation of phospholipase C via α_q. In contrast, activation of G_i or G_o can influence multiple second messenger systems, including inhibition of adenylate cyclase, activation of cyclic GMP phosphodiesterase, and regulation of potassium and calcium channels. Pertussis toxin is commonly used as a diagnostic tool for identifying pathways involving $G\alpha_{i1}$, $G\alpha_{i2}$, $G\alpha_{i3}$, $G\alpha_{o1}$, $G\alpha_{o2}$, and $G\alpha_t$ since it catalyses the irreversible ADP-ribosylation of their C-terminal cysteine resulting in inactivation of the G protein. While these are general rules, GPCR can also activate other signalling pathways such as tyrosine kinases (Bence *et al.* 1997; Diverse-Pierluissi *et al.* 1997), protein kinase B, and MAPkinase (Gutkind 1998; Lopez-Ilasaca 1998; Gudermann *et al.* 2000) and small, monomeric G proteins (Hall *et al.* 1999; Seasholtz *et al.* 1999).

9.1.4 Non G protein-mediated pathways

In the last few years, evidence has been accumulating for the existence of a number of G protein-independent signalling pathways (reviewed in Heuss and Gerber 2000). Many of these involve arrestin binding subsequent to receptor phosphorylation (see Section 5.2.2). In this paradigm arrestin acts as an adaptor protein, linking the receptor to activation of the non-receptor tyrosine kinases of the SRC family (Luttrell *et al.* 1999) and thus to regulation of MAPkinase and nuclear transcription. GPCR can interact with PDZ domain-containing proteins such as PSD-95 through variants of the T/S-x-V motif in their C-terminal tails (Hall *et al.* 1999). For example, the β_2-adrenergic receptor binds the Na^+/H^+ exchange factor (NHERF) in an agonist-dependent fashion (Hall *et al.* 1998). mGluR bind members of the Homer family of EVH domain-containing proteins through a polyproline rich region in the C-terminal tail and this facilitates a functional interaction between mGluRs and ER-based IP_3 receptors (Brakeman *et al.* 1997; Bockaert and Pin 1999). Other candidate proteins for direct receptor interaction are eNOS, calmodulin, and the small G proteins ARF and RhoA (Heuss and Gerber 2000).

9.2 Overall structural features

The defining characteristics of this large superfamily of receptors are the seven transmembrane domain structure with an extracellular N-terminus and intracellular C-terminus and an ability to activate heterotrimeric G proteins. Until recently, structural modelling has been based on the structure of bacteriorhodopsin, a light-activated proton pump found in photosynthetic bacteria, which also has a seven transmembrane structure. Although GPCR and bacteriorhodopsin appear to share a similar overall morphology, bacteriorhodopsin is not a GPCR and there is little actual sequence homology. More recently, the crystal structure of rhodopsin has been determined to be 2.8 Å resolution (Palczewski *et al.* 2000) and this structure, combined with the results of site-directed mutagenesis studies (e.g. Meng and Bourne 2001; Savarese and Fraser 1992), and extensive molecular modelling (Ballesteros *et al.* 2001; Gershengorn and Osman 2001; Lu *et al.* 2002) has refined our understanding of the molecular function of GPCR (see also www.gpcr.org/7tm). As a result there is considerably more known about the structure of family 1 GPCR than for the other GPCR families. In general, the sequence similarity between the different GPCRs is in the transmembrane segments while most of the variability resides in the size of the N-terminal extracellular domain, the third intracellular loop, and the C-terminal domain.

9.2.1 **Family 1 — rhodopsin family**

The main structural features of a 'generic' G protein-coupled receptor of family 1 are illustrated in Fig. 9.2. Remarkably, even within a family, GPCR share little sequence homology but there are a small number of highly conserved residues such as the 'DRY' (Asp-Arg-Tyr) motif found in all family 1 GPCR which is important in the conformational changes which lead to G protein activation (see below). It is thought that, at least for rhodopsin, the helical structure of transmembrane domains five and six (TM5 and TM6) may extend into the cytoplasm from the membrane (Ballesteros *et al.* 2001). The three-dimensional shape (Fig. 9.2B) resembles that of a distorted, pear-shaped barrel lined by TM3, 4, 5, 6, and 7 and closed off on the intracellular side by the tilted TM3. The seven transmembrane domains are linked by H-bonding networks and Van-der-Waals contacts which stabilize the ground-state structure (Lu *et al.* 2002). The transmembrane helices are arranged in a counterclockwise fashion (looking from the extracellular side) with TM3 almost in the centre (Fig. 9.2B).

All GPCR contain many post-translational modifications. Two cysteine residues in extracellular loops 1 and 2 are highly conserved and are important for stabilization of receptor conformation (Bockaert and Pin 1999). In many family 1 GPCR, there is a fourth intracellular loop, consisting of an amphiphilic helical structure, linking the cytoplasmic end of TM7 to a point where it is anchored to the membrane by palmitoylated cysteine residues (Lu *et al.* 2002). Agonist-driven depalmitoylation has been demonstrated for some receptors but it has been suggested that repalmitoylation may be affected by desensitization mechanisms (Loisel *et al.* 1999 and references therein). Palmitoylation is thought to be important in regulating G protein coupling, receptor phosphorylation, and internalization (*ibid*). There are glycosylation sites on the N-terminal domain, some of which are important for efficient expression at the plasma membrane (Benya *et al.* 2000; Deslauiers *et al.* 1999). There are a number of potential phosphorylation sites which are generally located in the C-terminal tail and the second and third intracellular loops. These phosphorylation sites are important in regulating receptor responsiveness (see below).

9.2.1.1 Heterodimerization

Traditionally GPCR have been viewed as single entities which interact with G proteins in a 1 : 1 ratio. However, since the first demonstration of dimer formation of β_2-adrenergic receptors in 1996 (Hebert *et al.* 1996), there has been accumulating evidence for the formation of both homo- and heterodimers from cross-linking studies (e.g. AbdAlla *et al.* 1999), co-immunoprecipitation experiments (e.g. Jordan and Devi 1999;

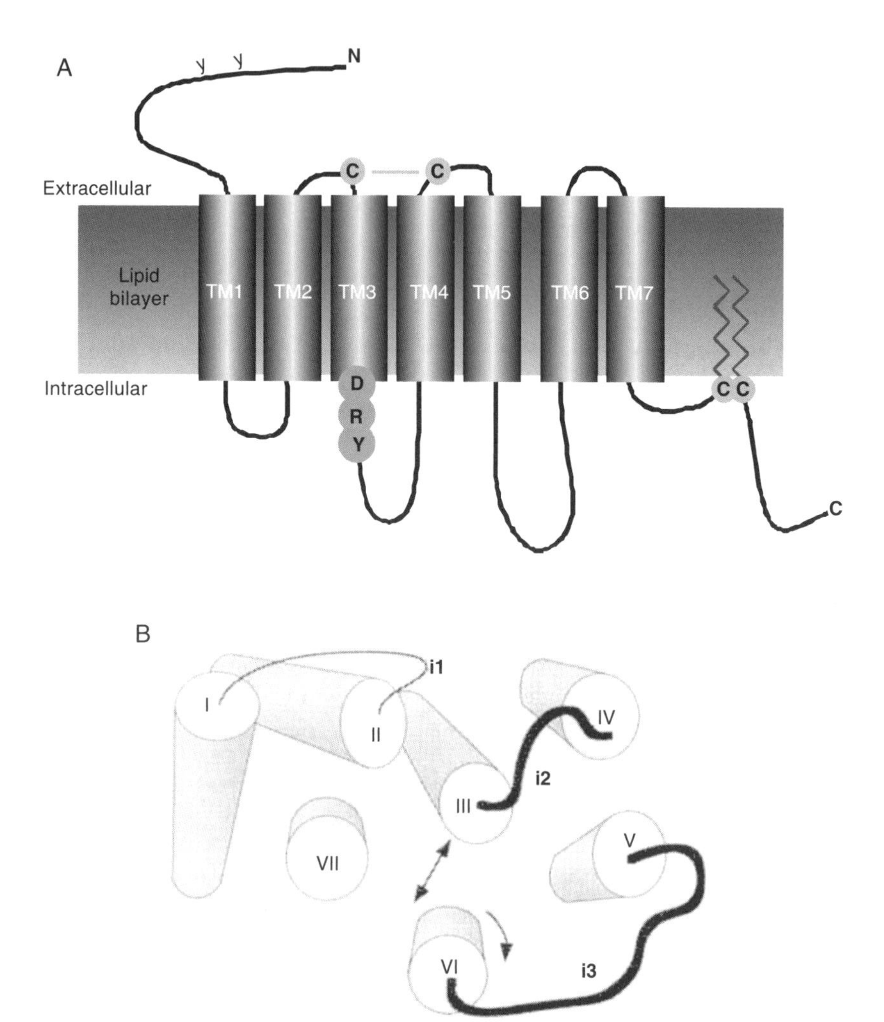

Fig. 9.2 A diagrammatic view of a 'generic' family 1 GPCR. (A) A two-dimensional view shows the major conserved features within this family. The seven transmembrane domains (TM1–TM7) are shown in grey in the lipid bilayer shown in blue. The N-terminus is extracellular and the C-terminus is intracellular. There are glycosylation sites (shown as y) on the N-terminal domain and multiple potential phosphorylation sites on the C-terminal domain as well as the 2nd and 3rd intracellular loops. There is a highly conserved cysteine disulfide bridge connecting extracellular loops one and two. Many receptors contain cysteine residue(s) in the C-terminal domain which are palmitoylated and thus provide for a potential fourth intracellular loop. The highly conserved 'DRY' motif is important for the conformational change leading to G protein activation. (B) A three-dimensional view looking from the cytoplasmic side (reproduced from Bockaert and Pin (1999) with permission) shows the arrangement of helices. Upon agonist activation the conformational change involves tilting of TM3 and TM6 away from each other and rotation of TM6. This allows the exposure of previously hidden sequences in the intracellular loops and possibly within the transmembrane helical bundle which are required for G protein interactions.

Zeng and Wess 1999), and FRET (fluorescence resonance energy transfer, e.g. Rocheville *et al.* 2000) or BRET (bioluminescence resonance energy transfer, e.g. Angers *et al.* 2000). In some cases, disulfide bond formation is thought to play a key role in dimer formation (Jordan and Devi 1999; Zeng and Wess 1999) but in many cases the molecular mechanism is unclear since various regions of the GPCR structure can be involved (e.g. the amino terminus for the B2 bradykinin receptor, AbdAlla *et al.* 1999 or the seventh transmembrane domain for the β_2-adrenergic receptor, Hebert *et al.* 1996). The functional effects of dimerization vary between receptor combinations. For example, co-expression of β_2-adrenergic receptors with δ-opioid or κ-opioid receptors does not affect ligand binding or G protein coupling but does alter the ability of the receptors to be internalized (Jordan *et al.* 2001). In contrast, D_2-dopamine receptor and $sstr_5$ somatostatin receptor heterodimers showed unusual pharmacological properties with somatostatin and dopamine agonists acting synergistically in both ligand binding and second messenger activation (Rocheville *et al.* 2000).

9.2.1.2 Genetic polymorphisms

Recently an increasing number of genetic polymorphisms in genes encoding GPCRs have been identified and this area is expanding rapidly (reviewed in Rana *et al.* 2001). In such cases multiple allelic forms occur in the population at frequencies above the background mutation rate, indicating that functional differences between the encoded proteins may have selective significance at the population level. The encoded variations in receptor structure can reflect subtle but important changes in ligand recognition (Bond *et al.* 1998; Compton *et al.* 2000), G protein signalling (Small and Liggett 2001), and regulation of plasma membrane receptor density (Green *et al.* 1994). Associations have been made between receptor gene polymorphic alleles and the probability of developing specific disease states. For example, the human D4 receptor has two high frequency alleles in the population, differing in the number of repeats of a sequence that encodes a 16 amino acid segment in the 3rd intracellular loop (Sealfon and Olanow 2000). This repeat number has been correlated with novelty-seeking personality traits and increased delusions in psychotic individuals. Mutation in the orexin B receptor gene in dogs is related to canine narcolepsy (Howard *et al.* 2001).

9.2.2 Family 2 — glucagon-like family

These receptors show some overall morphological similarity to family 1 receptors but have a much larger N-terminal domain which contains multiple potential disulfide bridges (Gether 2000). Except for the conserved disulfide between the 1st and 2nd extracellular loops, this family does not contain any of the structural features described for family 1 (Gether 2000). Most of the ligands acting at these receptors are peptides or glycoprotein hormones, typically 30–40 amino acids, and are likely to interact with the receptor over large surface areas. These receptors are often described as having two halves, an exodomain that consists of the N-terminal extracellular hormone binding domain and an endodomain that is the C-terminal, seven transmembrane signal generating domain (Zeng *et al.* 2001). The exodomain is entirely responsible for hormone binding and the resulting hormone–exodomain complex adjusts its conformation and interacts with the endodomain.

9.2.3 Family 3 — mGluR/GABA$_B$ family

The major characteristic of this family is the extremely large N-terminal extracellular domain which bears some resemblance to bacterial periplasmic binding proteins. This domain is thought to contain the binding site for transmitter amino acids glutamate and GABA (Gether 2000). The third intracellular loop is very short and highly conserved within this family. The seven transmembrane domain shows very little sequence similarity with receptors of family 1 (~12%) but it is thought that the overall topology is probably similar since family 3 receptors contain the conserved cysteines making the disulfide bond between extracellular loops one and two (DeBlasi *et al.* 2001).

Family 3 receptors appear to form dimers but do so in quite different ways. The GABA$_B$ receptor forms heterodimers between GABA$_B$R1 and GABA$_B$R2 through coiled-coil regions in the C-terminal tails and this dimerization is required for efficient cell surface expression and signalling (Marshall *et al.* 1999). In contrast, metabotropic glutamate receptor dimerization is stabilized by disulfide bonds in the amino-terminal extracellular domain (DeBlasi *et al.* 2001).

9.3 Receptor–Ligand interactions

9.3.1 Rhodopsin-like subfamily

The ligand binding site for small molecule neurotransmitters is in a pocket in the transmembrane region and involves amino acids from helices 3, 5, 6, and 7 (reviewed by Savarese and Fraser 1992; Strader *et al.* 1994; Gether and Kobilka 1998; Ji *et al.* 1998; Gether 2000; Lu *et al.* 2002). While the actual amino acid residues involved in ligand binding are different for the different small molecule receptors, the topological location of these residues shows remarkable similarity. In the case of the β-adrenergic receptor, mutagenesis studies have identified a small number of transmembrane residues that are important in catecholamine binding (shown diagrammatically in Fig. 9.3) including Asp113 in TM3, two serines in TM5 (Ser204 and Ser207), and Phe290 in the TM6. These serines and phenylalanine residues are present in all GPCRs that bind catecholamines but not in other GPCRs. In contrast, the histamine receptor contains an Asp186 and Thr190 in analogous positions to the Ser204 and Ser207 of the β-adrenergic receptor and an interaction of these residues with the imidazole ring of histamine has been suggested. The muscarinic receptor contains Thr231 and Thr234 in the analogous part of the TM5 which might interact with the ester group of acetylcholine. The aspartate in TM3 is conserved among the biogenic amine receptors and is also important in forming

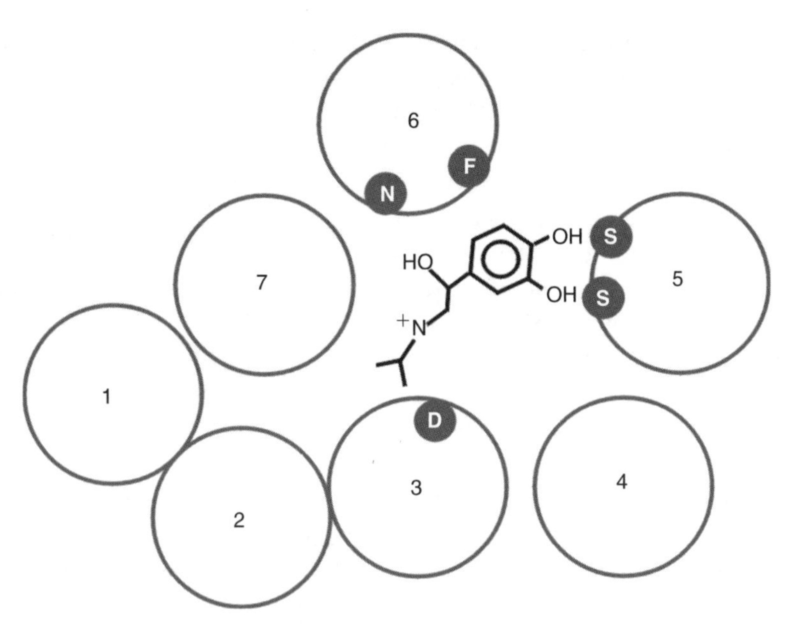

Fig. 9.3 The ligand binding site for isoproterenol (isoprenaline) at the β_2-adrenergic receptor. There are five major points of interaction. The aspartate (D) in TM3 forms an ionic interaction with the protonated amine group of the agonist. Two serines (S) in TM5 form hydrogen bonding interactions with the hydroxyl groups of the catechol moiety. A phenylalanine (F) in TM6 has hydrophobic interactions with the benzene ring. An asparagine (N) in TM6 has hydrogen-bonding interactions with the aliphatic hydroxyl group.

salt-bridge interactions with the positively charged head groups of dopamine, serotonin, histamine, and acetylcholine.

Although this simplistic picture holds for the well-studied β-adrenergic and muscarinic receptors, the binding sites for other neurotransmitters and peptides clearly do not involve the same distribution of amino acids (Savarese and Fraser 1992; Strader *et al.* 1994; Gether and Kobilka 1998; Ji *et al.* 1998; Gether 2000). In the neurokinin receptor (NK1), peptide agonists bind to residues in the amino terminus, extracellular loops 1 and 2, and a residue at the top of the TM7 and as yet there is little evidence for interaction of peptide agonists with regions deep in the transmembrane domains (Gether 2000). Other peptide hormones, however, bind both with the extracellular domains and in the transmembrane cleft (e.g. angiotensin II, endothelin, somatostatin, opioids, and bradykinin). Although the actual residues for interaction differ between the peptides, they all appear to be located on the surface of the predicted binding crevice illustrating the concept that there is a high degree of structural similarity of the receptors even though they bind quite chemically diverse ligands using chemically different interacting amino acids.

Non-peptide agonists generally bind within the transmembrane domains and do not require residues in the extracellular loops. Thus the subtype selectivity of the dopamine D2 and D4 receptors is determined by residues in TM2, 3, and 7 (Simpson *et al.* 1999). However, mutagenesis studies have shown that the binding sites for a number of different classes of drugs at the same receptor do not necessarily include the same set of amino acids. In some cases, a single amino acid substitution can decrease the binding affinity for one class of drugs but not another (Strader *et al.* 1994). This implies that several different combinations of interactions with different parts of the transmembrane region can collectively induce a conformational change sufficient to activate a G protein.

Similarly, there are no hard and fast rules as to which residues are important in antagonist binding. By definition, antagonists bind to the receptor and prevent agonist binding but do not cause a conformational change required for G protein activation. For the β-adrenergic receptor, the aryloxyalkyamine antagonists such as alprenolol and propranolol bind to some of the same residues as the agonists but have an additional contact with an asparagine in TM7 (Gether 2000). In the case of peptide agonists and non-peptide antagonists of the NK1 receptor and the angiotensin II receptor, agonists and antagonists do not interact with the same set of amino acids (Strader *et al.* 1994; Ji *et al.* 1998; Gether 2000). Given that at least part of the agonist binding site is located in a pocket in the bilayer region, it is likely that binding of antagonists while not inducing a conformational change, could sterically block access of the agonist to its binding site since they often bind with residues closer to the surface of the membrane (Gether 2000).

9.3.2 mGluR family — receptor ligand interactions

Modelling, mutagenesis, and more recently crystallographic studies of the extracellular domain revealed that the binding site for glutamate was within a cleft between the two globular lobes of the large extracellular domain (DeBlasi *et al.* 2001). Exactly how the binding of agonist in the extracellular domain can be transmitted to cause G protein activation is still unclear but it has been proposed that dimerization may be important (*ibid*). There are a number of compounds which are structurally related to the amino acid agonist transmitters which bind competitively within the extracellular domain. In contrast, a group of novel allosteric antagonists has been identified which bind entirely within the transmembrane domains, particularly interacting with residues in TM3 and TM7 of the mGlu5 receptors (Spooren *et al.* 2001).

9.4 Receptor-G protein interactions

9.4.1 How are receptor-G protein interactions measured?

Receptor activation of G protein-mediated pathways can be measured by assaying for accumulation of second messengers such as cAMP, inositol phosphates and calcium, or modulation of ion channels using electrophysiological techniques. It is also possible to measure further 'upstream' using the binding of ^{35}S-GTPγS and recent developments have used antibodies selective for the different Gα subunits to determine incorporation of ^{35}S-GTPγS into particular subtypes and isoforms of Gα (e.g. Carruthers *et al.* 1999).

Another method for examining receptor–G protein interactions is to measure the effect of G proteins on the binding of agonist to its receptor (reviewed in Birnbaumer *et al.* 1990). When the G protein is bound to the receptor, the receptor–G protein complex has a high affinity for agonist. Once the G protein is activated and released from the receptor, the receptor converts to a conformation with low affinity for agonists. This conversion requires the presence of magnesium ions and GTP to allow activation of the G protein (Birnbaumer *et al.* 1990). Ligand binding assays are usually performed with well-washed membrane preparations under equilibrium conditions so that, in the absence of GTP, both high- and low-affinity states are measured. In the presence of GTP and magnesium, only the low-affinity state is measured since binding of agonist rapidly induces the change from high to low affinity. The interaction of G proteins with the β-adrenergic receptor can be mimicked by the action of synthetic peptides based on regions of the α subunit of G_S but not with peptides derived from G_i or G_o (Rasenick *et al.* 1994). Binding of agonists to intact cells generally shows only low-affinity characteristics since intracellular levels of GTP are naturally high. Antagonist binding generally does not show this sensitivity to G protein binding since antagonists do not promote the active conformation of the receptor and do not cause dissociation of the receptor–G protein complex.

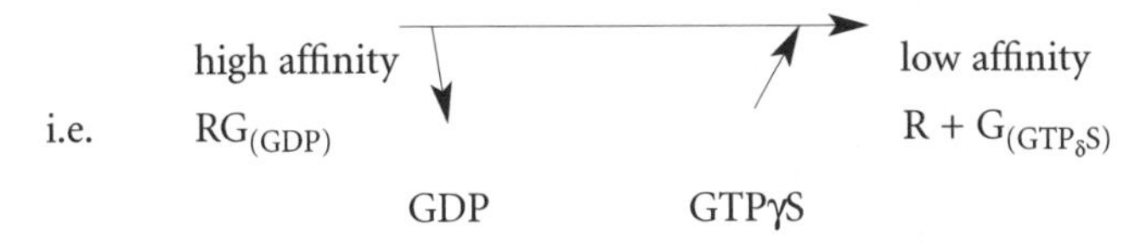

9.4.2 Structural features of receptors involved in G protein activation

9.4.2.1 How does agonist binding cause receptor conformational change?

Historically there has been a clear distinction made between the concepts of affinity and efficacy. The affinity of the drug–receptor interaction describes how tightly the drug binds and efficacy describes how well the drug causes the conformational change required to activate G proteins. The simplest models of receptor function incorporate two receptor states, an inactive conformation which is in equilibrium with an active conformation. Although it is likely that there may be multiple active conformations (see below), only one will be assumed for the purpose of this discussion. Thus a strong agonist will bind and cause a conformational change leading to maximal active state probability. A partial agonist may bind with high affinity but will only be able to promote a submaximal active state probability. In contrast, an antagonist may bind with high affinity but will cause no conformational change, i.e. have no efficacy.

How do we turn this theoretical description into a 'visual' representation of what is happening at the molecular level? From Section 9.3 above, it is clear that agonists can interact with receptors in many different ways to cause G protein activation but it seems that the mechanism of the conformational change may be highly conserved. There are constraining intramolecular interactions which keep receptors preferentially silent in the absence of agonist and stabilizing interactions have been proposed particularly between TM5 and TM6 as well as between TM3 and TM7 (Bourne 1997; Gether and Kobilka 1998; Gether 2000; Lu *et al.* 2002). For members of the rhodopsin family, conformational changes upon agonist activation are thought to be transmitted by the highly conserved 'DRY' motif at the cytoplasmic side of TM3 (Fig. 9.2) (Bourne 1997; Gudermann *et al.* 1997; Gether and Kobilka 1998; Gether 2000). The arginine in this motif is constrained within a hydrophilic pocket formed by residues from other transmembrane helices. Upon receptor activation, the aspartate is protonated causing movement of adjacent residues and tilting of the transmembrane helices which leads to exposure of previously hidden sequences in the intracellular loops which interact with G proteins. The exact molecular mechanism and the role of specific amino acid residues is still a matter of debate and at least two molecular modelling solutions have been proposed (Gether 2000; Lu *et al.* 2002).

There is, however, considerable experimental evidence for the tilting of the transmembrane helices. Site-directed mutagenesis has been used to engineer either cysteine–cysteine or zinc–histidine bridges between the helices in strategic locations which could prevent receptor activation (Meng and Bourne 2001). Another experimental approach has been to introduce site-specific pairs of sulfhydryl-reactive spin labels or fluorophores and to measure helical interactions by either EPR spectroscopy or fluorescence spectroscopy respectively (Gether and Kobilka 1998). It has been proposed that activation of the receptor requires the cytoplasmic ends of TM3 and TM6 to move away from each other and TM6 to rotate slightly (Bockaert and Pin 1999; Gether 2000). These movements of the helices causes conformational changes in the intracellular loops leading to exposure of previously hidden sequences which can then interact with G proteins. A major interaction site on the G protein is in the last seven amino acid stretch of the C-terminal and it is this region which is thought to predominantly determine receptor specificity (Bourne 1997; Kostenis 2001) although other regions of the G protein, including the N-terminus, may also play an important role (Lu *et al.* 2002). Other contact points, particularly those with the $\beta\gamma$ subunit are less well established (Bourne 1997; Hamm 1998). It is clear though, that the nucleotide binding domain of the α subunit is some distance away (~30 Å) from the membrane and therefore the receptor must act at a distance to induce GDP release. Molecular modelling has been employed to explore potential hypothetical mechanisms and one proposal has been that movement of the helices may open a crevice in the receptor into which the C-terminal portion of the G protein may move (Iiri *et al.* 1998) but this field remains speculative (Bourne 1997; Lu *et al.* 2002).

Considerable effort has been put into identifying the exact sequences in the receptor which are necessary for directing specificity of receptor–G protein interaction. Generation of chimeric β-adrenergic and muscarinic receptors has identified multiple separate small stretches of amino acid residues in the second intracellular loop and in the N- and C-terminal portion of the third intracellular loop which act in a co-operative fashion to dictate G protein coupling specificity (Gudermann *et al.* 1997; Wess *et al.* 1997). In the muscarinic receptors, a four amino acid motif has been identified ('VTIL') which is located at the third intracellular loop – TM6 junction which can direct G protein specificity (Wess *et al.* 1997). However, despite extensive mutagenesis work, it has proven difficult to predict general rules for G protein specificity of receptor from primary structure. This has led Bourne (1997) to propose that activated receptors can take on a restricted set of conformations that can mould to interact with certain G protein subtypes. In contrast, an alternative approach incorporating data-mining and pattern recognition has identified some sequences from receptors with known G protein coupling which can predict the coupling specificity of newly-cloned receptor sequences (Moller *et al.* 2001).

9.4.2.2 Constitutive activity

Until recently, models of receptor function assumed that, in the absence of agonist, all receptors existed in the inactive conformation and that addition of agonist shifted the equilibrium towards the active conformation (Leff *et al.* 1997; Strange 2002). The increasing use of transfected cell systems containing high receptor density led to the realization that basal levels of receptor activity could be measured in the absence of agonist and that antagonists were actually able to decrease this basal activity (deLigt *et al.* 2000; Gether 2000; Newman-Tancredi *et al.* 2000). It has since been realized that some receptors do show constitutive activity even when expressed at 'physiological' levels and these include rat dopamine D1, rat and human histamine H2, human dopamine D3, human serotonin 5HT1a, human cannabinoid CB1 and CB2 (deLigt *et al.* 2000; Pan *et al.* 1998). Many antagonists have been renamed 'inverse agonists' to reflect their ability to decrease basal activity, since by definition an antagonist is a drug which binds but has no efficacy. Several antipsychotic drugs have been reclassified as inverse agonists rather than antagonists (Weiner *et al.* 2001) and now much drug discovery work is focused upon distinguishing between antagonist and inverse agonist drugs since it is likely that this will be of great therapeutic importance.

A number of mutations were identified which served to increase the level of basal activity without affecting the ability of agonists to further activate the receptors (e.g. Greasley *et al.* 2001). Most of these mutations were initially identified in the C-terminal end of the third cytoplasmic loop but more recently mutations in other regions of the structure can cause constitutive activity. This underlines the idea that there are many stabilizing interactions between helices that hold the receptor in an inactive state and that interfering with these interactions, either by mutation or by agonist activation can lead to activation of G proteins.

9.4.2.3 Multiple active conformations — stimulus trafficking

Although it is widely accepted that proteins can take on multiple conformations, most of the available experimental data suggests that these multiple conformations are indistinguishable experimentally and models describing receptor function with just one active and one inactive state are sufficient. However, there are some reports which suggest that there may be multiple active conformations and that certain drugs can induce formation of certain conformations: this has been termed 'stimulus trafficking' (Kenakin 1995; Watson *et al.* 2000). This means that different drugs are able to promote distinct receptor conformations which interact with different G proteins resulting in activation of distinct signalling pathways. Experimentally this is observed as a reversal in the order of potency of agonists. For example, $PACAP_{27}$ is more potent than $PACAP_{38}$ in stimulating cAMP production via the PACAP receptor but less potent in stimulating inositol phosphate accumulation (Spengler *et al.* 1993). Differential stimulation of inositol phosphate and arachidonic acid second messenger systems has also been demonstrated for a series of partial agonists at the human $5HT_{2A}$ and $5HT_{2C}$ receptors (Berg *et al.* 1998).

9.4.3 Cell-type specific factors

The central question considered here is how cells, which might contain multiple receptor subtypes and G proteins (α and $\beta\gamma$) control the ultimate functional response which is produced. At first glance, one could expect to answer this by finding out which receptor subtypes activate which G proteins and which G proteins activate which effectors. However, this approach was stymied by the finding that in purified systems, the receptor–G protein interactions are remarkably non-specific. This has led to the realization that there are more subtle controls within the cell that determine the signalling pathway(s) that respond to activation of a given receptor and that these controls will vary from one cell to another. Some examples to illustrate this are given below and are most certainly not exhaustive.

9.4.3.1 Receptor splice variants

A number of receptors show functional diversity due to the existence of splice variants which exhibit distinct patterns of tissue distribution. Most of these receptor isoforms have structural differences in the intracellular loops and C-terminal tail and therefore show different patterns of G protein coupling (Kilpatrick *et al.* 1999). For example, the two isoforms of the dopamine D2 receptor differ by a 29 amino acid insert in the third intracellular loop and show signalling through distinct G proteins (Monsma *et al.* 1989). Sometimes the effect is less clear cut and may be observed as differences in the strength of coupling to G proteins as shown in the four isoforms of the prostanoid receptor EP_3 which differ in their C-terminal tail (Sugimoto *et al.* 1993).

9.4.3.2 Levels of receptor expression and signal amplification

Signals generated by activation of GPCRs undergo varying degrees of amplification depending upon the nature of the system under study. In a system with high receptor density which shows strong coupling to a particular G protein pathway, the concentration of drug required to generate second messengers may be very much less than the concentration required to occupy a significant fraction of receptors and this system will show a large amount of signal amplification. This is also termed 'receptor reserve', 'spare receptors', or 'strong coupling'. This signal amplification has important physiological consequences illustrated in Fig. 9.4. For a response to be generated quickly upon release of neurotransmitter, the binding interaction needs to reach equilibrium quickly. The rate at which a reaction reaches equilibrium depends upon the rate constants for association and dissociation and for low-affinity interactions such as those of acetylcholine or noradrenaline (binding

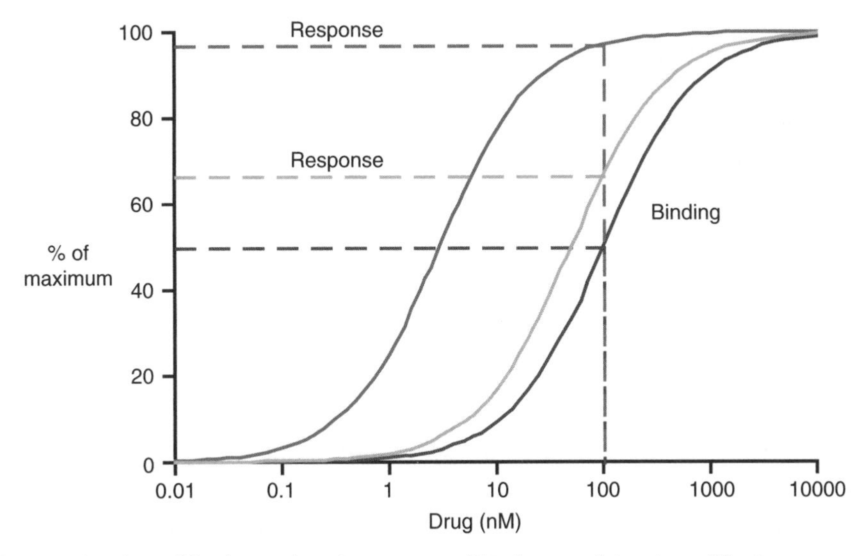

Fig. 9.4 Signal amplification and receptor reserve. The degree of signal amplification or receptor reserve can vary widely depending on the nature of the system under study. Consider a receptor where a drug concentration of 100 nM is sufficient to occupy half the receptors (the K_D is 100 nM). In a strongly coupled system (also said to have large signal amplification or large receptor reserve), this concentration causes enough signal so that the response is effectively maximal. This is reflected by a considerable shift to the left of the dose-response curve compared to the binding curve. In contrast, the same receptors in this cell may also signal through another, less well coupled pathway having less signal amplification and less receptor reserve. Now, the drug concentration of 100 nM is only able to produce 67% of the maximal response and 2 μM is required to produce a maximal response.

affinities in the μM to mM range) equilibrium can be reached very quickly. Furthermore, once the transmitter is removed from the synapse, dissociation occurs equally quickly. The amount of transmitter released into the synapse may be significantly lower than the concentration required to reach high levels of receptor occupancy. Therefore, if there is a large amount of signal amplification, a low concentration of transmitter can bind and reach equilibrium quickly but also create a strong signal within the cell. Peptide hormones generally have much higher affinities (in the nM–pM range) and take much longer to reach equilibrium and to dissociate. Thus the temporal characteristics of their responses will be quite different to those of neurotransmitters with lower affinities.

Levels of receptor expression can vary widely between cell types and typically the receptor density in a stably transfected cell line may be ten to a thousand times higher than that found in a cultured cell line endogenously expressing the receptor (Kenakin 1997). The coupling strength of two different G protein-mediated pathways activated by the same receptor may differ widely. For example, muscarinic m_2 receptors generally couple only in a negative manner to adenylate cyclase via G_i. However, when the receptor is expressed at high levels, stimulation of phospholipase C is also observed (Vogel *et al.* 1995 and references therein). The concentration of agonist required to activate the G_i pathway may be much lower than that required to activate the phospholipase C pathway. This suggests that, although the receptor is actually able to couple to both, at normal expression levels only the stronger signal is observed.

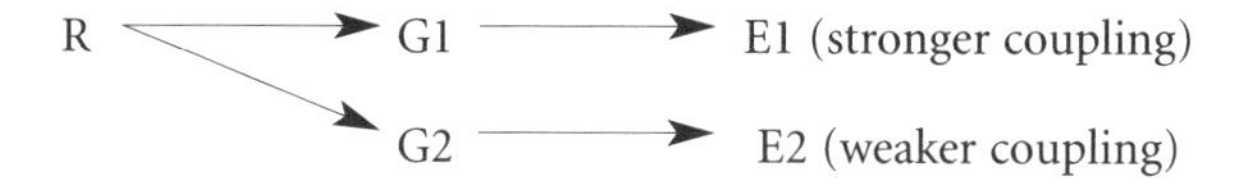

The observed strength of coupling could also be determined by the efficacy of the agonist. If a receptor has a greater ability to stimulate one G protein (G1) than another (G2) then a full agonist could theoretically produce enough activated G1 and G2 to activate both pathways whereas a partial agonist might only produce enough activated G1 to produce a measurable response. An example of this occurs with the μ-opioid receptor where morphine is a partial agonist and etorphine is a full agonist. In transfected cells, there is strong coupling of the μ-opioid receptor to inhibition of adenylyl cyclase via Gi and in this system, both morphine and etorphine appear to show similar maximal responses — i.e. they both behave as if they are full agonists (Arden *et al.* 1995; Blake *et al.* 1997; Keith *et al.* 1996). However, receptor phosphorylation and internalization are much less well coupled and morphine appears to have no activity while etorphine does. Overexpression of GPCR kinase reveals that morphine can indeed promote receptor phosphorylation but requires a higher proportion of kinase relative to receptor density (Zhang *et al.* 1998). This underlines the concept that an agonist may be able to promote formation of the active conformation of a receptor, but being able to observe a resulting response depends upon the levels of expression of all the components in the signalling pathways and their affinities for each other. A quantitative analysis of the relationship between receptor density and agonist efficacy, and potency has been developed and may be used to eliminate apparent discrepancies due to different expression levels (Whaley *et al.* 1994).

9.4.3.3 Specificity of receptors for G protein subtypes

Some receptors can show selectivity for a certain α subtype and if a cell line does not contain the subtype required then the appropriate response will not be seen even though it might be present in another cell type. For example, somatostatin receptors transfected into CHO-DG44 cells do not show inhibition of adenylate cyclase whereas their transfection in CHO-K1 cells does reveal the appropriate response (Patel *et al.* 1994). There are other examples where receptors appear to show a great degree of selectivity even though several G protein isoforms may be present. For example, both somatostatin and muscarinic receptors regulate calcium channel function in GH_3 cells via different variants of the same G protein subtype (Kleuss *et al.* 1992). The exact mechanism underlying this selectivity is not yet clear.

R (muscarinic) → G ($\alpha_{01}/\beta_3/\gamma$) → E (calcium channels)

R (somatostatin) → G($\alpha_{02}/\beta_1/\gamma$) ↗

9.4.3.4 Restricted localization

For many years, various models of GPCR signalling have visualized the receptor, G proteins, and effectors floating in a sea of membrane and that agonist-occupied receptors couple transiently to and activate G proteins which are then released to interact with effectors (Chidiac 1998; Ostrom *et al.* 2000). More recently, developments in experimental technologies for measuring protein–protein interactions, such as yeast two-hybrid and immunoprecipitation of fusion proteins, has led to an accumulating mass of information that suggests that signalling molecules are found in complexes held together by scaffolding proteins (reviewed in Pawson and Scott 1997; Ranganathan and Ross 1997; Ostrom *et al.* 2000; Sheng and Sala 2001; Michel and Scott 2002). The true state of affairs is probably somewhere between the two extremes. Compartmentation of signalling pathways is an attractive explanation for the apparent specificity of interactions that occurs within intact cells but not with purified proteins *in vitro.*

Receptors, G proteins, and effectors are often localized to particular parts of the cell membrane and are generally not found to be uniformly distributed (Hopkins 1992; Mellman *et al.* 1993; Wozniak and Limbird 1996; Yeaman *et al.* 1996). The mechanisms involved in this targetting are often not clear and may involve interaction with specific protein partners such as scaffolding proteins or may result from continual trafficking of the protein between the plasma membrane and internal compartments such as endosomes. A review of protein trafficking in neurons is given by Connolly (this volume, Chapter 4). Targetting motifs are beginning to be identified, for example domains within the intracellular loops of the mGlu7 receptor are thought to provide a signal for axonal targetting in neurons (Dev *et al.* 2001). Several GPCR contain PDZ domains at their C-terminus and this is thought to be important for interaction with NHERF (Na^+–K^+ exchange regulatory factor) and endosomal sorting of receptors (Cao *et al.* 1999; Tsao *et al.* 2001). In Drosophila eye, the scaffolding protein INAD associates with PDZ domains on a calcium channel (TRP), phospholipase C β, and protein kinase C (PKC) and this helps to organize the signalling pathways downstream of rhodopsin activation (Pawson and Scott 1997). Although AKAPs and RACKs are important scaffolding proteins for protein kinase A (PKA)- (Michel and Scott 2002) and PKC- (Mochly-Rosen and Kauvar 1998) mediated signalling respectively, they do not appear to hold the GPCR or G protein in these supermolecular complexes.

The mGlu receptors (mglu1 and mglu5) contain proline-rich sequences (PPxxFR) at their extreme C-termini which interact with the Homer class of proteins. Some Homer proteins can dimerize via coiled-coil domains and thus may link the mGlu receptors to IP_3 receptors which also bind Homer via their PDZ domains (Dev *et al.* 2001). The intermediate early gene Homer 1a, which is induced upon excitatory synaptic activity, can disrupt this interaction since it does not allow dimerization. This may provide a mechanism for temporal regulation in response to activity (Xiao *et al.* 1998).

Organization of signalling pathways within membrane microdomains or lipid rafts has received much recent attention (Ostrom 2002; Zajchowski and Robbins 2002). The best studied form of lipid raft are caveolae which are flask-like invaginations of the plasma membrane enriched in the protein caveolin, cholesterol, and sphingolipids (Ostrom *et al.* 2000; Shaul and Anderson 1998). Caveolae are thought to be involved in endocytosis and transcytosis. They are also enriched in many signalling molecules such as GPCR, G proteins, tyrosine kinase receptors as well as downstream effectors such as adenylyl cyclase, PKA, PKC, nitric oxide synthase, mitogen-activated protein kinase and the small G proteins, Ras, Rac, Raf, and Rho. There is increasing evidence for a role of caveolae in the formation of microdomains within the plasma membrane that concentrate the components of the signalling pathways.

9.5 Regulation of G protein-coupled receptor function

9.5.1 Desensitization/resensitization

Desensitization can be manifest as a decrease in responsiveness during continuous drug application (e.g. adrenergic receptor regulation of calcium channels, Diverse-Pierluissi *et al.* 1996) or as a shift in the concentration–effect curve such that higher concentrations for drug are required to achieve the same response (e.g. adrenergic receptor regulation of cAMP formation, Oakley *et al.* 1999). After removal of drug, the receptor activity recovers, although the speed and extent of this resensitization can depend on the duration of agonist activation. Rapid desensitization, which occurs within seconds or minutes, is thought to result from receptor phosphorylation, arrestin binding, and receptor internalization into an intracellular compartment. Long-term desensitization, or down-regulation, may involve changes in receptor and/or G protein levels. There may also be changes in the expression and stability of mRNA (reviewed in Morris and Malbon 1999). Long-term changes in the levels of GPCRs and their accessory proteins are known to be induced by chronic drug treatment and are involved in a number of pathological states (e.g. Aguirre *et al.* 1997; Chen *et al.* 1997; Bloch *et al.* 1999).

9.5.1.1 Phosphorylation

Receptors can be phosphorylated either by a second messenger kinase (protein kinase A, PKA; protein kinase C, PKC) or by a G protein receptor kinase (GRK, previously known as βARK) (reviewed in Morris and Malbon 1999; Ferguson 2001). For receptors coupled to G_s, such as the β-adrenergic receptor, agonist binding results in an increase in cAMP levels which activates PKA which, in turn, phosphorylates the receptor on serine residues located in the third intracellular loop and the C-terminal tail (Clark *et al.* 1989; Hausdorff *et al.* 1989). The concentration of agonist required to phosphorylate a substantial proportion of receptors will reflect the concentration required to maximally activate adenylate cyclase (Hausdorff *et al.* 1989), which will depend on both the receptor density and the coupling efficiency (reviewed in Clark *et al.* 1999). This concentration may be quite low. PKA-mediated desensitization is an example of heterologous desensitization since stimulation of other receptors that activate G_s will also activate PKA. An analogous mechanism can occur for receptors coupled to $G_{q/11}$ which ultimately activate PKC and can be phosphorylated by PKC at quite low concentrations of agonist (e.g. Balmforth *et al.* 1997).

In contrast, the GRKs specifically recognize the agonist-occupied form of the receptor. The agonist-dependent nature of this phosphorylation could explain homologous desensitization. There are at least six distinct isoforms of GRK (reviewed in Carman and Benovic 1998; Bunemann and Hosey 1999; Ferguson 2001) with GRK1 being mainly responsible for phosphorylation of rhodopsin. GRK2 and 3 were previously known as βARK1 and 2 and contain a βγ-binding domain and a PH, PIP_2-binding domain which are both implicated in targetting to the plasma membrane. GRK4 and 6 are palmitoylated and this is thought to aid in membrane targetting. Most GRKs show fairly wide tissue distribution with the exception of GRK1 which is restricted to photoreceptor cells and GRK4 which is mainly found in the testis.

9.5.1.2 Arrestin

Mutagenesis of phosphorylation sites on the intracellular loops and C-terminal domain and the use of inhibitors of PKA and GRK have demonstrated that phosphorylation plays an important role in desensitization of many GPCR including the β-adrenergic receptor (Hausdorff *et al.* 1989; Lohse *et al.* 1990; Premont *et al.* 1995). However, *in vitro* studies using purified proteins have shown that phosphorylation alone does not decrease the GTPase activity stimulated by activated receptor (Lohse *et al.* 1992) and that β-arrestin is also required. This suggested that agonist-occupation of the receptor allows phosphorylation of the receptor by GRK, and that this enhances the binding of β-arrestin to the receptor complex thereby preventing interaction between the receptor and G protein. Therefore, the arrestin-bound phosphorylated receptor is thought to represent a desensitized form. This has been demonstrated for β_2-adrenergic and muscarinic acetylcholine receptors and it is assumed that the same mechanism applies for other GPCR. There is some receptor selectivity for the different isoforms of arrestin. Visual arrestin showed the greatest affinity for phosphorylated rhodopsin with very little binding to the adrenergic or muscarinic receptors; β-arrestin had greater affinity for the adrenergic than the muscarinic receptor while arrestin-3 had similar affinity for these two receptors (reviewed in Ferguson 2001).

The absolute requirement of phosphorylation for desensitization is still a matter of controversy. Several studies have demonstrated that mutation of many if not all phosphorylation sites or truncation of the C-terminal tail does not prevent desensitization (e.g. Seibold *et al.* 1998). This may result from the observation that arrestin can bind to unphosphorylated receptors but often with lower affinity (Gurevich *et al.* 1995). Therefore, the effect observed experimentally is likely to depend upon the relative concentrations of kinases and arrestins expressed in each cell type and agonist-independent β-arrestin association may be observed (Anborgh *et al.* 2000). Alternatively, the sites of phosphorylation identified *in vitro* may not reflect those which are important *in vivo* and it is possible that phosphorylation of other sites may compensate for the loss of sites removed by mutagenesis.

A major difficulty in elucidating the relative importance of GRKs is the lack of selective inhibitors. Heparin has been used to inhibit GRK (e.g. Lohse *et al.* 1990) and has been shown to decrease agonist-induced desensitization of the α_{2A}-adrenergic receptor (Liggett *et al.* 1992). However, heparin is not very selective for GRKs and is believed to interact with a number of growth factor receptors and extracellular matrix proteins (among others) due to its strongly anionic nature (reviewed in Tyrrell *et al.* 1995). Overexpression of GRKs results in increased desensitization of many receptors including adenosine receptors (Palmer *et al.* 1996), the α_{1A}-adrenergic receptor (Diviani *et al.* 1996) and the dopamine 1A receptor (Tiberi *et al.* 1996). A decrease in desensitization after overexpression of a dominant negative mutant of GRK2 is often used as a diagnostic for a GRK-mediated desensitization mechanism (e.g. Mundell *et al.* 1997). Injection of purified GRKs into sensory neurones enhances the desensitization of α_2-adrenergic receptor mediated calcium channel inhibition (Diverse-Pierluissi *et al.* 1996).

9.5.2 **Receptor trafficking**

Most GPCRs undergo agonist-induced internalization of receptor (and possibly ligand) into an endosomal compartment (reviewed in Bohm *et al.* 1997; Koenig and

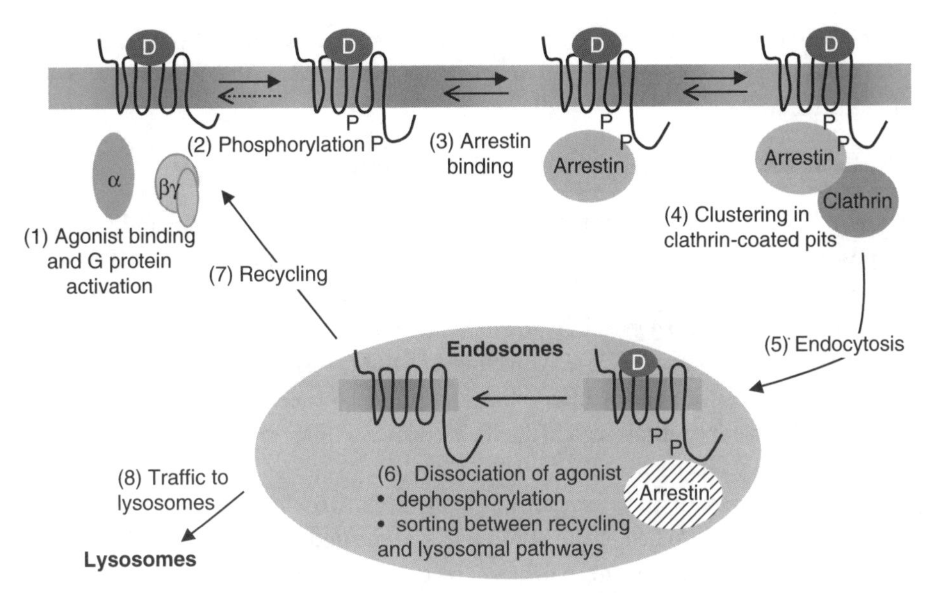

Fig. 9.5 Mechanisms of receptor regulation. Agonist activation of receptors results in G protein activation (1) and phosphorylation (2) of the receptor at sites on the intracellular loops and C terminal domain. Arrestin then binds to the phosphorylated form of the receptor (3) preventing G protein binding and thus this represents the desensitized form of the receptor. Arrestin also binds clathrin and the receptors cluster in clathrin-coated pits (4). The receptors are endocytosed into clathrin-coated vesicles (5) and the vesicles uncoat and fuse with endosomes. In the acidic environment of endosomes (6) agonist dissociates, the receptor is dephosphorylated and sorting events occur to determine whether the receptor is to be returned to the plasma membrane via the recycling pathway (7) or sent to lysosomes for degradation (8).

Edwardson 1997, shown diagrammatically in Fig. 9.5). For small molecule GPCR such as the adrenergic and muscarinic receptors, internalization is usually measured as a loss of hydrophilic antagonist ligand binding (e.g. Koenig and Edwardson 1994). On the other hand, the internalization of peptide receptors is usually demonstrated as internalization of a labelled agonist ligand into an acid-wash resistant internal compartment (e.g. Koenig *et al.* 1997). It is assumed that the receptor and ligand are internalized together in a one-to-one ratio, but this has not been directly demonstrated. Significant reduction in surface receptor number can occur within a few minutes for some receptors, while other receptors show much slower kinetics. The rate and extent of internalization can depend as much on the cell type as the receptor type (Koenig and Edwardson 1996). After removal of agonist, receptors are generally recycled back to the plasma membrane. However, there are a few exceptions for some peptide hormone receptors (see below). Kinetic analysis of the trafficking of muscarinic receptors in neuroblastoma cell lines (Koenig and Edwardson 1994) has shown that 12–30% of surface receptors are internalized in the first minute.

After 20–30 min of incubation with agonist, when a steady state has been reached, the rates of endocytosis and recycling almost balance each other, with a turnover of 5–7% of receptors/min.

9.5.2.1 Molecular mechanisms of internalization

Arrestin has been shown to interact with both receptors and clathrin and has been proposed to function as an adaptor protein in endocytosis. Clathrin-mediated endocytosis is a widely used mechanism and is blocked by agents which inhibit the formation of clathrin-coated pits such as hyperosmolar sucrose (e.g. Garland *et al.* 1994; Pippig *et al.* 1995; Koenig *et al.* 1997). Involvement of the cytoskeleton also appears to be important since, in some cells, internalization can be blocked by agents which disrupt the cytoskeleton for example, colchicine (e.g. substance P receptor, Garland *et al.* 1994 and references therein).

Some receptors can be internalized via both clathrin-coated and smooth pits although this is likely to be dependent on the cell type under study (e.g. CCK receptor, Roettger *et al.* 1995). After internalization receptors have been observed in the same endosomal compartment as transferrin (Garland *et al.* 1994; Ashworth *et al.* 1995; Fonseca *et al.* 1995) suggesting that GPCRs undergo similar trafficking to a number of other cell surface proteins.

9.5.2.2 Physiological role of receptor trafficking

The question of whether internalization can account partly or fully for the observed rapid or long-term desensitization has been hotly debated for a number of years. The short answer is probably that it is partly responsible for short-term desensitization for some receptors under certain conditions but it is not the primary mechanism in most cases. A more complete answer involves understanding some of the following issues. Is the receptor separated from G protein upon internalization? Under what circumstances is the ligand internalized with the receptor and, if the ligand is internalized with the receptor, does the receptor continue to generate a signal whilst in endosomes? How quickly is the internalized receptor recycled? When the recycled receptor reaches the cell surface again, is there a delay before recoupling to G protein and effector?

Answers to most of these questions are at best sketchy. Recent attention has focused on the observation that there is greater phosphatase activity in isolated light vesicles containing sequestered receptor and that the phosphatase activity is increased by the acid environment of the endosome (Sibley *et al.* 1986; Krueger *et al.* 1997). A number of groups have now shown that receptors which are not internalized, as a result of either mutation or prevention of clathrin-coated pit formation, also show an inability to recover from desensitization (β_2-adrenergic receptor, Yu *et al.* 1993; α_{1B}-adrenergic receptor, Fonseca *et al.* 1995; neurokinin receptors, Garland *et al.* 1996). Further, inhibition of phosphatases with calyculin A blocked β_2-adrenergic receptor resensitization (Pippig *et al.* 1995).

It remains debatable whether the internalized receptor does still generate second messengers in endosomes. There is evidence that neuropeptide receptor/ligand complexes (e.g. Beaudet *et al.* 1994 and references therein) and G proteins (e.g. Hendry and Crouch 1991) undergo retrograde axonal transport but the functional significance remains unknown. In the case of neuropeptides, a perinuclear localization of the ligand has been demonstrated (Beaudet *et al.* 1994) which may reflect accumulation in late endosomes (Gruenberg and Maxfield 1995). A role in the nucleus has been suggested and $GABA_B$ receptors have been found to associate directly with nuclear transcription factors (reviewed in Milligan and White 2001). Internalization may be involved in the removal and degradation of neuropeptide transmitters (Beaudet *et al.* 1994 and references therein).

The fate of endocytosed peptide hormone agonists is important in determining the subsequent activation of receptors. Although several peptide hormones have been shown to colocalize with lysosomal markers in confocal microscopy studies (e.g. angiotensin II, Hein *et al.* 1997) quantitative studies with radiolabelled metabolically stable analogues of somatostatin have shown that the agonist is recycled, but then rebinds and re-endocytoses with its receptor unless measures are taken to prevent rebinding (such as extensive washing or the inclusion of saturating concentrations of unlabelled agonist or antagonist) (Koenig *et al.* 1998). Sufficient agonist can be recycled to continue to activate second messenger responses and prevent recovery of surface receptor number (*ibid*).

9.5.3 Mechanisms of long-term down regulation

It is clear that long-term (>1 h) treatment with agonist induces the loss of total cellular receptor number in addition to the decrease in surface receptor number. This is an important mechanism in the response to chronic drug treatment. For example, chronic treatment with antidepressants such as fluoxetine, which elevate synaptic levels of 5HT, induces a decrease in the density of 5HT receptors (Sleight *et al.* 1995). Amino-terminal (i.e. extracellular) polymorphisms of the human β_2-adrenergic receptor lead to increased down-regulation (Green *et al.* 1994). This was proposed to be due to an increased degradation rate rather than effects on synthesis or agonist-induced internalization.

The most widely accepted current model is that when receptors are endocytosed, there are specific interactions between proteins involved in sorting and motifs in the C-terminal tail of the receptors which determine whether receptors enter the recycling pathway or the lysosomal pathway. Two distinct motifs have so far been identified and these are both located right at the end of the C-terminal tail. One is a PDZ-domain which interacts with NHERF (also known as EBP50) in a phosphorylation dependent manner (Cao *et al.* 1999). The other motif is a short sequence which regulates interaction with NSF (*N*-ethylmaleimide sensitive factor, Cong *et al.* 2001). Arrestin binding is also thought to be important for recycling since the V2 vasopressin receptor, which continues to bind arrestin whilst in endosomes, does not show recovery of receptors back to the plasma membrane (Oakley *et al.* 1999).

5.4 Regulation at the level of the G protein

The regulator of G protein signalling (RGS) family of proteins, which contains more than 20 members, are responsible for regulating the rate of hydrolysis of GTP in the $G\alpha$ subunit of the G protein and are thus sometimes known as GAPs or GTPase activating proteins (reviewed in Berman and Gilman 1998; Hepler 1999; DeVries *et al.* 2000). RGS proteins can also attenuate G protein actions that are mediated by $\beta\gamma$ subunits since they may alter the number of $\beta\gamma$ subunits available by enhancing the affinity of $G\alpha$ subunits for $\beta\gamma$ subunits after GTP hydrolysis thus accelerating the reformation of the heterotrimer (*ibid*). The action of RGS proteins is important in regulating the temporal characteristics of G protein actions. For example, RGS proteins play a role in accelerating the decay of agonist-induced activation of GIRK (G protein regulated inward rectifying potassium channels, Doupnik *et al.* 1997) and accelerate desensitization of adrenergic receptor induced N-type calcium channel currents (Diverse-Pierlussi *et al.* 1999). Effectors such as phospholipase $C\beta$ may act as GTPase activating proteins and decrease the lifetime of the activated G protein (Rhee and Bae 1997).

9.6 Conclusions

G protein-coupled receptors transduce extracellular signals into a multitude of intracellular changes including changes in electrical activity, levels of second messengers, secretion, morphology, growth, and differentiation. Understanding how the common structural features of these receptors allow such controlled yet complex signalling patterns is a fundamental question in neurobiology. Over the last few years most of these receptors and their intracellular effector enzymes have been identified and cloned. Now the challenge is two-fold. First to understand how the receptor changes its conformation in order to transmit the signal. This will await a full three-dimensional structure of a number of different receptors along with further mutagenesis experiments and structural modelling. The second, and arguably more difficult challenge, is to understand the cell biology involved. We need to know how the direction and amplification of signal transmission is controlled by different cells under different conditions and how the receptors interact with the rest of the cellular machinery.

References

AbdAlla, S., Zaki, E., Lother, H., and Quitterer, U. (1999) Involvement of the amino terminus of the B2 bradykinin receptor in agonist-induced receptor dimerization. *J. Biol. Chem.*, **274**, 26079–84.

Aguirre, N., Frechilla, D., Garcia-Osta, A., Lashera, B., and Rio, J. D. (1997) Differential regulation by methylenedioxymethamphetamine of 5-hydroxytryptamine1A receptor density and mRNA expression in rat hippocampus, frontal cortex and brainstem: the role of corticosteroids. *J. Neurochem.*, **68**, 1099–105.

Anborgh, P. H., Seachrist, J. L., Dale, L. B., and Ferguson, S. S. G. (2000) Receptor/β-arrestin complex formation and the differential trafficking and resensitization of β_2-adrenergic and angiotensin II type 1A receptors. *Mol. Endocrinol.*, **14**, 2040–53.

Angers, S., Salahpour, A., Joly, E., Hilairet, S., Chelsky, D., Dennis, M., *et al.* (2000) Detection of β_2-adrenergic receptor dimerization in living cells using bioluminescence resonance energy transfer (BRET) *Proc. Natl. Acad. Sci.*, **97**, 3684–9.

Arden, J. R., Segredo, V., Wang, Z., Lameh, J., and Sadee, W. (1995) Phosphorylation and agonist-specific intracellular trafficking of an epitope-tagged μ-opioid receptor expressed in HEK 293 cells. *J. Neurochem.*, **65**, 1636–45.

Ashworth, R., Yu, R., Nelson, E. J., Dermer, S., and Gerschegorn, M. C. (1995) Visualization of the thyrotropin-releasing hormone receptor and its ligand during endocytosis and recycling. *Proc. Natl. Acad. Sci.*, **92**, 512–16.

Ballesteros, J. A., Shi, L., and Javitch, J. A. (2001) Structural mimicry in G protein-coupled receptors: implications of the high-resolution structure of rhodopsin for structure-function analysis of rhodopsin-like receptors. *Mol. Pharmacol.*, **60**, 1–19.

Balmforth, A. J., Shepherd, F. H., Warburton, P., and Ball, S. G. (1997) Evidence of an important and direct role for protein kinase C in agonist-induced phosphorylation leading to desensitization of the angiotensin AT1A receptor. *Br. J. Pharmacol.*, **122**, 1469–77.

Beaudet, A., Mazella, J., Nouel, D., Chabry, J., Castel, M. N., Laduron, P., *et al.* (1994) Internalization and intracellular mobilization of neurotensin in neuronal cells. *Biochem. Pharmacol.*, **47**, 43–52.

Bence, K., Ma, W., Kozasa, T., and Huang, X. Y. (1997) Direct stimulation of Bruton's tyrosine kinase by G_q protein α-subunit. *Nature*, **389**, 296–9.

Benya, R. V., Kusui, T., Katsuno, T., Tsuda, T., Mantey, S. A., Battey, J. F., *et al.* (2000) Glycosylation of the gastrin-releasing peptide receptor and its effect on expression, G protein coupling and receptor modulatory processes. *Mol. Pharmacol.*, **58**, 1490–501.

Berg, K. A., Maayani, S., Goldfarb, J., Scaramellini, C., Leff, P., and Clarke, W. P. (1998) Effector pathway dependent relative efficacy at serotonin type 2A and 2C receptors: evidence for agonist-directed trafficking of receptor stimulus. *Mol. Pharmacol.*, **54**, 94–104.

Berman, D. M. and Gilman, A. G. (1998) Mammalian RGS proteins: barbarians at the gate. *J. Biol. Chem*, **273**, 1269–72.

Birnbaumer, L., Abramowitz, J., and Brown, A. M. (1990) Receptor-effector coupling by G proteins. *Biochim. Biophys. Acta*, **1031**, 163–224.

Blake, A. D., Bot, G., Freeman, J. C., and Reisine. T. (1997) Differential opioid agonist regulation of the mouse μ opioid receptor. *J. Biol. Chem.*, **272**, 782–90.

Bloch, B., Dumartin, B., and Bernard, V. (1999) In vivo regulation of intraneuronal trafficking of G protein-coupled receptors for neurotransmitters. *Trends Pharmacol. Sci*,. **20**, 315–19.

Bockaert, J. and Pin, J. P. (1999) Molecular tinkering of G protein coupled receptors: an evolutionary success. *EMBO J.*, **18**, 1723–9.

Bohm, S. K., Grady, E. F., and Bunnett, N. W. (1997) Regulatory mechanisms that modulate signalling by G protein-coupled receptors. *Biochem. J.*, **322**, 1–18.

Bond, C., LaForge, K. S., Tian, M., Melia, D., Zhang, S., Borg, L., *et al.* (1998) Single-nucleotide polymorphism in the human μ opioid receptor gene alters β-endorphin binding and activity: possibly implications for opiate addiction. *Proc. Natl. Acad. Sci.*, **95**, 9608–13.

Bourne, H. R. (1997) How receptors talk to trimeric G proteins. *Curr. Opin. Cell Biol.*, **9**, 134–42.

Brakeman, P. R., Lanahan, A. A., O'Brien, R., Roche, K., Barnes, C. A., Huganir, R. L., *et al.* (1997) Homer: a protein that selectively binds metabotropic glutamate receptors. *Nature*, **386**, 284–8.

Bunemann, M. and Hosey, M. M. (1999) G protein coupled receptor kinases as modulators of G protein signalling. *J. Physiol.*, **517**, 5–23.

Cao, T. T., Deacon, H. W., Reczek, D., Bretscher, A., and Zastrow, M. V. (1999) A kinase-regulated PDZ domain interaction controls endocytic sorting of the β_2-adrenergic receptor. *Nature*, **401**, 286–90.

Carman, C. V. and Benovic, J. L. (1998) G protein-coupled receptors: turn-ons and turn-offs. *Curr. Opin. Neurobiol.*, **8**, 335–44.

Carruthers, A., Warner, A., Michel, A., Feniuk, W., and Humphrey, P. (1999) Activation of adenylate cyclase by human recombinant sst(5) receptors expressed in CHO-K1 cells and involvement of G(alpha s) proteins. *Br. J. Pharmacol.*, **126**, 1221–9.

Chen, J. J., Dymshitz, J., and Vasko, M. R. (1997) Regulation of opioid receptors in rat sensory neurons in culture. *Mol. Pharmacol.*, **51**, 666–73.

Chidiac, P. (1998) Rethinking receptor-G protein-effector interactions. *Biochem. Pharmacol.*, **55**, 549–56.

Clark, R. B., Friedman, J., Dixon, R. A. F., and Strader, C. D. (1989) Identification of a specific site required for rapid heterologous desensitization of the β-adrenergic receptor by cAMP-dependent protein kinase. *Mol. Pharmacol.*, **36**, 343–8.

Clark, R. B., Knoll, B. J., and Barber. R. (1999) Partial agonists and G protein-coupled receptor desensitization. *Trends Pharmacol. Sci.*, **20**, 279–86.

Compton, S. J., Cairns, J. A., Palmer K.-J., Al-Ani, B., Hollenberg, M. D., and Walls, A. F. (2000) A polymorphic protease-activated receptor 2 (PAR2) displaying reduced sensitivity to trypsin and differential responses to PAR agonists. *J. Biol. Chem.*, **275**, 39207–12.

Cong, M., Perry, S. J., Hu, L. Y. A., Hanson, P. I., Claing, A., and Lefkowitz, R. J. (2001) Binding of the β_2-adrenergic receptor to *N*-ethylmaleimide-sensitive factor regulates receptor recycling. *J. Biol. Chem.*, **276**, 45145–53.

DeBlasi, A., Conn, P. J., Pin J.-P., and Nicoletti, F. (2001) Molecular determinants of metabotropic glutamate receptor signaling. *Trends Pharmacol. Sci.*, **22**, 114–20.

deLigt, R. A. F., Kourounakis, A. P., and Ijzerman, A. P. (2000) Inverse agonism at G protein-coupled receptors: (patho)Physiological relevance and implications for drug discovery. *Br. J. Pharmacol.*, **130**, 1–12.

Deslauiers, B., Ponce, C., Lombard, C., Larguier, R., Bannafous, J.-C., and Marie, J. (1999) *N*-glycosylation requirements for the AT1a angiotensin II receptor delivery to the plasma membrane. *Biochem. J.*, **339**, 397–405.

Dev, K. K., Nakanishi, S., and Henley, J. M. (2001) Regulation of mglu7 receptors by proteins that interact with the intracellular C terminus. *Trends Pharmacol. Sci.*, **22**, 365–1.

DeVries, L., Zheng, B., Fischer, T., Elenko, E., and Farquhar, M. G. (2000) The regulator of G protein signaling family. *Annu. Rev. Pharmacol. Toxicol.*, **40**, 235–71.

Diverse-Pierluissi, M., Inglese, J., Stoffel, R. H., Lefkowitz, R. J., and Dunlap, K. (1996) G protein-coupled receptor kinases mediates desensitization of norepinephrine-induced Ca channel inhibition. *Neuron*, **16**, 579–85.

Diverse-Pierluissi, M., Remmers, A. E., Neubig, R. R., and Dunlap. K. (1997) Novel form of crosstalk between G protein and tyrosine kinase pathways. *Proc. Natl. Acad. Sci.*, **94**, 5417–21.

Diverse-Pierlussi, M. A., Fischer, T., Jordan, J. D., Schiff, M., Ortiz, D. F., Farguhas, M. G., *et al.* (1999) Regulators of G protein signaling proteins as determinants of the rate of desensitization of presynaptic calcium channels. *J. Biol. Chem.*, **274**, 14490–4.

Diviani, D., Lattion, A-L., Larbi, N., Kunapuli, P., Pronin, A., Benovic, J. L., *et al.* (1996) Effect of different G protein-coupled receptor kinases on phosphorylation and desensitization of the α1B-adrenergic receptor. *J. Biol. Chem.*, **271**, 5049–58.

Doupnik, C. A., Davidson, N., Lester, H. A., and Kofuji, P. (1997) RGS proteins reconstitute the rapid gating kinetics of G$\beta\gamma$-activated inwardly rectifying K^+ channels. *Proc. Natl. Acad. Sci.*, **94**, 10461–6.

Ferguson, S. S. G. (2001) Evolving concepts in G protein-coupled receptor endocytosis: the role in receptor desensitization and signaling. *Pharmacol. Rev.*, **53**, 1–24.

Flower, D. R. (1999) Modeling G protein-coupled receptors for drug design. *Biochim. Biophys. Acta*, **1422**, 207–34.

Fonseca, M. I., Button, D. C., and Brown, R. D. (1995) Agonist regulation of α_{1b}-adrenergic receptor subcellular distribution and function. *J. Biol. Chem.*, **270**, 8902–9.

Garland, A. M., Grady, E. F., Lovett, M., Vigna, S. R., Frucht, M. M., Krause, J. E., *et al.* (1996) Mechanisms of desensitization and resensitization of the G protein coupled neurokinin1 and neurokinin2 receptors. *Mol. Pharmacol.*, **49**, 438–46.

Garland, E. M., Grady, E. F., Payan, D. G., Vigna, S. R., and Bunnett, N. W. (1994) Agonist induced internalization of the substance P (NK1) receptor expressed in epithelial cells. *Biochem. J.*, **383**, 177–86.

Gershengorn, M. C. and Osman, R. (2001) Insights into G protein-coupled receptor function using molecular models. *Endocrinology*, **142**, 2–10.

Gether, U. (2000) Uncovering molecular mechanisms involved in activation of G protein coupled receptors. *Endocrine Rev.*, **21**, 90–113.

Gether, U. and Kobilka, B. K. (1998) G protein-coupled receptors. II Mechanism of agonist activation. *J. Biol. Chem.*, **273**, 17979–82.

Greasley, P. J., Fanelli, F., Scheer, A., Abuin, L., Nenniger-Tosato, M., De Benedetti, P. G., *et al.* (2001) Mutational and computational analysis of the α_{1b}-adrenergic receptor. *J. Biol. Chem.*, **276**, 46485–94.

Green, S. A., Turki, J., Innis, M., and Liggett, S. B. (1994) Amino-terminal polymorphisms of the human β_2 adrenergic receptor impart distinct agonist-promoted regulatory properties. *Biochem.*, **33**, 9414–19.

Gruenberg, J. and Maxfield, F. R. (1995) Membrane transport in the endocytic pathway. *Curr. Opin. Cell Biol.*, **7**, 552–63.

Gudermann, T., Grosse, R., and Schultz, G. (2000) Contribution of receptor/G protein signaling to cell growth and transformation. *Naunyn-Schiedeberg's Arch. Pharmacol.*, **361**, 345–62.

Gudermann, T., Schoneberg, T., and Schultz, G. (1997) Functional and structural complexity of signal transduction via G protein-coupled receptors. *Annu. Rev. Neurosci.*, **20**, 399–427.

Gurevich, V. V., Dion, S. B., Onorato, J. J., Ptasienski, J., Kim, C. M., Sternemarr, R., *et al.* (1995) Arrestin interactions with G protein coupled receptors — direct binding studies of wild-type and mutant arrestins with rhodopsin, beta(2)-adrenergic and m2 muscarinic cholinergic receptors. *J. Biol. Chem.*, **270**, 720–31.

Gutkind, J. S. (1998) The pathways connecting G protein-coupled receptors to the nucleus through divergent mitogen-activated protein kinase cascades. *J. Biol. Chem.*, **273**, 1839–42.

Hall, R. A., Premont, R. T., Chow, C.-W., Blitzer, J. T., Pitcher, J. A., Claing, A., *et al.* (1998) The β_2-adrenergic receptor interacts with the Na^+/H^+ exchanger regulatory factor to control Na^+/H^+ exchange. *Nature*, **392**, 626–30.

Hall, R. A., Premont, R. T., and Lefkowitz, R. J. (1999) Heptahelical receptor signaling: beyond the G protein paradigm. *J. Cell Biol.*, **145**, 927–32.

Hamm, H. E. (1998) The many faces of G protein signaling. *J. Biol. Chem.*, **273**, 669–72.

Hausdorff, W. P., Bouvier, M., O'Dowd, B. F., Irons, G. P., Caron, M. G., and Lefkowitz. R. J. (1989) Phosphorylation sites on two domains of the β_2-adrenergic receptor are involved in distinct pathways of receptor desensitization. *J. Biol. Chem.*, **264**, 12657–65.

Hebert, T. E., Moffett, S., Morello, J.-P., Loisel, T. P., Bichet, D. G., Barret, C., *et al.* (1996) A peptide derived from a β_2-adrenergic receptor transmembrane domain inhibitis both receptor dimerization and activation. *J. Biol. Chem.*, **271**, 16384–92.

Hein, L., Meinel, L., Pratt, R. E., Dzau, V. J., and Kobilka, B. K. (1997) Intracellular trafficking of angiotensin II and its AT1 and AT2 receptors: evidence for selective sorting of receptor and ligand. *Mol. Endocrinol.*, **11**, 1266–77.

Hendry, I. and Crouch, M. (1991) Retrograde axonal-transport of the GTP-binding protein-Gi-alpha — a potential neutrophic intraaxonal messenger. *Neurosci. Lett.*, **133**, 29–32.

Hepler, J. R. (1999) Emerging roles for RGS proteins in cell signalling. *Trends Pharmacol. Sci.*, **20**, 376–82.

Heuss, C. and Gerber, U. (2000) G protein-independent signaling by G protein coupled receptors. *Trends Neurosci.*, **233**, 469–75.

Hopkins, C. R. (1992) Selective membrane protein trafficking: vectorial flow and filter. *Trends Biochem. Sci.*, **17**, 27–36.

Howard, A. D., McAllister, G., Feighner, S. D., Liu, Q., Nargund, R. P., Van der Ploeg, L. H., *et al.* (2001) Orphan G protein-coupled receptors and natural ligand discovery. *Trends Pharmacol. Sci.*, **22**, 132–40.

Iiri, T., Farfel, Z., and Bourne, H. R. (1998) G protein diseases furnish a model for the turn-on switch. *Nature*, **394**, 35–38.

Ji, T. H., Grossman, M., and Ji, I. (1998) G protein-coupled receptors: 1. diversity of receptor-ligand interactions. *J. Biol. Chem.*, **273**, 17299–302.

Jordan, B. A. and Devi, L. A. (1999) G protein-coupled heterodimerization modulates receptor function. *Nature*, **399**, 697–700.

Jordan, B. A., Trapaidze, N., Gomes, I., Nivarthi, R., and Devi. L. A. (2001) Oligomerization of opioid receptors with β_2-adrenergic receptors: a role in trafficking and mitogen-activated protein kinase activation. *Proc. Natl. Acad. Sci.*, **98**, 343–8.

Keith, D. E., Murray, S. R., Zaki, P. A.,Chu, P. C., Lissin, D. V., Kang, L., *et al.* (1996) Morphine activates opioid receptors without causing their rapid internalization. *J. Biol. Chem.*, **271**, 19021–4.

Kenakin T (1995) Agonist-receptor efficacy II: agonist trafficking of receptor signals. *Trends Pharmacol. Sci.*, **16**, 232–8.

Kenakin, T. (1997) Differences between natural and recombinant G protein-coupled receptor systems with varying receptor/G protein stoichiometry. *Trends Pharmacol. Sci.*, **18**, 456–64.

Kilpatrick, G. J., Dautzenberg, F. M., Martin, G. R., and Eglen, R. M. (1999) 7TM receptors: the splicing on the cake. *Trends Pharmacol. Sci.*, **20**, 294–301.

Kleuss, C., Scherubl, H., Hescheler, J., Schultz, G., and Wittig, B. (1992)Different beta-subunits determine G protein interaction with transmembrane receptors. *Nature*, **358**, 424–6.

Koenig, J. A. and Edwardson, J. M. (1994) Kinetic analysis of the trafficking of muscarinic acetylcholine receptors between the plasma membrane and intracellular compartments. *J. Biol. Chem.*, **269**, 17174–82.

Koenig, J. A. and Edwardson, J. M. (1996) Intracellular trafficking of the muscarinic acetylcholine receptor: importance of subtype and cell type. *Mol. Pharmacol.*, **49**, 351–9.

Koenig, J. A. and Edwardson, J. M. (1997) Endocytosis and recycling of G protein-coupled receptors. *Trends Pharmacol. Sci.*, **18**, 276–87.

Koenig, J. A., Edwardson, J. M., and Humphrey, P. P. A. (1997) Somatostatin receptors in Neuro-2a cells: 2, ligand internalisation. *Br. J. Pharmacol.*, **120**, 52–9.

Koenig, J. A., Kaur, R., Dodgeon, I., Edwardson, J. M., and Humphrey, P. P. A. (1998) Fates of endocytosed somatostatin receptors and associated agonists. *Biochem. J.*, **336**, 291–8.

Kostenis, E. (2001) Is $G\alpha_{16}$ the optimal tool for fishing ligands of orphan G protein-coupled receptors. *Trends Pharmacol. Sci.*, **22**, 560–4.

Krueger, K. M., Daaka, Y., Pitcher, J. A., and Lefkowitz, R. J. (1997) The role of sequestration in G protein-coupled receptor resensitization. *J. Biol. Chem.*, **272**, 5–8.

Leff, P., Scaramellini, C., Law, C., and McKechnie, K. (1997) A three-state receptor model of agonist action. *Trends Pharmacol. Sci.*, **18**, 355–62.

Liggett, S., Ostrowski, J., Chesnut, L., Kurose, H., Raymond, J. R., Caron, M. G., *et al.* (1992) Sites in the 3rd intracellular loop of the alpha-2A adrenergic receptor confer short-term agonist-promoted desensitization—evidence for a receptor kinase-mediated mechanism. *J. Biol. Chem.*, **267**, 4740–6.

Lohse, M. J., Andexinger, S., Pitcher, J., Truckawinski, S., Codina, J., Faure, J. P., *et al.* (1992) Receptor-specific desensitization with purified proteins. *J. Biol. Chem.*, **267**, 8558–64.

Lohse, M. J., Benovic, J. L., Caron, M. G., and Lefkowitz, R. J. (1990) Multiple pathways of rapid-β_2-adrenergic receptor desensitization. *J. Biol. Chem.*, **265**, 3202–9.

Loisel, T. P., Ansanay, H., Adam, L., Marullo, S., Seifert, R., Lagace, M.,*et al.* (1999) Activation of the β_2-adrenergic receptor-$G\alpha_2$ complex leads to rapid depalmitoylation and inhibition of repalmitoylation of both the receptor and Gas. *J. Biol. Chem.*, **274**, 31014–4019.

Lopez-Ilasaca, M. (1998) Signaling from G protein-coupled receptors to mitogen-activated protein (MAP) kinase cascades. *Biochem. Pharmacol.*, **56**, 269–77.

Lu, Z-L., Saldanha, J. W., and Hulme, E. C. (2002) Seven-transmembrane receptors: crystals clarify. *Trends Pharmacol. Sci.*, **23**, 140–6.

Luttrell, L. M., Ferguson S. S. G., Daaka, Y., Miller, W. E., Mausdsley, S., Della Rocca, G. J., *et al.* (1999) β-arrestin-dependent formation of β_2 adrenergic receptor-src protein kinase complexes. *Science*, **283**, 655–61.

Marchese, A., George, S. R., Kolakowski, L. F., Lynch, K. R., and O'Dowd, B. F. (1999) Novel GPCRs and their endogenous ligand: expanding the boundaries of physiology and pharmacology. *Trends Pharmacol. Sci.*, **20**, 370–5.

Marshall, F. H., Jones, K. A., Kaupmann, K., and Bettler, B. (1999) $GABA_B$ receptors — the first 7TM heterodimers. *Trends Pharmacol. Sci.*, **20**, 396–9.

Mellman, I., Yamamoto, E., Whitney, J. A., Kim, M., Hunziker, W., and Matter, K. (1993) Molecular sorting in polarized and non-polarized cells: common problems, common solutions. *J. Cell Sci.*, **supplement 17**, 1–7.

Meng, E. C. and Bourne, H. R. (2001) Receptor activation: what does the rhodopsin structure tell us? *Trends Pharmacol. Sci.*, **22**, 587–93.

Michel, J. J. C. and Scott, J. D. (2002) AKAP mediated signal transduction. *Annu. Rev. Pharmacol. Toxicol.*, **42**, 235–57.

Milligan, G. and White, J. H. (2001) Protein-protein interactions at G protein-coupled receptors. *Trends Pharmacol. Sci.*, **22**, 513–18.

Mochly-Rosen, D. and Kauvar, L. M. (1998) Modulating protein kinase C signal transduction. *Adv. Pharmacol.*, **44**, 91–145.

Moller, S., Vilo, J., and Croning, M. D. R. (2001) Prediction of the coupling specificity of G protein coupled receptors to their G proteins. *Bioinformatics*, **17 suppl 1**, S174–S181.

Monsma, F. J., McVittie, L. D., Gerfen, C. R., Mahan, L. C., and Sibley, D. R. (1989) Multiple D2 dopamine receptors produced by alternative RNA splicing. *Nature*, **342**, 926–9.

Morris, A. J. and Malbon, C. C. (1999) Physiological regulation of G protein-linked signaling. *Physiol. Rev.*, **79**, 1373–413.

Mundell, S. J., Benovic, J. L., and Kelly, E. (1997) A dominant negative mutant of the G protein-coupled receptor kinase 2 selectively attenuates adenosine A2 receptor desensitization. *Mol. Pharmacol.*, **51**, 991–8.

Newman-Tancredi, A., Audinot, V., Moreira, C., Verreile, L., and Millan, M. J. (2000) Inverse agonism and constitutive activity as functional correlates of serotonin h5-HT_{1B} receptor/G protein stoichiometry. *Mol. Pharmacol.*, **58**, 1042–9.

Oakley, R. H., Laporte, S. A., Holt, J. A., Barak, L. S., and Caron, M. G. (1999) Association of β-arrestin with G protein-coupled receptors during clathrin-mediated endocytosis dictates the profile of receptor resensitization. *J. Biol. Chem.*, **274**, 32248–57.

Ostrom, R. S. (2002) New determinants of receptor-effector coupling: trafficking and compartmentation in membrane microdomains. *Mol. Pharmacol.*, **61**, 473–6.

Ostrom, R. S., Post, S. R., and Insel, P. A. (2000) Stoichiometry and compartmentation in G protein-coupled receptor signaling: implications for therapeutic intervention involving Gs. *J. Pharmacol. Exp. Ther.*, **294**, 407–12.

Palczewski, K., Kumasaka, T., Hori, T., Behnke, C. A., Motoshima, H., Fox, B. A., *et al.* (2000) Crystal structure of rhodopsin: A G protein coupled receptor. *Science*, **289**, 739–45.

Palmer, T. M., Benovic, J. L., and Stiles, G. L. (1996) Molecular basis for subtype specific desensitization of inhibitory adenosine receptors. *J. Biol. Chem.*, **271**, 15272–8.

Pan, X., Ikeda, S. R., and Lewis, D. L. (1998) SR141716A acts as an inverse agonist to increase neuronal voltage-dependent calcium currents by reversal of tonic CB1 cannabinoid receptor activity. *Mol. Pharmacol.*, **54**, 1064–72.

Patel, Y. C., Greenwood, M. T., Warszynska, A., Panetta, R., and Srikant, C. B. (1994) All 5 cloned human somatostatin receptors (hsstr1-5) are functionally coupled to adenylyl-cyclase. *Biochem. Biophys. Res. Commun.*, **198**, 605–12.

Pawson, T. and Scott, J. D. (1997) Signaling through scaffold, anchoring and adaptor proteins. *Science*, **278**, 2075–80.

Pippig, S., Andexinger, S., and Lohse, M. J. (1995) Sequestration and recycling of β2-adrenergic receptors permit receptor resensitization. *Mol. Pharmacol.*, **47**, 666–76.

Premont, R. T., Inglese, J., and Lefkowitz, R. J. (1995) Protein kinases that phosphorylate activated G protein-coupled receptors. *FASEB J.*, **9**, 175–82.

Rana, B. K., Shiina, T., and Insel, P. A. (2001) Genetic variations and polymorphisms of G protein-coupled receptors: functional and therapeutic implications. *Annu. Rev. Pharmacol. Toxicol.*, **41**, 593–624.

Ranganathan, R. and Ross, E. M. (1997) PDZ domain proteins: scaffolds for signalling complexes. *Curr. Biol.*, **7**, R770–3.

Rasenick, M. M., Watanabe, M., Lazarevic, M. B., Hatta, S., and Hamm, H. E. (1994) Synthetic peptides as probes for G protein function. *J. Biol. Chem.*, **269**, 21519–25.

Rhee, S. G. and Bae, Y. S. (1997) Regulation of phosphoinositide-specific phospholipase C isozymes. *J. Biol. Chem.*, **272**, 15045–8.

Rocheville, M., Lange, D. C., Kumar, U., Patel, S. C., Patel, R. C., and Patel, Y. C. (2000) Receptors for dopamine and somatostatin: formation of hetero-oligomers with enhanced functional activity. *Science*, **288**, 154–7.

Roettger, B. F., Rentsch, R. U., Pinon, D., Holicky, E., Hadac, E., Larkin, J. M., *et al.* (1995) Dual pathways of internalization of the cholecystokinin receptor. *J. Cell Biol.*, **128**, 1029–41.

Savarese, T. M. and Fraser, C. M. (1992) In vitro mutagenesis and the search for structure function relationships among G protein coupled receptors. *Biochem. J.*, **283**, 1–19.

Sealfon, S. C. and Olanow, C. W. (2000) Dopamine receptors: from structure to behaviour. *Trends Neurosci.*, **23**, S34–S40.

Seasholtz, T. M., Majumdar, M., and Brown, J. H. (1999) Rho as a mediator of G protein-coupled receptor signaling. *Mol. Pharmacol.*, **55**, 949–56.

Seibold, A., January, B. G., Friedman, J., Hipkin, R. W., and Clark, R. B. (1998) Desensitization of β_2-adrenergic receptors with mutations of the proposed G protein-coupled receptor kinase phosphorylation sites. *J. Biol. Chem.*, **273**, 7637–42.

Shaul, P. W. and Anderson, R. G. W. (1998) Role of plasmalemmal caveolae in signal transduction. *Am. J. Physiol.*, **275**, L843–L851.

Sheng, M. and Sala, C. (2001) PDZ domains and the organization of supramolecular complexes. *Annu. Rev. Neurosci.*, **24**, 1–29.

Sibley, D. R., Strasser, R. H., Benovic, J. L., Daniel, K., and Lefkowitz, R. J. (1986) Phosphorylation/dephosphorylation of the b-adrenergic receptor regulates its functional coupling to adenylate cyclase and subcellular distribution. *Proc. Natl. Acad. Sci.*, **83**, 9408–12.

Simpson, M. M., Ballesteros, J. A., Chiappa, V., Chen, J., Suehiro, M., Hartman, D. S., *et al.* (1999) Dopamine D4/D2 receptor selectivity is determined by a divergent aromatic microdomain contained within the second, third and seventh membrane spanning segments. *Mol. Pharmacol.*, **56**, 1116–26.

Sleight, A., Carolo, C., Petit, N., Zwingelstein, C., and Bourson, A. (1995) Identification of 5-hydroxytryptamine(7) receptor-binding sites in rat hypothalamus — sensitivity to chronic antidepressant treatment. *Mol. Pharmacol.*, **47**, 99–103.

Small, K. M. and Liggett, S. B. (2001) Identification and functional characterization of α_2-adrenoceptor polymorphisms. *Trends Pharmacol. Sci.*, **22**, 471–7.

Spengler, D., Waeber, C., Pantaloni, C., Holsboer, F., Bockaert, J., Seeburg, P. H., *et al.* (1993) Differential signal transduction by five splice variants of the PACAP receptor. *Nature*, **365**, 170–5.

Spooren, W. P. J. M., Gasparini, F., Salt, T. E., and Kuhn, R. (2001) Novel allosteric antagonists shed light on mglu5 receptors and CNS disorders. *Trends Pharmacol. Sci.*, **22**, 331–7.

Stadel, J. M., Wilson, S., and Bergsma, D. J. (1997) Orphan G protein-coupled receptors: a neglected opportunity for pioneer drug discovery. *Trends Neurosci.*, **18**, 430–7.

Strader, C. D., Fong, T. M., Toda, M. R., Underwood, D., and Dixon, R. A. F. (1994) Structure and function of G protein coupled receptors. *Ann. Rev. Biochem.*, **63**, 101–32.

Strange, P. G. (2002) Mechanisms of inverse agonism at G protein-coupled receptors. *Trends Pharmacol. Sci.*, **23**, 89–95.

Sugimoto, Y., Negishi, M., Hayashi, Y., Namba, T., Honda, A., Watabe, A., *et al.* (1993) Two isoforms of the EP3 receptor with different carboxyl-terminal domains. Identical ligand binding properties and different coupling properties with Gi proteins. *J. Biol. Chem.*, **268**, 2712–18.

Tiberi, M., Nash, S. R., Bertrand, L., Lefkowitz, R. J., and Caron, M. G. (1996) Differential regulation of dopamine D_{1A} receptor responsiveness by various G protein-coupled receptor kinases. *J. Biol. Chem.*, **271**, 3771–8.

Tsao, P., Cao, T., and Zastrow, M. V. (2001) Role of endocytosis in mediating downregulation of G protein-coupled receptors. *Trends Pharmacol. Sci.*, **22**, 91–6.

Tyrrell, D., Kilfeather, S., and Page, C. (1995) Therapeutic uses of heparin beyond its traditional role as an anticoagulant. *Trends Pharmacol. Sci.*, **16**, 198–204.

Vogel, W. K., Mosser, V. A., Bulseco, D. A., and Schimerlik, M. I. (1995) Porcine m_2 muscarinic acetylcholine receptor-effector coupling in Chinese-Hamster Ovary cells. *J. Biol. Chem.*, **270**, 15485–93.

Watson, C., Chen, G., Irving, P., Way, J., Chen W.-J., and Kenakin, T. (2000) The use of stimulus-biased assay systems to detect agonist-specific receptor active states: implications for the trafficking of receptor stimulus by agonists. *Mol. Pharmacol.*, **58**, 1230–8.

Weiner, D. M., Burstein, E. S., Nash, N., Croston, G. E., Currier, E. A., Vanover, K. E., *et al.* (2001) 5-hydroxytryptamine$_{2A}$ receptor inverse agonists as antipsychotics. *J. Pharmacol. Exp. Ther.*, **299**, 268–76.

Wess, J., Liu, J., Blin, N., Yun, J., Lerche, C., and Kostenis, E. (1997) Structural basis of receptor/G protein coupling selectivity studied with muscarinic receptors as model systems. *Life Sci.*, **60**, 1007–14.

Whaley, B. S., Yuan, N., Birnbaumer, L., Clark, R. B., and Barber, R. (1994) Differential expression of the β-adrenergic receptor modifies agonist stimulation of adenylyl cyclase: a quantitative evaluation. *Mol. Pharmacol.*, **45**, 481–9.

Wozniak, M. and Limbird, L. E. (1996) The three α_2-adrenergic receptor subtypes achieve basolateral localization in Madin-Darby Canine Kidney cells via different targeting mechanisms. *J. Biol. Chem.*, **271**, 5017–24.

Xiao, B., Tu, J. C., Petralia, R. S., Yuan, J. P., Doan, A., Breder, C. D., *et al.* (1998) Homer regulates the association of group 1 metabotropic glutamate receptors with multivalent complexes of homer-related synaptic proteins. *Neuron*, **21**, 707–16.

Yeaman, C., Heinflink, M., Falck-Pedersen, E., Rodriguez-Boulan, E., and Gerschengorn, M. C. (1996) Polarity of TRH receptors in transfected MDCK cells is independent of endocytosis signals and G protein coupling. *Am. J. Physiol.*, **270**, C753–62.

Yu, S. S., Lefkowitz, R. J., and Hausdorff, W. P. (1993) β-adrenergic receptor sequestration: a potential mechanism of receptor resensitization. *J. Biol. Chem.*, **268**, 337–41.

Zajchowski, L. D. and Robbins, S. M. (2002) Lipid rafts and little caves. *Eur. J. Biochem.*, **269**, 737–52.

Zeng, F-Y. and Wess, J. (1999) Identification and molecular characterization of m_3 muscarinic receptor dimers. *J. Biol. Chem.*, **274**, 19487–97.

Zeng, H., Phang, T., Song, Y. S., Ji, I., and Ji, T. H. (2001) The role of the hinge region of the luteinizing hormone receptor in hormone interaction and signal generation. *J. Biol. Chem.*, **276**, 3451–8.

Zhang, J., Ferguson, S. S. G., Barak, L. S., Bodduluri, S. R., Laporte, S. P., Law, P. Y., *et al.* (1998) Role for G protein-coupled receptor kinase in agonist-specific regulation of μ-opioid receptor responsiveness. *Proc. Natl. Acad. Sci.*, **95**, 7157–62.

Chapter 10

Synapse-to-nucleus calcium signalling

Giles E. Hardingham

10.1 Introduction

Many extracellular stimuli result in an elevation of intracellular calcium concentration. Calcium ions act as messengers, coupling many external events or stimuli to the cell's responses to those stimuli. Calcium has a central role to play in the nervous system, as well as mediating other important processes such as activation of the immune system and fertilization.

For a variety of reasons, the calcium ion has been 'selected' by evolution as an intracellular messenger in preference to other monatomic ions that are abundant in the cellular environment. The reasons why this is the case essentially centre around the need for an intracellular messenger to bind tightly and with high specificity to downstream components of the signalling cascade (often enzymes) and for the capacity for the concentration of the messenger to vary considerably between elevated and basal levels in a manner that is as energetically efficient as possible. The doubly charged Ca^{2+} ion binds strongly to target proteins and is thus a more appropriate messenger than the monovalent sodium, potassium, and chloride ions. It is also more flexible than the smaller divalent magnesium ion and so is able to coordinate more effectively with protein-binding sites. A complete discussion of these issues can be found elsewhere (Carafoli and Penniston 1985).

In addition, it is energetically favourable to utilize calcium as a second messenger. Basal levels of free calcium in the cell are necessarily very low ($\sim 10^{-7}$ M) as higher levels would combine with phosphate ions in the cell to form a lethal precipitate. The very low basal levels of intracellular calcium compared to other ions (e.g. $\sim 10^{-3}$ M for magnesium) make it energetically efficient to use it as a second messenger — a relatively small amount of calcium needs to pass in to the cytoplasm to increase the concentration of the ion severalfold and similarly, relatively little energy need be spent pumping it out again to return the concentration to basal levels.

This review will address how calcium acts as a second messenger in mammalian neurons to couple synaptic activity to gene transcription. Such new gene expression has an important role to play in triggering long-term changes to neuronal physiology, function, and fate.

10.2 Calcium as an intracellular second messenger

Many cell types rely on an elevation of intracellular calcium to activate essential biological functions. This elevation can occur via either influx of calcium through proteinaceous channels into the cell from the extracellular medium, or through the release of calcium from internal stores (typically the endoplasmic reticulum).

Calcium influx is a critical step in communication between neurons. An action potential, travelling the length of a neuron, will arrive at the axon terminal and trigger calcium entry into the terminal through voltage-dependent calcium channels. This in turn results in calcium-dependent neurotransmitter release into the synaptic cleft. This neurotransmitter causes an electrical change in the post-synaptic neuron through the activation of neurotransmitter-gated ion channels. Thus, calcium is responsible for coupling action potentials to neurotransmitter release and enabling information to be passed on from neuron to neuron.

However, as well as contributing to the nuts and bolts of inter-neuronal communication, synaptically evoked cellular calcium transients activate signalling pathways in the cell and so are responsible for much *intra*-cellular communication as well. The predominant excitatory neurotransmitter in the central nervous system is glutamate and when released at the synapse it acts on glutamate receptors located on the post-synaptic membrane. Calcium influx is mainly mediated by the NMDA subtype of ionotropic glutamate receptors. Although some forms of non-NMDA receptors also pass calcium, it is more often than not that NMDA receptors mediate the calcium influx at the post-synaptic membrane that activate intracellular signalling pathways. This calcium influx can be augmented by release from intracellular stores, for example release from inositol triphosphate-sensitive stores via the activation of certain metabotropic glutamate receptors or via simple calcium-induced calcium release via ryanodine receptors (Berridge 1993*a*, *b*, Bockaert and Pin 1997; Verkhratsky and Petersen 1998; Emptage *et al.* 1999*a*, *b*).

10.3 Synaptic plasticity in the nervous system

An important characteristic of an animal's nervous system is that it adapts in a structural and functional way in response to certain patterns of synaptic stimulation (Bliss and Collingridge 1993). The mature animal depends on this activity-dependent plasticity to change neuronal connectivity and strength in ways that enable the process of learning and memory (Malenka 1994). It is therefore a fundamental goal of neurobiologists to understand how electrical activity results in these long-lasting changes.

Synaptic plasticity can be split into three phases (see Chapter 14). During the early phase, seconds to minutes after electrical activity, changes in neuronal connections take place via the modification of existing proteins, particularly ion channels, for example by phosphorylation or delivery to the post-synaptic membrane (Robertson *et al.* 1996; Braithwaite *et al.* 2000; Malinow *et al.* 2000). In the later stages (minutes to hours), synthesis of new proteins (phase II) or new gene expression and subsequent protein synthesis (phase III) converts these initial transient changes into long-lasting ones. In the mammalian brain these changes in gene expression are primarily triggered by calcium influx into neurons and involve the activation of intracellular signalling pathways (Bading *et al.* 1993).

The hippocampus has long been the focus of studies into memory formation in mammals since clinicians observed that patients with hippocampal lesions could not form new memory, suffered anterograde and retrograde amnesia, and were deficient in spatial learning tasks (Milner *et al.* 1968; Nadel and Moscovitch 1997; Whishaw *et al.* 1997). The phenomenon of hippocampal long-term potentiation (LTP) is an extensively studied model for learning and memory. LTP is an activity-dependent increase in synaptic efficacy that can last for days to weeks in intact animals (Bliss and Lomo 1973; Bliss and Collingridge 1993). It is induced in the postsynaptic neuron by repeated, high-frequency stimulation of presynaptic afferents. LTP is characterized by an early, protein synthesis independent phase and later phases whose establishment is blocked by protein synthesis inhibitors (Frey *et al.* 1993) and, for the longest-lasting phase, requires a critical period of transcription after the LTP-inducing stimuli have been applied (Nguyen *et al.* 1994). Critically, its induction was found to be dependent on an elevation of post-synaptic calcium (Lynch *et al.* 1983).

The changes in calcium following LTP-inducing stimuli elicit the rapid induction of a number of immediate early genes (IEGs—see also Chapter 14). Wisden *et al.* (1990) and Cole *et al.* (1989) showed a correlation between LTP and the induction of *zif268* transcription. Later Worley *et al.* (1993) showed that stronger stimuli could also induce *c-fos* and *c-jun* and that such stimulus-induced gene expression was dependent on activation of NMDA receptors. IEGs are genes whose transcription can be triggered in the absence of *de novo* protein synthesis and many are transcription factors. These transcription factors likely contribute to secondary waves of transcription, leading to the structural and functional changes to the neuron required for the maintenance of LTP, although the exact mechanisms underlying this are unclear. Since these early studies, many genes up-regulated by LTP-inducing stimuli have been implicated in the maintenance phase of LTP such as tissue plasminogen activator (Baranes *et al.* 1998) and activity-regulated cytoskeleton-associated protein (Guzowski *et al.* 2000). These aspects of synaptic plasticity are discussed in Chapter 14.

Thus, the activation of gene expression in electrically excitable cells has been the subject of much recent research. Below is a brief overview of the essentials of transcriptional activation — the point at which gene expression is most often regulated.

10.4.1 **Control of gene expression**

The control of gene expression (at the protein level) can occur at many stages in the process; at transcription initiation and elongation, RNA processing (including alternative splicing), mRNA stability, control of translation, and of protein degradation. By far the commonest point of regulation is in transcription initiation. The synthesis of mRNA is catalysed by RNA polymerase (pol) II but a large number of additional proteins are needed to direct and catalyse initiation at the correct place.

A DNA sequence near the transcription start site, called the core promoter element, is the site for the formation of the pre-initiation complex (PIC), a complex of RNA pol II and proteins called basal transcription factors (Roeder 1996). RNA pol II and the basal transcription factors are sufficient to facilitate a considerable amount of transcription *in vitro* (called basal transcription). However, *in vivo*, basal transcription levels are often extremely low, reflecting the fact that *in vivo* the DNA containing the core promoter is associated with histones and subsequently less accessible to incoming factors. For transcription to take place, other accessory factors, called activating transcription factors (hereafter known as transcription factors) are required. These factors bind to specific DNA promoter elements, located upstream of the core promoter and enhance the rate of PIC formation by contacting and recruiting the basal transcription factors, either directly, or indirectly via adapters or coactivators (Ptashne and Gann 1997). They can also modify or disrupt the chromatin structure (for example by histone acetylation) to make it easier for other factors to come in and bind.

The ability of many transcription factors to influence the rate of transcription initiation can be regulated by signalling pathways. This provides a mechanism whereby a stimulus applied to the cell that activates a signalling pathway can result in the specific activation of a subset of transcription factors. These signalling mechanisms often involve regulatory phosphorylation events at the transcription factor level that control, for example, DNA binding affinity, subcellular localization, or its interactions with the basal transcription machinery (Hunter and Karin 1992; Hill and Treisman 1995*a*; Whitmarsh and Davis 2000). Genes whose promoters contain binding sites for these signal-inducible transcription factors are transcribed as a result of signal-activating stimuli. There are several well-characterized DNA elements that act as binding sites for transcription factors that are regulated by calcium-activated signalling pathways, some examples of which are listed below.

10.4.2 Calcium-responsive DNA regulatory elements and their transcription factors

The **cyclic-AMP Response Element (CRE)** was first identified in the promoter of the somatostatin gene as the element required to confer cAMP inducibility on the gene (Comb *et al.* 1986; Montminy *et al.* 1986). The CRE was subsequently found in a number of other genes and is a 8 bp palindromic sequence, 5′-TGACGTCA-3′. The calcium-inducibility of the CRE was demonstrated first in PC12 cells (Sheng *et al.* 1988; Sheng *et al.* 1990). Bursts of synaptic activity strongly activate CREB by triggering synaptic NMDA receptor dependent calcium transients (Hardingham *et al.* 2002) and the CRE is activated by stimuli that generate long-lasting LTP in area CA1 of the hippocampus (Impey *et al.* 1996). The transcription factor that can mediate activation via the CRE, CRE binding protein (CREB) was isolated as a phosphoprotein that bound the CRE on the mouse somatostatin gene (Montminy and Bilezikjian 1987) and was cloned by Hoeffler *et al.* (1988). In neurons, calcium activation of CREB is mediated by the CaM kinase and Ras-ERK1/2 signalling pathways (see below).

The **Serum Response Element (SRE)** was identified as an element centred at –310 bp required for serum induction of *c-fos* in fibroblasts (Treisman 1985). The SRE comprises a core element 5′-CC[A/T]$_6$GG-3′ that is the binding site for serum response factor, SRF (Treisman 1987; Schröter *et al.* 1987; Norman *et al.* 1988). In addition, the SRE contains a ternary complex factor (TCF) binding site, 5′-CAGGAT-3′, situated immediately 5′ to the core SRE, which is bound by TCF. TCF is an umbrella name for a group of Ets domain proteins, SAP-1, Elk-1, and SAP-2 (Price *et al.* 1995). These proteins cannot bind the SRE on their own but recognize the SRE/SRF complex. Like the CRE, the SRE is a target for calcium signalling pathways, and can confer calcium inducibility onto a minimal *c-fos* promoter in response to activation of L-type calcium channels (Bading *et al.* 1993; Misra *et al.* 1994; Johnson *et al.* 1997) and NMDA receptors (Bading *et al.* 1993). Calcium-dependent synaptic activation of the SRE in hippocampal neurons is mediated by the ERK1/2 pathway (Hardingham *et al.* 2001*a*).

The **Nuclear Factor of Activated T cells (NFAT) Response Element** is another well-characterized calcium-response element (Rao *et al.* 1997). NFAT activity is regulated by the calcium-activated phosphatase calcineurin at the level of subcellular localization (Crabtree and Clipstone 1994; Rao *et al.* 1997). Calcineurin dephosphorylates the normally cytoplasmic NFAT, which exposes a nuclear localization signal and leads to its active transport into the nucleus. In the absence of continuing elevated levels of calcium (and calcineurin activity), NFAT becomes rephosphorylated by glycogen synthase kinase 3 (GSK3) and is re-exported to the cytoplasm. While these mechanisms were primarily characterized in T-cells, they also apply to neurons (Graef *et al.* 1999).

The next section of this review will focus on the mechanism of calcium-dependent CREB activation. The importance of CREB-dependent transcription on various aspects of neuronal physiology make it an extensively studied transcription factor.

10.5 The physiological importance of CREB

The study of the calcium activation of CREB-mediated gene expression bears considerable neurophysiological relevance. CREB seems to have an important role in the establishment of long-term memory in a variety of organisms (Silva *et al.*, 1998). Genetic and molecular studies of learning paradigms in the marine snail, *Aplysia californica* and the fruit fly, *Drosophila melanogaster* have shown that modulating CREB levels or affecting CREB-dependent transcription severely affects the long-term, protein synthesis-dependent phase of the learning paradigm studied (Dash *et al.* 1990; Bailey and Kandel 1994; Tully *et al.* 1994; Yin *et al.* 1995*a*, *b*). In the mammalian central nervous system CREB was also found to play a role in information storage. The intrahippocampal perfusion of antisense oligonucleotides designed to bind and trigger degradation of CREB mRNA achieved a transient decrease in CREB levels in the hippocampus, an area of the brain needed for certain spatial memory tasks. This strategy blocked the animal's long-term memory of these spatial tasks without affecting short-term memory (Guzowski and McGaugh 1997). Mice deficient in alpha and delta forms of CREB have defective long-term (but not short-term) memory (Bourtchuladze *et al.* 1994).

There is also considerable evidence that CREB has a role in other aspects of neuronal physiology including drug addiction (Blendy and Maldonado 1998), circadian rhythmicity (King and Takahashi 2000; Gau *et al.* 2002) and neuronal survival (Walton and Dragunow 2000). Mice deficient in CREB exhibit excess apoptosis in sensory neurons (Lonze *et al.* 2002) and CREB mediates many of the pro-survival effects of neurotrophins (Riccio *et al.* 1999; Finkbeiner 2000; Walton and Dragunow 2000).

One mechanism by which the calcium-mediated activation of CREB modulates neuronal functions may involve BDNF, the activation of which is controlled at least in part by a CRE/CREB-dependent mechanism (Tao *et al.* 1988; Shieh *et al.* 1998). BDNF plays an important role in the survival and differentiation of certain classes of neurons during development (Ghosh *et al.* 1994; Schwartz *et al.* 1997) and is also implicated in the establishment of neuronal plasticity (Thoenen 1995). Other CREB-regulated genes that are implicated in maintaining changes in synaptic strength and efficacy include nNOS (Sasaki *et al.* 2000) and tissue plasminogen activator (Baranes *et al.* 1998). Apart from BDNF, CREB-dependent pro-survival genes include bcl-2, mcl-1, and vasoactive intestinal peptide (Wilson *et al.* 1996; Hahm and Eiden 1998; Wang *et al.* 1999).

10.6 **The mechanism of CREB activation**

10.6.1 **CREB activation requires a crucial phosphorylation event**

CREB can bind to the CRE even prior to the activation of CRE-dependent gene expression, indicating that regulation of its activity is not via the control of its DNA-binding activity (Sheng and Greenberg 1990; Sheng *et al.* 1990). CREB binds the CRE as a dimer, mediated by a leucine zipper motif (Yamamoto *et al.* 1988). To activate CREB-mediated transcription, CREB must become phosphorylated on serine 133 (Gonzalez and Montminy 1989).

Sheng *et al.* (1990) showed that elevation of intracellular calcium, following depolarization of PC12 cells, resulted in CREB phosphorylation on serine 133 and activation of CREB-mediated gene expression (Sheng *et al.* 1991). CREB-mediated gene expression was abolished by mutating serine 133 to an alanine, underlining the importance of the site as a point of control by calcium signalling pathways. These results showed that CREB is a calcium-responsive transcription factor and led to the assumption that CREB-mediated gene expression was triggered solely by phosphorylation of CREB on serine 133. Calcium-activated phosphorylation of CREB was subsequently shown in neurons of several types (Ginty *et al.* 1993; Bito *et al.* 1996, 1997; West *et al.* 2001).

10.6.1.1 Calcium-dependent signalling molecules capable of phosphorylating CREB on serine 133

10.6.1.1.1 **CaM kinases and their role in calcium-activated, CRE-dependent gene expression** CREB phosphorylation on serine 133 can be mediated by a number of protein kinases, including the multifunctional calcium/calmodulin dependent protein kinases (CaM kinases) II, IV and the less well studied CaM kinase I (Sheng *et al.* 1991; Matthews *et al.* 1994; Sun *et al.* 1994). CaM kinases play a role in diverse biological processes such as secretion, gene expression, LTP, cell cycle regulation, and translational control (for a review, see Schulman 1993). A role for CaM kinases in the calcium activation of *c-fos* expression (which contains a CRE and a SRE) was indicated by the attenuation of L-type calcium channel-activated *c-fos* expression in neurons by the CaM kinase inhibitor, KN-62 (Bading *et al.* 1993), and by the blocking of calcium-dependent *c-fos* expression by the calmodulin antagonist, calmidazolium (Morgan and Curran 1986; Bading *et al.* 1993). CaM kinase II is a protein highly expressed in the nervous system (Lin *et al.* 1987). CaM kinase IV is similar in sequence to CaM kinase II's catalytic domain (Ohmstede *et al.* 1989). It is expressed in some cells of the immune system but also in neuronal cells, including the cerebellum and the hippocampus and has been shown to be mainly localized to the nucleus (Jensen *et al.* 1991; Bito *et al.* 1996).

Regulation and structural organization of CaM kinases II and IV are broadly similar (Schulman 1993; Ghosh and Greenberg 1995; Heist and Schulman 1998; Hook and Means 2001; Soderling and Stull 2001). Both have a N-terminal catalytic domain and a central calcium/calmodulin binding regulatory domain. Note that the kinase itself does not bind calcium, activation of the enzyme occurs when calcium complexed with a small protein, calmodulin, binds and displaces an auto-inhibitory domain that otherwise occludes the catalytic site. Despite their structural similarities and their ability to phosphorylate CREB on serine 133, CaM kinases II and IV have very different effects on CRE/CREB-mediated gene expression. Matthews *et al.* (1994) showed that a constitutively active form of CaM kinase IV, but not an active form of CaM kinase II could activate CRE-dependent transcription. This is because CaM kinase II also phosphorylates CREB at an inhibitory site, serine 142 (Sun *et al.* 1994).

Thus CaM kinase IV appears to be a prime candidate for the activation of CREB-mediated gene expression by nuclear calcium signals, being located largely in the nucleus and able to efficiently activate CREB. Indeed, antisense oligonucleotide-mediated disruption of CaM kinase IV expression suppressed calcium-activated CREB phosphorylation in hippocampal neurons (Bito *et al.* 1996) and calcium-activated CREB phosphorylation is impaired in neurons cultured from mice deficient in CaM kinase IV (Ho *et al.* 2000; Ribar *et al.* 2000; Kang *et al.* 2001). As would be predicted from the importance of CREB in long-term synaptic plasticity, CaM kinase IV is critical for long-term hippocampal LTP (Ho *et al.* 2000; Kang *et al.* 2001). Mice deficient in CaM Kinase IV exhibit defective cognitive memory related to a noxious shock (fear conditioning, (Wei *et al.* 2002)) and while the same mice showed no obvious defects in spatial learning (Ho *et al.* 2000), a study that employed a more subtle technique (expressing a dominant interfering mutant of CaM kinase IV only in the post-natal forebrain) did find defective consolidation/retention of hippocampus-dependent long-term memory. In another well-studied paradigm for synaptic plasticity, cerebellar long-term depression (LTD), both inhibition of either CREB function or CaM kinase IV blocked the late phase (but not induction) of long-term depression in the cerebellum (Ahn *et al.* 1999; Kang *et al.* 2001).

10.6.1.1.2 **The Ras-ERK1/2 (MAP kinase) cascade** The role of the ERK1/2 pathway in signalling to CREB was characterized first in the context of growth factor stimulation. Growth factors such as nerve growth factor (NGF) can activate CREB phosphorylation by a mechanism mediated by the Ras-ERK1/2 pathway. NGF treatment of PC12 cells resulted in the Ras-ERK1/2 dependent activation of a CREB kinase (Ginty *et al.* 1994) found to be a member of the previously identified pp90 RSK family, RSK2 (Xing *et al.* 1996). RSK2 was able to mediate CREB phosphorylation *in vivo* and *in vitro*. The fact that NGF cannot efficiently activate CRE-dependent transcription (Bonni *et al.* 1995; Johnson *et al.* 1997) demonstrates that CREB phosphorylation on serine 133 is not sufficient to activate CREB-mediated gene expression — additional activating steps are required (see below).

The Ras/MAP kinase (ERK1/2) pathway is also activated by calcium (Bading and Greenberg 1991; Rosen *et al.* 1994) and so RSK2 is activated by calcium signals as well as growth factors. This pathway is involved in both the induction and maintenance of LTP and other memory paradigms (Impey *et al.* 1999; Adams and Sweatt 2002).

10.6.1.2 Parallel activation of CaM kinase and ERK1/2 pathways by synaptic activity

At first glance it may appear that there is a certain degree of redundancy in the parallel activation of both the CaM kinase and ERK1/2 pathways when it comes to phosphorylating CREB. However, both pathways have critical roles to play. While CaM kinase IV itself is calcium dependent, ERK1/2 and RSK2 are not (the calcium-dependent activation step is far upstream) and so they are activated slower than CaM kinases and, importantly, their activity remains long after synaptic activity has ceased. Thus, while the CaM kinase pathway mediates CREB phosphorylation within the first few seconds of calcium influx, and both pathways contribute to CREB phosphorylation at intermediate time points, the ERK1/2 pathway is needed to prolong CREB phosphorylation after activity has ceased (Hardingham *et al.* 1999; Hardingham *et al.* 2001*a*; Impey and Goodman 2001; Wu *et al.* 2001) which is important for robust activation of CREB-dependent gene expression (Bito *et al.* 1996; Impey and Goodman 2001).

The importance of CaM kinase IV comes from its role in not only phosphorylating CREB, but in carrying out a 2nd critical activating step.....

10.6.1.3 Uncoupling of CREB phosphorylation from CREB-mediated transcription

CREB phosphorylation on serine 133, while necessary for CREB to function as a transcriptional activator, is not sufficient for full induction of gene expression. A wide variety of extracellular signals lead to CREB phosphorylation on serine 133, but many of them, including stimulation with nerve growth factor (NGF) or epidermal growth factor (EGF), which rely solely on the ERK1/2 pathway to trigger phosphorylation, are poor activators of CRE-mediated transcription (Xing *et al.* 1996). Further experiments demonstrate that serine 133 phosphorylation is not sufficient for calcium-induced CREB-mediated transcription: CaM kinase inhibition blocks CRE-mediated gene expression without inhibiting CREB phosphorylation (Chawla *et al.* 1998; Hardingham *et al.* 1999) showing that the remaining 'CREB kinase' pathway (the ERK1/2 pathway) is unable to activate CREB-dependent gene expression.

Thus, CRE-dependent transcription requires additional activation events that are provided by CaM kinase activity but not, for example, by NGF or EGF treatment, activators of the Ras-ERK1/2 kinase pathway. This fact is further reinforced by the observation that both acutely activated CaM kinase IV and activated Ras trigger CREB phosphorylation but that only CaM kinase IV could activate CREB-mediated transcription (Chawla *et al.* 1998).

10.6.2 The role of CREB Binding Protein (CBP) in CREB-mediated transcription

10.6.2.1 Phosphorylated CREB activates transcription by recruiting its coactivator CREB binding protein, CBP

As stated earlier, CREB is regulated by modification of its transactivation domain, rather than its subcellular localization or DNA-binding activity. For CREB to activate transcription it must be associated with its coactivator, CBP, via the inducible part of the CREB transactivation domain, the kinase inducible domain (KID). CBP and p300 (a closely related protein) function as coactivators for many signal-dependent transcription factors such as c-Jun (Arias *et al.* 1994; Bannister *et al.* 1995), interferon-α signalling through STAT2 (Bhattacharya *et al.* 1996), Elk-1 (Janknecht and Nordheim 1996*a*, *b*), p53 (Gu *et al.* 1997), and nuclear hormone receptors (Chakravarti *et al.* 1996). The association of CBP with CREB is dependent on CREB being phosphorylated on serine 133 (Chrivia *et al.* 1993; Parker *et al.* 1996).

CBP's ability to stimulate transcription may be due to its ability to recruit to the promoter components of the basal transcription machinery — it has been reported to associate with TFIIB, TATA-binding protein (TBP), and RNA polymerase II complex (Kwok *et al.* 1994; Swope *et al.* 1996; Nakajima *et al.* 1997). In addition, CBP has an intrinsic histone acetyl transferase (HAT) activity (Ogryzko *et al.* 1996; Bannister and Kouzarides 1996) as well as being able to associate with other HAT proteins, p/CAF and SRC1 (Yao *et al.* 1996; Chen *et al.* 1997).

10.6.2.2 A Model for nuclear calcium-regulated transcription: regulation of CBP

The purpose of CREB phosphorylation on serine 133 appears to be to recruit the transcriptional coactivator, CREB binding protein (CBP) to the promoter (Chrivia *et al.* 1993). However, evidence described above suggests that this is insufficient to activate transcription fully. This pointed to the possibility that the second regulatory event critical for transcriptional activation may involve the activation of CBP. This was indeed found to be the case. CBP's transactivating potential is positively regulated by calcium, acting via CaM kinase IV (Chawla *et al.* 1998; Hardingham *et al.* 1999; Hu *et al.* 1999). CaM kinase IV-dependent enhancement of CBPs activity occurs predominantly via a phosphorylation event on serine-301 (Impey *et al.* 2002). In addition, recent use of a mutant form of CREB that constitutively binds CBP showed that in hippocampal neurons this mutant, while modestly active, is nowhere near as active as wild-type CREB activated by calcium signals and could itself be further activated by calcium signals and CaM kinase IV (Impey *et al.* 2002). Thus, the critical role for CaM kinase IV in calcium-activation of CREB-dependent transcription is in activating CBP, while the ERK1/2 pathway is responsible for ensuring prolonged CREB phosphorylation (and thus association of CREB with CBP, see Fig. 10.1).

As mentioned earlier, CBP/p300 acts as a coactivator for a large number of transcription factors. Thus, the fact that CBP is subject to calcium-dependent regulation means that calcium could potentially regulate transcription mediated by many of these factors, either on its own or in conjunction with other signals. Indeed, the calcium-dependent activation of the CBP-interacting transcription factor c-Jun has been reported (Cruzalegui *et al.* 1999) that can occur independently of what were hitherto thought to be crucial regulatory phosphorylation sites (targets of stress activated protein kinases).

10.7 **Decoding the calcium signal**

Until a few years ago there was little evidence to suggest anything other than the idea that elevated levels of intracellular calcium activate a specific set of 'calcium-responsive genes' to a greater or lesser extent, depending on the level of calcium concentration in the cell. This was in contrast to the apparent complexity of the processes that calcium was mediating. However, evidence is mounting that intracellular signalling pathways in neurons are laid down in a very sophisticated manner to enable cells to distinguish between calcium signals of differing properties. These properties include the amplitude of the signal (Hardingham *et al.* 1997), its temporal properties (including oscillatory frequency (Hardingham *et al.* 2001*b*), its spatial properties (Hardingham *et al.* 1997; Hardingham and Bading 1998; Hardingham *et al.* 2001*a*), and its site of entry (Bading *et al.* 1993; Hardingham *et al.* 1999, 2002; Sala *et al.* 2000). The subcellular localization of calcium-responsive signalling molecules points to different spatial requirements for calcium. For example, CaM kinase IV is predominantly nuclear,

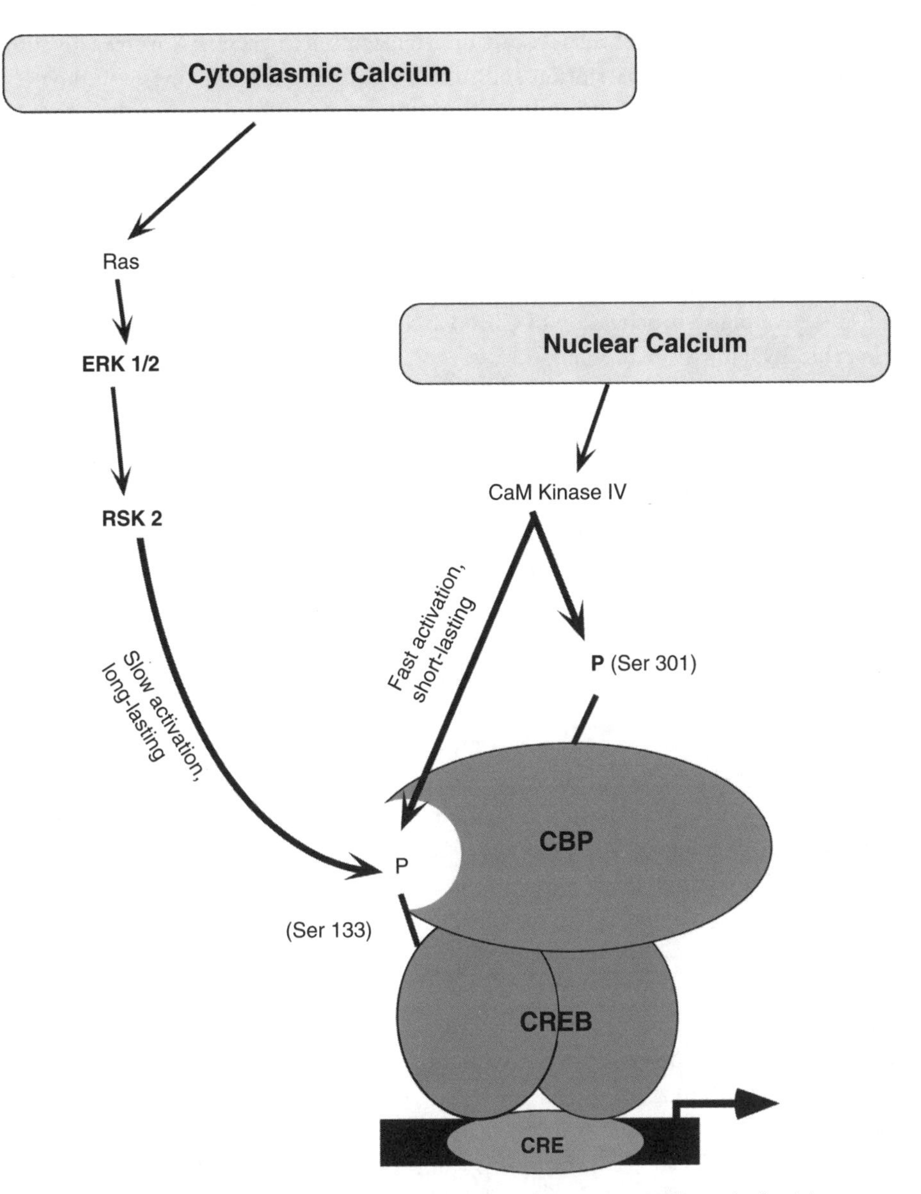

Fig. 10.1 Calcium-activation of the transcription factor CREB. Synaptically evoked calcium transients trigger two parallel pathways that result in CREB phosphorylation on serine 133 (necessary for CBP recruitment); the Ras-ERK1/2 pathway and the CaM Kinase IV pathway. The Ras-ERK1/2 pathway that activates the CREB kinase RSK2 is in particular responsible for the prolonged phosphorylation of CREB. CaM kinase IV carries out the second critical activation step targeting CBP; this includes a phosphorylation event on serine 301. Note: the spatial requirements for calcium activation of these pathways is different (see Section 10.7).

which reflects the fact that nuclear calcium transients are necessary for the activation of CREB-dependent transcription and sufficient to activate CaM kinases in the neuronal nucleus (Hardingham *et al.* 1997, 2001*b*). As well as elevation of nuclear calcium being triggered by electrical activity, translocation of the calcium-sensing protein calmodulin into the nucleus has also been described (Deisseroth *et al.* 1998; Mermelstein *et al.* 2001). Were this to happen in neurons where nuclear calmodulin concentration is limiting this would assist in the activation of nuclear CaM kinases. Thus, an elevation of nuclear calcium is necessary for triggering the crucial 2nd CREB-activating step; activation of CBP (Chawla *et al.* 1998; Hardingham *et al.* 1999; Hu *et al.* 1999; Impey *et al.* 2002).

In sharp contrast, many components of the Ras-ERK1/2 pathway (including putative calcium-dependent activators of the pathway, PYK2 and synGAP) are contained within a protein complex with the NMDA receptor at the membrane (Husi *et al.* 2000). Calcium requirements for activation of this pathway are very different; increased calcium levels just under the membrane near the site of entry are sufficient to activate the ERK1/2 pathway (Hardingham *et al.* 2001*a*). Thus, relatively weak or spatially restricted calcium signals would be able to activate this pathway, which acts on the SRF/TCF transcription factor complex (Janknecht *et al.* 1993) and is responsible for prolonged CREB phosphorylation (though is unable to carry out the crucial 2nd activation step).

Many proteins located at the cytoplasm are of course not tethered near the membrane. For the freely diffusible cytoplasmic signalling molecule calcineurin, submembranous calcium elevation is not enough — it requires a global increase in cytoplasmic calcium concentration to be appreciably activated in order to trigger NFAT nuclear translocation (Hardingham *et al.* 2001*a*). Continued nuclear localization would then rely on active calcineurin in the nucleus (and therefore an elevation in nuclear calcium).

The ability of spatially distinct calcium signals to differentially activate transcription naturally only has relevance if scenarios exist in the neuron whereby nuclear, cytoplasmic, and submembranous calcium levels change to differing degrees. In neurons, isolated synaptic inputs can yield extremely spatially restricted calcium transients (Emptage *et al.* 1999*a*). However, where synaptic inputs are stronger or repetitive, global calcium transients can result (Alford *et al.* 1993). Also, where synaptic inputs contribute to causing the postsynaptic cell to fire action potentials, synaptic inputs can co-operate with back-propagating action potentials to yield global calcium transients (Magee and Johnston 1997; Emptage 1999*b*). Thus, differing patterns of electrical activity can in theory recruit different calcium-dependent signalling modules, which would have a qualitative effect on the resulting transcriptional output (Fig. 10.2).

Differing buffering capacities or calcium clearance mechanisms of different areas of the cell can also lead to transient subcellular differences in calcium. The nucleus appears to be particularly suited to the propagation and prolongation of calcium signals.

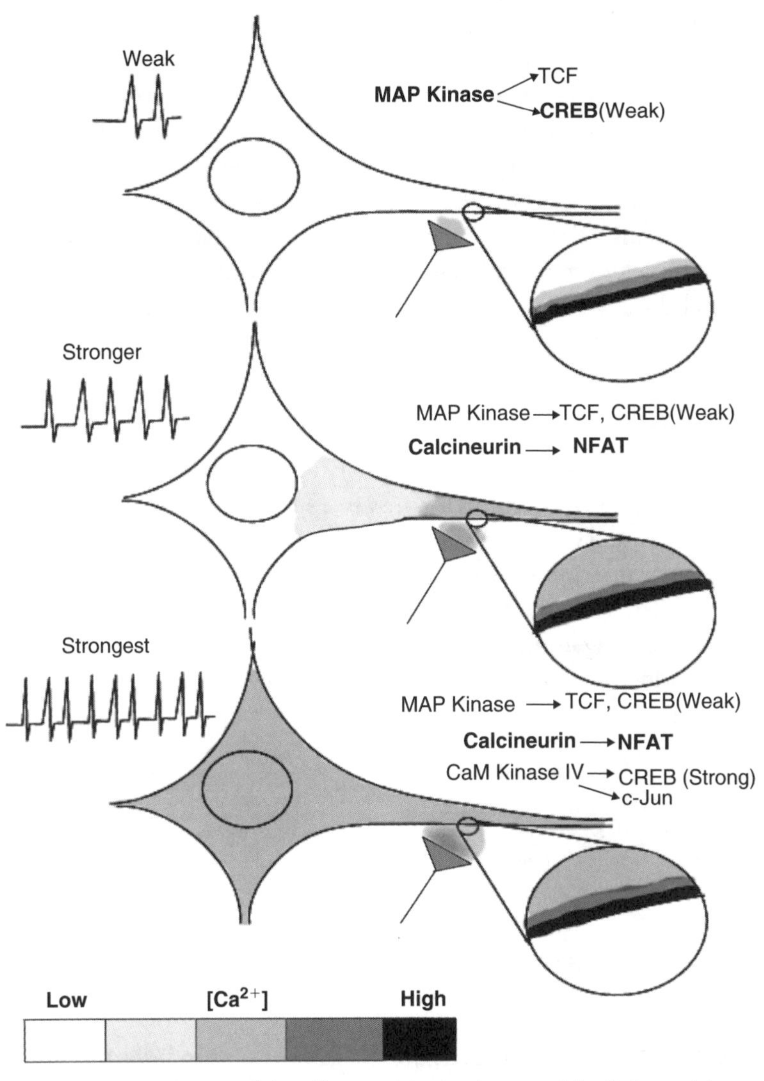

Fig. 10.2 Successive recruitment of signalling modules by three spatially distinct calcium pools may underline differential gene expression by synaptic activity. The figure shows three independent calcium-regulated signalling pathways, each with distinct functions in gene regulation, that are differentially controlled by spatially distinct calcium pools. Submembranous calcium influx via synaptic NMDA receptors is sufficient to activate the Ras-ERK1/2 (MAP Kinase)-RSK2 cascade, which induces TCF/SRF-mediated transcription and is important in prolonging the activation state of CREB. Synaptically evoked calcium influx can further invade the cytoplasm, recruiting the calcineurin signalling module that triggers NFAT nuclear translocation. Nuclear calcium elevation is sufficient for CaM Kinase IV-dependent CREB/CBP activation. Successive recruitment of signalling pathways allows different electrical impulse patterns to be converted into unique transcriptional responses providing a means for storage of activity pattern-specific information in the nervous system. The electrical inputs are coming in via a schematic glutamatergic synapse releasing glutamate (marked red) onto the postsynaptic neuron.

The absence of calcium buffering ATPases (which take calcium up into the ER/inter membrane space) on the inner nuclear membrane (Humbert *et al.* 1996) is thought to be behind the striking ability of elementary calcium release events proximal to the nucleus to trigger global increases in nuclear calcium concentration which last long after the cytoplasmic elementary 'trigger' has died away (Gerasimenko *et al.* 1996; Bootman *et al.* 1997; Lipp *et al.* 1997; Nakazawa and Murphy 1999). This property of nuclei applies to primary neurons where the nucleus is able to integrate synaptically evoked global calcium transients of quite low frequencies to give an elevated calcium 'plateau' (Hardingham *et al.* 2001*b*) ideal for the activation of nuclear events (for example via CaM kinase IV).

10.8 **Concluding remarks**

Activation of transcription by synaptic activity is an important adaptive response in the mammalian CNS. The neuron is wired in a sophisticated way to respond differently to different calcium signals. While much remains undiscovered, the molecular mechanisms of transcription factor activation and the basis for differential genomic responses to different calcium signals are becoming clearer. Such knowledge, coupled with an understanding of the physiological role of synaptically evoked transcription, means that a bridge between molecular and cellular events, and physiological behaviour, is slowly being built.

References

Adams, J. P., and Sweatt, J. D. (2002) Molecular psychology: roles for the ERK MAP kinase cascade in memory. *Annu. Rev. Pharmacol. Toxicol.*, **42**, 135–63.

Ahn, S., Ginty, D. D., and Linden, D. J. (1999) A late phase of cerebellar long-term depression requires activation of CaMKIV and CREB. *Neuron*, **23**, 559–68.

Alford, S., Frenguelli, B. G., Schofield, J. G., and Collingridge, G. L. (1993) Characterization of Ca2+ signals induced in hippocampal CA1 neurones by the synaptic activation of NMDA receptors. *J. Physiol.*, **469**, 693–716.

Arias, J., Alberts, A. S., Brindle, P., Claret, F. X., Smeal, T., Karin, M., *et al.* (1994) Activation of cAMP and mitogen responsive genes relies on a common nuclear factor. *Nature*, **370**, 226–9.

Bading, H., Ginty, D. D., and Greenberg, M. E. (1993) Regulation of gene expression in hippocampal neurons by distinct calcium signaling pathways. *Science*, **260**, 181–6.

Bading, H. and Greenberg, M. E. (1991) Stimulation of protein tyrosine phosphorylation by NMDA receptor activation. *Science*, **253**, 912–14.

Bailey, C. H. and Kandel, E. R. (1994) Structural changes underlying long-term memory storage in *Aplasia*: a molecular prospective. *Neuroscience*, **6**, 35–44.

Bannister, A. and Kouzarides, T. (1996) The CBP co-activator is a histone acetyltransferase. *Nature*, **384**, 641–3.

Bannister, A. J., Oehler, T., Wilhelm, D., Angel, P., and Kouzarides, T. (1995) Stimulation of c-Jun activity by CBP: c-Jun residues Ser63/73 are required for CBP induced stimulation in vivo and CBP binding *in vitro*. *Oncogene*, **11**, 2509–14.

Baranes, D., Lederfein, D., Huang, Y. Y., Chen, M., Bailey, C. H., and Kandel, E. R. (1998) Tissue plasminogen activator contributes to the late phase of LTP and to synaptic growth in the hippocampal mossy fiber pathway. *Neuron*, **21**, 813–25.

Berridge, M. J. (1993*a*) Inositol trisphosphate and calcium signalling. *Nature*, **361**, 315–25.

Berridge, M. J. (1993*b*) Cell signalling. A tale of two messengers. *Nature*, **365**, 388–9.

Bhattacharya, S., Eckner, R., Grossman, Oldread, E., Arany, Z., D., Andrea, A., *et al.* (1996) Cooperation of Stat2 and p300//CBP in signalling introduced by interferon-alpha. *Nature*, **383**, 226–8.

Bito, H., Deisseroth, K., and Tsien, R. W. (1996) CREB phosphorylation and dephosphorylation: a Ca(2+)- and stimulus duration-dependent switch for hippocampal gene expression. *Cell*, **87**, 1203–14.

Bito, H., Deisseroth, K., and Tsien, R. W. (1997) Ca2+-dependent regulation in neuronal gene expression. *Curr. Opin. Neurobiol.*, **7**, 419–29.

Blendy, J. A. and Maldonado, R. (1998) Genetic analysis of drug addiction: the role of cAMP response element binding protein. *J. Mol. Med.*, 104–10.

Bliss, T. V. P. and Collingridge, G. L. (1993) A synaptic model of memory: long-term potentiation in the hippocampus. *Nature*, **361**, 31–9.

Bliss, T. V. P. and Lomo, T. (1973) Long-lasting potentiation of synaptic transmission in the dentate area of the anaesthetized rabbit following stimulation of the perforant path. *J. Physiol.*, **232**, 331–56.

Bockaert, J. and Pin, J. (1997) In *Recombinant cell surface receptors: focal point for therapeutic intervention* (ed. M. Browne), pp. 75–102, Landes Bioscience publishers, Georgetown, USA.

Bonni, A., Ginty, D. D., Dudek, H., and Greenberg, M. E. (1995) Serine 133-phosphorylated CREB induces transcription via a cooperative mechanism that may confer specificity to neurotrophin signals. *Mol. Cell. Neurosci.*, **6**, 168–83.

Bootman, M. D., Berridge, M. J., and Lipp, P. (1997) Cooking with calcium: the recipes for composing global signals from elementary events. *Cell*, **91**, 367–73.

Bourtchuladze, R., Frenguelli, B., Blendy, J., Cioffi, D., Schutz, G., and Silva, A. J. (1994) Deficient long-term memory in mice with a targeted mutation of the cAMP-responsive element-binding protein. *Cell*, **79**, 59–68.

Braithwaite, S. P., Meyer, G., and Henley, J. M. (2000) Interactions between AMPA receptors and intracellular proteins. *Neuropharmacology*, **39**, 919–30.

Carafoli, E. and Penniston, J. (1985) The calcium signal. *Scientific American*, **253**, 70–8.

Chakravarti, D., LaMorte, V. J., Nelson, M. C., Nakajima, T., Schulman, I.G., Juguilon, H., *et al.* (1996) Role of CBP/P300 in nuclear receptor signalling. *Nature*, **383**, 99–102.

Chawla, S., Hardingham, G. E., Quinn, D. R., and Bading, H. (1998) CBP: a signal-regulated transcriptional coactivator controlled by nuclear calcium and CaM kinase IV. *Science*, **281**, 1505–9.

Chen, H., Lin, R., Schiltz, R., Chakravati, D., Nash, A., Nagy, L., *et al.* (1997) Nuclear receptor coactivator ACTR is a novel histone acetyltransferase and forms a multimeric activation complex with P/CAF and CBP/p300. *Cell*, **90**, 569–80.

Chrivia, J. C., Kwok, R. P. S., Lamb, N., Hagiwara, M., Montminy, M. R., and Goodman, R. H. (1993) Phosphorylated CREB binds specifically to the nuclear protein CBP. *Nature*, **365**, 855–9.

Cole, A. J., Saffen, D. W., Baraban, J. M., and Worley, P. F. (1989) Rapid increase of an immediate early gene messenger RNA in hippocampal neurons by synaptic NMDA receptor activation. *Nature*, **340**, 474–6.

Comb, M., Birnberg, N. C., Seasholtz, A., Herbert, E., and Goodman, H. M. (1986) A cyclic AMP- and phorbol ester-inducible DNA element. *Nature*, **323**, 353–6.

Crabtree, G. R. and Clipstone, N. A. (1994) Signal transmission between the plasma membrane and nucleus of T lymphocytes. *Annu. Rev. Biochem.*, **63**, 1045–83.

Cruzalegui, F. H., Hardingham, G., and Bading, H. (1999) c-Jun functions as a calcium-regulated transcriptional activator in the absence of JNK/SAPK1 activation. *EMBO J.*, **18**, 1335–44.

Dash, P. K., Hochner, B., and Kandel, E. R. (1990) Injection of the cAMP-responsive element into the nucleus of Aplysia sensory neurons blocks long-term facilitation. *Nature*, **345**, 718–21.

Deisseroth, K., Heist, E. K., and Tsien, R. W. (1998) Translocation of calmodulin to the nucleus supports CREB phosphorylation in hippocampal neurons. *Nature*, **392**, 198–202.

Emptage, N., Bliss, T. V. P., and Fine, A. (1999*a*) Calcium on the up: supralinear calcium signaling in central neurons. *Neuron*, **22**, 115–24.

Emptage, N. (1999*b*) Single synaptic events evoke NMDA receptor-mediated release of calcium from internal stores in hippocampal dendritic spines. *Neuron*, **24**, 495–7.

Finkbeiner, S. (2000) CREB couples neurotrophin signals to survival messages. *Neuron*, **25**, 11–14.

Frey, U., Huang, Y.-Y., and Kandel, E. R. (1993) Effects of cAMP simulate a late stage of LTP in hippocampal CA1 neurons. *Science*, **260**, 1661–4.

Gau, D., Lemberger, T., Von Gall, C., Kretz, O., Le Minh, N., Gass, P., *et al.* (2002) Phosphorylation of CREB Ser142 regulates light-induced phase shifts of the circadian clock. *Neuron*, **34**, 245–252.

Gerasimenko, O. V., Gerasimenko, J. V., Tepikin, A. V., and Petersen, O. H. (1996) Calcium transport pathways in the nucleus. *Pflugers Archiv-European Journal Of Physiology*, **432**, 1–6.

Ghosh, A., Carnahan, J., and Greenberg, M. E. (1994) Requirement for BDNF in activity-dependent survival of cortical neurons. *Science*, **263**, 1618–23.

Ghosh, A. and Greenberg, M. E. (1995) Calcium signaling in neurons: molecular mechanisms and cellular consequences. *Science*, **268**, 239–47.

Ginty, D. D., Bonni, A., and Greenberg, M. E. (1994) Nerve growth factor activates a Ras-dependent protein kinase that stimulates c-fos transcription via phosphorylation of CREB. *Cell*, **77**, 713–25.

Ginty, D. D., Kornhauser, J. M., Thompson, M. A., Bading, H., Mayo, K. E., Takahashi, J. S., *et al.* (1993) Regulation of CREB phosphorylation in the suprachiasmatic nucleus by light and a circadian clock. *Science*, **260**, 238–41.

Gonzalez, G. A. and Montminy, M. R. (1989) Cyclic AMP stimulates somatostatin gene transcription by phosphorylation of CREB at serine 133. *Cell*, **59**, 675–80.

Graef, I. A., Mermelstein, P. G., Stankunas, Neilson, J. R., Deisseroth, K., Tsien, R. W., *et al.* (1999) L-type calcium channels and GSK-3 regulate the activity of NF-ATc4 in hippocampal neurons. *Nature*, **401**, 703–8.

Gu, W., Shi, X., and Roeder, R. (1997) Synergistic activation of transcription by CBP and p53. *Nature*, **387**, 819–22.

Guzowski, G. A. and McGaugh, J. L. (1997) Antisense oligodeoxynucleotide-mediated disruption of hippocampal cAMP response element binding protein levels impairs consolidation of memory for water maze training. *Proc. Natl. Acad. Sci. USA*, **94**, 2693–8.

Guzowski, J. F., Lyford, G. L., Stevenson, G. D., Houston, F. P., McGaugh, J. L., Worley, P. F., *et al.* (2000) Inhibition of activity-dependent arc protein expression in the rat hippocampus impairs the maintenance of long-term potentiation and the consolidation of long-term memory. *J. Neurosci.*, **20**, 3993–4001.

Hahm, S. H. and Eiden, L. E. (1998) Cis-regulatory elements controlling basal and inducible VIP gene transcription. *Ann. NY Acad. Sci.*, **865**, 10–26.

Hardingham, G. E., Arnold, F., and Bading, H. (2001*a*) A calcium microdomain near NMDA receptors: on switch for ERK-dependent synapse-to-nucleus communication. *Nat. Neurosci.*, **4**, 565–6.

Hardingham, G. E., Arnold, F. A., and Bading, H. (2001*b*) Nuclear calcium signaling controls CREB-mediated gene expression triggered by synaptic activity. *Nat. Neurosci.*, **4**, 261–7.

Hardingham, G. E. and Bading, H. (1998) Nuclear calcium: a key regulator of gene expression. *Biometals*, **11**, 345–58.

Hardingham, G. E., Chawla, S., Cruzalegui, F. H., and Bading, H. (1999) Control of recruitment and transcription-activating function of CBP determines gene regulation by NMDA receptors and L-type calcium channels. *Neuron*, **22**, 789–98.

Hardingham, G. E., Chawla, S., Johnson, C. M., and Bading, H. (1997) Distinct functions of nuclear and cytoplasmic calcium in the control of gene expression. *Nature*, **385**, 260–5.

Hardingham, G. E., Fukunaga, Y., and Bading, H. (2002) Extrasynaptic NMDARs oppose synaptic NMDARs by triggering CREB shut-off and cell death pathways. *Nat. Neurosci.*, **5**, 405–14.

Heist, E. and Schulman, H. (1998) The role of Ca2+/calmodulin-dependent protein kinases within the nucleus. *Cell Calcium*, **23**, 103–14.

Hill, C. S. and Treisman, R. (1995*a*) Transcriptional regulation by extracellular signals: mechanisms and specificity. *Cell*, **80**, 199–211.

Ho, N., Liauw, J. A., F. Blaeser, F. W., Wei, F., Hanissian, S., Muglia, L. M., *et al.* (2000) Impaired synaptic plasticity and cAMP response element-binding protein activation in Ca2+/calmodulin-dependent protein kinase type IV/Gr-deficient mice. *J. Neurosci.*, **20**, 6459–72.

Hoeffler, J., Meyer, T., Yun, Y., Jameson, J., and Habener, J. (1988) Cyclic AMP-responsive DNA-binding protein: structure based on a cloned placental cDNA. *Science*, **242**, 1430–3.

Hook, S. S. and Means, A. R. (2001) Ca(2+)/CaM-dependent kinases: from activation to function. *Annu. Rev. Pharmacol. Toxicol.*, **41**, 471–505.

Hu, S. C., Chrivia, J., and Ghosh, A. (1999) Regulation of CBP-mediated transcription by neuronal calcium signaling. *Neuron*, **22**, 799–808.

Humbert, J. P., Matter, N., Artault, J. C., Koppler, P., and Malviya, A. N. (1996) Inositol 1,4,5-trisphosphate receptor is located to the inner nuclear membrane vindicating regulation of nuclear calcium signaling by inositol 1,4,5-trisphosphate. Discrete distribution of inositol phosphate receptors to inner and outer nuclear membranes. *J. Biol. Chem.*, **271**, 478–85.

Hunter, T. and Karin, M. (1992) The regulation of transcription by phosphorylation. *Cell*, **70**, 375–87.

Husi, H., Ward, M. A., Choudhary, J. A., Blackstock, W. P., and Grant, S. G. N. (2000) Proteomic analysis of NMDA receptor-adhesion protein signaling complexes. *Nat. Neurosci.*, **3**, 661–9.

Impey, S., Fong, A. L., Wang, Y., Cardinaux, J. R., Fass, D. M., Obrietan, K., *et al.* (2002) Phosphorylation of CBP mediates transcriptional activation by neural activity and CaM kinase IV. *Neuron*, **34**.

Impey, S. and Goodman, R. H. (2001) CREB signaling — timing is everything. *Science- STKE*, **PE1**.

Impey, S., Mark, M., Villacres, E. C., Poser, S., Chavkin, C., and Storm, D. R. (1996) Induction of CRE-mediated gene expression by stimuli that generate long-lasting LTP in area CA1 of the hippocampus. *Neuron*, **16**, 973–82.

Impey, S., Obrietan, K., and Storm, D. R. (1999) Making new connections: role of ERK/MAP kinase signaling in neuronal plasticity. *Neuron*, **23**, 11–14.

Janknecht, R., Ernst, W. H., Pingoud, V., and Nordheim, A. (1993) Activation of ternary complex factor Elk-1 by MAP kinases. *EMBO J.*, **12**, 5097–104.

Janknecht, R. and Nordheim, A. (1996*a*) Regulation of the c-fos promoter by the ternary complex factor Sap-1a and its coactivator CBP. *Oncogene*, **12**, 1961–9.

Janknecht, R. and Nordheim, A. (1996*b*) MAP kinase-dependent transcriptional coactivation by Elk-1 and its cofactor CBP. *Biochem. Biophys. Res. Commun.*, **228**, 831–7.

Jensen, K. F., Ohmstede, C.-A., Fisher, R. S., and Sayhoun, N. (1991) Nuclear and axonal localization of Ca2+/calmodulin-dependent protein kinase type Gr in rat cerebellar cortex. *Proc. Natl. Acad. Sci. USA*, **88**, 2850–3.

Johnson, C. M., Hill, C. S., Chawla, S., Treisman, R., and Bading, H. (1997) Calcium controls gene expression via three distinct pathways that can function independently of the Ras/mitogen-activated protein kinases (ERKs) signaling cascade. *J. Neurosci.*, **17**, 6189–202.

Kang, H., Sun, L. D., Atkins, C. M., Soderling, T. R., Wilson, M. A., and Tonegawa, S. (2001) An important role of neural activity-dependent CaMKIV signaling in the consolidation of long-term memory. *Cell*, **106**, 771–83.

King, D. P. and Takahashi, J. S. (2000) Molecular genetics of circadian rhythms in mammals. *Annu. Rev. Neurosci.*, **23**, 713–42.

Kwok, R. P. S., Lundblad, J. R., Chrivia, J. C., Richards, J. P., Bachinger, H. P., Brennan, R. G., *et al.* (1994) Nuclear protein CBP is a coactivator for the transcription factor CREB. *Nature*, **370**, 223–6.

Lin, C., Kapiloff, M., Durgerian, S., Tatemoto, K., Russo, A., Hanson, P., *et al.* (1987) Molecular cloning of a brain-specific calcium/calmodulin-dependent protein kinase. *Proc. Natl. Acad. Sci. USA*, **84**, 5962–6.

Lipp, P., Thomas, D., Berridge, M. J., and Bootman, M. D. (1997) Nuclear calcium signalling by individual cytoplasmic calcium puffs. *EMBO J.*, **16**, 7166–73.

Lonze, B. E., Riccio, A., Cohen, S. and Ginty, D. D. (2002) Apoptosis, axonal growth defects, and degeneration of peripheral neurons in mice lacking CREB. *Neuron*, **34**, 371–85.

Lynch, G., Larson, J., Kelso, S., Barrionuevo, G., and Schottler, F. (1983) Intracellular injections of EGTA block induction of hippocampal long-term potentiation. *Nature*, **305**, 719–21.

Magee, J. C. and Johnston, D. (1997) A synaptically controlled, associative signal for Hebbian plasticity in hippocampal neurons. *Science*, **275**, 209–13.

Malenka, R. (1994) Synaptic plasticity in the hippocampus: LTP and LTD. *Cell*, **78**, 535–8.

Malinow, R., Mainen, Z. F., and Hayashi, Y. (2000) LTP mechanisms: from silence to four-lane traffic. *Curr. Opin. Neurobiol.*, **10**, 352–7.

Matthews, R. P., Guthrie, C. R., Wailes, L. M., Zhao, X., Means, A. R., and McKnight, G. S. (1994) Calcium/calmodulin-dependent protein kinase types II and IV differentially regulate CREB-dependent gene expression. *Mol. Cell. Biol.*, **14**, 6107–16.

Mermelstein, P. G., Deisseroth, K., Dasgupta, N., Isaksen, A. L., and Tsien, R. W. (2001) Calmodulin priming: nuclear translocation of a calmodulin complex and the memory of prior neuronal activity. *Proc. Natl. Acad. Sci. USA*, **98**, 15342–7.

Milner, B., Corkin, S., and Teurber, H. (1968) Further analysis of the hippocampal amnesic syndrome: 14 year follow-up study of H.M. *Neurophysiologia*, **6**, 215–34.

Misra, R. P., Bonni, A., Miranti, C. K., Rivera, V. M., Sheng, M., and Greenberg, M. E. (1994) L-type voltage-sensitive calcium channel activation stimulates gene expression by a serum response factor-dependent pathway. *J. Biol. Chem.*, **269**, 25483–93.

Montminy, M. R. and Bilezikjian, L. M. (1987) Binding of a nuclear protein to the cyclic-AMP response element of the somatostatin gene. *Nature*, **328**, 175–8.

Montminy, M. R., Sevarino, K. A., Wagner, J. A., Mandel, G., and Goodman, R. H. (1986) Identification of a cyclic-AMP-responsive element within the rat somatostatin gene. *Proc. Natl. Acad. Sci. USA*, **83**, 6682–6.

Morgan, J. I. and Curran, T. (1986) Role of ion flux in the control of c-fos expression. *Nature*, **322**, 552–5.

Nadel, L. and Moscovitch, M. (1997) Memory consolidation, retrograde amnesia and the hippocampal complex. *Curr. Opin. Neurobiol.*, **7**, 217–27.

Nakajima, T., Uchida, C., Anderson, S. E., Parvin, J. D., and Montminy, M. (1997) Analysis of a cAMP-responsive activator reveals a two-component mechanism for transcriptional induction via signal-dependent factors. *Genes Dev.*, **11**, 738–47.

Nakazawa, H. and Murphy, T. H. (1999) Activation of nuclear calcium dynamics by synaptic stimulation in cultured cortical neurons. *J. Neurochem.*, **73**, 1075–83.

Nguyen, P. V., Abel, T., and Kandel, E. R. (1994) Requirement of a critical period of transcription for induction of a late phase of LTP. *Science*, **265**, 1104–7.

Norman, C., Runswick, M., Pollock, R., and Treisman, R. (1988) Isolation and properties of cDNA clones encoding SRF, a transcription factor that binds to the c-fos serum response element. *Cell*, **55**, 989–1003.

Ogryzko, V. V., Schiltz, R. L., Russanova, V., Howard, B. H., and Nakatani, Y. (1996) The transcriptional coactivators p300 and CBP are histone acetyltransferases. *Cell*, **87**, 953–9.

Ohmstede, C.-A., Jensen, K. F., and Sahyoun, N. E. (1989) Ca2+/calmodulin-dependent protein kinase enriched in cerebellar granule cells. Identification of a novel neuronal calmodulin-dependent protein kinase. *J. Biol. Chem.*, **264**, 5866–75.

Parker, D., Ferreri, K., Nakajima, T., LaMorte, V. J., Evans, R., Koerber, S. C., *et al.* (1996) Phosphorylation of CREB at Ser-133 induces complex formation with CREB-binding protein via a direct mechanism. *Mol. Cell. Biol.*, **16**, 694–703.

Price, M. A., Rogers, A. E., and Treisman, R. (1995) Comparative analysis of the ternary complex factors Elk-1, SAP-1a and SAP-2 (ERP/NET). *EMBO J.*, **14**, 2589–601.

Ptashne, M. and Gann, A. (1997) Transcriptional activation by recruitment. *Nature*, **386**, 569–77.

Rao, A., Luo, C., and Hogan, P. G. (1997) Transcription factors of the NFAT family: regulation and function. *Annu. Rev. Immunol.*, **15**, 707–47.

Ribar, T. J., Rodriguiz, R. M., Khiroug, L., Wetsel, W. C., Augustine, G. J., and Means, A. R. (2000) Cerebellar defects in Ca2+/calmodulin kinase IV-deficient mice. *J. Neurosci.*, **20**, RC107.

Riccio, A., Ahn, S., Davenport, C. M., Blendy, J. A., and Ginty, D. D. (1999) Mediation by a CREB family transcription factor of NGF-dependent survival of sympathetic neurons. *Science*, **286**, 2358–61.

Roberson, E. D., English, J. D., and Sweatt, J. D. (1996) A biochemist's view of long-term potentiation. *Learning and Memory*, **3**, 1–24.

Roeder, R. G. (1996) The role of zgeneral initiation factors in transcription by RNA polymerase II. *Trends Biochem. Sci.*, **21**, 327–34.

Rosen, L. B., Ginty, D. D., Weber, M. J., and Greenberg, M. E. (1994) Membrane depolarization and calcium influx stimulate MEK and MAP kinase via activation of Ras. *Neuron*, **12**, 1207–21.

Sala, C., Rudolph-Correia, S., and Sheng, M. (2000) Developmentally regulated NMDA receptor-dependent dephosphorylation of cAMP response element-binding protein (CREB) in hippocampal neurons. *J. Neurosci.*, **20**, 3529–36.

Sasaki, M., Gonzalez-Zulueta, M., Huang, H., Herring, W. J., Ahn, S., Ginty, D. D., *et al.* (2000) Dynamic regulation of neuronal NO synthase transcription by calcium influx through a CREB family transcription factor-dependent mechanism. *Proc. Natl. Acad. Sci. USA*, **97**, 8617–22.

Schröter, H., Shaw, P. E., and Nordheim, A. (1987) Purification of intercalator-released p67, a polypeptide that interacts specifically with the c-fos serum response element. *Nucleic Acids Res.*, **15**, 10145–57.

Schulman, H. (1993) The multifunctional Ca2+/calmodulin-dependent protein kinases. *Curr. Opin. Cell Biol.*, **5**, 247–53.

Schwartz, P. M., Borghesani, P. R., Levy, R. L., Pomeroy, S. L., and Segal, R. A. (1997) Abnormal cerebellar development and foliation in BDNF-/- mice reveals a role for neurotrophins in CNS patterning. *Neuron*, **19**, 269–81.

Sheng, M., Dougan, S. T., McFadden, G., and Greenberg, M. E. (1988) Calcium and growth factor pathways of c-fos transcriptional activation require distinct upstream regulatory sequences. *Mol. Cell. Biol.*, **8**, 2787–96.

Sheng, M. and Greenberg, M. E. (1990) The regulation and function of c-fos and other immediate early genes in the nervous system. *Neuron*, **4**, 477–85.

Sheng, M., McFadden, G., and Greenberg, M. E. (1990) Membrane depolarization and calcium induce c-fos transcription via phosphorylation of transcription factor CREB. *Neuron*, **4**, 571–82.

Sheng, M., Thompson, M. A., and Greenberg, M. E. (1991) CREB: a Ca(2+)-regulated transcription factor phosphorylated by calmodulin-dependent kinases. *Science*, **252**, 1427–30.

Shieh, P. B., Hu, S.-C., Bobb, K., Timmusk, T., and Ghosh, A. (1998) Identification of a signaling pathway involved in calcium regulation of BDNF expression. *Neuron*, **20**, 727–40.

Silva, A. J., Kogan, J. H., Frankland, P. W., and Kida, S. (1998) CREB and memory. *Annu. Rev. Neurosci.*, **21**, 127–48.

Soderling, T. R. and Stull, J. T. (2001) Structure and regulation of calcium/calmodulin-dependent protein kinases. *Chem. Rev.*, **101**, 2341–51.

Sun, P., Enslen, H., Myung, P. S., and Maurer, R. A. (1994) Differential activation of CREB by Ca2+/calmodulin-dependent protein kinases type II and type IV involves phosphorylation of a site that negatively regulates activity. *Genes Dev.*, **8**, 2527–39.

Swope, D., Mueller, C., and Chrivia, J. (1996) CREB-binding protein activates transcription through multiple domains. *J. Biol. Chem.*, **271**, 28138–45.

Tao, X., Finkbeiner, S., Arnold, D. B., Shaywitz, A. J., and Greenberg, M. E. (1988) Ca2+ influx regulates BDNF transcription by a CREB family transcription factor-dependent mechanism. *Neuron*, **20**, 709–26.

Thoenen, H. (1995) Neurotrophins and neuronal plasticity. *Science*, **270**, 593–8.

Treisman, R. (1985) Transient accumulation of c-fos RNA following serum stimulation requires a conserved 5' element and c-fos 3' sequences. *Cell*, **42**, 889–902.

Treisman, R. (1987) Identification and purification of a polypeptide that binds to the c-fos serum response element. *EMBO J.*, **6**, 2711–17.

Tully, T., Preat, T., Boynton, S. C., and Del Vecchio, M. (1994) Genetic dissection of consolidated memory in Drosophila. *Cell*, **79**, 35–47.

Verkhratsky, A. J. and Petersen, O. H. (1998) Neuronal calcium stores. *Cell Calcium*, **24**, 333–43.

Walton, M. R. and Dragunow, M. (2000) Is CREB a key to neuronal survival? *Trends Neurosci.*, **23**, 48–53.

Wang, J. M., Chao, J. R., Chen, W., Kuo, M. L., Yen, J. J., and Yang-Yen, H. F. (1999) The antiapoptotic gene mcl-1 is up-regulated by the phosphatidylinositol 3-kinase/Akt signaling pathway through a transcription factor complex containing CREB. *Mol. Cell. Biol.*, **19**, 6195–206.

Wei, F., Qiu, C.-S., Liauw, J., Robinson, D. A., Ho, N., Chatila, T., ***et al*** (2002) Calcium calmodulin-dependent protein kinase IV is required for fear memory. *Nat. Neurosci.*, **5**, 573–9.

West, A. E., Chen, W. G., Dalva, M. B., Dolmetsch, R. E., Kornhauser, J. M., Shaywitz, A. J., ***et al.*** (2001) Calcium regulation of neuronal gene expression. *Proc. Natl. Acad. Sci. USA*, **98**, 11024–31.

Whishaw, I., McKenna, J., and Maaswinkel, H. (1997) Hippocampal lesions and path integration. *Curr. Opin. Neurobiol.*, **7**, 228–34.

Whitmarsh, A. J. and Davis, R. (2000) Regulation of transcription factor function by phosphorylation. *Cell. Mol. Life Sci.*, **57**, 1172–83.

Wilson, B. E., Mochon, E., and Boxer, L. M. (1996) Induction of bcl-2 expression by phosphorylated CREB proteins during B-cell activation and rescue from apoptosis. *Mol. Cell. Biol.*, **16**, 5546–56.

Wisden, W., Errington, M. L., Williams, S., Dunnett, S. B., Waters, C., Hitchcock, D., *et al.* (1990) Differential expression of immediate early genes in the hippocampus and spinal cord. *Neuron*, **4**, 603–14.

Worley, P. F., Bhat, R. V., Baraban, J. M., Erickson, C. A., McNaughton, B. L., and Barnes, C. A. (1993) Thresholds for synaptic activation of transcription factors in hippocampus: correlation with long-term enhancement. *J. NeuroSci.*, **13**, 4776–86.

Wu, G. Y., Deisseroth, K., and Tsien, R. W. (2001) Activity-dependent CREB phosphorylation: convergence of a fast, sensitive calmodulin kinase pathway and a slow, less sensitive mitogen-activated protein kinase pathway. *Proc. Natl. Acad. Sci. USA*, **98**, 2808–13.

Xing, J., Ginty, D. D., and Greenberg, M. E. (1996) Coupling of the RAS-MAPK pathway to gene activation by RSK2, a growth factor-regulated CREB kinase. *Science*, **273**, 959–63.

Yamamoto, K. K., Gonzalez, G. A., Biggs 3rd, W. H., and Montminy, M. R. (1988) Phosphorylation-induced binding and transcriptional efficacy of nuclear factor CREB. *Nature*, **334**, 494–8.

Yao, T. T., Ku, G., Zhou, N., Scully, R., and Livingston, D. M. (1996) The nuclear hormone receptor coactivator SRC-1 is a specific target of p300. *Proc. Natl. Acad. Sci. USA*, **93**, 10626–31.

Yin, J., Wallach, J. S., Wilder, E. L., Klingensmith, J., Dang, D., Perrimon, N., *et al.* (1995*a*) A Drosophila CREB/CREM homolog encodes multiple isoforms, including a cyclic AMP-dependent protein kinase-responsive transcriptional activator and antagonist. *Mol. Cell Biol.*, 5123–30.

Yin, J. C., Vecchio, M. D., Zhou, H., and Tully, T. (1995*b*) CREB as a memory modulator: induced expression of a dCREB2 activator isoform enhances long-term memory in Drosophila. *Cell*, **81**, 107–15.

Chapter 11

Signalling by tyrosine phosphorylation in the nervous system

Jean-Antoine Girault

Introduction

Phosphorylation is a universal mode of regulation of protein properties, which plays a central role in all living cells. This reversible post-translational modification is a key part of virtually all intracellular signalling pathways (Fischer 1993; Krebs 1993). In vertebrates the amino acid residues that are the substrates of regulatory phosphorylation are serine, threonine, and tyrosine. In normal cells phosphotyrosine accounts for less than 1% of all the phosphorylated residues in proteins. This explains why tyrosine phosphorylation was discovered much later than serine and threonine phosphorylation. The first protein tyrosine kinases to be identified were v-Src, the product of an avian oncogene, coded by the Rous sarcoma virus, and its cellular counterpart, c-Src (Martin 2001). At about the same time the ability of the epidermal growth factor receptor (EGFR) to phosphorylate tyrosine was discovered (Ushiro and Cohen 1980). Thus, from the onset it appeared that tyrosine phosphorylation was involved in the control of cell growth, either in the context of tumor formation or in the action of growth factors. In fact, subsequent investigations have confirmed this initial impression, but have also substantially widened the scope of cellular functions in which tyrosine phosphorylation is involved. Studies in many different species have shown that, although phosphotyrosine can be found in bacteria as well as in eukaryotic cells, specific regulated tyrosine phosphorylation appears to be an evolutionary 'invention' of metazoans, i.e. multicellular animals. Generally, tyrosine phosphorylation is involved in cell–cell interactions, regulating cell growth, migration or shape in response to cell contacts or to soluble extracellular messengers. Because of its close relationship with cell growth, tyrosine phosphorylation initially did not attract much interest from investigators concerned with the mature nervous system. However, this interest changed when it was shown that neurotrophin receptors were tyrosine kinases (the Trk family), or associated with tyrosine kinases (e.g. receptors for glial cell line-derived neurotrophic factor (GDNF) and ciliary neurotrophic factor (CNTF)). Moreover, work over the last ten years or so has shown that non-receptor tyrosine kinases play a critical role in the nervous system, including during adulthood, regulating functions that are often seen as specific for neurons such as synaptic plasticity, learning, and memory. In this chapter we will first describe the most important features of the structure and function of the protein tyrosine kinases and phosphatases and we will then examine their role in neurons at different stages of development, in the adult as well as their role in glial cells and in pathological conditions. All tyrosine kinases are characterized by a conserved catalytic domain that is also highly related to that of serine/threonine kinases (Hanks and Hunter 1995). However, the rest of the peptide chain is composed of very diverse domains that provide specific means of targeting, regulation and interactions. Tyrosine kinases can be conveniently divided into receptor tyrosine kinases, which have an extracellular region, and non-receptor tyrosine kinases.

11.1 **Receptor tyrosine kinases**

A recent survey of putative tyrosine kinases encoded in the human genome identified 90 unique kinase genes and 5 pseudogenes (Robinson *et al.* 2000). Of the 90 tyrosine kinases, 58 are receptor tyrosine kinases (RTKs) distributed into 20 subfamilies. RTKs are type 1 transmembrane proteins, i.e. with an extracellular N-terminus, a single transmembrane domain and an intracellular C-terminus. The tyrosine kinase catalytic core corresponds to all or most of the intracellular domain. The extracellular region is much more diverse and encompasses various domains, depending on the receptor, which include the ligand-binding domain. RTKs are classified into several groups, according to the nature of their extracellular domains (Fig. 11.1).

The epidermal growth factor receptor (EGFR) group is characterized by two cysteine-rich domains in the extracellular region. It contains the receptor for EGF itself, and the ErbBs which are receptors for the neuregulins. Neuregulins (NRG-1–4), coded by at least four different genes, encompass EGF-like domains and exist under various forms due to alternative splicing and promoter usage (Buonanno and Fischbach 2001). Neuregulins 1–3 are expressed in the nervous system and play a role in neuronal development and in the adult. ErbB 2–4 are enriched at neuromuscular junctions, and at the level of postsynaptic densities in adult brain. ErbB receptors are also involved in myelinating cell differentiation.

The insulin and IGF-1 receptors have a distinct organization among RTKs inasmuch as the precursor peptide chain is cleaved during its processing into two different chains (α and β). The mature receptor is an heterotetramer ($\alpha_2\beta_2$) in which the β subunits are linked with each other and with α chain by disulfide bridges (Fig. 11.1). The extracellular α chain contains a cysteine-rich domain involved in ligand binding and the β chain crosses the membrane and encompasses the intracellular catalytic domain.

Several subfamilies of RTKs are characterized by an extracellular region encompassing several immunoglobulin-like (Ig) domains. The kinase domain of these receptors is interrupted by a peptide termed 'kinase insert'. These subfamilies include those of PDGF-R, FGF-R, and Flt-1, which differ by the number of Ig domains. The extracellular

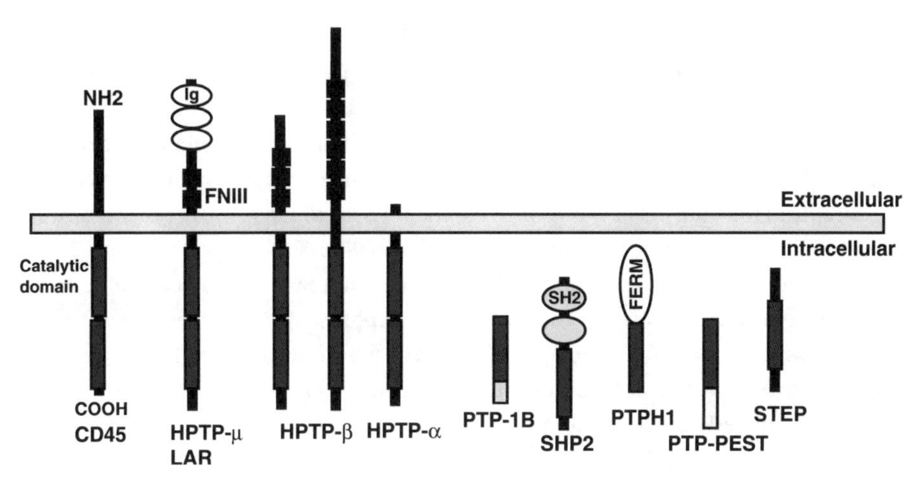

Fig. 11.1 Examples of receptor tyrosine kinases. The conserved catalytic tyrosine kinase domain is in black. KI: kinase insert, Fibro III: type III fibronectin domain.

region of the FGF-R group is interrupted by a sequence rich in acidic residues. The 3-D structure of the FGF-R has been determined and has revealed that this acidic region is involved in the binding of a mucopolysaccharide, such as heparin, which forms a ternary complex with the FGF and its receptor (Pellegrini 2001).

The Trk subfamily plays a particularly important role in neurons since it encodes the high-affinity receptors for neurotrophins. NGF activates TrkA, BDNF and NT-4/5 activate trk B, and NT-3 activates TrkC. Their extracellular region contains cell adhesion motifs with three tandem leucine-rich motifs flanked by two cysteine clusters, as well as two Ig domains (Patapoutian and Reichardt 2001). The Ig domain proximal to the membrane is essential for the binding of the neurotrophin (Wiesmann *et al.* 1999). Trk can associate with another protein, named $p75^{NTR}$, which belongs to the tumor necrosis factor (TNF) receptor family. $p75^{NTR}$ is a low-affinity receptor for neurotrophins, capable to mediate its own signal transduction, but which further increases the affinity of Trk for neurotrophins (Hempstead 2002). Since the signals mediated by Trk or by the Trk- $p75^{NTR}$ complex lead to cell survival, whereas those generated by $p75^{NTR}$ alone can be pro-apoptotic, the ratio of the expression of these molecules is clearly an important parameter for the cellular responses to neurotrophins.

One transmembrane tyrosine kinase deserves special attention since it plays an important role in the action of some neurotrophic factors. It is the product of the proto-oncogene c-Ret. While Ret by itself has no ligand-binding domain, it interacts with a group of proteins called GDNF-family receptors-α (GFRα), which are receptors for glial cell line-derived neurotrophic factor (GDNF) and related factors (neurturin, artemin and persephin) (Airaksinen and Saarma 2002). The GFRα do not possess a transmembrane domain but are attached to the membrane by a glycosylphosphatidyl-inositol

(GPI) that concentrates them in specific membrane regions that are rich in sphingolipids and cholesterol, and are called lipid rafts. Each subtype of GFRα interacts with Ret and the complex forms an active GDNF receptor.

The Eph receptors constitute the largest subfamily of RTKs and are particularly important in the nervous system, both during development and adulthood. Eph receptors are receptors for ephrins, a group of GPI-anchored (ephrins A) or transmembrane (ephrins B) ligands (Mellitzer *et al.* 2000; Klein 2001; Kullander and Klein 2002). There are at least 14 Eph receptors and 8 ephrins in humans. In general, EphA receptors bind ephrins A, whereas EphB receptors bind ephrins B. Eph receptor extracellular domain encompass a short cysteine-rich region and fibronectin type III domains. Two Eph receptors bind two ephrin molecules forming a functional tetramer (Himanen *et al.* 2001). Interestingly both the ephrins and their receptors are capable of transducing signals, providing a bidirectional exchange of information at zones of cell contacts where these two proteins interact.

There are several additional classes of RTKs among which several are expressed in the nervous system. Among these we can mention the Axl-UFO-Tyro 3 subfamily which is characterized by the presence of both Ig and fibronectin type III domains. These kinases are receptors for protein S and Gas6. Receptors and ligands are expressed in many cell types including those in the nervous system, but their function remains to be determined.

11.2 **Mechanisms of activation and signalling of receptor tyrosine kinases**

How can the binding of a ligand to the extracellular domain alter the activity of the intracellular tyrosine kinase? This question was initially very puzzling when one considers that the two domains of an RTK are linked by a single transmembrane region. Studies from many groups have shown that activation of such receptors involves dimerization (Weiss and Schlessinger 1998; Schlessinger 2000). Depending on the receptors, this may take different paths. In some cases, such as the insulin receptor, the receptor is already an oligomer in the absence of ligand. Insulin binding is thought to trigger a conformational change of the oligomer which results in a modification of the respective position of the two catalytic cores, and allows them to phosphorylate each other. In other cases, such as the FGF receptor, the receptor is thought to preexist as a monomer. In that case binding of the ligand, in interaction with a third molecule, a mucopolysaccharide (see above), promotes the formation of a dimer of two ternary complexes. Whatever the mechanism of interaction between the ligand and the extracellular region of the receptor, the formation of a dimer triggered on the extracellular side of the membrane, alters the position of the intracellular catalytic domains relative to each other and enables them to phosphorylate each other (trans-phosphorylation) on tyrosine residues. This reaction is the trigger for the signalling cascades activated by the receptor.

One major effect of the tyrosine phosphorylation is that it allows the recruitment of proteins that possess domains that bind to specific peptide sequences that encompass a phosphorylated tyrosine (Fig. 11.2). Such domains include Src-homology 2 (SH2) and phosphotyrosine-binding (PTB) domains. The important point is that these domains bind to the growth factor receptor only when it is phosphorylated on tyrosine. The proteins that contain SH2 domains belong to several categories. Some are themselves enzymes such as phospholipase Cγ (PLCγ) and phosphatidyl-inositol 3-kinase (PI3-K). Their binding to the receptor brings them in the vicinity of their substrates that are membrane lipids. PLCγ cleaves phosphatidylinositol 4,5 diphosphate (PIP2) into diacylglycerol (DAG) and inositol triphosphate (IP3). IP3 releases Ca^{2+} from intracellular stores, which in combination with DAG activates protein kinases C (PKC), a group of enzymes that phosphorylate their substrates on serines or threonines. PI3-K, on the other hand, phosphorylates the inositol moiety of PIP2, generating PIP3 (phosphatidylinositol 3,4,5 trisphosphate). PIP3 provides an anchoring site at the membrane for the pleckstrin homology (PH) domain of two serine threonine kinases, PKB/Akt and its activating kinase PDK1. This translocation results in the activation of Akt, which is a major effector of growth factor receptors.

A different type of proteins that are recruited on tyrosine-phosphorylated growth factor receptors are adaptor molecules. One such protein, Grb-2, has a SH2 domain that binds to the receptor, and two Src homology 3 (SH3) domains that bind to specific proline-rich sequences (usually two proline residues separated by two other amino acids). Through its SH3 domains Grb2 is associated with an enzyme called SOS that belongs to the category of the guanine nucleotide exchange factors. These enzymes act on small GTP binding proteins of the Ras family and help them to release GDP and to exchange it for GTP, putting them in an active conformation. Since Ras is itself attached to the inner face of plasma membrane through a lipid anchor, the recruitment of Grb-2-SOS to the activated receptor places it in the vicinity of its substrate, and

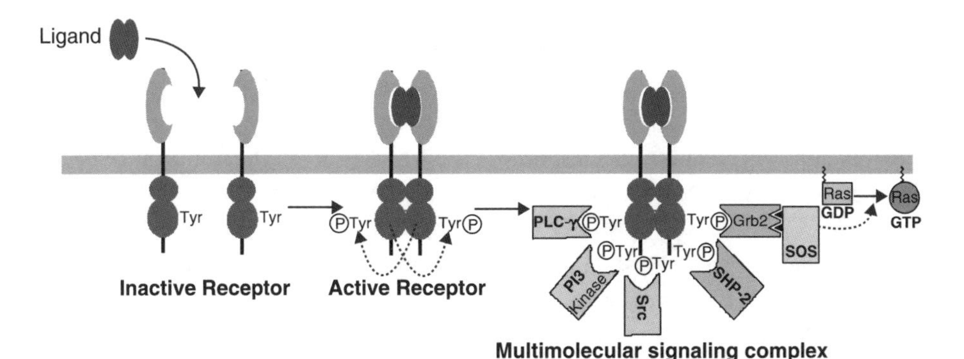

Fig. 11.2 Schematic representation of the activation of a tyrosine kinase growth factor receptor. Some of the proteins that can be recruited to the phosphorylated receptor through their SH2 domain are schematically indicated.

triggers the formation of Ras-GTP. Ras-GTP is recognized by proteins which have a Ras-binding domain, a major one being the protein kinase Raf-1. In fact there are several different isoforms of Raf, one of which, B-Raf, is highly enriched in neurons and has properties slightly different from those of Raf-1. Raf-1 is activated by its binding to Ras-GTP and by concomitant phosphorylation, and becomes capable of activating the extracellular signal-regulated kinase (ERK) pathway, also called mitogen-activated kinase (MAP-kinase) pathway (Derkinderen *et al.* 1999). This is achieved by phosphorylation and activation by Raf-1 or B-Raf of a protein kinase termed MEK (MAP-kinase/ERK kinase). MEK is an unusual kinase which is capable of phosphorylating ERK on both threonine and tyrosine residues. This results in the activation of ERK, a serine/threonine kinase.

In the case of some RTKs, Grb-2 does not bind directly to the phosphorylated receptor and additional adaptor molecules are involved, such as insulin receptor substrate (IRS) and Shc, which bind to the phosphorylated receptors by a PTB domain, becomes phosphorylated on tyrosine and allow the recruitment of Grb-2 or of other proteins. An additional important type of enzyme that can be recruited to activated RTKs are tyrosine kinases of the Src family (see below) that phosphorylate neighboring proteins, including Shc, and amplify the signals generated by the receptor. Interestingly, protein tyrosine phosphatases (the enzymes that dephosphorylate tyrosine residues) that possess SH2 domains, such as SHP-2, can also be recruited to the activated receptors (Tonks and Neel 2001). These phosphatases can have both a positive and negative feedback role.

The important point in understanding the signalling generated by RTKs is that their activation triggers directly or indirectly the recruitment to the membrane of a number of enzymes which are capable of activating a variety of signalling pathways. Many of these enzymes are physically associated, directly or indirectly, with the receptor, forming large multimolecular signalling complexes. Each RTK can bind a specific combination of signalling molecules and, although there is a considerable overlap between receptors, each of them activates a selective combination of signalling pathways. Besides these binding specificities, another important difference between receptors is their different kinetics of activation. Upon ligand binding some RTKs remain active for a long time, while others are only transiently activated. This can have important consequences on the type of cellular responses that are generated. The importance of the duration of activation has been well demonstrated in the case of PC12 cells differentiation, a classical model of study of NGF (Traverse *et al.* 1992). In this model, although both EGF and NGF activate the ERK pathway, only stimulation by NGF is sufficiently prolonged to trigger differentiation. Interestingly, in neurons the signalling complexes including Trk can be endocytosed and appear to remain active at the surface of vesicles that are retrogradely transported to the cell body (Ginty and Segal 2002). This is likely to be one of the mechanisms by which survival signals generated at the contact of specific targets are retrogradely conveyed to the neuronal cell body.

A remarkable feature of the various signalling pathways triggered by RTKs is that they lead to the activation of several serine/threonine protein kinases with a wide variety of substrates, such as PKC, Akt, and ERK. These kinases can phosphorylate substrates in the vicinity of their site of activation and exert local effects. This is probably an important aspect in the action of growth factors when they are located on dendritic spines or shafts or on nerve terminals. But the activated serine/threonine kinases have also the important capacity to alter gene expression. For example active ERK can translocate to the nucleus where it phosphorylates and activates a transcription factor named Elk which is part of a ternary complex that controls the transcription of specific genes. Although RTKs are usually low-abundance proteins, their activation exerts major effects due to the simultaneous activation of several signalling pathways that are often synergistic and result in enhanced survival and growth. Of course cells normally possess several means to control the level of activation of these pathways and limit in time their action. These include several mechanisms including desensitization and degradation of receptors, and dephosphorylation of tyrosines, as well as serines and threonines under the action of protein phosphatases (see below). However, sometimes molecules in the pathways summarized above escape these controls because of mutation and become constitutively activated. This is a major factor in the apparition of cancer (Hunter 1997).

11.3 **Non-receptor tyrosine kinases**

Non-receptor tyrosine kinases (NRTK) are a rather heterogeneous group of enzymes that have in common a conserved tyrosine kinase catalytic domain and a lack of extracellular ligand-binding domain. There are at least 32 non-receptor tyrosine kinases in humans distributed into 10 families (Robinson *et al.* 2000). These families have various domains in addition to their tyrosine kinase domain (Fig. 11.3) and their functions in cells are also diverse.

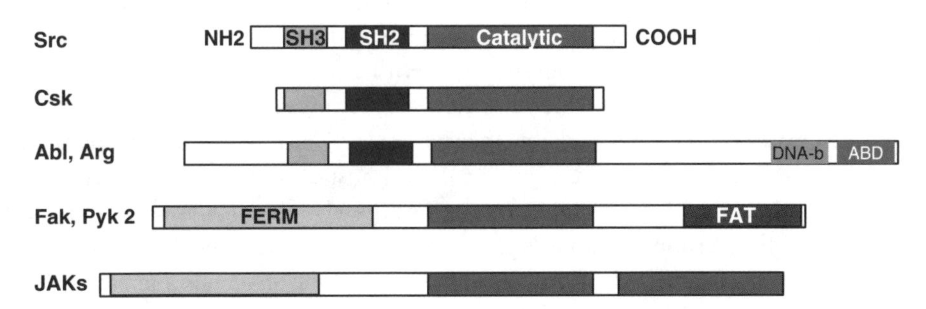

Fig. 11.3 Examples of non-receptor tyrosine kinases. The conserved catalytic tyrosine kinase domain is in black. SH2: Src-homology 2, SH3: Src-homology 3, DNA-b: DNA binding, ABD: actin-binding domain, FERM: four-point-one, ezrin, radixin, moesin, FAT: focal adhesion targeting. Note that in JAKs, only the C terminal tyrosine kinase domain is catalytically active.

Janus kinases (JAKs) are so named because they have two kinase domains like the two-faced Roman God. Only one of these domains is active, however. The N-terminus of JAKs is related to FERM (four-point-one, ezrin, radixin, moesin) domains, which are present in a number of proteins capable of interacting reversibly with membrane proteins (Girault *et al.* 1999*a*). The JAKs do have the very important property of interacting with a group of transmembrane proteins known as the cytokine receptors. When the cytokine receptors bind their cognate ligands (cytokines), they dimerize and activate the JAKs that phosphorylate themselves and the associated receptors on tyrosines. Phosphorylation of the receptor provides a docking site for transcription factors of the STAT family (signal transducer and activator of transcription) which possess an SH2 domain. The STAT molecules are then phosphorylated by JAKs and dimerize (Fig. 11.4). The resulting STAT dimers are no longer bound to the receptors and translocate to the nucleus where they activate directly the transcription of specific genes. The complex between JAKs and cytokine receptors closely resemble RTKs in their ability to transduce signals in response to direct activation by extracellular messengers. However, the cytokine receptors/JAKs complex have an original signalling mechanism, which involves a direct control of gene transcription through STATs. Cytokine receptors that play an important role in nerve cells include the receptor for ciliary-derived neurotrophic factor (CNTF) and the receptor of leptin, a polypeptide involved in the control of food intake and body weight.

The Src family is the archetype of NRTKs, which comprises at least 8 members in humans. These kinases encompass two conserved domains in addition to their catalytic domain (Fig. 11.3), referred to as Src-homology 2 and 3 (SH2, SH3), capable of binding specific peptides containing a phosphorylated tyrosine or a proline-rich sequence, respectively (Thomas and Brugge 1997). Src family kinases are usually membrane attached, due to the presence, at their N-terminus, of a covalently bound myristic acid, a 14-carbon saturated fatty acid. Myristoylation enriches Src kinases in membrane rafts. Src kinases are maintained in an inactive state by several intramolecular interactions: the SH2 domain binds to a phosphorylated tyrosine located at the C-terminus, while the SH3 domain interacts with a specific sequence in the linker region between the catalytic and SH2 domains. This organization allows to understand how Src kinases are regulated and how they become activated following disruption of their intramolecular interactions (Fig. 11.5). The phosphorylation of the C-terminal tyrosine which is the intramolecular ligand of the SH2 domain, is an important parameter. The C-terminal tyrosine is phosphorylated by a specific enzyme, termed C-terminal Src kinase (Csk), which acts on all members of the Src family and has a critical regulatory role. Conversely, dephosphorylation of this tyrosine is catalysed by several tyrosine phosphatases whose action result in Src activation. This is an unusual situation in which a tyrosine phosphatase can activate a tyrosine phosphorylation pathway. Another means of activation of Src is the displacement of its SH2 domain from the C-terminal phosphorylated tyrosine by a competing phosphopeptide.

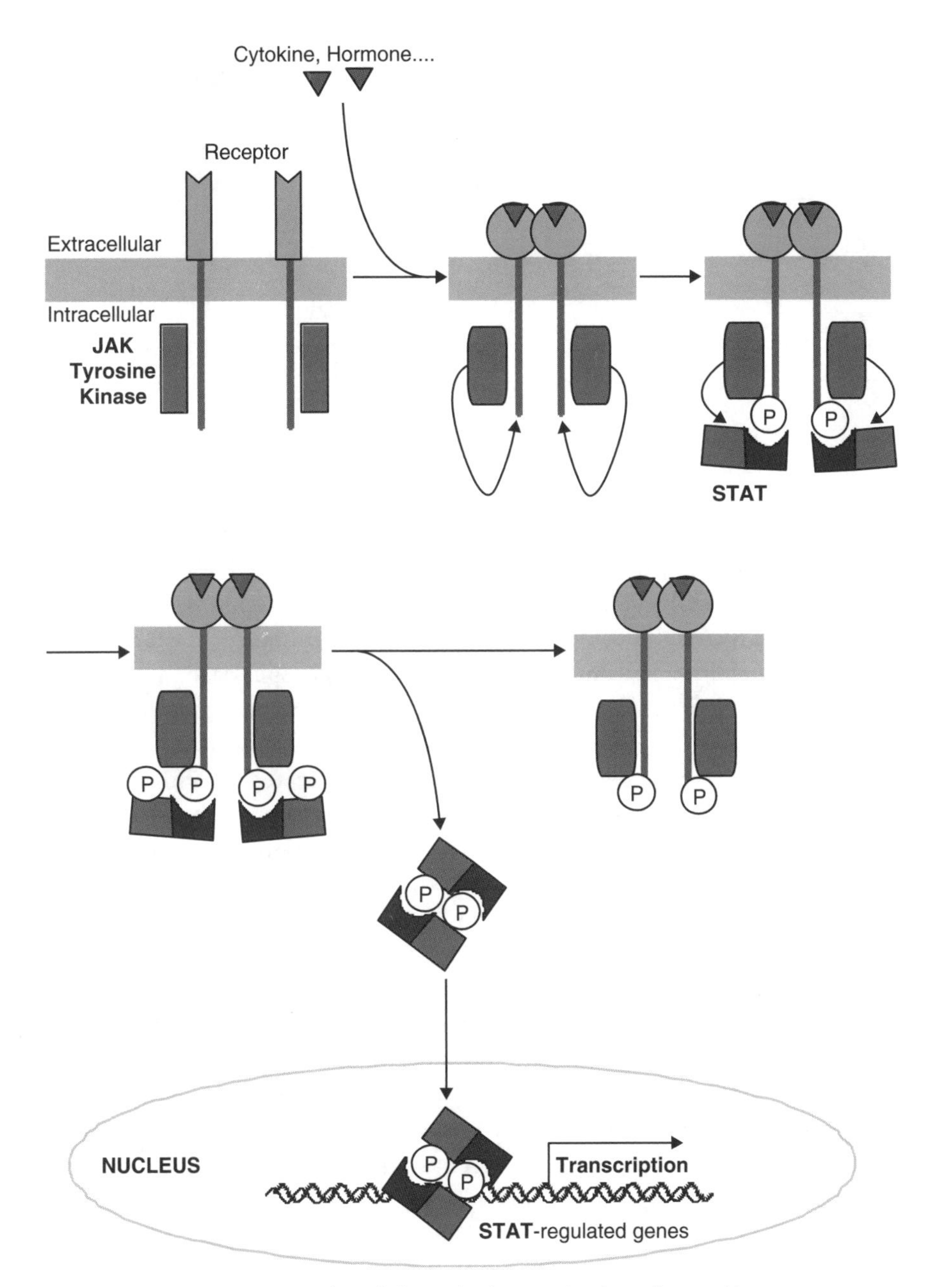

Fig. 11.4 Schematic representation of the activation mechanism of a cytokine receptor coupled to JAK tyrosine kinases and of STAT.

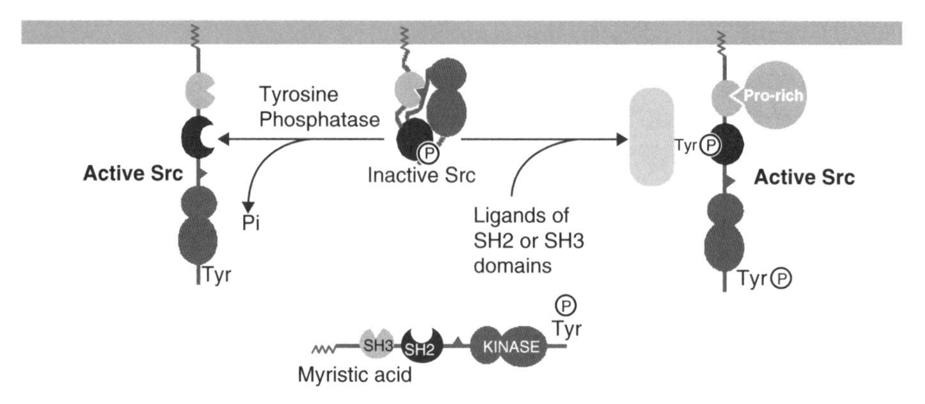

Fig. 11.5 Schematic representation of the regulation of Src family kinases. These kinases are attached to the membrane through an N-terminal myristic acid. They are maintained in an inactive state by intramolecular interactions that can be alleviated by the indicated mechanisms.

This occurs for instance when Src binds to a phosphorylated RTK. Similarly Src activation can be prompted by the binding of its SH3 domain to a proline-rich sequence in an interacting protein. In addition, it should be mentioned that, like many other tyrosine kinases, Src is capable of autophosphorylation on a tyrosine residue located in the activation loop of the kinase domain, and that this autophosphorylation enhances its catalytic activity. Neurons express several members of the Src family such as c-Src or its neuronal splice variant n-Src, Fyn, Yes, and Lck. Using various combinations of the mechanisms described above, these kinases can be activated in neurons through several pathways including recruitment to activated growth factor receptors through its SH2 domain, or in response to G protein-coupled receptors (Pierce and Lefkowitz 2001).

Focal adhesion kinase (FAK) and the closely related PYK2 (proline-rich tyrosine kinase-2 also termed CAKβ or CADTK or RAFTK) form a distinct group of NRTK (Girault *et al.* 1999*b*). In cells in culture FAK is enriched at focal adhesions, where integrins attached to extracellular matrix components are concentrated and form points of anchorage for the actin cytoskeleton. FAK contains a central kinase domain, a C-terminal focal adhesion targeting (FAT) domain, and an N-terminal FERM domain, related to that of JAKs (Fig. 11.3). FAK mediates adhesion-dependent survival signals. It is also involved in cell motility by controlling the turnover of focal adhesions. FAK is activated following integrin engagement and in response to G protein-coupled receptors. Its activation corresponds to its autophosphorylation on a tyrosine located at the junction between the N-terminal domain and the kinase domain. This provides a binding site for Src and Fyn, as well as for PI3-kinase. Recruitment of these proteins triggers signalling pathways reminiscent of those activated by growth factor receptors. Neurons express alternatively spliced isoforms of FAK with an increased autophosphorylation activity. In hippocampal slices

FAK can be activated by various neurotransmitters, including glutamate, and by neuromodulators, such as endocannabinoids (Girault *et al.* 1999*b*). PYK2 has properties quite similar to FAK, although it generally does not respond to cell adhesion, but is rather activated in response to signals that raise intracellular Ca^{2+} through a mechanism that is not fully understood. PKC seems to be involved in PYK2 regulation in some cells but not in others. In neurons PYK2 is exquisitely sensitive to depolarization and stimuli which increase Ca^{2+}. Both FAK and PYK2 provide means for activating Src family kinases and other signalling pathways in response to neurotransmitters in brain.

Other NRTK well studied in neurons include Abl, and Arg, the product of the Abl-related gene. Abl has attracted attention as a clinically efficient target for anticancer drugs (Druker and Lydon 2000), but it is also important for normal cell functions including during the development of the nervous system (Van Etten 1999). In addition to SH3, SH2, and kinase domains Abl contains an actin-binding domain and a DNA-binding domain at its C-terminus (Fig. 11.3). These features account for the nuclear and cytoplasmic locations of the enzyme, and its involvement in the control of actin cytoskeleton.

11.4 **Tyrosine phosphatases**

A major feature of phosphorylation reactions is that they are reversible in physiological conditions. Phosphotyrosines are dephosphorylated by several classes of tyrosine phosphatases, including classical protein tyrosine phosphatases (PTPs), low molecular weight PTPs, and two groups of dual specificity phosphatases (Fauman and Saper 1996; Tonks and Neel 2001). In spite of a lack of sequence similarities in the catalytic domains of these four classes of phosphatases, they share a common catalytic mechanism involving a critical cysteine residue and a similar structural organization. In general PTPs have a high intrinsic activity and their inhibition in living cells leads to a marked increase in total protein tyrosine phosphorylation. The catalytic cysteine residue is very sensitive to oxidation and it is thought that its oxidation by H_2O_2 is a regulatory mechanism of some PTPs. Classical PTPs are subdivided into receptor-like PTPs (RPTPs) and non-receptor-like PTPs (Fig. 11.6). Although many RPTPs have two catalytic domains, the C-terminal one has usually little or no catalytic activity, and its function is unclear. In contrast to RTK little is known about possible ligands that might regulate the activity of RPTPs. An important feature in their regulation appears to be their precise localization. Extracellular domains of RPTPs may concentrate them at specific locations, through poorly understood interactions with proteins of other cells or of extracellular matrix. This class of enzymes appears to play an important role in axon guidance during development (Van Vactor 1998). Among the non-receptor-like PTPs, some possess a SH2 domain which targets them to activated growth factor receptors (Fig. 11.6) (see above). Although in general PTPs tend to oppose tyrosine kinase signalling, in some cases they can activate specific tyrosine kinases, as discussed above in the case of Src.

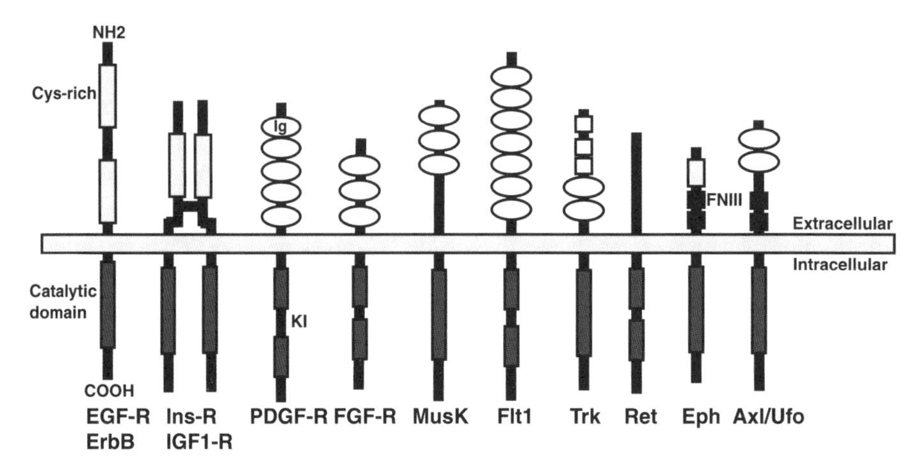

Fig. 11.6 Examples of tyrosine phosphatases. The conserved catalytic tyrosine kinase domain is in green. SH2: Src-homology 2, Ig: immunoglobulin-like domain, FERM: four-point-one, ezrin, radixin, moesin, PEST: proline, glutamate, serine threonine-rich domain. STEP is a striatal enriched tyrosine phosphatase. Note that in RPTPs with tandem catalytic domains, the C terminal phosphatase domain, is catalytically inactive.

11.5 **Role of protein tyrosine phosphorylation during development of the nervous system**

Since receptors for virtually all growth factors acting on neurons are either RTKs or associated with tyrosine kinases, it is not surprising that tyrosine phosphorylation is a major signalling pathway during neuronal development. Growth factors that are active on neurons include molecules that are also well characterized in other cells, such as EGF, FGF, or IGF-1, and factors that are specific to neurons. Among the latter category neurotrophic factors which act on Trk receptors have been extensively studied (see above). The signalling mechanisms of Trk receptors include the activation of PLCγ, the Ras-ERK pathway, and PI3-kinase (Huang and Reichardt 2001). Another type of RTK plays a specific role in synapse formation at the neuromuscular junction (Sanes and Lichtman 2001). During development the clustering of nicotinic acetylcholine receptors is due to agrin, a large heparan sulfate proteoglycan secreted by nerve terminals. Agrin binds to MusK, a tyrosine kinase present at the plasma membrane of muscle cells. Both agrin and MusK are necessary for the clustering of acetylcholine receptors, which is mediated by its interaction with a cytoplasmic protein, rapsyn.

Protein tyrosine phosphorylation is involved in axon guidance throughout development. The first evidence originated from studies in Drosophila (Van Vactor 1998), but similar results have now been obtained in other species including mammals and Aplysia (Lanier and Gertler 2000; Jay 2001). It is well established that actin polymerization is an important aspect of growth cone progression. Tyrosine phosphorylation controls this process, by directly or indirectly activating proteins involved in actin polymerization/depolymerization, such as ENA/Vasp. Tyrosine phosphorylation also participates in the regulation of small GTPases of the Rho family, including Cdc42, Rac, and Rho A that promote the extension of filopodia and lamellipodia, and microfilaments bundling, respectively. Tyrosine phosphorylation can also control cell adhesion molecules such as cadherins. Enzymes that are important in the regulation of tyrosine phosphorylation during neurite extension include protein tyrosine kinases of the Abl and Src families, and several RPTPs. Thus, increased tyrosine phosphorylation in response to specific extracellular clues which activate tyrosine kinases facilitates growth cone progression, whereas decreased phosphorylation through activation or enrichment of tyrosine phosphatases have the opposite effect. Very localized regulation of this balance will have steering effects, inducing the turning of the cone.

Ephrins and their cognate Eph receptors also play an important role during the development of the nervous system, by generating bidirectional signalling between adjacent cells. These proteins mediate contact-dependent cell interactions that regulate the repulsion mechanism involved in the guidance of growth cones (Mellitzer *et al.* 2000; Klein 2001; Kullander and Klein 2002). They are also critical for topographic mapping, as illustrated in the case of the retinotectal projections (Mellitzer *et al.* 2000). Repulsive effects seem to result from the fact that activation of Eph receptors inhibits several pathways including those which involve ERK, Abl, Cdc42, and Rac, and antagonizes the effects of integrins engagement. Reverse signalling by ephrins involve phosphorylation by Src family kinases and recruitment of adaptor proteins, such as Grb4, that may allow regulation of the actin cytoskeleton. Interestingly, ephrins can be shed from cell surface allowing cells or their processes to detach from each other, and repulsion can be turned into adhesion by alternative splicing of Eph receptors (Klein 2001).

11.6 **Role of tyrosine phosphorylation in the regulation of ion channels and receptors**

The role of protein tyrosine phosphorylation in the control of cell growth and differentiation, as well as in the regulation of cell shape and movement is to some extent not specific to nerve cells. On the other hand, work over the last ten years has revealed that tyrosine phosphorylation is also involved in the regulation of ion channels and neurotransmitter receptors that are either specific of neurons or highly enriched in these cells (Smart 1997; Boxall and Lancaster 1998; Swope *et al.* 1999). Regulation of ligand-gated ion channels by tyrosine phosphorylation has been extensively investigated. As mentioned above, tyrosine phosphorylation is the trigger of nicotinic acetylcholine receptor clustering at the neuromuscular junction. However, this clustering does not result from direct phosphorylation of the receptor (Sanes and Lichtman 2001). Instead, tyrosine phosphorylation of the nicotinic receptor controls the rate of receptor desensitization (Hopfield *et al.* 1988). The NMDA glutamate receptor (NMDA-R) is also regulated by tyrosine phosphorylation (Ali and Salter 2001). This receptor comprises a combination of NR1 and NR2 subunits. The NR2 subunits

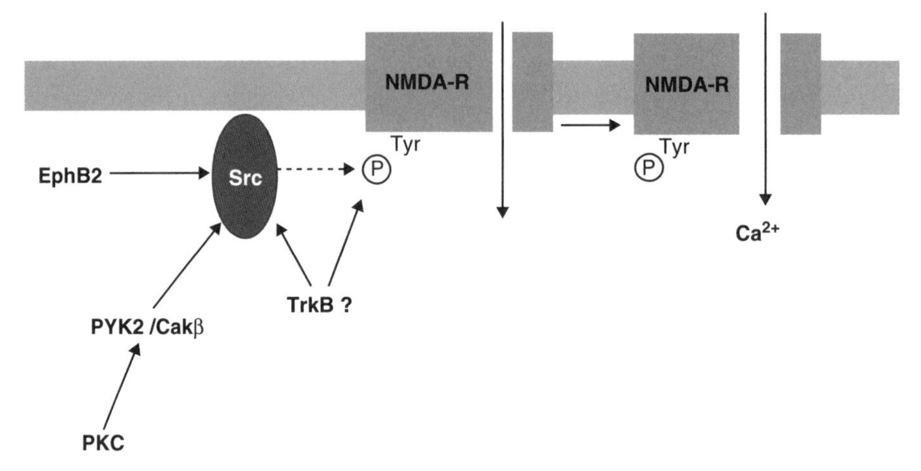

Fig. 11.7 Regulation of the NMDA receptor by tyrosine phosphorylation. The NR2A and NR2B are phosphorylated by Src or Fyn. This phosphorylation results in an increased function of the receptor, including an enhancement of glutamate-triggered Ca^{2+} influx. Src kinase may be activated through several pathways, including PYK2/Cakβ that can be itself activated by PKC and by EphB2 that is directly associated with NMDA receptors. In addition, NMDA-R are regulated by TrkB and neurotrophins.

A and B are substrates of Src family kinases, Src and Fyn, which are physically associated with the NMDA-R multiprotein complex. Tyrosine phosphorylation results in an upregulation of the NMDA-R function. Interestingly, protein kinase C also activates NMDA receptors, and this activation appears to be mediated through PYK2/Cakβ and Src family kinases (Fig. 11.7). The GABA-A receptor, a chloride channel which is activated by the major inhibitory neurotransmitter in the adult brain, is also phosphorylated on tyrosine and this phosphorylation increases its function (Moss *et al.* 1995). Voltage-gated ion channels are also substrates for tyrosine kinases of the Src family and for specific tyrosine phosphatases (Levitan 1999). Several types of K^+ channels are regulated by tyrosine phosphorylation, and depending on the channels considered, the effects of phosphorylation can be either an increase or a suppression of their function.

11.7 **Role of protein tyrosine phosphorylation in synaptic plasticity**

As we have seen above, tyrosine phosphorylation in neurons is controlled by two major types of mechanisms: growth factors directly activate tyrosine kinase receptors, whereas depolarization, classical neurotransmitters, as well as other types of intercellular messengers are capable of enhancing tyrosine phosphorylation through non-receptor tyrosine kinases (Girault *et al.* 1999*b*). On the other hand, tyrosine phosphorylation is known to regulate directly or indirectly a variety of cellular functions ranging from ion channel properties to the organization of cytoskeletal proteins and gene expression. Thus, tyrosine phosphorylation is strategically located at crossroads of signalling mechanisms, and it comes as no surprise that it is implicated in synaptic plasticity (Boxall and Lancaster 1998; Girault *et al.* 1999*b*). The first clue came from experiments which demonstrated that non-specific tyrosine kinase inhibitors were capable of preventing long-term potentiation in the CA1 region of hippocampus (O'Dell *et al.* 1991). Subsequently it was found that LTP was blunted in hippocampus of Fyn knockout mice, whereas mice devoid of Src, Yes, or Abl were indistinguishable from their wild-type counterparts (Grant *et al.* 1992). The impairment of synaptic plasticity in Fyn −/− mice is unlikely to be secondary to the developmental defects observed in these animals, since they were rescued by transient re-expression of Fyn (Kojima *et al.* 1997). Interestingly, Fyn knockout mice display other deficits thought to be linked to NMDA-R dysfunction including an hypersensitivity to hypnotic effects of ethanol (Miyakawa *et al.* 1997). Although these data suggest a more critical role of Fyn than of Src *in vivo*, other types of studies have implicated Src and the regulation of NMDA-R (Ali and Salter 2001). Indeed, NMDA-R tyrosine phosphorylation is increased in the course of LTP, and PYK2/Cakβ together with Src exert a facilitatory effect on LTP. Therefore it has been suggested that PYK2/Cakβ and Src play a critical role in LTP induction by enhancing NMDA-R function, a process that may be triggered by Ca^{2+} influx and/or PKC activation (see also Chapter 14).

Receptor tyrosine kinases have also been implicated in the induction of LTP. Most noticeably hippocampal LTP is impaired by reagents that alter BDNF action, as well as in BDNF and TrkB knockout mice (Huang and Reichardt 2001). These effects are observed in the absence of detectable alterations of neuronal circuitry and basal synaptic transmission, strongly indicating that neurotrophins and their receptors are directly involved in the activity-dependent mechanisms that regulate synaptic efficacy. In fact, recent work from several laboratories has demonstrated that the relative contribution of neurotrophins to synaptic plasticity depends on the type of stimuli used, and that it involves pre and post-synaptic actions, as well as immediate and delayed effects (Schuman 1999; Patterson *et al.* 2001; Kovalchuk *et al.* 2002). It is remarkable in this context that BDNF and NT-4/5 appear to have excitatory effects similar to those of glutamate and that these effects are blocked by a tyrosine kinase inhibitor (Kafitz *et al.* 1999).

Ephrins and Eph receptors are also important players in synapse formation and plasticity (Kullander and Klein 2002). In hippocampal neurons in culture, EphB2 interacts with a cell surface proteoglycan, syndecan-2, to induce the formation of dendritic spines, associates directly with NMDA receptors and promotes its clustering and synapse formation upon binding of EphrinB. Moreover, EphB activation increases NMDA-R tyrosine phosphorylation and glutamate-induced Ca^{2+} transients, through phosphorylation by Src kinases. In spite of these striking effects, in EphB2 −/− mice, hippocampal synapses appear normal indicating the existence of redundant pathways for development. However, several forms of synaptic plasticity are impaired in these animals. Thus, receptor and non-receptor tyrosine kinases appear to be critical modulators of synaptic plasticity in hippocampus, and they appear to exert many of their effects at the level of the NMDA-R.

11.8 **Other roles of protein tyrosine phosphorylation in the normal and diseased nervous system**

Tyrosine kinases have other roles in the nervous tissue besides those described above. Many of their functions are similar to those they have outside of the nervous system and do not need to be specifically addressed here. However, they are also involved in additional processes that are specific of the nervous system. One striking example is the requirement of Fyn for normal myelin formation (Umemori *et al.* 1994). This function is specific for Fyn and depends on its kinase activity (Sperber *et al.* 2001). Myelin defects have also been observed in mice lacking PTPε (Peretz *et al.* 2000). Thus tyrosine phosphorylation pathways are important regulators of myelin formation.

The involvement of tyrosine phosphorylation in the course of diseases of the nervous system has been also investigated and interesting results have been recently reported. One concerns the mechanism of brain damage in stroke. Src is a mediator of the vascular permeability triggered by vascular endothelial growth factor (VEGF) that is produced in response to ischemic injury (Paul *et al.* 2001). This response contributes significantly to the formation of edema. Importantly, the infarct size in experimental stroke was decreased in Src −/− mice, or following administration of Src inhibitors, suggesting that these compounds could be interesting therapeutic leads. The role of non-receptor tyrosine kinases has also been studied in the context of Alzheimer's disease. It has been shown that paired helical filaments of tau proteins are tyrosine phosphorylated, and, interestingly, that fibrillar amyloid beta-peptides are capable of enhancing phosphorylation on tyrosine of tau and other neuronal proteins via a pathway that includes FAK and Src (Williamson *et al.* 2002). At present, it is not known whether this mechanism is involved in the pathogenesis of the disease. Finally, it should be mentioned that neurotrophic factors are the subject of intense research, with the hope to prevent cell death or to increase recovery in a variety of neurological conditions. Since the actions of virtually all these factors involve protein tyrosine phosphorylation, molecules that could stimulate specific protein kinases or inhibit specific protein tyrosine phosphatases could be potential therapeutic agents for human diseases. The better knowledge of the proteins and pathways controlling tyrosine phosphorylation is important in this prospect.

11.9 **Conclusion**

Protein tyrosine phosphorylation is a universal signalling mechanism among multicellular animals to control cell–cell interactions. In the nervous system, in addition to its general role in cell growth and survival, tyrosine phosphorylation is important for specific functions. Through conserved molecular mechanisms, receptor and non-receptor tyrosine kinases appear to function in a coordinate manner in processes that include neurite extension and guidance, topographical organization of brain projections, target-dependent survival, synapse formation, and plasticity. Recent findings also indicate that some of these pathways may be interesting targets for treatment of human diseases. Since a growing number of drugs that inhibit specific protein kinases are being synthesized by the pharmaceutical industry, this area of investigation will certainly be expanding at a rapid pace.

References

Note that generally the bibliographic list corresponds to reviews, references to original publications can be found therein. Original references are indicated only when such reviews were not available.

Airaksinen, M. S. and Saarma, M. (2002) The GDNF family: signalling, biological functions and therapeutic value. *Nat. Rev. Neurosci.*, 383–94.

Ali, D. W and Salter, M. W. (2001) NMDA receptor regulation by Src kinase signalling in excitatory synaptic transmission and plasticity. *Curr. Opin. Neurobiol.*, **11**, 336–42.

Boxall, A. R. and Lancaster, B. (1998) Tyrosine kinases and synaptic transmission. *Eur J. Neurosci*, **10**, 2–7.

Buonanno, A. and Fischbach, G. D. (2001) Neuregulin and ErbB receptor signalling pathways in the nervous system. *Curr. Opin. Neurobiol.*, **11**, 287–96.

Derkinderen, P., Enslen, H., and Girault, J. A. (1999) The ERK/MAP–kinase cascade in the nervous system. *Neuroreport*, **10**, R24–34.

Druker, B. J. and Lydon, N. B. (2000) Lessons learned from the development of an abl tyrosine kinase inhibitor for chronic myelogenous leukemia. *J. Clin. Invest.*, **105**, 3–7.

Fauman, E. B. and Saper, M. A. (1996) Structure and function of the protein tyrosine phosphatases. *Trends Biochem. Sci.*, **21**, 413–17.

Fischer, E. H. (1993) Protein phosphorylation and cellular regulation II (Nobel lecture). *AngewChem(Engl)*, **32**, 1130–7.

Ginty, D. D. and Segal R.A. (2002) Retrograde neurotrophin signalling: Trk–ing along the axon. *Curr. Opin. Neurobiol.*, **12**, 268–74.

Girault, J. A., Labesse, G., Mornon, J.-P., and Callebaut, I. (1999*a*) The N–termini of FAK and JAKs contain divergent band 4.1 domains. *TIBS*, **24**, 54–7.

Girault, J. A., Costa, A., Derkinderen, P., Studler, J. M., and Toutant, M. (1999*b*) FAK and PYK2/CAK in the nervous system, a link between neuronal activity, plasticity and survival? *Trends Neurosci.*, **22**, 257–63.

Grant, S. G. N., O'Dell, T. J., Karl, K. A., Stein, P. L., Soriano, P., and Kandel, E. R. (1992) Impaired long-term potentiation, spatial learning, and hippocampal development in fyn mutant mice. *Science*, **258**, 1903–10.

Hanks, S. K. and Hunter, T. (1995) Protein kinases 6: The eukaryotic protein kinase superfamily: Kinase (catalytic) domain structure and classification. *FASEB J.*, **9**, 576–96.

Hempstead, B. L. (2002) The many faces of p75(NTR). *Curr. Opin. Neurobiol.*, **12**, 260–7.

Himanen, J. P., Rajashankar, K. R., Lackmann, M., Cowan, C. A., Henkemeyer, M., and Nikolov, D. B. (2001) Crystal structure of an Eph receptor–ephrin complex. *Nature*, **414**, 933–8.

Hopfield, J. F., Tank, D. W., Greengard, P., and Huganir, R. L. (1988) Functional modulation of the nicotinic acetylcholine receptor by tyrosine phosphorylation. *Nature*, **336**, 677–80.

Huang, E. J. and Reichardt, L. F. (2001) Neurotrophins: roles in neuronal development and function. *Annu. Rev. Neurosci.*, **24**, 677–736.

Hunter, T. (1997) Oncoprotein networks. *Cell*, **88**, 333–46.

Jay, D. G. (2001) A Src-astic response to mounting tension. *J. Cell. Biol.*, **155**, 327–30.

Kafitz, K. W., Rose, C. R., Thoenen, H., and Konnerth, A. (1999) Neurotrophin-evoked rapid excitation through TrkB receptors. *Nature*, **401**, 918–21.

Klein, R. (2001) Excitatory Eph receptors and adhesive ephrin ligands. *Curr. Opin. Cell. Biol.*, **13,** 196–203.

Kojima, N., Wang, J., Mansuy, I. M., Grant, S. G. N., Mayford, M., and Kandel, E. R. (1997) Rescuing impairment of long-term potentiation in fyn-deficient mice by introducing Fyn transgene. *Proc. Natl. Acad. Sci. USA*, **94**, 4761–5.

Kovalchuk, Y., Hanse, E., Kafitz, K. W., and Konnerth, A. (2002) Postsynaptic Induction of BDNF-Mediated Long-Term Potentiation. *Science*, **295**, 1729–34.

Krebs, E. G. (1993) Protein phosphorylation and cellular regulation I (Nobel lecture). *AngewChem(Engl)*, **32**, 1122–9.

Kullander, K. and Klein, R. (2002) Mechanisms and functions of eph and ephrin signalling. *Nat. Rev. Mol. Cell Biol.*, **3**, 475–86.

Lanier, L. M. and Gertler, F. B. (2000) From Abl to actin: Abl tyrosine kinase and associated proteins in growth cone motility. *Curr. Opin. Neurobiol.*, **10**, 80–7.

Levitan, I. B. (1999) Modulation of ion channels by protein phosphorylation. How the brain works. *Adv. Second Messenger Phosphoprotein Res.*, **33**, 3–22.

Martin, G. S. (2001) The hunting of the Src. *Nat Rev Mol Cell Biol*, **2**, 467–75.

Mellitzer, G., Xu, Q., and Wilkinson, D. G. (2000) Control of cell behaviour by signalling through Eph receptors and ephrins. *Curr. Opin. Neurobiol.*, **10**, 400–8.

Miyakawa, T., Yagi, T., Kitazawa, H., Yasuda, M., Kawai, N., Tsuboi, K., *et al.* (1997) Fyn-kinase as a determinant of ethanol sensitivity: relation to NMDA-receptor function. *Science*, **278**, 698–701.

Moss, S. J., Gorrie, G. H., Amato, A., and Smart, T. G. (1995) Modulation of $GABA_A$ receptors by tyrosine phosphorylation. *Nature*, **377**, 344–8.

O'Dell, T. J., Kandel, E. R., and Grant, S. G. N. (1991) Long-term potentiation in the hippocampus is blocked by tyrosine kinase inhibitors. *Nature*, **353**, 558–60.

Patapoutian, A. and Reichardt, L. F. (2001) Trk receptors: mediators of neurotrophin action. *Curr. Opin. Neurobiol.*, **11**, 272–80.

Patterson, S. L., Pittenger, C., Morozov, A., Martin, K. C., Scanlin, H., Drake, C., *et al.* (2001) Some forms of cAMP-mediated long-lasting potentiation are associated with release of BDNF and nuclear translocation of phospho-MAP kinase. *Neuron*, **32**, 123–40.

Paul, R., Zhang, Z. G., Eliceiri, B. P., Jiang, Q., Boccia, A. D., Zhang, R. L., *et al.* (2001) Src deficiency or blockade of Src activity in mice provides cerebral protection following stroke. *Nat. Med.*, **7**, 222–7.

Pellegrini, L. (2001) Role of heparan sulfate in fibroblast growth factor signalling: a structural view. *Curr. Opin. Struct. Biol.*, **11**, 629–34.

Peretz, A., Gil-Henn, H., Sobko, A., Shinder, V., Attali, B., and Elson, A. (2000) Hypomyelination and increased activity of voltage-gated K(+) channels in mice lacking protein tyrosine phosphatase epsilon. *EMBO J.*, **19**, 4036–45.

Pierce, K. L. and Lefkowitz, R. J. (2001) Classical and new roles of beta-arrestins in the regulation of G-protein-coupled receptors. *Nat. Rev. Neurosci.*, **2**, 727–33.

Robinson, D. R., Wu, Y. M., and Lin, S. F. (2000) The protein tyrosine kinase family of the human genome. *Oncogene*, **19**, 5548–57.

Sanes, J. R. and Lichtman, J. W. (2001) Induction, assembly, maturation and maintenance of a postsynaptic apparatus. *Nat. Rev. Neurosci.*, **2**, 791–805.

Schlessinger, J. (2000) Cell signalling by receptor tyrosine kinases. *Cell*, **103,** 211–25.

Schuman, E. M. (1999) Neurotrophin regulation of synaptic transmission. *Curr. Opin. Neurobiol.*, **9**, 105–9.

Smart, T. G. (1997) Regulation of excitatory and inhibitory neurotransmitter-gated ion channels by protein phosphorylation. *Curr. Opin. Neurobiol.*, **7**, 358–67.

Sperber, B. R., Boyle–Walsh, E. A., Engleka, M. J., Gadue, P., Peterson, A. C., Stein, P. L., ***et al.*** (2001) A unique role for Fyn in CNS myelination. *J. Neurosci.*, **21**, 2039–47.

Swope, S. L., Moss, S. I., Raymond, L. A., and Huganir, R. L. (1999) Regulation of ligand-gated ion channels by protein phosphorylation. *Adv. Second Messenger Phosphoprotein Res.*, **33**, 49–78.

Thomas, S. M. and Brugge, J. S. (1997) Cellular functions regulated by Src family kinases. *Annu. Rev. Cell Dev. Biol.*, **13**, 513–609.

Tonks, N. K. and Neel, B. G. (2001) Combinatorial control of the specificity of protein tyrosine phosphatases. *Curr. Opin. Cell. Biol.*, **13**, 182–95.

Traverse, S., Gomez, N., Paterson, H., Marshall, C., and Cohen, P. (1992) Sustained activation of the mitogen-activated protein (MAP) kinase cascade may be required for differentiation of PC12 cells. Comparison of the effects of nerve growth factor and epidermal growth factor. *Biochem. J.*, **288**, 351–5.

Umemori, H., Sato, S., Yagi, T., Aizawa, S., and Yamamoto, T. (1994) Initial events of myelination involve Fyn tyrosine kinase signalling. *Nature*, **367**, 572–6.

Ushiro, H. and Cohen, S. (1980) Identification of phosphotyrosine as a product of epidermal growth factor receptor-activated protein kinase in A–431 cells. *J. Biol. Chem.*, **255**, 8363–5.

Van Etten, R. A. (1999) Cycling, stressed-out and nervous: cellular functions of c-Abl. *Trends Cell. Biol.*, **9**, 179–86.

Van Vactor, D. (1998) Protein tyrosine phosphatases in the developing nervous system. *Curr. Opin. Cell Biol.*, **10**, 174–81.

Weiss, A. and Schlessinger, J. (1998) Switching signals on or off by receptor dimerization. *Cell*, **94**, 277–80.

Wiesmann, C., Ultsch, M. H., Bass, S. H., and de Vos, A. M. (1999) Crystal structure of nerve growth factor in complex with the ligand- binding domain of the TrkA receptor. *Nature*, **401**, 184–8.

Williamson, R., Scales, T., Clark, B. R., Gibb, G., Reynolds, C. H., Kellie, S., ***et al.*** (2002) Rapid tyrosine phosphorylation of neuronal proteins including tau and focal adhesion kinase in response to amyloid–beta peptide exposure: involvement of Src family protein kinases. *J. Neurosci.*, **22**, 10–20.

Chapter 12

Mature neurons: Signal transduction-serine/threonine kinases

Renata Zippel, Simona Baldassa, and Emmapaola Sturani

12.1 Introduction

Protein phosphorylation is the most important reversible covalent modification utilized by the cells in response to extracellular signals to regulate biochemical and physiological processes (Hemmings *et al.* 1989). Many signaling pathways activated by neurotransmitters, hormones, growth factors, cytokines etc. converge on the phosphorylation of downstream protein effectors or other kinases to integrate the stimuli and propagate them in the various cellular compartments.

Phosphorylation is catalysed by a family of enzymes called kinases that transfer the terminal (gamma) phosphate group of ATP to the hydroxyl moieties of serine (Ser), threonine (Thr), or tyrosine (Tyr) residues at specific sites on target protein. This reaction adds negative charges to the protein that can alter its conformation and ultimately its functional activity (Edelman *et al.* 1987).

However, the phosphorylation state of a protein depends not only on the activity of kinases but also phosphatases that hydrolyse the phosphoester bond. Regulation of these two classes of enzymes and hence, the phosphorylation state of the different substrates (enzymes, structural proteins, ion channels, transcriptional factors etc.) is one of the most widely used mechanism to achieve fine-tuning of the cellular signaling both in the non-neuronal and in neuronal systems (Hunter 1995).

The kinases can be divided into two large families, Ser/Thr kinases and Tyr kinases, that originate from a common ancestral gene. In this chapter we will describe some of the principal classes of Ser/Thr kinases relevant for neuronal function mainly considering their role in mature neurons rather than during neuronal development. We will also provide information on some of the neuronal proteins, which are substrates of these enzymes. In the second part of the chapter we will describe some recent researches on some Ser/Thr phosphatases and their role in neuronal signaling.

12.2 Ser/Thr protein kinases

All protein kinases share a common structural design in their catalytic domain that extends also at the three-dimensional structure (Johnson *et al.* 1996). The cAMP-dependent protein kinase (PKA) catalytic subunit was the first protein kinase characterized at the three-dimensional structure and serves as a prototype for all the protein kinases (Knighton *et al.* 1991). The PKA kinase core comprises a bilobal scaffold that has an N-terminal lobe mainly composed of β sheets and a C-terminal lobe in which α helixes dominate. A hinge region allows the two lobes to articulate. The catalytic site is at the interface of the two lobes: the Mg-ATP binding site spans both lobes, while the substrate binding site (in the crystal structure identified by a specific PKA inhibitor PKI) is associated mostly with the C-terminal lobe. The two lobes can exist in an open or closed conformation. The open form allows the access of ATP to the catalytic site, while the closed form brings residues in the correct conformation to promote catalysis. The catalysis is carried out by an invariant aspartate present in all kinases (asp166 for PKA), which acts as a base via the hydroxyl group of the amino acid. For many protein kinases (but not all) the activation of the catalysis requires phosphorylation of specific residues in the activation segment, also called the activation loop, that in PKA spans amino acid 184–208 and that is differently positioned in the three-dimensional structure of the active and inactive kinase. In PKA the phosphate of threonine 197 interacts with basic amino acids maintaining the correct orientation and electrostatic environment for the catalytic base to act (Johnson *et al.* 1996).

Kinases can be divided into dedicated enzymes, if they have a narrow specificity toward substrates, or multifunctional ones, if they have a broad specificity. However, the latter kinases are by no means promiscuous for the different substrates. Recognition of the correct substrate and the appropriate amino acid on the polypeptide chain depends upon characteristic consensus sequences. Usually the amino acid to be phosphorylated is located in loops or in regions that are not organized in secondary structures.

Many different types of Ser/Thr kinases are expressed in the nervous system (see Table 12.1). We will subdivide these enzymes into two major categories:

1. Second messenger-dependent protein kinases;
2. Second messenger-independent protein kinases.

12.2.1 Second messenger-dependent protein kinases

Neurotransmitters, cytokines, hormones, odorants, light, and other extracellular signals induce the formation in the target cell of a second messenger that transmits the external signal. cAMP, calcium, diacylglycerol (DAG), and cGMP are the most prominent. These messengers activate specific kinases that decode the signal into the phosphorylation of proteins that integrate and propagate the external stimuli, although this is not the only mode of action.

Table 12.1 Major classes of protein serine–threonine kinases.

Second messenger-dependent protein kinases
cAMP-dependent protein kinase
cGMP-dependent protein kinase
Ca^{2+}/CaM-dependent protein kinases
Protein kinase C
Second messenger-independent protein kinases
MAP kinase cascade and target kinases
Raf, MEK kinases, MEKs, SEKs, ERKs, JNKs, SAPKs, RSKs
Cyclin dependent kinases (CDKs) and CDK regulating kinases
Cdc-2, CAK, CAK kinase
G protein-coupled receptor kinases
GRK2 (betaARK1), GRK3 (betaARK2), GRK5, and GRK6.
***P21 activated kinases* PAK**
Kinases involved in cytoskeletal organization and development
ROCK1/Rho kinase
Transmembrane receptor protein serine/threonine kinases
TGFβ receptor protein kinases
Casein kinases
CK1, CK2

This list is not intended to be complete. MAP kinase, mitogen activated protein kinase; ERK, extracellular signal-regulated kinase; JNK, Jun kinase; SAPK, stress-activated protein kinase; MEK, MAPK/ERK kinases; SEK, SAPK kinase; RSK, ribosomal S6 kinase; CDK, cyclin-dependent kinase; Cdc, cell division cycle; CAK, CDK-activating kinase; GRK, G protein receptor kinase; Rho-associated kinase1 (ROCK1).

All the kinases of this category share a common design (Fig. 12.1): they consist of several functional domains that can reside either on the same polypeptide chain or on separate ones. Each kinase has a catalytic domain, which is intrinsically active, but is kept in the inactive state by a regulatory domain. The regulatory domain consists of an auto-inhibitory region, and of the binding sites for the second messengers. In each case, interaction with the second messenger dissociates the auto-inhibitory site from the catalytic domain thus dis-inhibiting it. Additional regions of the kinases may be responsible for oligomerization or for targeting the kinases to distinct cellular localizations.

12.2.1.1 cAMP-dependent protein kinase (protein kinase A; PKA)

Signaling molecules that stimulate the synthesis of cAMP exert their intracellular effects primarily by activating PKA (Edelman *et al.* 1987). The holoenzyme, which consists of a tetramer of two catalytic (C) and two regulatory (R) subunits, is inactive. As each R subunit binds two molecules of cAMP its affinity for the C subunit is greatly reduced so that the C subunits dissociate as a free active enzyme. The steady-state level of cAMP determines the fraction of PKA that is in the dissociated active form.

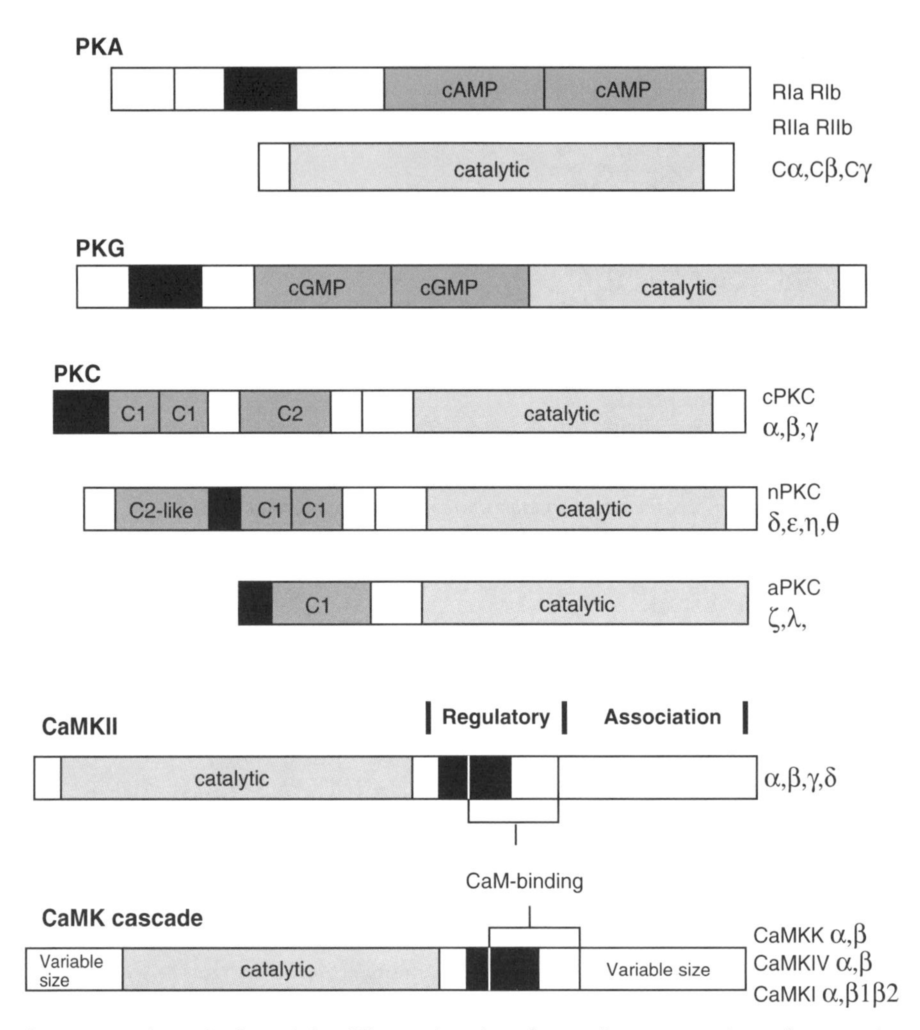

Fig. 12.1 A schematic view of the different domains of second messenger-dependent protein kinases. The different kinases contain a regulatory domain that encodes specialized functional domains and a conserved catalytic domain kept inactive by the presence of an auto-inhibitory region (black). For details see text.

Three isoforms of the C subunit, each of about 40 kDa, and four isoforms of the R subunit, each of 50–55 kDa, have been cloned from mammalian tissues. The three C subunits, designated Cα, Cβ, and Cγ, exhibit a similar and broad substrate specificity.

The four R subunits consist of two forms (α and β) each of type I and type II proteins. Their common modular structure contains an N-terminal dimerization domain that allows dimerization between the two R subunits, an auto-inhibitory site, and two cAMP binding domains. The holoenzymes always contain two identical regulatory subunits (PKAI contains RI while PKAII contains RII) that are joined together by two disulfide bonds. The α and β isoforms of the C subunit and all isoforms of the R subunits are expressed in the central nervous system, although with a differential pattern in various areas of the brain (Cadd and McKnight 1989).

Regulatory subunits differ in their regulation and biochemical properties. RII, but not RI subunits, undergo autophosphorylation in the inhibitory domain. This phosphorylation reduces the affinity of RII subunits for the catalytic subunit resulting in a more rapid dissociation and activation upon exposure to cAMP, as well as in a slower reassociation and inactivation. The binding affinity of the regulatory subunits to cAMP is in the order RIIβ < RIIα < RIα < RIβ (Edelman *et al.* 1987; Taylor *et al.* 1992). This implies that holoenzymes containing RI or RII subunits decode cAMP signals that differ in intensity and duration: PKAI is activated transiently by weak cAMP signals, whereas PKAII responds to high and persistent cAMP levels. Neurons, which predominantly express PKAII, are adapted to persistent high concentrations of cAMP.

The composition and biochemical properties of PKA isoenzymes account, in part, for differential cellular responses to discrete extracellular signals that activate adenylate cyclase.

Several studies suggest that RIα acts as a buffer for PKA activity (Brandon *et al.* 1997). For example, when C subunits are overexpressed in cultured cells, RIα increases while, when RII subunits are overexpressed, RIα decreases. In addition when one of the other R subunit types is removed by gene targeting, the RIα shows a compensatory increase. For example there is an increase in RIα level in the nervous system of mice lacking RIβ or RIIβ, due to a reduced rate of RIα degradation. The association with the C subunits protects RIα from proteolytic degradation (Amieux *et al.* 1997).

Protein kinase A activity is present throughout the neuronal cell. However, the kinase is highly compartmentalized within the cell, largely via several forms of anchoring proteins, termed A kinase anchor proteins (AKAPs) (Rubin 1994; Colledge and Scott 1999; Feliciello *et al.* 2001). Through an amphipatic helix AKAP tightly binds a large hydrophobic surface of the RII dimers and thereby tethers the inactive PKA holoenzyme to specific subcellular sites. A single species of neuronal AKAP is abundantly expressed in the brain and is highly conserved in mice (150 kDa), bovines (75 kDa), and humans (79 kDa). This prototypic neuronal AKAP targets PKAII to the postsynaptic densities keeping the protein kinase in close proximity to the signal-transduction proteins it phosphorylates to regulate synaptic transmission (Colledge and Scott 1999). The important role played by AKAPs is indicated by experiments showing that peptides corresponding to the RII-binding site of AKAP disrupt PKA-mediated regulation of AMPA-type glutamate receptors in neurons and of L-type voltage dependent calcium channels (VDCC) in skeletal muscle cells (Sim and Scott 1999).

The RI subunits are generally found in the cytosolic fraction although this assignment is not absolute. While anchoring of PKA to AKAPs via the RII subunit represents an extremely strong association, some AKAPs have been shown to anchor PKA through association with RI. Although the binding affinity of RI is much lower than that of RII, the dissociation constant (K_d) is still within the physiological concentrations of RI and AKAPs, so that anchoring PKAI may occur in regions of the cell where PKAII concentration is limited (Sim and Scott 1999).

12.2.1.2 Calcium/calmodulin dependent kinases (Ca^{2+}/CaM kinases; CaMKs)

Physiological elevation of intracellular free calcium is regulated by a diversity of Ca^{2+}-linked receptors and signal transduction systems that either increase influx of extracellular calcium or cause redistribution from intracellular stores. Most of the characterized Ca^{2+}-dependent kinases utilize the calcium-binding protein called calmodulin (CaM) as their calcium sensor.

Three major multifunctional Ca/CaM kinases, termed Ca/CaM kinase I, II, and IV (CaMKs), play a crucial role as second messenger-responsive kinases in neuronal signal transmission. The best characterized both at the structural and functional level is CaMKII (Hanson and Schulman 1992).

CaMKII is highly expressed in the brain where it represents 0.25% of total proteins, accounting for 1% in the cerebral cortex and 2% in the hippocampus. It is highly enriched at synaptic structures and particularly in the postsynaptic densities (PSD). CaMKII consists of two major subunits of 52 and 60 kDa encoded by α and β genes respectively. Additional isoforms are generated by alternative splicing as well as by γ and δ CaMKII genes. All the subunits share a common structure. At the N-terminal end there is the catalytic domain linked to the regulatory domain that partially overlaps the calmodulin-binding region. The association domain that helps to assemble the subunits into holoenzyme is located at the C-terminus of the protein. In this region we find major differences between the various isoforms. A short amino acid insert in some isoforms introduces a nuclear localization signal that targets the holoenzyme to the nucleus.

The holoenzyme consists of 6–12 subunits that through the association domain assemble into a central globular structure, from which the N-terminal catalytic/regulatory domains radially extend (Hanson and Schulman 1992) (Fig. 12.2A).

In the basal state the auto-inhibitory domain renders the kinase inaccessible to the substrates. Binding of calcium/calmodulin displaces the inhibitory domain from the active site, deinhibiting the kinase. Once activated by Ca/CaM, CaMKII can lock itself into the activated state by autophosphorylation on a conserved Thr residue present in the auto-inhibitory domain of all isoforms (Thr^{286} in α or Thr^{287} in β). The phosphorylation of $Thr^{286/287}$ occurs by an intersubunit reaction within each holoenzyme. The catalytic domain of one activated subunit in the holoenzyme phosphorylates $Thr^{286/287}$ in the regulatory domain of a neighbouring subunit that must also have calmodulin bound (Fig. 12.2B). Therefore, assembly of CaMKII into holoenzyme concentrates the subunits and positions them for autophosphorylation.

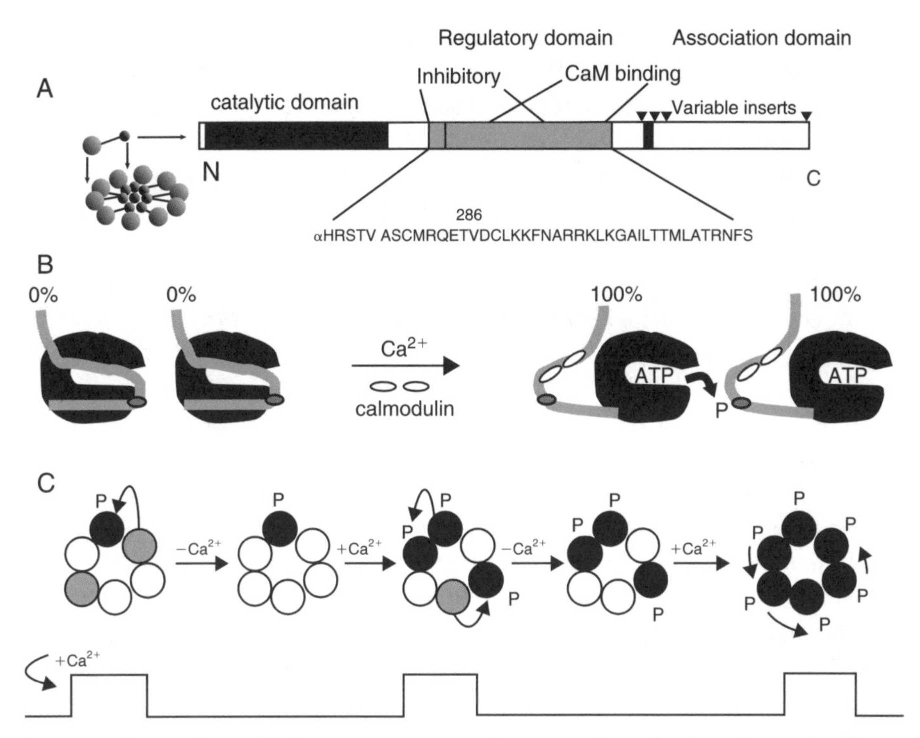

Fig. 12.2 A) Domains and holoenzyme structure of CaMKII; B) Mechanism of autophosphorylation. (Threonine 286 is indicated in red). Displacement of the auto-inhibitory domain by calmodulin exposes Thr286, which can then be phosphorylated if its proximate neighbour is active; C) Trapping of calmodulin (gray colour) to the CaMKII subunits increases the probability of autophosphorylation (black) during successive Ca^{2+} spikes at high frequency. For simplicity only a reduced number of subunits have been designed. Redrawn from Hanson and Schulman (1992) with permission from Annual Review of Biochemistry 61, 1994 by Annual Reviews.

Autophosphorylation of Thr $^{286/287}$ has two consequences: (*a*) calmodulin remains bound to the phosphorylated subunit for extended periods of time even at low Ca^{2+} concentrations (trapped state) because the autophosphorylation greatly reduces the calmodulin dissociation (Schulman *et al.* 1992; Singla *et al.* 2001); (*b*) autophosphorylated α and β subunits are rendered Ca/CaM-independent (autonomous) but still retain substantial kinase activity (Hanson and Schulman 1992; Schulman *et al.* 1992). Both autonomy and calmodulin trapping enable the phosphorylated kinase subunits to remain active beyond the limited duration of a Ca^{2+} spike. The multimeric structure of CaMKII, the intersubunit phosphorylation of $Thr^{286/287}$, and the resulting calmodulin trapping and autonomous activity are crucial elements that enable the kinase to act as a Ca^{2+} spike frequency detector (De Koninck and Schulman 1998). At low stimulus frequency, the time between stimuli is sufficient for calmodulin to dissociate and the kinase to be dephosphorylated. However at higher frequencies, some subunits will remain phosphorylated and bound to calmodulin, so that successive stimuli will result in more calmodulin bound per holoenzyme which will make autophosphorylation and subsequent calmodulin trapping more probable. At a threshold frequency, autophosphorylation and calmodulin trapping become more efficient and a higher level of activation and autonomy will be obtained (Fig. 12.2C).

The essential role of this autophosphorylation was demonstrated with the finding that mutant mice carrying Thr286Ala point mutation in αCaMKII do not exhibit long-term potentiation (LTP), are defective in spatial learning, and have unstable hippocampal place cells (Cho *et al.* 1998; Giese *et al.* 1998).

It is well known that activity-dependent stimuli promote translocation of CaMKII to the synapses and binding of the activated kinase to *N*-methyl-D-Aspartate (NMDA) receptors (Gardoni *et al.* 1999; Shen and Meyer 1999). Very recently, Bayer *et al.* (2001) have shown a Ca^{2+}/CaM-dependent association of the catalytic domain of CaMKII to NR2B subunits. This association allows some subunits of the holoenzymes to remain active even after the dissociation of Ca^{2+}/CaM. Moreover this interaction further facilitates autophosphorylation of CaMKII subunits leading to hyperphosphorylation of the holoenzyme. Interestingly, this manner of CaMKII activation generates an autonomous and CaM trapping state of CaMKII that cannot be reversed by phosphatases.

In addition to the multimeric CaMKII we can also find two other, closely related, monomeric proteins: CaMKI and IV (Hanson and Schulman 1992). There are at least three isoforms for CaMKI and two for CaMKIV. These kinases, too, require Ca/CaM binding to relieve intramolecular inhibition (disinhibition). A second auto-inhibitory mechanism unique to CaMIV is relaxed by the autophosphorylation of Ser12 and 13. However, for the full activation of these kinases, the phosphorylation by Ca/M kinase kinase (CaMKK) on a threonine present in the activation loop of the catalytic domain is also required. This fact introduces in this signaling pathway the concept of a cascade of kinases where all the members of the cascade respond to an elevation of calcium (Soderling 1999; Corcoran and Means 2001). Once activated by phosphorylation, CaMKIV acquires Ca/CaM independence whereas CaMKI remains dependent.

Both Ca/MKI and CaMKIV are highly expressed in the brain but, while CaMKIV is predominantly nuclear, CaMKI is primarily cytoplasmic (Soderling 1999) even if, given the size of the monomeric kinase, it is possible that it may have access to the nucleus by passive diffusion. Moreover, a translocation of this kinase to the nucleus after induction of LTP has been recently described in hippocampal neurons (Ahmed *et al.* 2000).

Two upstream activators of CaMKI and CaMKIV have been identified: CaMKKα and β which are encoded by two different genes (Soderling 1999). Both are highly expressed in the brain, share 80% similarity, and can phosphorylate CaMKI and CaMKIV *in vitro*. These kinases, too, are regulated by Ca/CaM, and the auto-inhibitory mechanism is similar to that of the other CaMKs. Given that CaMKK itself has potentially phosphorylatable residues in its activation loop we cannot exclude the possibility of a CaMK-kinase-kinase although at the moment no CaMKKK has been isolated (Corcoran and Means 2001).

While the ultrastructure of CaMKII allows this kinase to differentiate Ca^{2+} spike frequencies, the elaborate mechanisms for CaMK cascade activation, described above, probably render the CaMK cascade inadequate to fulfill a role as a neuronal Ca^{2+} signal detector. However, the inducible phosphorylation of CaMKI and CaMKIV can provide a mechanism for signal amplification.

12.2.1.3 Protein kinase C (PKC)

Neurotransmitters acting through G-coupled receptors activate phosphatidylinositol turnover, causing the production of diacylglycerol (DAG) and release of calcium from intracellular stores that act as second messengers to activate protein kinase C (PKC). The PKC family comprises at least 10 structurally related phospholipid-dependent protein kinases, and most of the various isoforms are expressed in the brain, albeit with a different distribution pattern that also depends on the developmental stage.

All PKC isoenzymes bind lipids with a remarkable selectivity for phophatidylserine. PKC isozymes have been grouped into three subclasses according to their regulatory properties, which are conferred by specific domains in the proteins (Mellor and Parker 1998). The 'conventional' PKCs (cPKCs; α, βI, βII, γ) are regulated by Ca^{2+}, DAG, or phorbol esters and phosphatidylserine. The 'novel' PKCs (nPKCs; δ, ε, θ, η) are activated by DAG or phorbol esters and phosphatidylserine but are Ca^{2+} independent. Finally, the 'atypical' PKCs (aPKC) are unresponsive to Ca^{2+}, DAG, and phorbol esters. A related enzyme, PKCμ or PKD (the mouse homolog), displays multiple unique features that make it a distant relative of the PKC isozymes. This enzyme also differs from other PKC isozymes in its regulation and substrate selectivity (Newton 1997).

Each of the PKCs isoforms consists of a single polypeptide chain that contains an N-terminal regulatory region and a C-terminal kinase domain (Parker *et al.* 1986). The regulatory region contains an auto-inhibitory or pseudosubstrate site and one or two membrane-targeting motifs, namely the C1 domain, which is present in all the isozymes and C2 domain which is present in conventional and novel PKCs. The C1 domain binds DAG or phorbol esters in all PKCs except for the aPKCs. The C2 region binds acidic phospholipids and in the cPKCs also Ca^{2+}. While the cPKCs and nPKCs have two copies of this motif in tandem, only a single copy is found in the aPKCs.

Regulation of PKCs isoenzymes requires the removal of the auto-inhibitory pseudosubstrate from the active site. This conformational change is achieved by highly specific binding of DAG and phosphatidylserine to the C1 and C2 domains. These interactions also allow the translocation of PKCs from the cytosol to the membrane causing maximal activation of the enzyme (Newton 1997). However both phosphorylation of PKCs and specific protein–protein interactions are also important for their activation.

Accumulating evidence suggests that PKC is phosphorylated in the activation loop by phosphoinositide-dependent protein kinase1 (PDK1)(Le Good *et al.* 1998). Phosphorylation by PDK1 is followed by autophosphorylation of two additional residues. All these phosphorylations are necessary for the newly synthesized PKC isoenzymes to obtain a catalytically competent conformation. Moreover, interaction of PKCs with other proteins plays a role in the localization and function of PKC isoenzymes. In the inactive conformation PKC can interact with AKAPs and 14-3-3 proteins, whereas in the active conformation it can bind to proteins called receptors for activated C kinase (RACKs) or to substrates that interact with C kinase (STICKs). Thus, while AKAPs can serve to localize inactive PKC isozymes in specific compartments of the cell, RACKs can function as shuttling proteins of the activated form. It is worthwhile recalling that in neurons one isoform of the AKAP proteins associates also with PKA and calcineurin (see below) and positions each enzyme just below the postsynaptic membrane where it can respond individually to activation signals such as calcium, or other second messengers (Ron and Kazanietz 1999).

In neurons an important role is played by PICK1, a scaffold protein that interacts with the AMPA-type glutamate receptors. This scaffold protein also binds to the C-terminal region of the activated PKCα. Thus in neurons PICK1 functions as a target and transport protein that directs the activated form of PKCα to GluR2 in the spines (Xia *et al.* 1999) (see below).

12.2.1.4 cGMP-dependent protein kinase (PKG)

Nitric oxide (NO), a very labile signaling molecule that can rapidly diffuse and affect targets in the same cell or in neighbouring neurons is the major activator of guanylate cyclase that produces cGMP. This second messenger activates cGMP-dependent protein kinase (PKG), a dimer composed of two identical subunits of about 75 kDa. PKG shows a more reduced cellular distribution and specificity than PKA and in fact the second messenger actions of cGMP in the regulation of neuronal function are very limited.

12.2.2 Second messenger-independent protein kinases

Many distinct families of protein kinases belong to this category. We will focus only on one specific family that comprises the enzymes involved in the mitogen-activated protein kinase (MAPK) cascade. These enzymes have acquired a particular importance in neuronal signaling since in recent years a crucial role in synaptic transmission and neuronal plasticity has been attributed to this cascade.

12.2.2.1 Mitogen-activated protein kinase and MAPK regulating kinases

The MAPK cascade contains at least three proteins that work in series. The last members of this cascade are MAPKs that were first characterized as mitogen-activated kinases and have been shown to play an important role in cell growth. The same enzymes have been described as microtubule-associated protein kinases for their ability to phosphorylate microtubule-associated proteins. The best-characterized family of MAPKs in the brain are the extracellular signal-regulated protein kinases (ERKs). Many reviews have been published on the molecular basis of ERK activation (Cobb and Goldsmith 1995; English *et al.* 1999).

Two isoforms of ERKs exist: ERK of 44 kDa and ERK of 42 kDa, both of which are highly expressed in the brain (Ortiz *et al.* 1995). These kinases are activated by the phosphorylation of both threonine and tyrosine in a conserved threonine–amino acid–tyrosine sequence (T-X-Y) present in the activation loop. This reaction is catalysed by the upstream ERK kinases (MEKs), dual-specificity kinases that phosphorylate ERKs in both residues. This dual phosphorylation is both necessary and sufficient for ERK activation, and ERKs are the only known substrates of MEKs. These properties have been used to create pharmacological tools to investigate the role of ERK cascade in neuronal functions (English *et al.* 1999).

Activation of MEK is accomplished by a third class of protein kinases, the Raf kinase family. Three members of this family are known; Raf-1 (c-Raf), ubiquitously expressed, B-Raf particularly enriched in brain and testes, and A-Raf which is not detected in the brain. The best characterized mechanism for Raf activation is that of Raf-1. This molecule contains two functional domains: the N-terminal regulatory domain and the C-terminal kinase domain. The small GTPase Ras in its GTP-bound activated form interacts with Raf-1, localizing it to the plasma membrane, where other events are required for full Raf-1 activation. Although at the moment the mechanisms of Raf-1 activation have still not been completely elucidated, phosphorylation on tyrosine residues by yet unidentified kinases (possibly src like kinases) and phosphorylation by PKC have been implicated. Moreover, interaction with the 14-3-3 proteins may play a role in facilitating and stabilizing the active Raf-1 conformation (Morrison and Cutler 1997).

The two isoenzymes expressed in the brain (Raf-1 and B-Raf) have similar regional distribution and are coexpressed in neurons. However their subcellular localization is different: while Raf-1 is mainly localized in the cytosol around the nucleus, B-Raf is widely distributed in the cell body and in neuritic processes suggesting that each one of them has a distinct regulatory function in the neuron (Morice *et al.* 1999).

The relevant feature of the MAPK cascade is the series of successive protein kinases involved that provide the basis for massive amplification of an initial signal. Moreover, Ras and the other closely related GTPase Rap, both implicated in the activation of this cascade in neurons (Grewal *et al.* 2000), can themselves be activated by a great variety of external stimuli. Not only neurotrophins acting through tyrosine kinase receptors but also neurotransmitters acting through G protein-coupled receptors and calcium activate the ERK cascade. In addition to the well-characterized SOS, Ras-GRFs, and G3C, that activate Ras or/and Rap, different guanine nucleotide exchange factors (GEFs) have been recently isolated (de Rooij *et al.* 1998, 1999; Kawasaki *et al.* 1998;

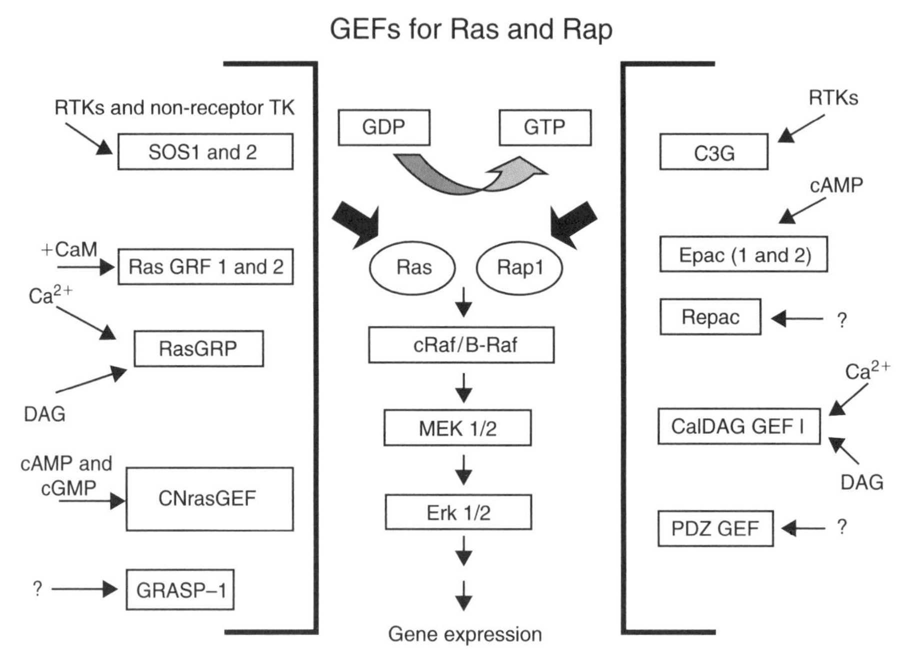

Fig. 12.3 Repertoire of neuronal guanine nucleotide exchange factors (GEFs) that convey on Ras and Rap the extracellular signals leading to MAPK activation and gene expression. RTK: receptor tyrosine kinases; Ras-GRP (Ebinu *et al.* 1998); Cyclic nucleotide rasGEF (CNrasGEF) (Pham *et al.* 2000); GRASP-1(Ye *et al.* 2000); Epac1,2 and Repac (de Rooij *et al.* 2000); CalDAG-GEF1 (Kawasaki *et al.* 1998); PDZ-GEF (de Rooij *et al.* 1999).

Pham *et al.* 2000). Most of these molecules are highly expressed in the brain and respond to Ca/CaM, cAMP or calcium, and DAG, allowing the integration of different second messenger systems in the activation of ERK signaling (Fig. 12.3). The relevance of the ERK cascade also depends on the fact that activated ERKs, which in the basal inactive state have a cytoplasmic localization, translocate to the nucleus and phosphorylate specific transcription factors either directly or through their downstream RSK2 kinase target (Xing *et al.* 1996). Thus the MAPK cascade not only amplifies extracellular stimuli but also integrates many signaling pathways and functions as a shuttle that imports the information into the nucleus.

12.3 **Neuronal substrates of kinases**

A great variety of neuronal proteins are substrates of the kinases described above and of other kinases. Phosphorylation of these many types of proteins allows the regulation of virtually every process in the nervous system. Many proteins are phosphorylated on more than one amino acid residue by different kinases. This may be important to integrate multiple intracellular pathways in order to achieve coordinated regulation of neuronal function.

I will only give few examples of substrates of kinases, but for more details many reviews can be consulted (Nestler and Greengard 1984).

12.3.1 Neurotransmitter release

At the presynaptic locus, protein phosphorylation regulates exocytosis, synapsins (synaptic vesicles-associated proteins) being key substrates of kinases in this process (Greengard *et al.* 1993). Synapsin I is phosphorylated by PKA and by CaMKII in the N-terminal region. In the C-terminus it is phosphorylated on two other serine residues by CaMKII and both at N- and C-terminus it is a substrate of ERKs. Under resting conditions, dephosphorylated synapsin is thought to anchor synaptic vesicles to the cytoskeleton. Following an action potential, the increase of intracellular calcium activates CaMKII, which phosphorylates synapsin at the C-terminus causing the release of neurotransmitters from synaptic vesicles of the reserve pool.

12.3.2 Ligand-gated ion channels and potassium channels

Protein kinases play a critical role in the induction and maintenance of LTP (see Chapter 14). A series of studies have shown that one particularly relevant substrate of CaMKII is the AMPA receptor that is potentiated during LTP. Phosphorylation of GluR1 subunits of AMPA receptors by CaMKII (either the endogenous receptor in cultured hippocampal neurons and in CA1 pyramidal neurons, or the overexpressed GluR1 subunit in HEK293 cells) results in potentiation of the whole-cell current mediated by AMPA receptor. The non-phosphorylated receptors exist predominantly in the lower conductance states whereas phosphorylation by CaMKII at serine 831 stabilizes the higher conductance (Derkach *et al.* 1999). Phosphorylation at serine 831 can also be mediated by PKC and it has been shown that AMPA receptor-mediated synaptic currents are similarly potentiated when the PKC catalytic domain is intracellularly perfused in CA1 hippocampal neurons.

GluR1 receptor subunits can be phosphorylated by PKA at serine 845 and their phosphorylation seems to regulate the open probability of AMPA receptors. Interestingly, it has been recently demonstrated that different kinases phosphorylate the GluR1 subunits depending on the past experience of the synapse (Lee *et al.* 2000).

Also trafficking of AMPA receptors is mediated in part by protein phosphorylation. It appears that phosphorylation of GluR2 subunits at serine 880 differentially regulates its interaction with the PDZ domain-containing proteins GRIP1 and PICK1. While phosphorylation of GluR2 subunits at serine 880 disrupts GRIP1 association, association with PICK1 is unaffected. PKC activation in neurons increases phosphorylation of GluR2 subunit at serine 880 and induces internalization of GluR2 subunits. PKC activation also induces a dramatic mobilization of PICK1 to synapses suggesting that PICK1 interaction with phosphorylated serine 880 of GluR2 may be involved in the internalization of AMPA receptor (Chung *et al.* 2000).

It has also been found that phosphorylation of ion channels regulates their ability to open or close, thereby mediating the synaptic response to neurotransmitters as well as more general changes in neuronal excitability. In the hippocampus voltage-dependent transient K^+ channels of the Shaker superfamily (Kv channels) can regulate neuronal excitability and also the magnitude of excitatory postsynaptic potential in response to synaptic activity (Dineley *et al.* 2001). One of the candidates for this function is the Shaker1-type K channel Kv 4.2, a voltage-dependent K channel localized at dendrites and cell bodies of hippocampal pyramidal neurons. This channel is a substrate of PKA, PKC, CaMKII and has been recently demonstrated to be also phosphorylated by ERKs in the C-terminal cytoplasmic domain both *in vitro* and *in vivo* (Adams *et al.* 2000). Although it is not clear what is the role of this modification, it has been hypothesized that ERK phosphorylation of the channel might lead to downregulation of hyperpolarizing K+ currents resulting in increased pyramidal neuron excitability. As support for a role of MAPKs in the regulation of Shaker-type current it has very recently been demonstrated that GDNF acutely modulates excitability by inhibiting transient Shaker type K channels in midbrain dopaminergic neurons and that this effect is mediated by MAPKs (Yang *et al.* 2001).

The inward rectifying potassium channel, IRK1, plays an important role in neuronal excitability and its function can be modulated by protein phosphorylation. It is a substrate of PKA and PKC, which can directly modulate its properties. Channel modulation can also be achieved by varying the amount of channel molecules present at the cell surface, and tyrosine phosphorylation of IRK1 has been implicated in the regulation of channel internalization (Wischmeyer *et al.* 1998). Moreover, we have recently shown that this channel is modulated by the ERK cascade: when ectopically expressed in HEK293 cells a consistent reduction of cell surface associated channels is caused by persistent activation of MAPKs (Giovannardi *et al.* 2002).

12.3.3 Transcription factors

Phosphorylation of transcription factors regulates the expression of specific genes in target neurons, this being the ultimate form of signal transduction and neuronal plasticity.

Activation of the transcription factor cAMP responsive element binding protein (CREB) is thought to be important in the formation of long-term memory in different species. Phosphorylation of serine 133 of CREB is critical for CREB activation allowing its association with the co-activator CREB binding protein (CBP) and thus promoting the transcription of genes with an upstream CRE element. Phosphorylation of CREB at serine 133 is mediated by different neuronal kinases the first one to be characterized being PKA. Also CaMKs and RSK2 phosphorylate CREB at the same serine (see also Chapter 10).

CREB has a nuclear localization, and translocation of the kinases to the nucleus is required to phosphorylate it. This, therefore, prompts the question of how cytoplasmic kinases can phosphorylate nuclear proteins. Persistent activation of PKA can cause translocation of its catalytic subunit to the nucleus (see below), while ERK translocation has been clearly demonstrated. More intriguing is the explanation of how CaMKs can phosphorylate CREB. CaMKII does not seem to translocate to the nucleus while CaMKIV has a nuclear localization. On the other hand, synaptic activity-dependent phosphorylation of CREB relies mainly on CaMK activity. Deisseroth *et al.* (1998) have shown that in hippocampal neurons brief bursts of activity cause a swift translocation of calmodulin from the cytoplasm to the nucleus and this event is important for rapid phosphorylation of CREB. A possible target of nuclear calmodulin increase might be the activation of CaMKIV. CREB can also be phosphorylated at serine 142 by CaMKII and phosphorylation of this residue has been shown to block activation of target genes and formation of the CREB–CBP complex *in vitro*. However substantial phosphorylation of this site has not been observed *in vivo* and it is not clear whether the inhibitory effect observed occurs *in vivo*.

12.4 Role of the kinases in synaptic transmission and cross-talk between the different kinase pathways

In mature neurons synaptic activity triggers a long-lasting cascade of events that activate the aforementioned kinases to transmit the signal from the postsynaptic membrane to dendrites, to the soma and to the nucleus and culminate in neuronal gene expression. A very complex network of neuronal kinases, which is often apparently redundant, is utilized to integrate the signal and converge it to final common effectors.

It is widely accepted that calcium is the key signal molecule at the excitatory synapse (Bito 1998). Calcium not only triggers release of neurotransmitters from presynaptic vesicles but is also one of the essential messengers regulating postsynaptic excitability as well as induction and expression of synaptic plasticity (Chawla and Bading 2001).

The majority of research on these aspects has been conducted on hippocampal or cortical cultures that represent a simple but bona fide cellular model to acquire biochemical informations on the various signaling pathways activated during synaptic activity.

In hippocampal neurons electric stimuli cause an influx of calcium mainly through L-type VDCC and NMDA receptors. Calcium can then associate with calmodulin and this complex activates CaMKII and/or the CaMK cascade. These kinases, which elicit changes in a variety of downstream effectors, can be sufficient to transfer information to the nucleus, where phosphorylation of CREB may induce gene expression. However, the Ca/CaM complex can also activate type 1 and type 8 adenylyl cyclases (ACs) leading to the production of cAMP and activation of PKA, whose catalytic subunits can migrate to the nuclei and activate transcription factors.

Postsynaptic calcium influx can activate additional signaling pathways among which is the ERK cascade. Multiple routes for the activity-dependent stimulation of ERKs have been demonstrated. The Ca/CaM complex can interact with Ras-GRF1 (Zippel *et al.* 2000) to activate Ras, the upstream regulator of Raf-1 in the ERK cascade. Moreover, several Ras-GEFs activated by calcium and DAG have been identified. It has also been demonstrated that in cortical neurons depolarization induces activation of the cytosolic tyrosine kinase Pyk2 in a calmodulin-dependent manner. Activation of Pyk2 may lead to the activation of the Ras exchange factor SOS.

The situation is even more complicated: in addition to the above-mentioned pathway Ras-Raf-1, Rap1, a small GTPase highly expressed in the brain, can activate the ERK cascade acting as upstream regulator of B-Raf, and many exchange factors for Rap1 directly modulated by second messengers have been recently identified.

In addition, PKA itself can activate the ERK cascade; it has been demonstrated that in PC12 cells, in hippocampal (Grewal *et al.* 2000) and cortical neurons (Baldassa *et al.* personal communication), PKA activity is required for the persistent ERK activation induced by depolarization. Moreover we have shown that Rap1 may mediate this effect (Baldassa *et al.* 2003).

It has previously been proposed that CaMKII positively acts on the ERK cascade: a synaptic GTPase activating protein for Ras (SynGAP), highly expressed in neurons and mainly localized at the postsynaptic densities, has been reported to be phosphorylated by CaMKII (Chen *et al.* 1998; Kim *et al.* 1998).While initially this phosphorylation seemed to inhibit GAP activity this does not seem to be the case (Chen *et al.* 2002).

Finally, stimulation of PKC produces a robust activation of ERKs in both hippocampal and cortical neurons.

All these results highlight the importance of the MAPK cascade and suggest that in neurons it can function as 'biochemical signal integrator and molecular coincidence detector' for coordinating responses to extracellular signals (Sweatt 2001*a*).

Cross-inhibitory interactions between signaling pathways have been demonstrated. It has been shown that *in vitro* PKA can phosphorylate CaMKK and suppress its ability to activate CaMKIV. Conversely, activation of CaMKIV can inhibit cAMP production by CaMKIV-mediated phosphorylation of type I adenylate cyclase.

A schematic diagram, which is undoubtedly reductive, illustrates the multiplicity and complexity of the signaling network that can be activated by calcium in neurons (Fig. 12.4). In this scheme we have also included calcineurin, a phosphatase activated by calcium/calmodulin and PP1. However for these details, see below.

12.5 **Role of the kinases in synaptic plasticity**

The ability of the above-mentioned kinases to initiate or maintain synaptic changes that underlie learning and memory might require that they themselves undergo some form of persistent change in activity. These kinases can be described as cognitive kinases because they remain in the activated state also after their second messengers have returned to basal level and because their target substrates modulate synaptic plasticity.

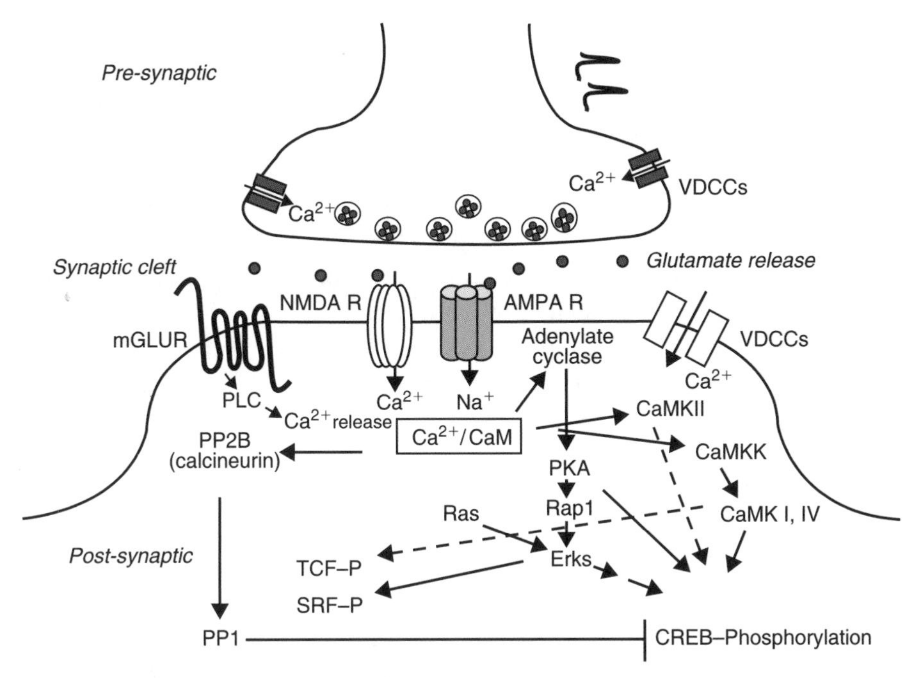

Fig. 12.4 Postsynaptic signaling induced by glutamate release in an excitatory synapse. A schematic diagram, undoubtedly reductive, illustrates the multiplicity and complexity of the signaling network that can be activated by calcium in a postsynaptic neuron. Serum response factor (SRF) and ternary complex factor (TCF) are other transcriptional factors that lead to gene expression.

12.5.1 **PKA**

In the pioneering works of Kandel an essential role for PKA has been established in long-term facilitation of the gill withdrawal reflex in *Aplysia* (Schacher *et al.* 1990). At the molecular level, release of serotonin from tail sensory neurons increases cAMP and activates PKA, producing a short-term increase in the phosphorylation of many proteins. However, prolonged or repeated exposure to serotonin leads to an increase in cAMP that persists for several minutes and causes a prolonged dissociation of the catalytic subunits of PKA from the holoenzyme. This allows the catalytic subunits to translocate into the nucleus and to activate several immediate response genes including ubiquitin hydrolase, which is required for the proteolysis of the regulatory subunit of PKA. Cleavage of the regulatory subunit results in persistent activation of PKA even when the second messenger is no longer present.

In higher organisms, particularly in the hippocampal area of the brain, PKA may also function as a cognitive kinase required for the maintenance of LTP (i.e. for the late phase — or phase III — of long-term potentiation, L-LTP — see also Chapter 14) rather than for the induction of LTP (Brandon *et al.* 1997; Dineley *et al.* 2001). In the

CA1 region, the NMDA receptor-dependent LTP is triggered by calcium influx either through NMDA receptors and/or VDCC. In these cells, calcium/calmodulin-dependent adenylate cyclases (AC1and AC8) generate cAMP waves critical for L-LTP (Poser and Storm 2001). Mice lacking both AC1 and AC8 do not exhibit L-LTP. In addition, mice with a targeted disruption of the regulatory subunit RIIB predominant in the striatum are severely deficient in PKA activity and exhibit changes in gene expression and locomotor behaviour.

PKA also seems to be involved in LTP at the CA3 region of the hippocampus. Tetanic stimulation of the mossy fibers induces a rise in presynaptic calcium and results in activation of Ca/CaM-dependent adenylate cyclase. The consequent activation of PKA elicits a long-lasting increase in transmitter release. Mutant mice lacking type I AC failed to induce mossy fibre LTP.

12.5.2 CaMKII

CaMKII is probably the most efficient cognitive kinase. The unique properties of molecular memory described above allow CaMKII to function as a memory molecule to decode the synaptic input. This kinase plays a key role during the induction of LTP rather than in the maintenance or consolidation of LTP (Fukunaga and Miyamoto 2000). It has been shown that mice lacking αCaMKII are deficient in LTP induction in the hippocampal CA1 region as well as in spatial learning. On the contrary, introduction of an active form of CaMKII into postsynaptic neurons induces potentiation in synaptic transmission that prevents subsequent LTP induction.

12.5.3 PKC

PKC can function as a cognitive kinase: it has been shown that in the CA1 region of rat hippocampal slices, tetanic stimulation triggers activation of PKC required for the induction of LTP. Moreover, it has been demonstrated that a persistent increase in PKC activity is associated with the maintenance phase of LTP. Either phosphorylation of PKC isoenzymes or proteolytic cleavage of the regulatory domains might be responsible for the persistent activation of PKCs in the absence of second messenger signals (Sweatt *et al.* 1998). Mice in which the brain-specific γ isoform of PKC has been deleted exhibited modest effects on memory, suggesting that other PKC isoforms are involved in mammalian learning and memory. Knock out mice for the β isoforms of PKC, which are highly expressed in the CA1 area of the hippocampus, did not exhibit deficits in hippocampal synaptic transmission or LTP. However, deletion of the PKCβ gene resulted in defects in amygdala-dependent functions such as clued and contextual fear conditioning (Dineley *et al.* 2001).

12.5.4 MAPKs

Accumulated data have identified a prominent role for ERKs in synaptic plasticity in a wide range of systems (Sweatt 2001*a*). Pharmacological drugs that specifically inhibit

MEK, the upstream activator of MAPK, have been extremely useful for these studies. Persistent activation of MAPKs and their translocation to the nucleus is required for plasticity-related changes in gene expression, suggesting that MAPKs might be required for the induction of L-LTP while they only modulate early-LTP.

Activation of the ERKs has been demonstrated to be necessary for the induction of both NMDA receptor-dependent and independent LTP in the CA1 area of the rat hippocampus, for the induction of LTP in the dentate gyrus and in the amygdala. ERK activation has also been shown to be required for long-term facilitation of the sensory-motor synapse in *Aplysia.* In addition, compelling evidence for ERK-dependent role in learning and memory comes from behavioral studies in rodents (Mazzucchelli and Brambilla 2000). In three different tasks (fear conditioning, aversive taste learning, and spatial learning), behavioral performance was associated with increased ERK activity while inhibition of ERK signaling specifically impaired learning. Taken together, these and many other findings provide strong evidence for an involvement of ERK activity in synaptic plasticity, learning, and memory. ERK1 knockout mice have been generated: in these mice a dramatic enhancement of ERK2-dependent signaling was observed and both electrophysiological and behavioral analyses revealed that altered modulation of ERK2 signalling affects neuronal plasticity in a region-specific manner. Tetanic stimulation elicits enhanced LTP in the nucleus accumbens but not in the hippocampus or in the basolateral nucleus of amygdala (Mazzucchelli *et al.* 2002).

12.6 Serine threonine phosphatases

For a long time, kinases have been regarded as the main 'switch on' molecules while phosphatases were simply considered housekeeping enzymes working constitutively to shut down signals.

It is now clear that in the neuronal system phosphatases play an active role in signaling and in synaptic plasticity. Although the general picture of how phosphates contribute to neuronal functions is not fully understood, it appears that phosphatases might provide an inhibitory constraint in the modulation of neuronal plasticity regulating the activity of a variety of kinases and in counteracting their effects.

The different phosphatases arise from different genes with no homology. In addition, it is impossible to identify consensus sequences for their specificity of substrate recognition. The same phosphatases can dephosphorylate substrates phosphorylated by different kinases. A limited number of multifunctional phosphatases account for most phosphatase activity. We will limit the description to two major phosphatases (PP1 and PP2B) whose role in the neuronal system has been characterized (Winder and Sweatt 2001).

12.6.1 Protein phosphatase2B (PP2B, Calcineurin)

PP2B (Calcineurin) is a calcium/calmodulin-dependent phosphatase highly enriched in the brain. It is a heterodimer composed of A and B subunits. The 60 kDa A subunit

(CnA) has an N-terminal catalytic domain and a C-terminal regulatory domain including an auto-inhibitory region, a calmodulin-binding domain, and a binding site for the 19 kDa regulatory B subunit (CnB). The latter is a calmodulin-like Ca binding protein. Calcineurin is activated by low concentrations of calcium: some activation of calcineurin is attained by the binding of calcium to CnB while stronger activation is obtained by the binding of Ca/CaM to CnA. Additional mechanisms exist for modulating and targeting PP2B activity. At least three PP2B binding proteins have been identified: the calcineurin inhibitor CAIN, AKAP 79, and FKBP12 which is required as an intermediate for FK506-mediated inhibition of calcineurin. FKBP12 promotes the association of PP2B to ryanodine and IP3 receptors allowing this phosphatase to regulate cytoplasmic free calcium. The interaction of CAIN with calcineurin potently inhibits its activity and has been implicated in the regulation of the exocytotic machinery. Finally AKAP 79 binds both PP2B and PKA, bringing them in close proximity for the regulation of receptors and ion channels.

The catalytic subunit of PP2B is expressed at a high level in the hippocampus and is enriched in dendritic spines while it is essentially absent from glia and interneurons in the hippocampus and is not readily detectable in presynaptic terminals. Its active role in hippocampal synaptic plasticity has been described. Bito *et al.* (1996) have shown that in hippocampal cultures a brief burst of synaptic activity causes a transient phosphorylation of CREB that is not sufficient to activate gene expression. However, if calcineurin activity is inhibited, the same stimuli can induce a persistent CREB phosphorylation similar to that obtained with a prolonged stimulation, leading to the activation of CRE-regulated genes. They suggest that while CaMK activity appears to play a critical role in the phosphorylation of CREB, the maintenance of this state is controlled by the regulation of PP2B. This phosphatase mediates nuclear PPI activation responsible for CREB dephosphorylation.

Further support for the role of PP2B in synaptic functions comes from studies on gene targeting. In mice overexpressing the auto-inhibitory domain of calcineurin LTP, elicited by subsaturating tetanization (but not saturating conditions), was enhanced. On the contrary, mice that overexpress an activated form of PP2B show suppression of LTP elicited by strong stimuli. These and other results suggest that inhibition of PP2B activity facilitates LTP formation (Sweatt 2001*b*). Interestingly, the manipulation of PP2B activity selectively interferes with forms of LTP that are sensitive to PKA inhibition suggesting that PP2B may compete with PKA for the regulation of specific substrates (see below).

12.6.2 **Protein phosphatase 1**

Several catalytic subunits of PP1, including α, β, $\gamma 1$, and $\gamma 2$ isoforms, are present in the brain where they interact with various other proteins acting in subcellular targeting and/or in the regulation of phosphatase activity (Winder and Sweatt 2001). For example, PP1 binds Yotiao, which in turn also binds NR1 receptors and the RII regulatory subunit of PKA, so that PP1 and PKA activity are targeted to the NMDA receptors.

PP1 activity can be regulated in various ways including direct inhibition. At least four proteins, known as PP1 inhibitor proteins [inhibitor 1 (I-1), inhibitor 2 (I-2) dopamine, and cAMP regulated phosphoprotein of 32 kDa (DARP-32) (Greengard *et al.* 1998), and nuclear inhibitor of PP1 (NIPP-1)], have been identified as regulating PP1 activity in neurons. This regulation depends on the phosphorylation of these inhibitors; however while phosphorylation of I-1 and DARP32 elicits the inhibition of PP1 activity, phosphorylation of I-2 and NIPP-1 has a positive effect on PP1 activity.

A 'gating' mechanism for PPI regulation has been proposed in which phosphorylation and dephosphorylation of an intermediary protein control the activity of downstream phosphatases (Winder and Sweatt 2001). I–1 and DARP32 are substrates of PKA and the specifically phosphorylated sites of I–1 (Thr35) and DARP32 (Thr34) are selectively dephosphorylated by PP2B (Greengard *et al.* 1998).

Thus, signals that activate the cAMP pathway would phosphorylate and activate DARP32 whereas signals that activate calcium pathways would dephosphorylate and inactivate DARP-32. Changes in DARP32 activity would then lead to altered activity of PP1 on target substrates. However, the situation is undoubtedly far more complicated since the type of second messenger and also the intensity, duration, and localization within the neurons may influence the balance between kinase and phosphatase activity.

CaMKII is among the relevant substrates of PP1 (and/or PP2A) activity. Based on *in vitro* studies, it has been proposed that small increases of intracellular calcium would preferentially activate calcineurin and induce PP1-mediated CaMKII dephosphorylation. However, strong calcium stimuli would activate not only calcineurin and CaMKII but also PKA causing I-1 mediated inhibition of PP1 activity on CaMKII (Greengard *et al.* 1998).

From the data available it appears that a dynamic interplay between phosphatases and kinases might set the thresholds that determine whether a given neuronal input triggers a long-lasting neuronal change. However, further information is required to fully understand this interplay.

Acknowledgments

We thank Marina Stone for the help in the preparation of the manuscript. The writing of this review was supported by MURST and CNR.

References

Adams, J. P., Anderson, A. E., Varga, A. W., Dineley, K. T., Cook, R. G., Pfaffinger, P.J ., *et al.* (2000) The A-type potassium channel Kv4.2 is a substrate for the mitogen-activated protein kinase ERK. *J. Neurochem.*, **75**, 2277–87.

Ahmed, B. Y., Yamaguchi, F., Tsumura, T., Gotoh, T., Sugimoto, K., Tai, Y., *et al.* (2000) Expression and subcellular localization of multifunctional calmodulin-dependent protein kinases–I, -II and -IV are altered in rat hippocampal CA1 neurons after induction of long-term potentiation. *Neurosci. Lett.*, **290**, 149–53.

Amieux, P. S., Cummings, D. E., Motamed, K., Brandon, E. P., Wailes, L. A., Le, K., *et al.* (1997) Compensatory regulation of RIalpha protein levels in protein kinase A mutant mice. *J. Biol. Chem.*, **272**, 3993–8.

Baldassa, S., Zippel, R., and Sturani, E. (2003) Depolarization-induced signaling to Ras, Rap1 and MAPKs in cortical neurons. *Mol. Brain Res.*, in press.

Bayer, K. U., De Koninck, P., Leonard, A. S., Hell, J. W., and Schulman, H. (2001) Interaction with the NMDA receptor locks CaMKII in an active conformation. *Nature*, **411**, 801–5.

Bito, H., Deisseroth K, and Tsien, R. W. (1996) CREB phosphorylation and dephosphorylation: a Ca(2+)- and stimulus duration-dependent switch for hippocampal gene expression. *Cell*, **87**, 1203–14.

Bito, H. (1998) The role of calcium in activity-dependent neuronal gene regulation. *Cell Calcium*, **23**, 143–50.

Brandon, E. P., Idzerda, R. L., and McKnight, G. S. (1997) PKA isoforms, neural pathways and behaviour: making the connection. *Curr. Opin. Neurobiol.*, **7**, 397–403.

Cadd, G., and McKnight, G. S. (1989) Distinct patterns of cAMP-dependent protein kinase gene expression in mouse brain. *Neuron*, **3**, 71–9.

Chawla, S. and Bading, H. (2001) CREB/CBP and SRE-interacting transcriptional regulators are fast on-off switches: duration of calcium transients specifies the magnitude of transcriptional responses. *J. Neurochem.*, **79**, 849–58.

Chen, H. J., Rojas-Soto, M., Oguni, A., and Kennedy, M. B. (1998) A synaptic Ras-GTPase activating protein (p135 SynGAP) inhibited by CaM kinase II. *Neuron*, **20**, 895–904.

Chen, H. J., Rojas-Soto, M., Oguni, A., and Kennedy, M. B. (2002) Erratum. *Neuron*, **33**, 151.

Cho, Y. H., Giese, K. P., Tanila, H., Silva, A. J., and Eichenbaum, H. (1998) Abnormal hippocampal spatial representations in alphaCaMKIIT286A and CREBalphaDelta-mice. *Science*, **279**, 867–9.

Chung, H. J., Xia, J., Scannevin, R. H., Zhang, X., and Huganir, R. L. (2000) Phosphorylation of the AMPA receptor subunit GluR2 differentially regulates its interaction with PDZ domain-containing proteins. *J. Neurosci.*, **20**, 7258–67.

Cobb, M. H. and Goldsmith, E. J. (1995) How MAP kinases are regulated. *J. Biol. Chem.*, **270**, 14843–6.

Colledge, M. and Scott, J. D. (1999) AKAPs: from structure to function. *Trends Cell Biol.*, **9**, 216–21.

Corcoran, E. E. and Means, A. R. (2001) Defining Ca^{2+}/calmodulin-dependent protein kinase cascades in transcriptional regulation. *J. Biol. Chem.*, **276**, 2975–8.

De Koninck, P. and Schulman, H. (1998) Sensitivity of CaM kinase II to the frequency of Ca^{2+} oscillations. *Science*, **279**, 227–30.

de Rooij, J., Zwartkruis, F. J., Verheijen, M. H., Cool, R. H., Nijman, S. M., Wittinghofer, A., *et al.* (1998) Epac is a Rap1 guanine-nucleotide-exchange factor directly activated by cyclic AMP. *Nature*, **396**, 474–7.

de Rooij, J., Boenink, N. M., van Triest, M., Cool, R. H., Wittinghofer, A., and Bos, J. L. (1999) PDZ-GEF1, a guanine nucleotide exchange factor specific for Rap1 and Rap2. *J. Biol. Chem.*, **274**, 38125–30.

de Rooij, J., Rehmann, H., van Triest, M., Cool, R. H., Wittinghofer, A., and Bos, J. L. (2000) Mechanism of regulation of the Epac family of cAMP-dependent RapGEFs. *J. Biol. Chem.*, **275**, 20829–36.

Deisseroth, K., Heist, E. K., and Tsien, R. W. (1998) Translocation of calmodulin to the nucleus supports CREB phosphorylation in hippocampal neurons. *Nature*, **392**, 198–202.

Derkach, V., Barria, A., and Soderling, T. R. (1999) Ca^{2+}/calmodulin-kinase II enhances channel conductance of alpha-amino-3-hydroxy-5-methyl-4-isoxazolepropionate type glutamate receptors. *Proc. Natl. Acad. Sci. USA*, **96**, 3269–74.

Dineley, K. T., Weeber, E. J., Atkins, C., Adams, J. P., Anderson, A. E., and Sweatt, J. D. (2001) Leitmotifs in the biochemistry of LTP induction: amplification, integration, and coordination. *J. Neurochem.*, **77**, 961–71.

Ebinu, J. O., Bottorff, D. A., Chan, E. Y., Stang, S. L., Dunn, R. J., and Stone, J. C. (1998) RasGRP, a Ras guanyl nucleotide- releasing protein with calcium- and diacylglycerol-binding motifs. *Science*, **280**, 1082–6.

Edelman, A. M., Blumenthal, D. K., and Krebs, E. G. (1987) Protein serine/threonine kinases. *Annu. Rev. Biochem.*, **56**, 567–613.

English, J., Pearson, G., Wilsbacher, J., Swantek, J., Karandikar, M., Xu, S., *et al.* (1999) New insights into the control of MAP kinase pathways. *Exp. Cell Res.*, **253**, 255–70.

Feliciello, A., Gottesman, M. E., and Avvedimento, E. V. (2001) The biological functions of A-kinase anchor proteins. *J. Mol. Biol.*, **308**, 99–114.

Fukunaga, K. and Miyamoto, E. (2000) A working model of CaM kinase II activity in hippocampal long-term potentiation and memory. *Neurosci. Res.*, **38**, 3–17.

Gardoni, F., Schrama, L. H., van Dalen, J. J., Gispen, W. H., Cattabeni, F., and Di Luca, M. (1999) AlphaCaMKII binding to the C-terminal tail of NMDA receptor subunit NR2A and its modulation by autophosphorylation. *FEBS Lett.*, **456**, 394–8.

Giese, K. P., Fedorov, N. B., Filipkowski, R. K., and Silva, A. J. (1998) Autophosphorylation at Thr286 of the alpha calcium-calmodulin kinase II in LTP and learning. *Science*, **279**, 870–3.

Giovannardi, S., Forlani, G., Balestrini, M., Bossi, E., Tonini, R., Sturani, E., *et al.* (2002) Modulation of the inward rectifier potassium channel IRK1 by the Ras signaling pathway. *J. Biol. Chem.*, **277**, 12158–63.

Greengard, P., Valtorta, F., Czernik, A. J., and Benfenati, F. (1993) Synaptic vesicle phosphoproteins and regulation of synaptic function. *Science*, **259**, 780–5.

Greengard, P., Nairn, A. C., Girault, J. A., Quimet, C. C., Snyder, G. L., Fisone, G., *et al.* (1998) The DARPP-32/protein phosphatase-1 cascade: a model for signal integration. *Brain Res. Brain Res. Rev.*, **26**, 274–84.

Grewal, S. S., Horgan, A. M., York, R. D., Withers, G. S., Banker, G. A., and Stork, P. J. (2000) Neuronal calcium activates a Rap1 and B-Raf signaling pathway via the cyclic adenosine monophosphate-dependent protein kinase. *J. Biol. Chem.*, **275**, 3722–8.

Hanson, P. I. and Schulman, H. (1992) Neuronal Ca^{2+}/calmodulin-dependent protein kinases. *Annu. Rev. Biochem.*, **61**, 559–601.

Hemmings Jr, H. C., Nairn, A. C., McGuinness, T. L., Huganir, R. L., and Greengard, P. (1989) Role of protein phosphorylation in neuronal signal transduction. *Faseb J.*, **3**, 1583–92.

Hunter, T. (1995) Protein kinases, and phosphatases: the yin, and yang of protein phosphorylation, and signaling. *Cell*, **80**, 225–36.

Johnson, L. N., Noble, M. E., and Owen, D. J. (1996) Active and inactive protein kinases: structural basis for regulation. *Cell*, **85**, 149–58.

Kawasaki, H., Springett, G. M., Toki, S., Canales, J. J., Harlan, P., Blumenstiel, J. P., *et al.* (1998) A Rap guanine nucleotide exchange factor enriched highly in the basal ganglia. *Proc. Natl. Acad. Sci. USA*, **95**, 13278–83.

Kim, J. H., Liao, D., Lau, L. F., and Huganir, R. L. (1998) SynGAP: a synaptic RasGAP that associates with the PSD-95/SAP90 protein family. *Neuron*, **20**, 683–91.

Knighton, D. R., Zheng, J. H., Ten Eyck, L. F., Ashford, V. A., Xuong, N. H., Taylor, S. S., *et al.* (1991) Crystal structure of the catalytic subunit of cyclic adenosine monophosphate-dependent protein kinase. *Science*, **253**, 407–14.

Le Good, J. A., Ziegler, W. H., Parekh, D. B., Alessi, D. R., Cohen, P., and Parker, P. J. (1998) Protein kinase C isotypes controlled by phosphoinositide 3-kinase through the protein kinase PDK1. *Science*, **281**, 2042–5.

Lee, H. K., Barbarosie, M., Kameyama, K., Bear, M. F., and Huganir, R. L. (2000) Regulation of distinct AMPA receptor phosphorylation sites during bidirectional synaptic plasticity. *Nature*, **405**, 955–9.

Mazzucchelli, C. and Brambilla, R. (2000) Ras-related, and MAPK signalling in neuronal plasticity and memory formation. *Cell Mol. Life Sci.*., **57**, 604–11.

Mazzucchelli, C., Vantaggiato, C., Ciamei, A., Fasano, S., Pakhotin, P., Krezel, W., *et al.* (2002) Knockout of ERK1 MAP kinase enhances synaptic plasticity in the striatum and facilitates striatal-mediated learning and memory. *Neuron*, **34,** 807–20.

Mellor, H. and Parker, P. J. (1998) The extended protein kinase C superfamily. *Biochem. J.*, **332(Pt 2)**, 281–92.

Morice, C., Nothias, F., Konig, S., Vernier, P., Baccarini, M., Vincent, J. D., *et al.* (1999) Raf-1 and B-Raf proteins have similar regional distributions but differential subcellular localization in adult rat brain. *Eur. J. Neurosci.*, **11**, 1995–2006.

Morrison, D. K. and Cutler, R. E. (1997) The complexity of Raf-1 regulation. *Curr. Opin. Cell Biol.*, **9**, 174–9.

Nestler, E. J. and Greengard, P. (1984) Neuron-specific phosphoproteins in mammalian brain. *Adv. Cyclic Nucleotide Protein Phosphorylation Res.*, **17**, 483–8.

Newton, A. C. (1997) Regulation of protein kinase C. *Curr. Opin. Cell Biol.*, **9**, 161–7.

Ortiz, J., Harris, H. W., Guitart, X., Terwilliger, R. Z., Haycock, J. W., and Nestler, E. J. (1995) Extracellular signal-regulated protein kinases (ERKs) and ERK kinase (MEK) in brain: regional distribution and regulation by chronic morphine. *J. Neurosci.*, **15**, 1285–97.

Parker, P. J., Coussens, L., Totty, N., Rhee, L., Young, S., Chen, E., *et al.* (1986) The complete primary structure of protein kinase C — the major phorbol ester receptor. *Science*, **233**, 853–9.

Pham, N., Cheglakov, I., Koch, C. A., de Hoog, C. L., Moran, M. F., and Rotin, D. (2000) The guanine nucleotide exchange factor CNrasGEF activates ras in response to cAMP and cGMP. *Curr. Biol.*, **10**, 555–8.

Poser, S. and Storm, D. R. (2001) Role of Ca^{2+}-stimulated adenylyl cyclases in LTP and memory formation. *Int. J. Dev. Neurosci.*, **19**, 387–94.

Ron, D. and Kazanietz, M. G. (1999) New insights into the regulation of protein kinase C and novel phorbol ester receptors. *Faseb J.*, **13**, 1658–76.

Rubin, C. S. (1994) A kinase anchor proteins and the intracellular targeting of signals carried by cyclic AMP. *Biochim. Biophys. Acta*, **1224**, 467–79.

Schacher, S., Glanzman, D., Barzilai, A., Dash, P., Grant, S. G., Keller, F., *et al.* (1990) Long-term facilitation in Aplysia: persistent phosphorylation and structural changes. *Cold Spring Harb. Symp. Quant. Biol.*, **55,** 187–202.

Schulman, H., Hanson, P. I., and Meyer, T. (1992) Decoding calcium signals by multifunctional CaM kinase. *Cell Calcium*, **13**, 401–11.

Shen, K. and Meyer, T. (1999) Dynamic control of CaMKII translocation and localization in hippocampal neurons by NMDA receptor stimulation. *Science*, **284**, 162–6.

Sim, A.T. and Scott, J. D. (1999) Targeting of PKA, PKC and protein phosphatases to cellular microdomains. *Cell Calcium*, **26**, 209–17.

Singla, S. I., Hudmon, A., Goldberg, J. M., Smith, J. L., and Schulman, H. (2001) Molecular characterization of calmodulin trapping by calcium/calmodulin–dependent protein kinase II. *J. Biol. Chem.*, **276**, 29353–60.

Soderling, T. R. (1999) The Ca-calmodulin-dependent protein kinase cascade. *Trends Biochem. Sci.*, **24**, 232–6.

Sweatt, J. D., Atkins, C. M., Johnson, J., English, J. D., Roberson, E. D., Chen, S. J., *et al.* (1998) Protected-site phosphorylation of protein kinase C in hippocampal long-term potentiation. *J. Neurochem.*, **71**, 1075–85.

Sweatt, J. D. (2001*a*) The neuronal MAP kinase cascade: a biochemical signal integration system subserving synaptic plasticity and memory. *J. Neurochem.*, **76**, 1–10.

Sweatt, J. D. (2001*b*) Memory mechanisms: the yin and yang of protein phosphorylation. *Curr. Biol.*, **11**, R391–4.

Taylor, S. S., Knighton, D. R., Zheng, J., Ten Eyck, L. F., and Sowadski, J. M. (1992) Structural framework for the protein kinase family. *Annu. Rev. Cell Biol.*, **8**, 429–62.

Winder, D. G. and Sweatt, J. D. (2001) Roles of serine/threonine phosphatases in hippocampal synaptic plasticity. *Nat. Rev. Neurosci.*, **2**, 461–74.

Wischmeyer, E., Doring, F., and Karschin, A. (1998) Acute suppression of inwardly rectifying Kir2.1 channels by direct tyrosine kinase phosphorylation. *J. Biol. Chem.*, **273**, 34063–8.

Xia, J., Zhang, X., Staudinger, J., and Huganir, R. L. (1999) Clustering of AMPA receptors by the synaptic PDZ domain-containing protein PICK1. *Neuron*, **22**, 179–87.

Xing, J., Ginty, D. D., and Greenberg, M. E. (1996) Coupling of the RAS-MAPK pathway to gene activation by RSK2, a growth factor-regulated CREB kinase. *Science*, **273**, 959–63.

Yang, F., Feng, L., Zheng, F., Johnson, S.W., Du, J., Shen, L., *et al.* (2001) GDNF acutely modulates excitability and A-type K(+) channels in midbrain dopaminergic neurons. *Nat. Neurosci.*, **4**, 1071–8.

Ye, B., Liao, D., Zhang, X., Zhang, P., Dong, H., and Huganir, R. L. (2000) GRASP-1: a neuronal RasGEF associated with the AMPA receptor/GRIP complex. *Neuron*, **26**, 603–17.

Zippel, R., Balestrini, M., Lomazzi, M., and Sturani, E. (2000) Calcium and calmodulin are essential for Ras-GRF1-mediated activation of the Ras pathway by lysophosphatidic acid. *Exp. Cell Res.*, **258**, 403–8.

Chapter 13

The cytoskeleton

Javier Díaz-Nido and Jesús Avila

13.1 Introduction

The intricate circuitry of the nervous system is built of billions of neurons, each of which has a complex morphology with numerous long processes that branch and interconnect through synaptic junctions. In this way, the highly asymmetrical shape of nerve cells is crucial for the functioning of brain.

At the beginning of the 20th century, neuroanatomists interested in the generation and maintenance of neuronal morphologies observed a 'neurofibril network' which arose in the cell body and extended into the axon and dendrites (Ramón y Cajal 1928). Electron microscopy later showed that 'neurofibrils' corresponded to bundles of cytoskeletal fibers similar to those found in all eukaryotic cells (Peters *et al.* 1976). Within the last 25 years, research has progressed impressively in the analysis of cytoskeletal components through the application of immunochemical, biochemical, and molecular genetic techniques. This has dramatically improved our knowledge of the cytoskeleton (Burgoyne 1991; Kreis and Vale 1993). In particular, the availability of complete genome sequences from a variety of organisms, and the recent technical advances to analyse the functions of many genes in parallel using *in vivo* or cell-based assays, are rapidly changing the field in a rather profound way (Martin and Drubin 2003).

The cytoskeleton consists of three types of filament structures: microfilaments, intermediate filaments, and microtubules. These fibrous structures are assembled from the polymerisation of certain protein subunits. A larger number of additional proteins associate with these filaments, modulating their structural stability and mediating their interaction with other cellular components. The dynamics of the assembly and disassembly of these protein polymers as well as their interactions with other cellular organelles and molecules provide the basis for the understanding of the physiological roles played by the cytoskeleton.

As the word 'cytoskeleton' implies, the skeleton of the cell is firstly required to sustain cell shape. The extraordinary morphologies of neurons consequently demand a highly developed cytoskeleton. Indeed, most cytoskeletal components are more abundant in neurons than in any other cell type. Furthermore, some proteins associated with the cytoskeleton appear to be specific for neurons.

Neuronal morphogenesis may thus be viewed as a process in which the relatively simple cytoskeleton of an undifferentiated neuroblast is progressively converted through a series of rearrangements into the complex cytoskeleton of a mature neuron. In this process, both changes in gene expression and post-translational modifications of cytoskeletal proteins take place.

Interestingly, the very sophisticated shapes of neurons are not unalterable during the entire life of an organism but they exhibit a noteworthy plasticity. Thus, neurons may undergo morphologic changes in response to their synaptic input, providing the nervous system with a flexibility of neuronal connectivity which might contribute to learning and memory mechanisms. Other neuron shape variation arise as the consequence of injury or denervation. Obviously, all these modifications in neuronal morphology are carried out through changes in the cytoskeleton.

With the availability of new and improved microscopy techniques that allow the imaging of cytoskeletal proteins within living cells, there has been a change from looking at the cytoskeleton as a rigid structure to seeing it as a dynamic network. Thus, the cytoskeletal system can grow and shrink within distinct cellular microcompartments as a result of the action of specific molecular regulatory elements. Furthermore, the entire cytoskeletal network is always mechanically tensed as a result of contractile forces which are generated by motor proteins. Thus, both changing the level of the tension in the cytoskeleton and modifying specific cytoskeletal proteins may significantly alter the morphology and physiology of cells. The regulation of proteins associated with the cytoskeleton by protein kinases and phosphatases activated in response to certain extracellular signals is particularly relevant in this respect.

In addition to its role in the development, maintenance and modification of neuronal morphology, the cytoskeleton is essential in the intracellular organization of the neuron. Thus, the sorting, distribution, transport, and anchoring of most cellular organelles depends on their interactions with cytoskeletal components. However, this does not merely constitute a passive support system, since the cytoskeletal scaffold is able to orient many of the enzymes and substrates that mediate critical cellular functions, including signal transduction, protein synthesis, transport, secretion, and turnover.

Finally, the significance of the cytoskeleton in neuronal physiology is highlighted by the neuropathological effects of agents that disturb the cytoskeleton. This may be the cause of several toxic neuropathies. Moreover, cytoskeletal abnormalities are also found in several naturally occurring neurodegenerative disorders.

This chapter summarizes the main features of the three major cytoskeletal polymers and reviews some aspects of their contribution to neuronal morphogenesis and plasticity, intraneuronal transport and neuropathology.

13.2 Components of the neuronal cytoskeleton

13.2.1 Microfilaments

Microfilaments are produced by the polymerization of a 43 kDa globular protein called actin. Two actin isoforms, β and γ actin, have been identified in neurons (Choo and Bray

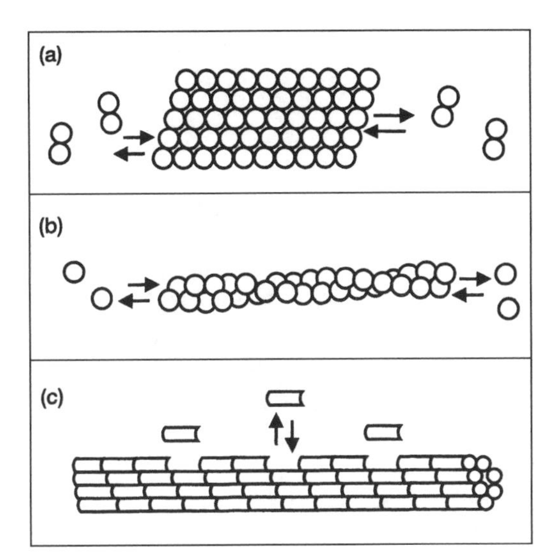

Fig. 13.1 Dynamics of cytoskeletal polymers. Microtubule (a) dynamics occur through the incorporation and release of tubulin heterodimers at the ends of the polymer. Microfilament (b) dynamics are also associated with the exchange of actin monomers at the polymer ends. Intermediate filament (c) dynamics mainly involve the lateral replacement of subunits. Reproduced from Atkinson *et al.* (1992) with permission from Current Biology Ltd.

1978). Actin is referred to as G-actin in its soluble form and once polymerized into microfilaments it is named F-actin. Actin monomers are arranged like two intertwined strings of beads giving a double helical filament of about 6 nm in diameter (see Fig. 13.1). Actin monomers are asymmetric and associate in a particular orientation. This results in the formation of polar microfilaments in which the two ends (the 'barbed' or 'plus' end and the 'pointed' or 'minus' end) are different. Actin is an ATP-binding protein that requires ATP in order to polymerize. Polymerized actin hydrolyses ATP rather slowly. One end of the microfilament contains ATP-actin and can incorporate new ATP-actin subunits. The other end contains ADP-actin, which can dissociate from the polymer. Thus, polymerization of ATP-actin occurs preferentially in one end of the microfilaments whereas depolymerization of ADP-actin takes place in the other. This dynamics of actin microfilaments is usually referred to as 'treadmilling' (Wegner 1985).

Actin microfilaments are found throughout the neuronal cytoplasm. Oligomers of actin are quite abundant immediately beneath the plasma membrane where they constitute the membrane skeleton (Luna and Hitt 1992). Actin polymers are mainly enriched in presynaptic terminals and in dendritic spines and postsynaptic densities (Ratner and Mahler 1983; Hirokawa *et al.* 1989; Fifkova and Morales 1992). Within axons, short filaments are associated with microtubules (Fath and Lasek 1988) and are also concentrated in the cortical region under the axonal plasma membrane (Tsukita *et al.* 1986). In developing neurons, long microfilaments are present within filopodia of nerve growth cones (Gordon-Weeks 1987; Smith 1988; Lewis and Bridgman 1992).

The variety of actin arrangements in different cellular locations as well as their dynamics are controlled by a number of actin-binding proteins (ABPs) (Dos Remedios *et al.* 2003). Table 13.1 summarizes the major classes of ABPs.

Table 13.1 Major actin-binding proteins

Protein	Properties
ADF/Cofilin	Sequesters G-actin. Severs F-actin. Abundant in growth cones and presynaptic terminals
Gelsolin	Severs and caps F-actin. Modulated by Ca^{2+} and PIP2. Present in growth cones
Profilin	Sequesters G-actin. Catalyses actin nucleotide exchange. Modulated by PIP_2. Abundant in growth cones.
WASP/SCAR	Activates Arp 2/3 complex. Promotes actin filament nucleation. Present in growth cones
Arp 2/3	Interacts with WASP/SCAR. Controls actin filament nucleation and branching. Present in growth cones
ENA/VASP	'Anti-capping' protein. Nucleates actin polymerization. Bundles and crosslinks actin filaments. Binds profilin. Present in growth cones
F-actin stabilizing proteins	
Tropomyosin	Dimer. Present both in dendrites and axons (including growth cones)
F-actin bundling and crosslinking proteins	
α-Actinin	Dimer. Modulated by Ca^{2+}. Binds vinculin, integrin and catenin. Present in growth cones
Amelin	Bundles F-actin and enhances spectrin binding. Abundant in soma and dendrites of mature neurons
Dystrophin	Dimer. Similar to spectrin
Filamin	Dimer. Crosslinks F-actin into networks. Present in growth cones.
Fimbrin	Bundles F-actin. Present in growth cones
MARCKS family	Crosslinks F-actin. Modulated by calmodulin binding and PKC phosphorylation
Spectrin ($\alpha\beta$)	Heterodimer and tetramer. Crosslinks F-actin into networks. Modulated by Ca^{2+}/calmodulin. Abundant in soma and dendrites of mature neurons. Enriched in dendritic spines
Fodrin/Spectrin ($\alpha\gamma$)	Homologous to spectrin. Present in growth cones. Abundant in axons
Synapsin I	Bundles F-actin and enhances spectrin binding. Associated with synaptic vesicles. Abundant in presynaptic terminals
Utrophin	Dimer. Similar to spectrin
Proteins anchoring F-actin to membranes	
Ankyrin	Anchors spectrin to membrane proteins
Catenin	Binds actin, α-actinin and cadherin
Talin	Binds vinculin and integrin. Present in growth cones
Vinculin	Binds actin, talin and α-actinin. Present in growth cones

Table 13.1 (continued) Major actin-binding proteins

Protein	Properties
Actin-activated ATPases	
Myosin I	
Myosin II	
Myosin V	

ADF, Actin depolymerizing factor; MARCKS, myristoylated alanine-rich C kinase substrate; PIP2, phosphatidylinositol-4, 5-bisphosphate; PKC, protein kinase C.

A class of ABPs regulates the assembly of actin by distinct mechanisms including the catalysis of actin ATP/ADP exchange, the sequestering of G-actin and the severing or capping of F-actin. Interestingly, the activities of some of these ABPs may be modulated by Ca^{2+} or phosphatidylinositol-4, 5-biphosphate. Thus, actin dynamics and organization may be modified in response to extracellular signals that induce changes in intracellular Ca^{2+} concentration or stimulate phosphatidylinositol biphosphate hydrolysis. As a case in point, receptors with tyrosine kinase activity phosphorylate phospholipase C-γ that is stimulated to hydrolyse phosphatidylinositol biphosphate tightly bound to profilin. As a consequence of phosphoinositide hydrolysis, profilin is released to the cytoplasm where it promotes actin polymerization (Aderem 1992).

The Arp 2/3 complex is composed of seven proteins and controls actin filament nucleation and branching. The ability of Arp 2/3 to initiate actin filament branching depends on its interaction with proteins of the WASP/SCAR family like N-WASP (Higgs and Pollard 2001). Triggered by the activation of small GTPases (like Rac or Cdc 42), and the binding of phosphatidylinositol-4, 5-biphosphate, proteins of the WASP family promote the nucleation of new actin filaments and the extension of existing filaments. Interestingly, the activation of N-WASP, which seems crucial for neurite outgrowth, is also regulated by protein tyrosine kinases (Suetsugu *et al.* 2002).

Profilin-binding protein of the VASP/Ena family are also key regulators of actin polymerization. VASP/Ena proteins associates with the barbed end of actin filaments, thereby preventing the binding of capping proteins and causing the formation of longer actin filaments (Krause *et al.* 2002). These 'anti-capping' proteins are abundant in growth cones and seem crucial for axonal growth. It is thought that VASP/Ena proteins cooperate with profilin to modulate changes in actin dynamics within growth cones in response to the extracellular signals that guide axons during development. Of relevance in this respect is the modulation of actin dynamics by the phosphorylation of VASP/Ena proteins (Loureiro *et al.* 2002).

The F-actin stabilizing proteins bind along actin filaments, blocking their interaction with other ABPs. For instance, F-actin bound to tropomyosin is resistant to the severing action of ADF (actin depolymerizing factor).

Other classes of ABPs are able to bundle and crosslink F-actin. For instance, fimbrin can make F-actin form bundles of uniform polarity such as those found within filopodia of growth cones.

The best-characterized F-actin crosslinking proteins are spectrins (Bennett and Baines 2001). These proteins crosslink actin oligomers, forming a network which is attached to the plasma membrane through the interaction of spectrin with ankyrin molecules. Spectrin can also bind directly to certain membrane proteins. Spectrin–actin complexes are the predominant form of the cytoskeleton immediately underneath the plasma membrane. Spectrin binding to actin can be modulated by Ca^{2+}/calmodulin. There are two major forms of spectrin in neurons: αβ heterodimers, which are mainly localized in dendrites (including dendritic spines) and αγ heterodimers, which are referred to as fodrin, and are predominantly localized in axons (Riederer *et al.* 1986). Fodrin appears first in developing neurons, whereas αβ spectrin is mainly expressed in mature neurons (Riederer *et al.* 1987). Other proteins somewhat similar to spectrins include filamin, dystrophin, and utrophin.

The MARCKS family of proteins, which are also referred to as GAP43-like proteins, are characterized by the presence of a basic domain that binds acidic phospholipids including phosphatidylinositol biphosphate, calcium/calmodulin, actin filaments and PKC in a mutually exclusive manner. These proteins include MARCKS (myristoylated alanine-rich C kinase substrate), GAP 43 (growth-associated protein 43, also known as B50, neuromodulin or F1) and CAP23 (Frey *et al.* 2000). In particular, GAP43 and CAP23 are abundant in axonal growth cones and are believed to be crucial for axonal growth and regeneration (Bomze *et al.* 2001).

Synapsins (I and II) are actin-bundling proteins. Synapsin I is very abundant in presynaptic terminals, and *in vitro* is able to bundle actin filaments, enhance the interaction of spectrin with actin and bind to synaptic vesicles. These activities are inhibited after synapsin phosphorylation by Ca^{2+}/calmodulin-dependent kinase. Synapsin I seems to favor the association of synaptic vesicles with the actin-rich cytoskeleton of the nerve terminal (see Chapter 6). After presynaptic membrane depolymerization, synapsin I becomes phosphorylated and synaptic vesicles are released from actin filaments. These synaptic vesicles are then free to move to the plasma membrane where they can undergo exocytosis to release their neurotransmitter content (Llinas *et al.* 1985). Additionally, synapsins also play significant roles during neuronal development (Ferreira and Rapoport 2002).

An important group of ABPs consists of proteins which anchor F-actin to membranes. These include ankyrins, vinculin, talin, and catenin. Spectrins, MARCKs, synapsins, and α-actinin have been mentioned as F-actin crosslinking and bundling proteins but could also be included in this group.

Ankyrins are protein linkers between integral membrane proteins and the spectrin-based cytoskeleton, playing important roles both in signal transduction and in the assembly of specialized membrane domains (Bennett and Chen 2001; Rubtsov and Lopina 2000). Ankyrins are the products of three genes: ANK 1 (ankyrin-R), ANK 2 (ankyrin-B) and ANK 3 (ankyrin-G), but multiple isoforms arise from the alternative splicing of the primary transcripts. There are two ankyrin-R isoforms that are generated by alternative splicing of the transcript of the ANK-1 gene. These proteins are present in the plasma membrane of cell bodies and dendrites of mature neurons. Two ankyrin-B isoforms arise from alternative splicing of the transcript of the ANK-2 gene. One 440 kDa isoforms appears in axons early during brain development and the other 220 kDa isoforms is mainly localized to neuronal cell bodies and dendrites in the adult brain. 440 kDa ankyrin-B is associated with the L1 CAM family of cell adhesion molecules and is essential for the survival of premyelinazed axons, at least in the case of the optic nerve. 480/270 kDa ankyrin-G isoforms are concentrated at axon initial segments and nodes of Ranvier in myelinazed axons. Ankyrin-G seems to associate with voltage-gated sodium channels as well as neurofascin and NrCAM cell adhesion molecules (Bennett and Lambert 1999).

Vinculin, talin, and α-actinin form a complex that anchors F-actin to integrins. Integrins are transmembrane proteins that function as receptors for extracellular matrix proteins. These actin-binding protein complexes also bind signalling proteins that are implicated in the signal transduction pathways responsible for integrin-induced changes in cell behavior (Clark and Brugge 1995).

Catenins are involved in the association of F-actin with the cytoplasmic domains of the Ca^{2+}-dependent cell adhesion molecules referred to as cadherins.

Finally, some ABPs are actin-activated ATPases that transduce the chemical energy of ATP into mechanical energy. These actin-dependent motor proteins are involved in growth cone motility (see Section 13.3.1) and in the intracellular transport of organelles along actin filaments (see Section 13.5.4.3).

Summarizing, a large variety of ABPs may modulate actin assembly and organization. Moreover, ABPs are regulated by second messengers, phospholipids, protein kinases, and other signalling proteins responding to specific extracellular signals. Thus, the actin cytoskeleton is tightly connected to signal transduction pathways (Meyer and Feldman 2002).

13.2.2 **Intermediate filaments**

Intermediate filament proteins (IFP) are a multigene family of polypeptides able to polymerize into filaments about 10 nm in diameter (Herrmann and Aebi 2000). These proteins share a similar three-domain molecular structure: a variable amino-terminal 'head' domain, a relatively conserved central 'rod' domain, and a variable carboxyl-terminal 'tail' domain. The 'rod' domain contains heptades of hydrophobic amino acids with a tendency to adopt a two-chain coiled-coil α-helical configuration. Based on their amino acid sequence homologies, IFP are catalogued into six classes as shown in Table 13.2.

Interestingly, these proteins show a cell type-specific expression. At different stages of differentiation, neuronal cells express distinct IFP. Neural stem cells of the central nervous system express nestin (a class VI IFP). Before differentiation, neuroblasts and neurons also express vimentin (a class III IFP), which is gradually replaced by neurofilaments.

Neurofilament triplet proteins (NF-L, NF-M, and NF-H, class IV IFP) are expressed by differentiating and mature neurons. Most neurons express relatively low levels of

Table 13.2 Intermediate filament proteins

Class	Proteins	Mass (kDa)	Distribution
I	Acidic cytokeratins	40–64	Epithelial cells
II	Basic cytokeratins	52–68	Epithelial cells
III	Vimentin	55	Mesenchymal cells Immature neuronal and glial cells
	Desmin	53	Muscle cells
	GFAP	51	Astroglia cells
	Peripherin	57	PNS neurons
IV	NF-L	68	Neurons
	NF-M	145	Neurons
	NF-H	200	Neurons
	α-internexin (NF-/66)	66	CNS neurons
V	Lamins	62–72	All cells
VI	Nestin	240	CNS neural stem cells

CNS, central nervous system; GFAP, glial fibrillary acidic protein; NF-L, neurofilament, low molecular weight; NF-M, neurofilament, middle molecular weight; NF-H, neurofilament, heavy molecular weight; PNS, peripheral nervous system.

NF-L and NF-M at the time of neurite outgrowth. After neuronal maturation, NF-H is expressed and a parallel up-regulation of NF-L and NF-M occurs. Neurofilaments are much more abundant in axons than in dendrites. In fact, neurofilaments are the most prominent cytoskeletal elements within large myelinated axons. Often the number of neurofilaments is an order of magnitude higher than the number of microtubules. Neurofilament numbers may determine the diameter of large myelinated axons and, consequently, the speed of nervous impulse conduction (Julien 1999). Ultrastructurally, neurofilaments consist of a core filament assembled from NF-L, with NF-M and NF-H co-assembling onto the core backbone. The 'tail' domains of NF-M and NF-H extend away from the filament surface (Nixon and Sihag 1991; Nixon and Shea 1992).

Some developing neurons in the central nervous system also express α-internexin (NF-66, a class IV IFP), which persists within small calibre axons in the adult brain. α-Internexin is also present within dendrites and dendritic spines within hippocampal neurons. Peripherin (a class III IFP) is expressed by some developing neurons of the peripheral nervous system and is down-regulated during maturation, although it is maintained at low levels in certain neurons (Nixon and Shea 1992).

Intermediate filaments are far more stable than microfilaments or microtubules. However, they exhibit a certain degree of dynamics which generally involves lateral exchanges of subunits (see Fig. 13.1). Class III intermediate filaments are more dynamic than class IV IFP. The transition from class III to class IV IFP during neuronal development therefore implies a stabilization of the cytoskeleton. The persistence of class III IFP in certain neurons may be correlated with their higher plasticity. For instance, axons from adult olfactory sensory neurons which are able to regenerate maintain the expression of vimentin as their major IF.

Dynamics of class III intermediate filaments is controlled by phosphorylation. Cyclic AMP-dependent protein kinase and PKC phosphorylate sites on the head domain of class III IFP impeding the polymerization of phosphorylated subunits and inducing the depolymerization of assembled filaments (Inagaki *et al.* 1989).

Several results indicate that neurofilaments are not entirely stable at least in immature growing axons. Thus, a slow lateral subunit exchange of NF-L has been observed in growing axons. This dynamic behavior declines after axonal maturation (Okabe *et al.* 1993).

Isolated NF-L proteins are able to polymerize *in vitro*. Subunit exchange is observed in these homopolymers. As for class III IFP, phosphorylation of the 'head' domain by cAMP-dependent protein kinase favors NF-L disassembly. However, homopolymerization of NF-L may not occur *in vivo*. Thus, NF-L is unable to form polymers in transfected cells lacking other intermediate filaments. NF-L can only polymerize if the cells express vimentin, NF-M or NF-H (Lee *et al.* 1993). In contrast, α-internexin (NF-66) is the only class IV IFP able to assemble in the absence of other intermediate filament proteins (Ching and Liem 1993). Accordingly, neurofilament triplet proteins are obligate heteropolymers *in vivo*. Perhaps NF-L/NF-M copolymers such as those found in growing axons still retain some dynamic behavior, whereas NF-L/NF-M/NF-H polymers such as those found in mature axons are almost completely stable.

Neurofilaments are extensively phosphorylated at sites on the 'tail' domains of NF-M and NF-H. Most of these sites correspond to repetitive Lys-Ser-Prof motifs and their phosphorylation may be catalysed by proline-directed protein kinases such as mitogen-activated protein kinases, stress-activated protein kinases, cyclin-dependent kinase 5, and glycogen synthase kinase 3. Phosphorylation mainly occurs when neurofilaments have entered the axon and continues throughout axonal transport. This gradual phosphorylation is associated with the slowing of neurofilament transport and with the integration of the neurofilaments into a stationary structure (see Section 13.5.2). The large number of phosphorylated residues on the tail domains of NF-M and NF-H possibly increases the electrostatic repulsion between filaments as a consequence of the huge number of negative charges. This extended conformation may be responsible for the increase in interfilament spacing and axonal calibre induced by myelination (see Section 13.3.2). Furthermore, phosphorylation protects neurofilaments against proteolysis by calcium-activated proteases like calpain. Thus, this type of neurofilament phosphorylation favors the stabilization of the axonal cytoskeleton (Grant and Pant 2000).

Curiously, mice with null mutations in neurofilament genes express subtle phenotypes but no gross abnormalities of the nervous system (Julien 1999). Mice with null mutations in the mid-sized neurofilament gene (NF-M) have axons with a diminished calibre but lack any overt behavioural phenotype (Elder *et al.* 1998). However, those animals suffer a progressive motor axonal atrophy in the lumbar ventral roots of the spinal cord after aging. Mice with null mutations in both NF-M and NF-H genes develop atrophy in ventral and dorsal roots as well as hindlimb paralysis with aging. This demonstrates that the presence of neurofilaments is crucial for the structural integrity of certain axons during aging (Elder *et al.* 1999).

13.2.3 Microtubules

Microtubules are hollow fibers with a diameter of 24 nm. They are formed by 13 longitudinal strands (protofilaments) arranged in a helical configuration. Each strand is composed of aligned globular heterodimers consisting of α-tubulin and β-tubulin subunits (Amos 2000). This leads to a polarized polymer with one end having exposed mainly α-subunits and the other end mainly β-subunits. As described for actin microfilaments, one end of the microtubule (the 'plus' end) polymerizes faster than the other (the 'minus' end) (Howard and Hyman 2003).

There are different genes coding for distinct isoforms of α- and β-tubulin. In mammals six genes for α- and six genes for β-subunits have been found. Some of these gene products are specific for neurons (Sullivan 1988). A γ-tubulin encoded by a different gene has also been described. This tubulin subunit is not found within microtubules but appears to be localized at the centrosome where it presumably functions as a seed for the initiation of microtubule polymerization (Oakley 1992).

Further tubulin isoforms are generated on assembled microtubules as the consequence of post-translational modifications. There is acetylation of α-tubulin, phosphorylation of a neuronal-specific β-tubulin, and polyglutamination of both α- and β-tubulin. These chemical modifications seem to occur preferentially on stable microtubules.

Tubulin is a GTP-binding protein. Soluble GTP-tubulin has the ability to polymerize into microtubules. Once tubulin is incorporated into the polymer, the GTP bound to β-tubulin is hydrolysed to GDP. When the proportion of GTP-tubulin at a microtubule end decreases below a threshold level, the microtubule starts to rapidly depolymerize. After this, there is the possibility that the microtubule may again incorporate GTP-tubulin and stop depolymerizing. This behavior of microtubules has been referred to as 'dynamic instability' and describes well the properties of polymers assembled from purified tubulin *in vitro* (Kirschner and Mitchison 1986). However, microtubules are less dynamic *in vivo* than they are *in vitro*. There are also differences in the behavior of microtubules in distinct cell types and at different stages of development. For instance, a progressive microtubule stabilization is attained during the maturation of axons and dendrites (Baas *et al.* 1991).

Microtubules are distributed as bundles throughout the neuronal cytoplasm, although their organization differs in axons and dendrites (see Section 13.3.3). Microtubules are crucial elements in the generation and maintenance of neuronal morphology. Thus, both axons and dendrites shrink back to the cell body after treatment of cultured neurons with microtubule-depolymerizing drugs. Microtubules serve also as tracks for bidirectional transport of various organelles and macromolecules between cell bodies and neurite tips (see Section 13.5.1).

The dynamics of microtubules as well as their interactions with other cellular components are regulated by microtubule-binding proteins. Table 13.3 shows the main types of microtubule-binding proteins.

Some microtubule-binding proteins promote the assembly of tubulin into microtubules and remain bound to the microtubule surface, resulting in microtubule stabilization. These proteins are quite abundant and were the first microtubule-binding proteins identified, being named microtubule-associated proteins (MAPs). MAPs bind to the carboxyl-terminal domain of tubulin, a domain implicated in the regulation of tubulin polymerization possibly through its interaction with the GTP-binding domain

Table 13.3 Major microtubule-binding proteins

Protein	Properties
Assembly-promoting and microtubule-stabilizing proteins	
HMW-Tau	Present in the peripheral nervous system
LMW-Tau	Abundant in axons of the central nervous system Contributes to microtubule stabilization
MAP-1A	Abundant in dendrites of mature neurons
MAP-1B (MAP5)	Present both in axons and dendrites. Contributes to neural migration and initial neurite outgrowth
MAP-2A, B	Present in cell bodies and dendrites (including dendritic spines)
MAP-2C, D	Present in axon, dendrites and glial cells
MAP-4	Present in glial cells and in immature neurons
DCX	Contributes to neuronal migration
LLS1	Contributes to neuronal migration
Microtubule end-binding protein	
CLIP-170	Attachment of microtubules to endosomes
APC	Attachment of microtubules to cell cortex
EB1	Attachment of microtubules to cell cortex
Microtubule-destabilizing proteins	
OP18/stathmin	Highly abundant. Favors microtubule destabilization
Microtubule-activated ATPases	
Dyneins	Move organelles from 'plus' to 'minus' ends of microtubules
Kinesins	Move organelles from 'minus' to 'plus' ends of microtubules
Proteins anchoring microtubules to membrane receptors	
Gephyrin	Binds glycine receptors

Table 13.3 (continued) Major microtubule-binding proteins

Protein	Properties
Signaling proteins interacting with microtubules	
JNK (SAPK)	Jun amino-terminal kinase (stress-activated protein kinase)
ERK (MAPK)	Extracellular regulated kinase (mitogen-activated protein kinase)
GSK-3	Glycogen synthase kinase-3
PP2A	Protein phosphatase 2A
Heterotrimeric G proteins	Bind tubulin
GEF-H1	Rho guanine nucleotide exchange factor. Regulates Rho protein and actin dynamics

of the tubulin molecule (Padilla *et al.* 1993). MAPs are a heterogeneous group of proteins that show developmental stage-specific expression as well as subcellular-specific compartmentalization. Interestingly, MAP functionality can be regulated by phosphorylation and dephosphorylation (Tucker 1990; Avila *et al.* 1994; Cassimeris and Spittle 2001). There are two major groups of MAPs: the MAP1 and the MAP2/MAP4/tau protein families.

MAP1A and MAP1B are encoded by different genes, but show extensive amino acid sequence homology. These proteins are much more abundant in neurons than in other cell types. MAP1B (also called MAP5) is the first MAP expressed by neurons during differentiation and is particularly abundant in growing axons. The expression of MAP1B diminishes after neuronal maturation, and the remaining protein is localized both in axons and dendrites of adult neurons. Null mutant mice lacking MAP1B die during the early embryonic stages, whereas hypomorphous mutant mice expressing low levels of MAP1B exhibit gross neuronal abnormalities including defects in neuronal migration and in axon formation (Edelmann *et al.* 1996; Gonzalez-Billault *et al.* 2000, 2001; Takei *et al.* 2000). Thus, MAP1B seems to play an essential role in neuronal morphogenesis.

MAP1B is highly phosphorylated *in vivo*. There are at least two modes of MAP1B phosphorylation. One is catalysed by proline-directed protein kinases including glycogen synthase kinase-3 (GSK-3), mainly takes place in growing axons, and practically disappears after axonal maturation (Mansfield *et al.* 1991). Interestingly axons from regenerating neurons express this phosphorylated MAPB1B (Dieterich *et al.* 2002). The other mode of MAP1B phosphorylation is catalysed by casein kinase 2 and is maintained both in axons and dendrites of mature neurons (Ulloa *et al.* 1994). As for MAP1A, its expression is up-regulated during neuronal maturation, and it is abundant in dendrites and is also *in vivo* phosphorylated.

MAP2, MAP4, and tau proteins share a homologous tubulin-binding domain that is different from that present in MAP1A or MAP1B. MAP2, MAP4, and tau are each encoded by a single gene, but considerable heterogeneity arises from alternative splicing of primary transcripts.

MAP2 isoforms include high molecular weight proteins MAP2A and MAP2B and low molecular weight proteins MAP2C and MAP2D. MAP2C is expressed in the embryonic brain and it is down-regulated during brain maturation. In contrast, MAP2A appears only after brain maturation. MAP2C is present in neuronal cell bodies, dendrites and axons as well as in glial cells, whereas high molecular weight MAP2 is a neuronal-specific protein selectively localized in dendrites and neuronal cell bodies (Tucker 1990). The compartmentalization of high molecular weight MAP2 into dendrites may be due to the selective transport of the corresponding mRNA into dendrites (Garner *et al.* 1988). In addition to its association with microtubules, MAP2 is associated with actin in dendritic spines (Fifkova and Morales 1992). Binding of MAP2 to tubulin and actin is regulated by phosphorylation. MAP2 can be modified by cAMP-dependent protein kinase, Ca^{2+}/calmodulin-dependent protein kinase, PKC and proline-directed protein kinases including mitogen-activated protein kinase (MAP kinases) and GSK-3 (Sanchez *et al.* 2000). MAP4 proteins are mainly expressed in nonneuronal tissues. In the brain, MAP4 is present in glial cells and very immature neurons.

Low molecular weight tau proteins are found in the central nervous system whereas additional high molecular weight tau proteins are present in the peripheral nervous system. In brain, tau proteins are particularly abundant within axons, although some tau is also present in neuronal cell bodies and dendrites. The association of tau with tubulin may also be regulated through phosphorylation by several protein kinases such as cAMP-dependent protein kinase, Ca^{2+}/calmodulin-dependent protein kinase, PKC, casein kinase 1, casein kinase 2, and proline-directed protein kinases including mitogen-activated protein kinases, stress-activated protein kinases, cyclin-dependent kinase-5, and glycogen synthase kinase-3 (Muñoz-Montaño *et al.* 1997; Shahani and Brandt 2002). *In vivo*, tau phosphorylation by GSK-3 is especially prominent in embryonic tau as compared with adult tau. It is plausible that tau dephosphorylation at these sites favors binding of tau to tubulin and therefore contributes to microtubule stabilization during axonal maturation (Ferreira *et al.* 1993). Interestingly, tau protein becomes hyperphosphorylated under certain pathological conditions (Buee *et al.* 2000; Shahani and Brandt 2002).

Interestingly, all these MAPs may play partially redundant and synergistic functions during neuronal development. Thus, suppression of the expression of one particular MAP has a relatively minor effect whereas the suppression of more than one MAP (MAP1B/MAP2 or MAP1B/Tau) has a dramatic effect on neuronal morphogenesis (Takei *et al.* 2000; Gonzalez-Billault *et al.* 2002).

In addition to these conventional and abundant MAPs, other microtubule-binding proteins have been described. These include doublecortin (DCX) and LIS1, which are the products of two genes involved in neuronal migration. Mutations in these genes give rise to neuronal migration arrest, cerebral cortex malformation, and disruption of neuronal morphology. DCX seems to be a classical MAP which is able to stabilize microtubules, and LIS1 might be a microtubule-end binding protein which also associates with the microtubule motor dynein (Feng and Walsh 2001).

Whereas conventional MAPs bind all along the microtubule surface, other microtubule-interacting proteins bind exclusively at microtubule ends, particularly at the 'plus' ends. These plus-end-binding proteins, also known as plus-end tracking proteins (+TIP), form multiprotein complexes that are thought to regulate microtubule dynamics, mediate the anchoring of microtubules to the cellular periphery and modulate the delivery of proteins to cell ends (Galjart and Perez 2003). Major +TIPs include CLIP-170 (cytoplasmic linker protein-170), APC (*adenomatous polyposis coli* gene product) and EB1 (end-binding protein 1). These +TIPs may also interact with other proteins including conventional MAPs, like MAP1B or tau, and microtubule-dependent motor proteins like dynein, facilitating the anchoring of microtubules to the cell cortex (Allan and Nathke 2001). These multiprotein complexes are also regulated by phosphorylation and the role of GSK-3 is quite significant in this respect (Allan and Nathke 2001).

A different type of microtubule-binding protein is Op18/stathmin. This 17 kDa protein was initially discovered as a protein which becomes up-phosphorylated in response to a variety of extracellular signals. It is now clear that Op 18/stathmin has the ability to bind tubulin and favor rapid microtubule depolymerization (Belmont and Mitchison 1996). Another type of microtubule-binding protein with microtubule severing activity has recently been described. It is not known whether these proteins are present in neurons (Shiina *et al.* 1995).

Other microtubule-binding proteins are microtubule-activated ATPases, also called motor proteins. Kinesin, kinesin-related proteins, and cytoplasmic dynein are involved in the transport of cell organelles and other macromolecular complexes along microtubules (see Section 4.5.4).

Microtubules are important not only for the organization of organelles within the cytoplasm but also for the clustering of protein complexes within membranes. Proteins anchoring membrane proteins to microtubules play an essential role in this respect. As a case in point, gephyrin links glycine receptors of postsynaptic membranes to microtubules and microfilaments (Kirsh and Betz 1996). MAP 1B seems to be responsible for the anchoring and clustering GABA-C receptors and this linkage to the cytoskeleton may modulate the sensitivity of the receptors (Hanley *et al.* 1999; Billups *et al.* 2000). Thus, the binding of membrane receptors to the cytoskeleton may control the functional properties of the receptors (see Section 13.4).

In a more general way, it is now considered that microtubules play a crucial role in the regulation of signal transduction. First, microtubules are affected by signalling pathways since MAPs, motors, and other microtubule-binding proteins are targets of signal transduction pathways. Second, microtubules seem to be critical for the spatial organization of signal transduction and also contribute to the transmission of intracellular signals to downstream targets. Thus, microtubules may sequester signalling factors from the cytoplasm and release them under certain conditions. Microtubules may also act as a scaffold to promote the interaction of certain factors, which would otherwise not interact. Finally, microtubules may additionally serve to deliver signalling factors to specific sites in the cell. Among the signalling proteins that interact with microtubules there are many protein kinases, phosphatases, and G proteins (Gundersen and Cook 1999). Of special relevance is the binding of Rho guanine nucleotide exchange factor (GEF-H1) since this protein modulates Rho GTPase activity and therefore actin dynamics, thus constituting a way of cross-talk between microtubules and the actin cytoskeleton (Krendel *et al.* 2002).

13.2.4 Interactions of cytoskeletal components

Microfilaments, intermediate filaments, and microtubules are not isolated from each other in the neuronal cytoplasm. Indeed, there are a large number of connections between the three types of cytoskeletal structures. Many proteins are involved in these interactions. For instance, spectrin not only binds actin but also ankyrin, membrane proteins, microtubules, and neurofilaments. Certain microtubule-associated proteins have also been reported to bind microfilaments and neurofilaments. Additionally, many membrane organelles and soluble proteins are anchored to the cytoskeleton. The clearest example is the plasma membrane, where the cytoplasmic 'tails' of many membrane proteins are bound to cytoskeletal proteins (Kirsh and Betz 1996).

Development of the quick-freeze, deep-etch techniques has allowed the observation of the three-dimensional architecture of the unfixed neuronal cytoskeleton. These images have provided evidence that there is a cytoskeletal network that should be considered as an integrated functional entity (Hirokawa 1982). Figure 13.2 is a schematic drawing of the cytoskeleton beneath the plasma membrane of the squid giant axon as visualized by the quick-freeze, deep-etch technique (Tsukita *et al.* 1986). The existence of connections between microtubules and microfilaments and between these cytoskeletal polymers and the plasma membrane is apparent.

13.3 The cytoskeleton in neuronal morphogenesis

13.3.1 Neurite growth

An essential event for neuronal morphogenesis is the extension of the neurites that are the hallmark of neuronal cell shape. All neurites, both axon and dendrites, grow at specialized terminal appendages called growth cones, which were first observed by Ramón y Cajal in the embryonic chick spinal cord. Growth cones are primarily involved in the extension of the neurite and in the guidance of the developing neurite to reach its target. Accordingly, growth cones function as sensory devices that decode extracellular signals in order to direct neurite growth through the regulation of intracellular cytoskeletal dynamics.

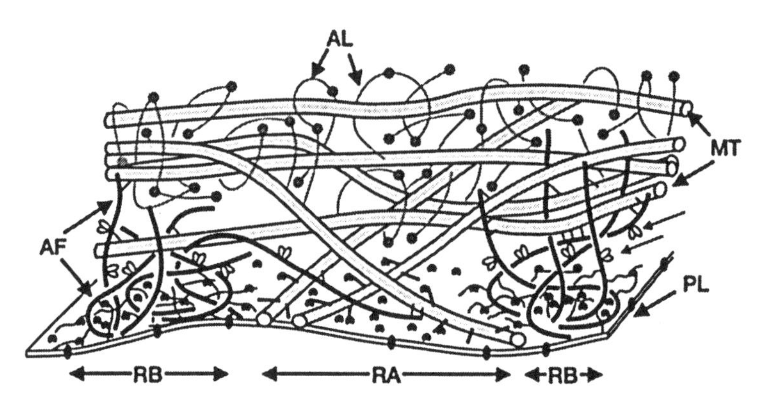

Fig. 13.2 Diagram showing the architecture of the cytoskeleton underneath the plasma membrane of a squid giant axon. The plasma membrane (PL) consists of actin microfilament associated (RB) and microtubule-associated (RA) domains. The cross-linkers between microtubules (MT) and actin filaments (AF), between MT and PL, and between AF and PL are apparent. Microtubules are surrounded by axolinin (AL) molecules. Axolinin is a squid giant axon MAP. Reproduced from *The Journal of Cell Biology* (1986), vol. 102, pp. 1710–25 by copyright permission of The Rockefeller University Press.

Typical growth cones appears as fan-shaped or leaf-shaped expansion of neurite tips. Growth cones are characterized by the presence of long thin projections called filopodia, which are embedded within broad expansions named lamellipodia. Microtubules are mostly located in the central portion of the growth cone. Bundled unipolar actin microfilaments constitute the cytoskeleton of filopodia and a network of crosslinked actin predominates within lamellipodia (Lewis and Bridgman 1992).

Neurite outgrowth occurs through the protrusion of filopodia and lamellipodia and the subsequent invasion of the expanded bases of filopodia and lamellipodia by microtubules. The bundling of the invading microtubules constitutes the consolidation of the growth of the neurite. The protrusive activity of lamellipodia and filopodia of cultured neurons is inhibited by F-actin depolymerizing drugs like cytochalasin B, while it persists in the presence of microtubule-depolymerizing drugs like nocodazole or colchicine. In contrast, neurite elongation is blocked in the presence of microtubule-depolymerizing drugs and continues in the presence of F-actin depolymerizing drugs. This indicates the existence of two linked processes during neurite outgrowth that can be uncoupled by different drugs. One is the growth cone motility, which is presumably involved in neurite guidance and depends on actin dynamics. The other process is neurite elongation which depends on microtubules. Under physiological conditions, both processes are connected (Bradke and Dotti 2000; Goldberg 2003).

The molecular mechanisms responsible for the protrusive activity of filopodia and lamellipodia are not entirely elucidated. However, a hypothetical description, known as the 'clutch' model, has been proposed (Mitchison and Kirschner 1988; Smith 1988; Stossel 1993; Lin *et al.* 1996; Jay 2000). Figure 13.3 shows a schematic drawing

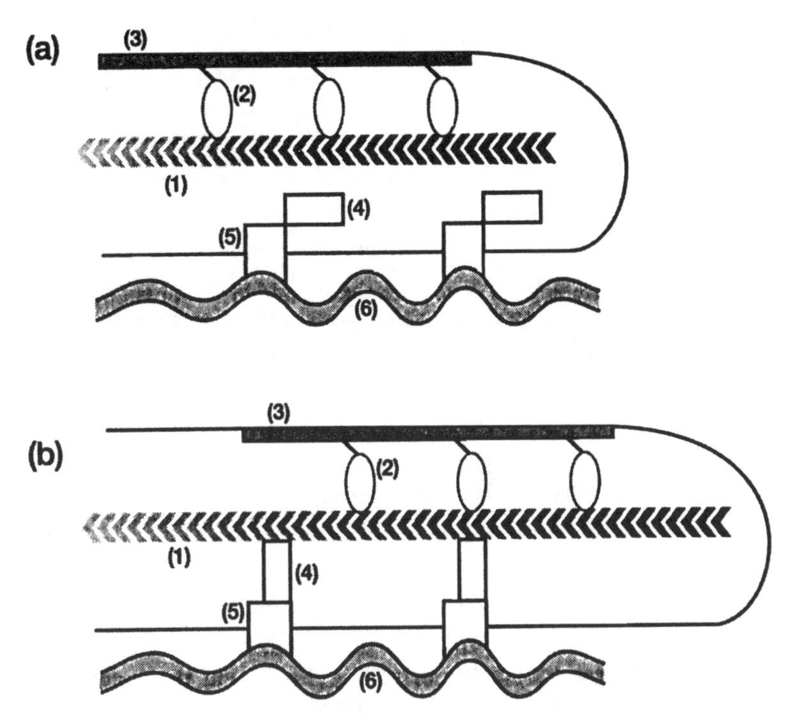

Fig. 13.3 Model showing how anterograde movement of the growth cone may occur when the retrograde flow actin microfilaments ceases as the consequences of anchoring to the extracellular substrate. In (a), actin microfilaments (1) are subjected to 'treadmilling': assembling at the 'barbed' end (shaded chevrons) and disassembling at the 'pointed' end (faded chevrons). There is a retrograde flow of actin microfilaments (1) driven by myosin molecules (2) anchored on to the membrane skeleton (3). In (b), actin filaments become stationary because of their association with anchoring proteins (4) bound to transmembrane protein (5) which are linked to extracellular matrix molecules (6). Addition of new actin monomers to the 'barbed' end also occur. Myosin (3) and the membrane skeleton (3) crawl to the leading edge of the growth cone. Reproduced from Mitchison and Kirschner (1988) with permission from Cell press.

that explains how forward movement of filopodia and lamellipodia might occur. It is well known that there is a retrograde flow of F-actin in filopodia and lamellipodia. This retrograde flow is particularly robust in immotile growth cones and it is easily observed when neurons are plated on a poly-L-lysine substrate lacking certain extracellular matrix proteins. When a growth cone is on a more permissive substrate, the retrograde F-actin flow slows down and the growth cone advances. This suggests that the advance of the growth cone depends on the coupling of the cytoskeleton to the substrate. Figure 13.3a depicts a growth cone in which the cytoskeleton is not linked with the extracellular substrate. In this situation, motor proteins (presumably myosin molecules anchored to the rigid membrane skeleton) can produce the retrograde movement of actin microfilaments that are undergoing 'treadmilling' (see Section 13.2.1). Figure 13.3b depicts the situation in which the cytoskeleton of the growth cone is firmly attached to the extracellular substrate. When this occurs, the actin microfilaments are stationary and the motor protein myosin and the membrane skeleton crawl to the edge of the growth cone. Actin polymerization at the 'plus' ends of microfilaments also occurs in parallel, thus leading to growth cone advance.

There is a large number of actin-binding proteins present at the growth cone (see Table 13.1) and their complex interactions make it difficult to understand how these proteins act in a concerted way. It has been suggested that myosin V functions as the motor protein anchored to the membrane skeleton. Talin seems to be involved in the anchoring of actin filaments to the membrane (the 'clutch' module), and zyxin promotes actin assembly at the tip, possibly by recruiting proteins from the Ena/VASP family (Jay 2000).

Extracellular signal molecules that control axon extension and guidance are able to modulate growth cone motility through the activation of intracellular signalling factors that regulate actin dynamics (Dickson 2001; Meyer and Feldman 2002; Goldberg 2003; Huber *et al.* 2003). In particular, proteins of the Rho subfamily of Ras-related GTPases (CDC42 Rac and Rho) are involved in the signal transduction pathways leading to the remodelling of the actin cytoskeleton within growth cones in response to certain extracellular signals. CDC42 primarily stimulates filopodia formation, Rac stimulates the formation of lamellipodia and Rho could participate in growth cone retraction (Dickson 2001; Meyer and Feldman 2002; Goldberg 2003; Huber *et al.* 2003).

The forward movement of the growth cone is, under physiological conditions, accompanied by neurite extension. This presumably occurs through the translocation and polymerization of microtubules from the central part of the growth cone to the expanded initial portions of the protrusive filopodia and lamellipodia. This microtubule extension may simply be favored by the fact that actin filament density is decreasing at the base of filopodia and lamellipodia when these are advancing. Alternatively, there might be molecules crosslinking actin microfilaments and microtubules that could pull microtubules to the base of filopodia and lamellipodia. Some evidence favors the first possibility, namely, that microtubules tend to extend when not restrained by the retrograde flow of actin microfilaments. Thus, treatment of growth cones with microfilament-depolymerizing drugs stimulates the extension of microtubules (Forscher and Smith 1988; Bradke and Dotti 2000). It has been suggested that the actin filament-severing protein gelsolin and the filament depolymerising family of ADF/cofilin proteins increase actin turnover at the transitional zone between actin filament 'pointed' ends and microtubule 'plus' ends.

As mentioned above, neurite elongation is completely dependent on microtubules. Both the transport of microtubules from the cell body and the elongation of microtubules contribute to neurite growth (Joshi and Baas 1993; Baas and Ahmad 2001). Figure 13.4 shows that short microtubules destined for the neurites are initiated at the centrosome within the cell body, after which they are released from the centrosome

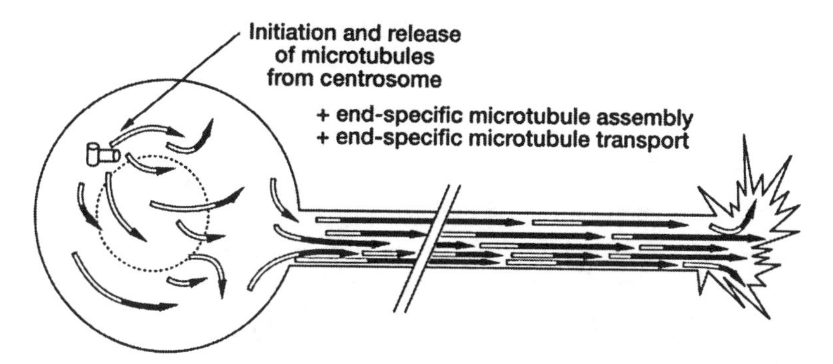

Fig. 13.4 Model describing the origin of axonal microtubules. Microtubules are nucleated at the centrosome, then released and transported into the axon where mictrotubule polymerization takes place at the 'plus' ends. The growth cone is a major site of microtubule assembly. Reproduced from *The Journal of Cell Biology* (1993), vol. 121, pp. 1191–6, by copyright permission of The Rockefeller University Press.

and transported into the neurite. It has been suggested that microtubule-dependent motor proteins like dynein are responsible for catalysing microtubule transport within growing neurites (Baas and Ahmad 2001). During their transit down the neurite, short microtubules are elongated. The central region of the growth cone is particularly enriched in microtubule 'plus' ends that are actively polymerizing by incorporation of tubulin subunits.

In addition to microtubule transport and elongation, the consolidation of the growing neurite is produced by the bundling of microtubules. Microtubule-associated proteins are possibly involved in the bundling of microtubules. In particular, an essential role for MAPs including MAP1B and tau or MAP2 has been demonstrated (Gonzalez-Billault *et al.* 2002).

Microtubules in distal neurite regions are more dynamic than those in the proximal domain, which suggests a progressive stabilization of microtubules during neurite extension. Different MAPs may stabilize microtubules in axons and dendrites. Thus, tau proteins are possibly implicated in a partial microtubule stabilization within growing axons (Harada *et al.* 1994) and high molecular weight MAP2 protein may perform a similar role in growing dendrites. Microtubules are not completely stable in extending neurites because these need a flexible cytoskeleton to allow growth. In particular, microtubule instability is required for new membrane insertion into the growing neurites (Zakharenko and Popov 1998). The precise degree of microtubule dynamics in developing axons may be controlled through the phosphorylation of MAPs including MAP1B (Ulloa *et al.* 1994) by kinases like GSK-3 (Goold *et al.* 1999). In mature neurites, the cytoskeleton becomes less plastic since the maintenance of morphology and the transport of organelles are its major functions. Thus, microtubule stabilization is increased with axonal and dendritic maturation (Baas *et al.* 1991). Dephosphorylation of certain sites on MAP1B and tau molecules can contribute to axonal microtubule stabilization (Ferreira *et al.* 1993; Ulloa *et al.* 1994), whereas the appearance of new MAPs like MAP1A may favor dendritic microtubule stabilization.

13.3.2 **Axonal maturation**

When axons reach their targets, the cytoskeleton of the growth cone is remodelled and converted into the cytoskeleton of a presynaptic terminal. Motility and extension cease, and an accumulation of synapsin, which crosslinks synaptic vesicles to actin microfilaments, is observed.

Myelination signals a new phase of axonal maturation, characterized by the radial growth of the axon. It is now clear that this increase in axonal diameter is due to the augmented expression and phosphorylation of neurofilaments. Thus, no radial growth of axons is observed in mice with mutant neurofilament genes (Julien 1999). A reduced axonal calibre is also observed in a mutant mouse called *trembler* in which myelination does not occur as a result of the mutation in the gene-encoding myelin

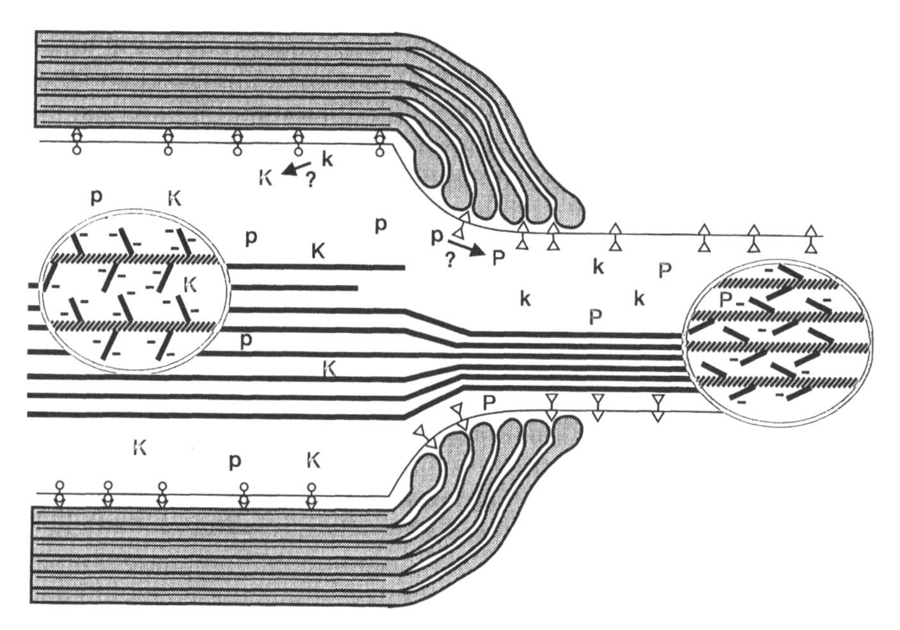

Fig. 13.5 Diagram showing the stimulation of axonal neurofilament phosphorylation by myelinating Schwann cells. Putative interactions between Schwann cell membrane and axonal membrane molecules trigger either the activation of a neurofilament kinase (K) or the inhibition of a phosphatase (P), thus leading to an enhanced phosphorylation of the 'tail' domains of NF-H and NF-M. These constitute the lateral projections of the neurofilament polymers and their high degree of phosphorylation may lead to electrostatic repulsion, resulting in a wide interfilament spacing and increased axonal calibre. In nonmyelinated axonal segments, the activity of the phosphatase is higher than the kinase activity, neurofilaments are therefore less phosphorylated, and a narrower interfilament spacing and a reduced axonal diameter are observed. Reproduced from De Waegh *et al.* (1992) with permission from Cell Press.

basic protein. *Trembler* axons have an increased density of neurofilaments that are closely spaced and underphosphorylated. This suggests that NF-M and NF-H tail domain phosphorylation augments interfilaments spacing, contributing to radial growth (see Section 13.2.2). Underphosphorylation results in narrow spacing, increased density, and reduced axonal calibre. Interestingly, this is not only observed in *trembler* axons but also in normal axons at the nodes of Ranvier. Thus, myelination seems to control NF-M and NF-H tail domain phosphorylation (see Fig. 13.5). This indicates that Schwann cell–axon interactions may trigger a signalling pathway that controls a neurofilament kinase or phosphatase (De Waegh *et al.* 1992).

13.3.3 Neuronal polarity

Axons and dendrites differ in their morphology, rate of growth, organelle content, and cytoskeletal composition and organization (see Table 13.4). It is plausible that the differences between the cytoskeleton of axons and dendrites are responsible for the differences in their morphology, rate of growth and organelle content (Craig and Banker 1994; Bradke and Dotti 2000; Scott and Luo 2001).

The formation of axons and dendrites follows a stereotyped pattern in cultured embryonic hippocampal neurons (Dotti *et al.* 1988; Bradke and Dotti 2000); after plating, the cells extend lamellipodia (stage 1). Several short neurites arise from these lamellipodia (stage 2). One of them elongates very rapidly, becoming the axon (stage 3). After a few days, the remaining short neurites begin to grow slowly to become dendrites (stage 4). Finally both axons and dendrites mature (stage 5).

The initiation of fast axonal growth (stage 3) marks the generation of neuronal polarity. A segregation of certain proteins into the nascent axon at this stage has been observed. One of these proteins is neuromodulin, which may have a regulatory function on axonal growth cone motility. Differential protein phosphorylation may also

Table 13.4 Major differences between axons and dendrites

Axons	Dendrites
Uniform calibre	Tapered morphology
Few branches	Highly branched
Lack of polyribosomes	Presence of polyribosomes
Fast growth	Slow growth
Abundance of neurofilaments	Abundance of microtubules
Uniform polarity of microtubules	Mixed polarity of microtubules
Narrow spacing between microtubules	Wide spacing between microtubules
Abundance of tau protein	Presence of MAP2A, B
Presence of $\alpha\gamma$ spectrin	Presence of $\alpha\beta$ spectrin
Highly phosphorylated NF-M and NF-H	Nonphosphorylated NF-M and NF-H

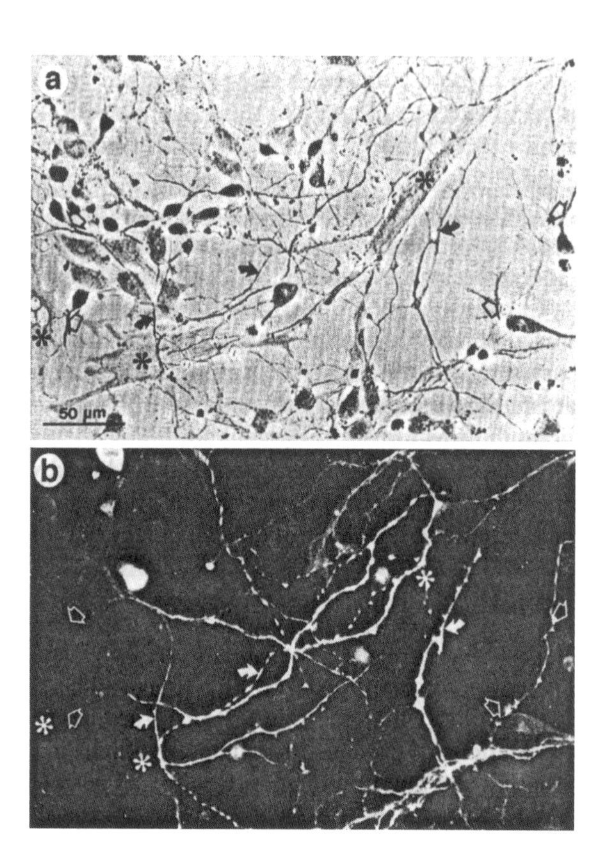

Fig. 13.6 Association of MAP1B phosphorylated by proline-directed protein kinases with developing axons. (a) Phase contrast micrograph of cortical neurons and glial cells in culture. (b) Immunofluorescence micrograph with an antibody that recognizes only MAP1B phosphorylated by proline-directed protein kinases. Axons (curved arrows) are intensely stained whereas glial cells (asterisks) and neuronal cell bodies and dendrites (open arrows) are not stained. Reproduced from Mansfield *et al.* (1991) with permission from Chapman and Hall Ltd.

contribute to the development of neuronal polarity. For example, MAP1B phosphorylated by GSK-3 is mainly localized to growing axons, as shown in Fig. 13.6 (Mansfield *et al.* 1991; Ulloa *et al.* 1994; Goold *et al.* 1999). Interestingly, laminin, an extracellular matrix protein that stimulates axonal growth, also promotes MAP1B phosphorylation (Di Tella *et al.* 1996).

An important event in the establishment of neuronal polarity is the appearance of microtubules with mixed polarity (see Fig. 13.7). Within the axon, all the microtubules have a uniform polar orientation with their 'plus' ends pointing toward the axon terminal. In contrast, dendritic microtubules are oriented with their 'plus' ends toward the dendrite tip or the cell body (Baas *et al.* 1989). All microtubules destined for axons or dendrites are probably initiated at the centrosome, released, and then transported into axons and dendrites. Thus, the uniform polarity of axonal microtubules and the mixed polarity of dendritic microtubules must arise from differences in their transport systems (Sharp *et al.* 1995; Baas 1999).

The distinct microtubule patterns within axons and dendrites may constitute a structural basis for organelle distribution in the neurons, as it has been hypothesized that certain organelles, including polyribosomes, are preferentially transported toward the 'minus' ends of the microtubules, which would facilitate their transport into dendrites.

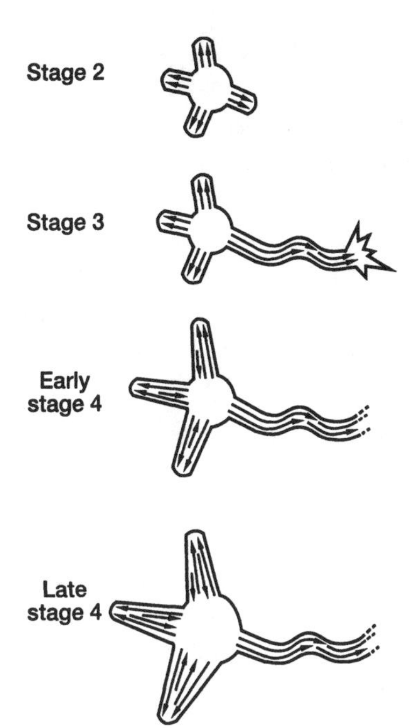

Fig. 13.7 Scheme depicting the appearance of microtubules with mixed polarity in developing dendrites of cultured rat hippocampal neurons. Reproduced from *The Journal of Cell Biology* (1989), vol. 109, pp. 3085–94, by copyright permission of The Rockefeller University Press.

13.4 The cytoskeleton in neuronal plasticity

Both in the developing and adult nervous systems, neurons can undergo structural modifications in response to certain extracellular signals. These modifications may range from the outgrowth of collateral ramifications from neurites, which establish new synapses, to alterations of pre-existing synapses, which include changes in the shape of dendritic spines and more subtle modifications of postsynaptic densities and presynaptic terminals. Some synaptic changes may be associated with learning (Greenough and Bailey 1987; Bailey and Kandel 1993; Montague 1993).

Rearrangements of the neuronal cytoskeleton have been proposed to play a crucial role in these synaptic remodelling events (Fifkova and Morales 1992). Such changes may be triggered by posttranslational modifications of cytoskeletal proteins, mainly through phosphorylation and dephosphorylation. According to this view, certain extracellular signals initiate transduction pathways that modulate cytoskeletal protein phosphorylation, which leads to the structural rearrangements underlying synaptic plasticity. In fact, some preliminary evidence has shown a correlation between MAP2 phosphorylation and certain examples of synaptic plasticity (Sanchez *et al.* 2000).

A great deal of attention has been focused on dendritic spines, as these specialized structures are postsynaptic sites. Actin provides the main structural basis for cytoskeletal organization within dendritic spines, which mostly lack microtubules and intermediate filaments (Halpain 2000; Matus 2000). Recent evidence has demonstrated a role for actin rearrangements in a form of synaptic plasticity (see also Chapter 14). Thus, long-term potentiation (LTP) of synapses in the hippocampal dentate gyrus is associated with the phosphorylation of cofilin, which gives rise to an increase in F-actin within spines, thus leading to the growth and strengthening of synapses (Fukazawa *et al.* 2003).

Cytoskeletal modifications may not only affect neuronal connectivity through morphological changes but might also control neuronal physiology through the modulation of neurotransmitter receptors and ion channels which are anchored to the membrane skeleton (Rosenmund and Westbrook 1993; Billups *et al.* 2000).

13.5 The cytoskeleton in intraneuronal transport

13.5.1 Axonal and dendritic transport

Although all eukaryotic cells need active systems to generate intracellular movement of organelles, protein complexes, and mRNAs, neurons have a much more pronounced demand and challenge since they are highly polarized cells with processes of significance length. All neuronal mRNA synthesis and most proteins synthesis takes place in the cell body. Specific transport systems allow the delivery of macromolecules from the cell body to axons and dendrites. The fact that axons and dendrites each receive a number of unique proteins (see Section 13.3.3) calls for the existence of molecular sorting mechanisms. Furthermore, certain mRNAs are transported into dendrites where they contribute to local protein synthesis (Steward and Banker 1992).

Dendritic transport has not been so thoroughly studied as axonal transport. For this reason we will focus on axonal transport. The necessity for an axonal transport mechanism was first appreciated by Ramón y Cajal (1928) who established that the integrity of the axon depends on the neuronal cell body. When a peripheral nerve axon is severed, the segment which is disconnected from the cell body degenerates, while the remaining portion of the axon regenerates. Both axon regeneration in peripheral nerves and axon outgrowth in immature neurons take place as a result of materials continuously supplied from the cell body to the axon.

Axonal transport has been mainly studied by radioisotopic labelling methods. For instance, in the visual system, radiolabeled amino acids injected into the vitreous humor are incorporated into proteins by the retinal ganglion neurons. Labeled proteins are then transported down the axons constituting the optic nerve. The movement of labelled proteins is analysed by removing the optic nerve, sectioning it into segments and measuring the distribution of radioactivity within these pieces at different times after injection.

The kinetics of axonal transport that emerge from these studies are rather complex, indicating the existence of distinct sets of conveyed proteins according to their rates of transport. These groups of proteins move as coherent waves down the axon at different rates, with no exchange of proteins between them. This has led to the proposal that transported proteins move as parts of cytological structures (Tytell *et al.* 1981).

Four main components of axonal transport have been identified:

(i) Slow component A. This group of proteins moves at a rate of 0.2–1 mm day^{-1} (0.002–0.01 μm sec^{-1}) and consists primarily of polypeptides associated with neurofilaments and microtubules.

(ii) Slow component B. This is a quite complex group of proteins comprising more than 100 polypeptides moving at 2–8 mm day^{-1} (0.02–0.08 μm sec^{-1}). It seems to correspond to the transport of microtubules and actin microfilaments including their associated proteins. Some of these are metabolic enzymes which bind to actin microfilaments.

(iii) Intermediate component. This corresponds to mitochondria, which are conveyed along microtubules at a rate of 50–100 mm day^{-1} (0.6–1.2 μm sec^{-1}).

(iv) Fast component. This is a rather complex group of membrane-associated proteins moving at a rate of 200–400 mm day^{-1} (2.4–4.8 μm sec^{-1}) and corresponds to the movement of most membrane organelles along microtubules.

Rates of axonal transport presumably depend on the specific mechanochemical properties of the motor proteins implicated and on the hindrance opposing the movement of organelles and other macromolecular complexes. This hindrance depends on the degree of cross-linking of cytoskeletal polymers and the presence of specific cytoskeletal-associated proteins . Thus, there are differences in the axonal transport components among distinct neuronal types. Curiously, the transport velocities of the slow components in axons from the peripheral nervous system are faster than those in axons from the central nervous system. Moreover, a novel component of axonal transport referred to a slow component C has recently been described in peripheral axons and consists of proteins moving at a rate of 7–9 mm day^{-1}. This corresponds to some proteins associated with tubulin and actin, including the mode-I phosphorylated isoforms of MAP1B. It has been suggested that this relatively less slow transport of phosphorylated MAP1B may account for the high concentration of this protein in the distal ends of growing axons (Ma *et al.* 2000).

13.5.2 **Slow axonal transport**

The molecular mechanisms of slow axonal transport are not entirely understood. It used to be generally accepted that cytoskeletal proteins assembled in the cell body and that cytoskeletal polymers were the moving elements in slow axonal transport. According to this 'structural hypothesis', cytoskeletal polymers entered the axon and there was a continuous movement of the cytoskeletal scaffolding along the axon to the synaptic terminal where microtubules and neurofilaments would be disassembled and proteolyzed. Alternatively to this model, cytoskeletal proteins may be transported down the axon as subunits, oligomers, or small polymers and serve as precursors for a stationary axonal cytoskeleton constituted by highly crosslinked polymers (Nixon 1991). Current evidence favors this latter view.

The existence of moving and stationary cytoskeletal elements is particularly clear for neurofilaments. Highly phosphorylated NF-H and NF-M isoforms (see Section 13.2.2) are associated with stationary neurofilaments. Thus, it is plausible that non-phosphorylated neurofilaments are assembled in the cell body and enter the axon as moving oligomers or polymers which become progressively phosphorylated and integrated into a stationary cytoskeleton (Nixon and Sihag 1991). Interestingly, it has been demonstrated that fluorescently tagged NF-M moves in a rapid, intermittent and highly asynchronous manner within axons of cultured neurons. This has raised the possibility that the slow rate of neurofilament tranport *in vivo* may merely be the result of rapid movements driven by a fast motor protein interrupted by prolonged pauses (Wang *et al.* 2000). Accordingly, all the components of the slow axonal transport would correspond to protein complexes exhibiting bursts of fast movement interspersed with variable periods of non-motility. This view is consistent with recent studies indicating an implication of motor proteins similar to those mediating fast organelle transport (see Section 13.5.4) in slow axonal transport (Shea and Flanagan 2001; Xia *et al.* 2003). It has been suggested that kinesin might be involved in the initial transport of neurofilament subunits into and along axons. As these neurofilament proteins undergo progressive phosphorylation during axonal transport, they might dissociate from kinesin and associate with other neurofilament subunits, eventually formed a large bundled macro-structure unable to translocate. The reversibility of the phosphorylation of neurofilament proteins as well as the reversibility of the interactions of these proteins with kinesin might contribute to the observed cycles of rapid movement and pauses (Shea and Flanagan 2001; Ackerley *et al.* 2003).

13.5.3 Fast axonal transport

Membrane organelles move along axonal cytoskeletal polymers (mainly microtubules but also actin microfilaments). Fast axonal transport actually comprises an anterograde (from the cell body to the axon tip) and a retrograde (from the axon tip to the cell body) component (Grafstein and Forman 1980). For instance, synaptic vesicles move anterogradely, while endocytotic and prelysosomal vesicles move retrogradely. The major function of the fast anterograde axonal transport is the conveying of membrane organelles required for presynaptic events at the axon ending. In developing neurites, anterogradely moving vesicles support the membrane extension which occurs at the growth cone. The retrograde component seems to perform two essential functions. First, it facilitates membrane recycling, since it conveys prelysosomal organelles from the nerve ending to the cell body. Second, it provides a retrograde pathway for the transmission of information through the transport of receptors associated with membrane vesicles. These receptors can convey signalling molecules from the axon tip to the neuronal cell body. For example, the retrograde transport of nerve growth factor and its receptor is well documented (Reynolds *et al.* 2000).

It now seems clear that axonal microtubules constitute the tracks for organelle transport over long distances whereas actin filaments provide movement to local sites along the axon and within the presynaptic terminal (Atkinson *et al.* 1992; Langford 1995). There is a bidirectional movement of organelles on microtubules, whereas transport on actin filaments is mainly (if not exclusively) unidirectional (toward the 'barbed' end). However, little is known about the coordination of the microtubule-mediated and microfilament-mediated movements of a given organelle (Huang *et al.* 1999).

13.5.4 **Molecular motors**

Membrane organelles and other macromolecular complexes (collectively referred as cargoes) move along microtubules or microfilaments through their association with molecular motors. These motor proteins are microtubule-dependent ATPases (kinesin, kinesin-related proteins, and cytoplasmic dynein; see Fig. 13.8) and actin-dependent ATPases (myosins) (Hirokawa 1998; Goldstein and Yang 2000; Miki *et al.* 2001; Karcher *et al.* 2002).

Conventional kinesin, also named as kinesin I, was discovered in the axoplasm of the squid giant axon as a microtubule-dependent organelle motor (Brady 1985). Kinesin I holoenzyme is a tetramer consisting of two heavy chains (110–130 kDa) and two light chains (60–80 kDa). Kinesin heavy chains have microtubule-activated ATPase activity and are able to move organelles from the 'minus' to the 'plus' ends of microtubules. The kinesin molecule appears as a long rod with a pair or globular 'head' domains at one end and a 'fan-shaped tail' at the opposite end. Kinesin binds ATP and microtubules through the 'head' domains and presumably interacts with specific protein cargoes through the 'tail' domain.

Molecular genetic techniques have allowed the identification of several kinesin-related proteins in different organisms. Kinesin-related proteins show homology to the 'head' domain of conventional kinesin and are highly divergent in the 'tail' domains. It is thus plausible that the different tails may be responsible for the interaction of different motors with distinct membrane organelles (or macromolecular complexes) (Terada and Hirokawa 2000).

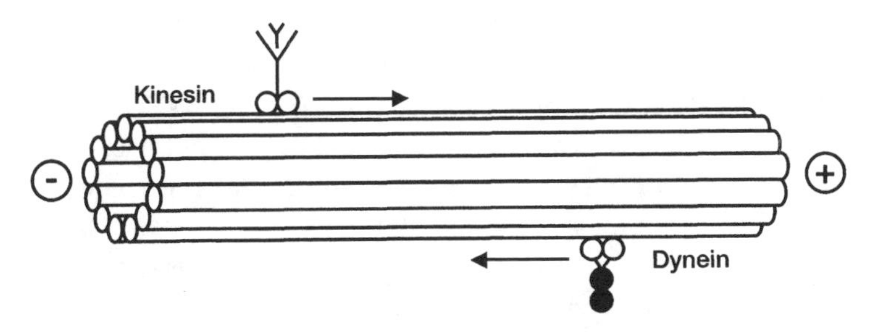

Fig. 13.8 Microtubule-dependent motor molecules. Scheme shows kinesin (a 'plus' end-directed motor protein) and dynein (a 'minus' end-directed motor protein).

Most kinesin holoenzymes are 500 kDa or smaller in size and, similarly to kinesin-I, contain between one and four copies of a heavy chain bearing the motor domain, as well as associated accessory subunits (Goldstein and Yang 2000). There are at least 45 genes in the human genome that code for kinesin-related heavy chains, which are referred to as kinesin superfamily proteins (KIFs). Of these, the expression of 38 KIFs has been detected in brain tissue (Miki *et al.* 2001). KIFs can be grouped according to the position of the globular motor domain: N-kinesins (NH_2-terminal motor domain type), M-kinesins (middle motor domain type), and C-kinesins (COOH-terminal motor domain type). Of the 45 human KIFs, there are 37 multimeric N-kinesins, 2 monomeric N-kinesins, 3 M-kinesins, and 3 C-kinesins.

Current evidence suggests that distinct KIF proteins are able to transport specific membrane organelles and macromolecular complexes. For instance, kinesin-I (KIF5B) can transport lysosomes, mitochondria and specific subsets of axonal vesicles bearing amyloid precursor protein (APP), GAP43 and low density lipoprotein (LDL) receptors. KIF1A transports a subset of synaptic vesicle precursors containing synaptophysin whereas the highly homologous KIF1B largely transports mitochondria. Kinesin-II (KIF3A/B) transports vesicles bearing fodrin as well as cytosolic choline acetyl transferase. KIF17 is able to move vesicles bearing NR2B, an NMDA-type glutamate receptor. In juvenile neurons, KIF4A translocates vesicles containing L1 cell adhesion molecule whereas KIF2A transports vesicles bearing βgc (an IGF-1 receptor relative).

The diversity of the non-motor tail domains of KIF proteins as well as the presence of accessory subunits within native kinesin holoenzymes are thought to give rise to specific interactions with transmembrane and scaffold proteins that might link kinesins to their particular cargoes. For instance, KIF13 binds directly to β1-adaptin, which interacts with clathrin and serves as a linking protein to vesicles bearing the mannose-6-phosphate receptor. KIF17 interacts through mLin-10 protein with vesicles bearing NMDA-type glutamate receptors. Curiously, one motor protein may bind to different cargoes through distinct linking proteins. Thus, kinesin-I light chains may bind to JIP, a group of proteins which forms a scaffold for c-Jun NH_2-terminal kinase (JNK, a member of the stress-activated protein kinase family) as well as serve as a link to membrane organelles bearing LDL receptors (ApoER2, LRP, and megalin). This linkage may serve to localize not only membrane proteins but also cytoplasmic signalling molecules, thus providing a targeting mechanism for the spatial regulation of signalling pathways at appropriate cellular domains (Verhey *et al.* 2001). On the other hand, kinesin-I light chains may also interact with APP, which links kinesin with membrane organelles bearing β-secretase, presenilin-1, synapsin-I, GAP-43, and the neurotrophin receptor Trk A (Kamal *et al.* 2001).

Interestingly, the association of kinesin with membrane organelles may be subjected to distinct regulatory mechanisms. Thus, either association with hsc70 chaperones or phosphorylation of kinesin light chains by glycogen synthase kinase-3 release kinesin-I from membrane organelles, thus inhibiting their anterograde axonal transport (Tsai *et al.* 2000; Morfini *et al.* 2002).

13.5.4.2 Dyneins

Dyneins (formerly referred to as cytoplasmic dyneins) are very large protein complexes of approximately 1–2 MDa which contain two or three dynein heavy chains of 500 kDa, and several intermediate chains of 74 dDa, light intermediate chains of 50–60 kDa and light chains of 8–29 kDa. Each dynein heavy chain bears a motor domain with microtubule-activated ATPase activity. All dyneins tested so far can convey organelles and other cargoes from the 'plus' ends to the 'minus' ends of microtubules (King 2000). Thus, dyneins may participate in retrograde axonal transport as well as in dendritic transport of organelles and other macromolecular complexes.

There are at least three dynein heavy chain genes in mammals. Whereas the cargo specificity of kinesin is believed to largely depend on 'tail' diversity encoded by different kinesin genes, dynein cargo interactions might be influenced by the diversity of dynein-associated polypeptides. In a similar way to kinesins, dynein complexes may also attach to cargoes by both binding to scaffolding proteins and by direct binding to transmembrane proteins.

Interestingly, an activator of dynein-catalysed organelle transport has been identified as dynactin, a 1.1 MDa protein complex consisting of 10–11 distinct polypeptides which include p150-Glued and the filament-forming actin-related protein ARP1 (Karki and Holzbaur 1999). It has been shown that the p150-Glued subunit binds to the 74 kDa dynein intermediate chain (DIC) and to microtubules. In this way, dynactin increases the run length of dynein-driven movements, acting as a processivity factor for the dynein motor on the microtubule (King and Schroer 2000). Furthermore, the ARP1 subunit of dynactin seems to interact with spectrin, which then binds to acidic phospholipids, thus providing a linkage between dynein and membrane vesicles (Muresan *et al.* 2001). Finally, dynactin may also play a crucial role in coordinating the bidirectional movement of organelles since p150-Glued not only binds DIC but also interacts with KAP, a protein which is a component of kinesin-II complex (Deacon *et al.* 2003; Gross 2003).

It has been shown that dynein, dynactin and kinesin-II may remain associated with membrane vesicles undergoing either anterograde or retrograde movements, indicating that their activities have to be properly controlled. For instance, dynein must be inactive during anterograde transport. However, little is known about the possible regulation of dynactin activity and its influence on kinesin and dynein motors and only a role for phosphorylation has been suggested (Reese and Haimo 2000).

A direct interaction of the dynein complex with transmembrane proteins may be important for the retrograde delivery of neurotrophic signals from the synaptic endings to neuronal cell bodies (Reynolds *et al.* 2000). Of interest in this regard is the reported association of Trk neurotrophin receptors with the 14 kDa light chain of dynein (Yano *et al.* 2001). Curiously, some neurotropic viruses like Herpes simplex 1 also use retrograde axonal transport to propagate the infection within the nervous system (Bearer *et al.* 2000) and this may also depend on the interaction of specific viral proteins with the dynein motor complex (Ye *et al.* 2000).

In addition to its role in intracellular transport of membrane organelles and other macromolecular complexes within mature neurons, dynein seems to play crucial roles during the development of the nervous system. Neuronal migration is an important event since neurons are formed in specialized proliferative zones but ultimately move to reside in distinct layered structures or organized nuclei (Rivas and Hatten 1995). During migration, a cell extends a process into which the nucleus is translocated. This nuclear movement, referred to as nucleokinesis, seems to be mediated by dynein in coordination with other microtubule-interacting proteins including LIS1 and NUDEL (Morris 2000; Sasaki *et al.* 2000; Feng and Walsh 2001). Finally, axonal extension and retraction, which are essential for proper axonal pathfinding, are essentially dependent on the movement of cytoskeletal polymers forward and backward. It has been suggested that dynein is also responsible for the transport of tubulin anterogradely down the axon (Baas and Ahmad 2001).

13.5.4.3 Myosins

Myosins are the motor molecules responsible for the movement of organelles along actin microfilaments. In particular, myosin V, which is very abundant in brain and has been localized in axons of cultured neurons seems to be implicated in the trafficking of synaptic vesicles containing synaptophysin (Prekeris and Terrian 1997).

13.6 The cytoskeleton in neurodegenerative diseases

Defects of the cytoskeleton may be a common feature contributing to neurodegeneration in many neurological diseases, which are characterized by the aberrant accumulation of certain cytoskeletal proteins. Disruption of the normal neuronal cytoskeleton may interfere with intraneuronal organelle transport and signal transduction, thus initiating a cascade of events including mitochondrial dysfunction and oxidative stress that ultimately leads to the loss of synapses and the death of neurons (McMurray 2000; Stamer *et al.* 2002). The contribution of cytoskeletal alterations to neurodegeneration has been thoroughly examined for some neurological disorders including Alzheimer's disease (AD), tauopathies, amyotrophic lateral sclerosis (ALS), and type II (axonopathy) of Charcot–Marie–Tooth disease (CMT).

13.6.1 Alzheimer's disease (AD)

Alzheimer's disease (AD) is the most prevalent cause of dementia in mid to old age. Clinically, patients with AD show a progressive deterioration of all cognitive functions. Histopathologically, AD is characterized by the presence of two kinds of abnormal protein deposits, amyloid plaques and neurofibrillary tangles, in specific areas of the patient brains, and finally by the atrophy of affected brain regions, which results from extensive losses of synapses and neurons (Ritchie and Lovestone 2002). Amyloid plaques are extracellular deposits of the Aβ peptide, which is derived through proteolysis from APP (amyloid precursor protein). Neurofibrillary tangles (NFT) are intraneuronal aggregates of fibrils constituted by hyperphosphorylated tau protein, which is also found in degenerating neurites (neuropil threads and dystrophic neurites).

The most widely accepted hypothesis on the pathogenesis of AD is the so-called 'amyloid cascade' hypothesis, which proposes that the processing of APP to give rise Aβ is the primary event in the pathogenic process. Then, the accumulation of Aβ peptide would lead to alterations in signal transduction, with the over-activation of protein kinases including GSK-3, CDK-5, and stress-activated protein kinases and the subsequent hyperphosphorylation of tau. Hyperphosphorylated tau would dissociate from microtubules and give rise to NFT formation, synaptic degeneration, and neuronal cell death (Alvarez *et al.* 2002; Mudher and Lovestone 2002). Supportive of this view is the fact that the overexpression of pseudohyperphosphorylated tau, a mutant tau protein in which phosphorylatable serine/threonine residues are substituted with glutamate to mimic hyperphosphorylation, leads to neuronal cell death (Fath *et al.* 2002). Furthermore, comparison of the effect of Aβ peptide on cultured neurons prepared from wild-type, tau knock-out and human tau transgenic mice suggests that tau plays a major role in Aβ-induced neurotoxicity (Rapoport *et al.* 2002). However, little is known about the mechanisms underlying tau-induced neurodegeneration. It has been speculated that a conformational change in tau that is induced by hyperphosphorylation promotes aberrant interactions with other proteins, thus causing cytoskeletal disruption and blockade of intraneuronal organelle transport (Alonso *et al.* 1997).

13.6.2 Tauopathies

Tauopathies comprise a group of heterogeneous dementias and movement disorders characterized by prominent intracellular fibrillar tau inclusions and neuronal degeneration in the absence of β-amyloid deposits. These tauopathies include progressive supranuclear palsy (PSP), corticobasal dementia with parkinsonism linked to chromosome 17 (FTDP-17), and amyotrophic lateral sclerosis/parkinsonism/dementia complex (ALS-PDC) (Lee *et al.* 2001). The identification of several autosomal dominant mutations in the tau gene which are responsible for some cases of inherited frontotemporal dementia (FTDP-17) has conclusively demonstrated that tau abnormalities can cause neurodegeneration. Whereas some mutations may decrease the association of protein phosphatase 2A with tau which thereby becomes hyperphosphorylated (Goedert *et al.* 2000), other mutations may directly cause a conformational change in tau protein which becomes 'neurotoxic' (Garcia and Cleveland 2001). Interestingly, it has been suggested that the 'neurotoxic' form of tau may be oligomeric since NFT or other large visible tau aggregates merely correspond to a late and severe stage in the neurodegenerative process (Wittmann *et al.* 2001). Thus, a 'toxic gain of function', presumably involving aberrant interactions with other proteins which would lead to cytoskeletal disruption, may be a common mechanism of tau-induced neurodegeneration both in AD and tauopathies.

13.6.3 **Amyotrophic lateral sclerosis (ALS)**

Amyotrophic lateral sclerosis (ALS) is characterized by the selective degeneration of motor neurons and the progressive atrophy of skeletal muscles resulting in eventual paralysis. Approximately 10% of ALS patients are inherited cases, whereas 90% of ALS cases are sporadic. Both sporadic and inherited cases share common pathological features such as the abnormal accumulation of neurofilaments within cell bodies and in the proximal part of axons of degenerating motor neurons (Julien 2001).

Some inherited cases of ALS are linked to autosomal dominant mutations in the gene coding for superoxide dismutase 1 (SOD1). Neurofilament inclusions within motor neurons have been described in ALS patients with SOD1 mutations and in transgenic mice expressing the mutant SOD1 forms. It has been suggested that mutant SOD1 may catalyse the oxidative damage of neurofilaments, thus driving neurofilament aggregation in motor neurons. Alternatively, neurofilament inclusions may merely arise from defects in slow axonal transport. Thus, mutant SOD1 may aggregate into insoluble protein complexes which interfere with slow axonal transport.

The appearance of neurofilament inclusions, the so-called spheroids, within axons may further block axonal transport, thus exacerbating the pathology. In contrast, large accumulations of neurofilaments in neuronal cell bodies that are induced in transgenic mice overexpressing NF-H may diminish the progression of the disease (Julien 2001). This protective effect of NF-H overexpression may be mediated by the formation of perikaryal neurofilament aggregates that act as phosphorylation sinks for CDK5, thus reducing the detrimental hyperphosphorylation of tau and other proteins (Nguyen *et al.* 2002). These results raise the possibility that tau abnormalities may also contribute to ALS pathogenesis. However, no filamentous tau aggregates have been isolated from transgenic mice overexpressing mutant SOD1.

13.6.4 **Charcot–Marie–Tooth disease type 2 (CMT-2)**

Charcot–Marie–Tooth disease (CMT) is an axonopathy primarily affecting motor neurons although sensory neurons may also degenerate. Some mutations in the gene coding for NF-L may lead to disrupted neurofilament assembly and axonal transport (Mersiyanova *et al.* 2000). Another form of CMT seems to be caused by mutation in the motor domain of KIF1Bβ (Zhao *et al.* 2001). This demonstrates that defects in axonal transport may underlie peripheral axonopathies.

It is tempting to speculate that the disorganization of the cytoskeleton and the subsequent failure of intraneuronal transport of organelles and other macromolecular complexes may be the proximal cause of synaptic dysfunction and neuronal cell death in many different neurodegenerative diseases. Of course, there might be many diverse molecular mechanisms triggering this cytoskeletal disruption in particular sets of neurons in distinct neurodegenerative disorders.

References

Ackerley, S., Thornhill, P., Grierson, A. J., Brownlees, J., Anderton, B. H., Leigh, P. N., *et al.* (2003) Neurofilament heavy chain side arm phosphorylation regulates axonal transport of neurofilaments. *J. Cell Biol.*, **161**, 489–95.

Aderem, A. (1992) Signal transduction and the actin cytoskeleton: the roles of MARCKS and profilin. *Trends Biochem. Sci.*, **17**, 438–43.

Allan, V. and Nathke, I. S. (2001) Catch and pull a microtubule: getting a grasp on the cortex. *Nat. Cell Biol.*, **3**, E226–8.

Alonso, A. D., Grundke-Iqbal, I., Barra, H. S., and Iqbal, K. (1997) Abnormal phosphorylation of tau and the mechanism of Alzheimer neurofibrillary degeneration: sequestration of microtubule-associated proteins 1 and 2 and the disassembly of microtubules by the abnormal tau. *Proc. Natl. Acad. Sci. USA*, **94**, 298–303.

Alvarez, G., Munoz-Montano, J. R., Satrustegui, J., Avila, J., Bogonez, E., and Diaz-Nido, J. (2002) Regulation of tau phosphorylation and protection against beta-amyloid-induced neurodegeneration by lithium. Possible implications for Alzheimer's disease. *Bipolar Disord.*, **4**, 153–65.

Amos, L. A. (2000) Focusing-in on microtubules. *Curr. Opin. Struct. Biol.*, **10**, 236–41.

Atkinson, S. J., Doberstein, S. K., and Pollard, T. D. (1992) Moving off the beaten tracks. *Curr. Biol.*, **2**, 326–8.

Avila, J., Dominguez, J., and Díaz-Nido, J. (1994) Regulation of microtubule dynamics by microtubule–associated protein expression and phosphorylation during neuronal development. *Int. J. Dev. Biol.*, **38**, 13–25.

Baas, P. W. (1999) Microtubules and neuronal polarity: lessons from mitosis. *Neuron*, **22**, 23–31.

Baas, P. W. and Ahmad, F. J. (2001) Force generation by cytoskeletal motor proteins as a regulator of axonal elongation and retraction. *Trends Cell Biol.*, **11**, 244–9.

Baas, P. W., Black, M. M., and Banker, G. A. (1989) Changes in microtubule polarity orientation during the development of hippocampal neurons in culture. *J. Cell Biol.*, **109**, 3085–94.

Baas, P. W., Slaughter, T., Brown, A., and Black, M. M. (1991) Microtubule dynamics in axons and dendrites. *J. Neurosci. Res.*, **30**, 134–53.

Bailey, C. H. and Kandel, E. R. (1993) Structural changes accompanying memory storage. *Annu. Rev. Physiol.*, **53**, 397–426.

Bearer, E. L., Breakefield, X. O., Schuback, D., Reese, T. S., and LaVail, J. H. (2000) Retrograde axonal transport of herpes simplex virus: evidence for a single mechanism and a role for tegument. *Proc. Natl. Acad. Sci. USA*, **97**, 8146–50.

Belmont, L. D. and Mitchison, T. J. (1996) Identification of a protein that interacts with tubulin dimers and increases the catastrophe rate of microtubules. *Cell*, **84**, 623–31.

Bennett, V. and Baines, A. J. (2001) Spectrin and ankyrin-based pathways: metazoan inventions for integrating cells into tissues. *Physiol. Rev.*, **81**, 1353–92.

Bennett, V. and Chen, L. (2001) Ankyrins and cellular targeting of diverse membrane proteins to physiological sites. *Curr. Opin. Cell Biol.*, **13**, 61–7.

Bennett, V. and Lambert, S. (1999) Physiological roles of axonal ankyrins in survival of premyelinated axons and localization of voltage-gated sodium channels. *J. Neurocytol.*, **28**, 303–18.

Billups, D., Hanley, J. G., Orme, M., Attwell, D., and Moss, S. J. (2000) GABAC receptor sensitivity is modulated by interaction with MAP1B. *J. Neurosci.*, **20**, 8643–50.

Bomze, H. M., Bulsara, K. R., Iskandar, B. J., Caroni, P., and Skene, J. H. (2001) Spinal axon regeneration evoked by replacing two growth cone proteins in adult neurons. *Nat. Neurosci.*, **4**, 38–43.

Bradke, F. and Dotti, C. G. (2000) Establishment of neuronal polarity: lessons from cultured hippocampal neurons. *Curr. Opin. Neurobiol.*, **10**, 574–81.

Brady, S. T. (1985) A novel brain ATPase with properties expected for the fast axonal transport motor. *Nature*, **317**, 73–5.

Buee, L., Bussiere, T., Buee-Scherrer, V., Delacourte, A., and Hof, P. R. (2000) Tau protein isoforms, phosphorylation and role in neurodegenerative disorders. *Brain Res. Brain Res. Rev.*, **33**, 95–130.

Burgoyne, R. D. (1991) The Neuronal Cytoskeleton. New York: Wiley–Liss.

Cassimeris, L. and Spittle, C. (2001) Regulation of microtubule-associated proteins. *Int. Rev. Cytol.*, **210**, 163–226.

Ching, G. Y. and Liem, R. K. (1993) Assembly of type IV neuronal intermediate filaments in nonneuronal cells in the absence of preexisting cytoplasmic intermediate filaments. *J. Cell Biol.*, **122**, 1323–35.

Choo, Q. L. and Bray, D. (1978) Two forms of neuronal actin. *J. Neurochem.*, **31**, 217–24.

Clark, E. A. and Brugge, J. S. (1995) Integrins and signal transduction pathways: the road taken. *Science*, **268**, 233–9.

Craig, A. M. and Banker, G. (1994) Neuronal polarity. *Annu. Rev. Neurosci.*, **17**, 267–310.

De Waegh, S. M., Lee, V. M., and Brady, S. T. (1992) Local modulation of neurofilament phosphorylation, axonal caliber, and slow axonal transport by myelinating Schwann cells. *Cell*, **68**, 451–63.

Deacon, S. W., Serpinskaya, A. S., Vaughan, P. S., Lopez Fanarraga, M., Vernos, I., Vaughan, K. T., *et al.* (2003) Dynactin is required for bidirectional organelle transport. *J. Cell Biol.*, **160**, 297–301.

Di Tella, M. C., Feiguin, F., Carri, N., Kosik, K. S., and Caceres, A. (1996) MAP–1B/TAU functional redundancy during laminin–enhanced axonal growth. *J. Cell Sci.*, **109**, 467–77.

Dickson, B. J. (2001) Rho GTPases in growth cone guidance. *Curr. Opin. Neurobiol.*, **11**, 103–10.

Dieterich, D. C., Trivedi, N., Engelmann, R., Gundelfinger, E. D., Gordon-Weeks, P. R., and Kreutz, M. R. (2002) Partial regeneration and long-term survival of rat retinal ganglion cells after optic nerve crush is accompanied by altered expression, phosphorylation and distribution of cytoskeletal proteins. *Eur. J. Neurosci.*, **15**, 1433–43.

Dos Remedios, C. G., Chhabra, D., Kekic, M., Dedova, I. V., Tsubakihara, M., Berry, D. A., *et al.* (2003) Actin binding proteins: regulation of cytoskeletal microfilaments. *Physiol. Rev.*, **83**, 433–73.

Dotti, C. G., Sullivan, C. A., and Banker, G. A. (1988) The establishment of polarity by hippocampal neurons in culture. *J. Cell Biol.*, **108**, 1507–16.

Edelmann, W., Zervas, M., Costello, P., Roback, L., Fischer, I., Hammarback, J. A., *et al.* (1996) Neuronal abnormalities in microtubule-associated protein 1B mutant mice. *Proc. Natl. Acad. Sci. USA*, **93**, 1270–5.

Elder, G. A., Friedrich Jr, V. L., Bosco, P., Kang, C., Gourov, A., Tu, P. H., *et al.* (1998) Absence of the mid-sized neurofilament subunit decreases axonal calibers, levels of light neurofilament (NF-L), and neurofilament content. *J. Cell Biol.*, **141**, 727–39.

Elder, G. A., Friedrich Jr, V. L., Margita, A., and Lazzarini, R. A. (1999) Age-related atrophy of motor axons in mice deficient in the mid-sized neurofilament subunit. *J. Cell Biol.*, **146**, 181–92.

Fath, K. R. and Lasek, R. J. (1988) Two classes of actin microfilaments are associated with the inner cytoskeleton of axons. *J. Cell Biol.*, **107**, 613–21.

Fath, T., Eidenmuller, J., and Brandt, R. (2002) Tau-mediated cytotoxicity in a pseudohyperphosphorylation model of Alzheimer's disease. *J. Neurosci.*, **22**, 9733–41.

Feng, Y. and Walsh, C. A. (2001) Protein-protein interactions, cytoskeletal regulation and neuronal migration. *Nat. Rev. Neurosci.*, **2**, 408–16.

Ferreira, A., Kincaid, R., and Kosik, K. S. (1993) Calcineurin is associated with the cytoskeleton of cultured neurons and has a role in the acquisition of polarity. *Mol. Biol. Cell*, **4**, 1225–38.

Ferreira, A. and Rapoport, M. (2002) The synapsins: beyond the regulation of neurotransmitter release. *Cell Mol. Life Sci.*, **59**, 589–95.

Fifkova, E. and Morales, M. (1992) Actin matrix of dendritic spines, synaptic plasticity, and long-term potentiation. *Int. Rev. Cytol.*, **139**, 267–307.

Forscher, P. and Smith, S. J. (1988) Actions of cytochalasins on the organization of actin filaments and microtubules in a neuronal growth cone. *J. Cell Biol.*, **107**, 1505–16.

Frey, D., Laux, T., Xu, L., Schneider, C., and Caroni, P. (2000) Shared and unique roles of CAP23 and GAP43 in actin regulation, neurite outgrowth, and anatomical plasticity. *J. Cell Biol.*, **149**, 1443–54.

Fukazawa, Y., Saitoh, Y., Ozawa, F., Ohta, Y., Mizuno, K., and Inokuchi, K. (2003) Hippocampal LTP is accompanied by enhanced F-actin content within the dendritic spine that is essential for late LTP maintenance in vivo. *Neuron*, **38**, 447–60.

Galjart, N. and Perez, F. (2003) A plus-end raft to control microtubule dynamics and function. *Curr. Opin. Cell Biol.*, **15**, 48–53.

Garcia, M. L. and Cleveland, D. W. (2001) Going new places using an old MAP: tau, microtubules and human neurodegenerative disease. *Curr. Opin. Cell Biol.*, **13**, 41–8.

Garner, C. C., Tucker, R. P., and Matus, A. (1988) Selective localization of messenger RNA for cytoskeletal protein MAP2 in dendrites. *Nature*, **336**, 674–7.

Goedert, M., Satumtira, S., Jakes, R., Smith, M. J., Kamibayashi, C., White 3rd, C. L., *et al.* (2000) Reduced binding of protein phosphatase 2A to tau protein with frontotemporal dementia and parkinsonism linked to chromosome 17 mutations. *J. Neurochem.*, **75**, 2155–62.

Goldberg, J. L. (2003) How does an axon grow? *Genes Dev.*, **17**, 941–58.

Goldstein, L. S. and Yang, Z. (2000) Microtubule-based transport systems in neurons: the roles of kinesins and dyneins. *Annu. Rev. Neurosci.*, **23**, 39–71.

Gonzalez-Billault, C., Avila, J., and Caceres, A. (2001) Evidence for the role of MAP1B in axon formation. *Mol. Biol. Cell*, **12**, 2087–98.

Gonzalez-Billault, C., Demandt, E., Wandosell, F., Torres, M., Bonaldo, P., Stoykova, A., *et al.* (2000) Perinatal lethality of microtubule-associated protein 1B-deficient mice expressing alternative isoforms of the protein at low levels. *Mol. Cell Neurosci.*, **16**, 408–21.

Gonzalez-Billault, C., Engelke, M., Jimenez-Mateos, E. M., Wandosell, F., Caceres, A. and Avila, J. (2002) Participation of structural microtubule-associated proteins (MAPs) in the development of neuronal polarity. *J. Neurosci. Res.*, **67**, 713–9.

Goold, R. G., Owen, R., and Gordon-Weeks, P. R. (1999) Glycogen synthase kinase 3beta phosphorylation of microtubule-associated protein 1B regulates the stability of microtubules in growth cones. *J. Cell Sci.*, **112 (Pt 19)**, 3373–84.

Gordon-Weeks, P. R. (1987) The cytoskeletons of isolated, neuronal growth cones. *Neuroscience*, **21**, 977–89.

Grafstein, B. and Forman, D. S. (1980) Intracellular transport in neurons. *Physiol. Rev.*, **60**, 1167–283.

Grant, P. and Pant, H. C. (2000) Neurofilament protein synthesis and phosphorylation. *J. Neurocytol.*, **29**, 843–72.

Greenough, W. T. and Bailey, C. H. (1987) The anatomy of a memory. *Trends Neurosci.*, **11**, 142–7.

Gross, S. P. (2003) Dynactin: coordinating motors with opposite inclinations. *Curr. Biol.*, **13**, R320–2.

Gundersen, G. G. and Cook, T. A. (1999) Microtubules and signal transduction. *Curr. Opin. Cell Biol.*, **11**, 81–94.

Halpain, S. (2000) Actin and the agile spine: how and why do dendritic spines dance? *Trends Neurosci.*, **23**, 141–6.

Hanley, J. G., Koulen, P., Bedford, F., Gordon-Weeks, P. R., and Moss, S. J. (1999) The protein MAP-1B links GABA(C) receptors to the cytoskeleton at retinal synapses. *Nature*, **397**, 66–9.

Harada, A., Oguchi, K., Okabe, S., Kuno, J., Terada, S., Ohshima, T., *et al.* (1994) Altered microtubule organization in small-calibre axons of mice lacking tau protein. *Nature*, **369**, 488–91.

Herrmann, H. and Aebi, U. (2000) Intermediate filaments and their associates: multi-talented structural elements specifying cytoarchitecture and cytodynamics. *Curr. Opin. Cell Biol.*, **12**, 79–90.

Higgs, H. N. and Pollard, T. D. (2001) Regulation of actin filament network formation through ARP2/3 complex: activation by a diverse array of proteins. *Annu. Rev. Biochem.*, **70**, 649–76.

Hirokawa, N. (1982) Cross-linker system between neurofilaments, microtubules, and membranous organelles in frog axons revealed by the quick-freeze, deep-etching method. *J. Cell Biol.*, **94**, 129–42.

Hirokawa, N. (1998) Kinesin and dynein superfamily proteins and the mechanism of organelle transport. *Science*, **279**, 519–26.

Hirokawa, N., Sobue, K., Kanda, K., Harada, A., and Yorifuji, H. (1989) The cytoskeletal architecture of the presynaptic terminal and molecular structure of synapsin 1. *J. Cell Biol.*, **108**, 111–26.

Howard, J. and Hyman, A. A. (2003) Dynamics and mechanics of the microtubule plus end. *Nature*, **422**, 753–8.

Huang, J. D., Brady, S. T., Richards, B. W., Stenolen, D., Resau, J. H., Copeland, N. G., *et al.* (1999) Direct interaction of microtubule- and actin-based transport motors. *Nature*, **397**, 267–70.

Huber, A. B., Kolodkin, A. L., Ginty, D. D. and Cloutier, J.-F. (2003) Signaling at the growth cone. *Annu. Rev. Neurosci.*, **26**, 509–63.

Inagaki, M., Nishi, Y., Nishizawas, K., Matsuyama, M., and Sato, C. (1989) Site-specific phosphorylation induces disassembly of vimentin *in vitro*. *Nature* **328**, 649–52.

Jay, D. G. (2000) The clutch hypothesis revisited: ascribing the roles of actin-associated proteins in filopodial protrusion in the nerve growth cone. *J. Neurobiol.*, **44**, 114–25.

Joshi, H. C. and Baas, P. W. (1993) A new perspective on microtubules and axon growth. *J. Cell Biol.*, **121**, 1191–6.

Julien, J. P. (1999) Neurofilament functions in health and disease. *Curr. Opin. Neurobiol.*, **9**, 554–60.

Julien, J. P. (2001) Amyotrophic lateral sclerosis. Unfolding the toxicity of the misfolded. *Cell*, **104**, 581–91.

Kamal, A., Almenar-Queralt, A., LeBlanc, J. F., Roberts, E. A., and Goldstein, L. S. (2001) Kinesin-mediated axonal transport of a membrane compartment containing beta-secretase and presenilin-1 requires APP. *Nature*, **414**, 643–8.

Karcher, R. L., Deacon, S. W., and Gelfand, V. I. (2002) Motor-cargo interactions: the key to transport specificity. *Trends Cell Biol.*, **12**, 21–7.

Karki, S. and Holzbaur, E. L. (1999) Cytoplasmic dynein and dynactin in cell division and intracellular transport. *Curr. Opin. Cell Biol.*, **11**, 45–53.

King, S. J. and Schroer, T. A. (2000) Dynactin increases the processivity of the cytoplasmic dynein motor. *Nat. Cell Biol.*, **2**, 20–4.

King, S. M. (2000) The dynein microtubule motor. *Biochim. Biophys. Acta*, **1496**, 60–75.

Kirschner, M. and Mitchison, T. (1986) Beyond self-assembly: from microtubules to morphogenesis. *Cell*, **45**, 329–42.

Kirsh, J. and Betz, H. (1996) The postsynaptic localization of the glycine receptor-associated protein Gephyrin is regulated by the cytoskeleton. *J. Neurosci.*, **15**, 4148–56.

Krause, M., Bear, J. E., Loureiro, J. J., and Gertler, F. B. (2002) The Ena/VASP enigma. *J. Cell Sci.*, **115**, 4721–6.

Kreis, T. and Vale, R. (1993) Guidebook to Cytoskeletal and Motor Proteins. Oxford: Oxford University Press.

Krendel, M., Zenke, F. T., and Bokoch, G. M. (2002) Nucleotide exchange factor GEF-H1 mediates cross-talk between microtubules and the actin cytoskeleton. *Nat. Cell Biol.*, **4**, 294–301.

Langford, G. M. (1995) Actin- and microtubule-dependent organelle motors: interrelationships between the two motility systems. *Curr. Opin. Cell Biol.*, **7**, 82–8.

Lee, M. K., Xu, Z., Wong, P. C., and Cleveland, D. W. (1993) Neurofilaments are obligate heteropolymers in vivo. *J. Cell Biol.*, **122**, 1337–50.

Lee, V. M., Goedert, M. and Trojanowski, J. Q. (2001) Neurodegenerative tauopathies. *Annu. Rev. Neurosci.*, **24**, 1121–59.

Lewis, A. K. and Bridgman, P. C. (1992) Nerve growth cone lamellipodia contain two populations of actin filaments that differ in organization and polarity. *J. Cell Biol.*, **119**, 1219–43.

Lin, C. H., Espreafico, E. M., Mooseker, M. S., and Forscher, P. (1996) Myosin drives retrograde F-actin flow in neuronal growth cones. *Neuron*, **16**, 769–82.

Llinas, R., McGuinness, T. L., Leonard, C. S., Sugimori, M., and Greengard, P. (1985) Intraterminal injection of synapsin I or calcium/calmodulin-dependent protein kinase II alters neurotransmitter release at the squid giant synapse. *Proc. Natl. Acad. Sci. USA*, **82**, 3035–9.

Loureiro, J. J., Rubinson, D. A., Bear, J. E., Baltus, G. A., Kwiatkowski, A. V., and Gertler, F. B. (2002) Critical roles of phosphorylation and actin binding motifs, but not the central proline-rich region, for Ena/vasodilator-stimulated phosphoprotein (VASP) function during cell migration. *Mol. Biol. Cell*, **13**, 2533–46.

Luna, E. J. and Hitt, A. L. (1992) Cytoskeleton–plasma membrane interactions. *Science*, **258**, 955–64.

Ma, D., Himes, B. T., Shea, T. B., and Fischer, I. (2000) Axonal transport of microtubule-associated protein 1B (MAP1B) in the sciatic nerve of adult rat: distinct transport rates of different isoforms. *J. Neurosci.*, **20**, 2112–20.

Mansfield, S. G., Díaz-Nido, J., Gordon-Weeks, P. R., and Avila, J. (1991) The distribution and phosphorylation of the microtubule-associated protein MAP 1B in growth cones. *J. Neurocytol.*, **20**, 1007–22.

Martin, A. C. and Drubin, D. G. (2003) Impact of genome-wide functional analyses on cell biology research. *Curr. Opin. Cell Biol.*, **15**, 6–13.

Matus, A. (2000) Actin-based plasticity in dendritic spines. *Science*, **290**, 754–8.

McMurray, C. T. (2000) Neurodegeneration: diseases of the cytoskeleton? *Cell Death Differ.*, **7**, 861–5.

Mersiyanova, I. V., Perepelov, A. V., Polyakov, A. V., Sitnikov, V. F., Dadali, E. L., Oparin, R. B., *et al.* (2000) A new variant of Charcot-Marie-Tooth disease type 2 is probably the result of a mutation in the neurofilament-light gene. *Am. J. Hum. Genet.*, **67**, 37–46.

Meyer, G. and Feldman, E. L. (2002) Signaling mechanisms that regulate actin-based motility processes in the nervous system. *J. Neurochem.*, **83**, 490–503.

Miki, H., Setou, M., Kaneshiro, K., and Hirokawa, N. (2001) All kinesin superfamily protein, KIF, genes in mouse and human. *Proc. Natl. Acad. Sci. USA*, **98**, 7004–11.

Mitchison, T. and Kirschner, M. (1988) Cytoskeletal dynamics and nerve growth. *Neuron*, **1**, 761–72.

Montague, P. R. (1993) Transforming sensory experience into structural change. *Proc. Natl. Acad. Sci. USA*, **90**, 6379–80.

Morfini, G., Szebenyi, G., Elluru, R., Ratner, N., and Brady, S. T. (2002) Glycogen synthase kinase 3 phosphorylates kinesin light chains and negatively regulates kinesin–based motility. *EMBO J.*, **21**, 281–93.

Morris, N. R. (2000) Nuclear migration. From fungi to the mammalian brain. *J. Cell Biol.*, **148**, 1097–101.

Mudher, A. and Lovestone, S. (2002) Alzheimer's disease — do tauists and baptists finally shake hands? *Trends Neurosci.*, **25**, 22–6.

Muñoz-Montaño, J. R., Moreno, F. J., Avila, J., and Díaz-Nido, J. (1997) Lithium inhibits Alzheimer's disease-like tau protein phosphorylation in neurons. *FEBS Lett.*, **411**, 183–8.

Muresan, V., Stankewich, M. C., Steffen, W., Morrow, J. S., Holzbaur, E. L., and Schnapp, B. J. (2001) Dynactin-dependent, dynein-driven vesicle transport in the absence of membrane proteins: a role for spectrin and acidic phospholipids. *Mol. Cell*, **7**, 173–83.

Nguyen, M. D., Lanviere, R. C., and Julien, J. P. (2002) Deregulation of Cdk5 in a mouse model of ALS: Toxicity alleviated by perikaryal neurofilament inclusions. *Neuron*, **8**, 135–47.

Nixon, R. A. (1991) Axonal transport of cytoskeletal proteins. In *The Neuronal Cytoskeleton*, (ed. R. D. Burgoyne), pp. 175–200. New York: Alan R. Liss.

Nixon, R. A. and Shea, T. B. (1992) Dynamics of neuronal intermediate filaments: a developmental perspective. *Cell Motil. Cytoskeleton*, **22**, 81–91.

Nixon, R. A. and Sihag, R. K. (1991) Neurofilament phosphorylation: a new look at regulation and function. *Trends Neurosci.*, **14**, 501–6.

Oakley, B. R. (1992) Gamma tubulin: the microtubule organizer? *Trends Cell Biol.*, **2**, 1–5.

Okabe, S., Miyasaka, H., and Hirokawa, N. (1993) Dynamics of the neuronal intermediate filaments. *J. Cell Biol.*, **121**, 375–86.

Padilla, R., López Otín, C., Serrano, L., and Avila, J. (1993) Role of the carboxy terminal region of beta tubulin on microtubule dynamics through its interaction with the GTP phosphate binding region. *FEBS Lett.*, **325**, 173–6.

Peters, A., Palay, S. L., and De Webster, H. (1976) The Fine Structure of the Nervous System (translated by RM May). London: Oxford University Press.

Prekeris, R. and Terrian, D. M. (1997) Brain myosin V is a synaptic vesicle-associated motor protein: evidence for a Ca^{2+}-dependent interaction with the synaptobrevin-synaptophysin complex. *J. Cell Biol.*, **137**, 1589–601.

Ramón y Cajal, S. (1928) Degeneration and Regeneration of the Nervous System (translated by RM May). London: Oxford University Press.

Rapoport, M., Dawson, H. N., Binder, L. I., Vitek, M. P., and Ferreira, A. (2002) Tau is essential to beta-amyloid-induced neurotoxicity. *Proc. Natl. Acad. Sci. USA*, **99**, 6364–9.

Ratner, N. and Mahler, H. R. (1983) Structural organization of filamentous proteins in postsynaptic density. *Biochemistry*, **22**, 2446–53.

Reese, E. L. and Haimo, L. T. (2000) Dynein, dynactin, and kinesin II's interaction with microtubules is regulated during bidirectional organelle transport. *J. Cell Biol.*, **151**, 155–66.

Reynolds, A. J., Bartlett, S. E., and Hendry, I. A. (2000) Molecular mechanisms regulating the retrograde axonal transport of neurotrophins. *Brain Res. Brain Res. Rev.*, **33**, 169–78.

Riederer, B. M., Zagon, I. S., and Goodman, S. R. (1986) Brain spectrin(240/235) and brain spectrin(240/235E): two distinct spectrin subtypes with different locations within mammalian neural cells. *J. Cell Biol.*, **102**, 2088–97.

Riederer, B. M., Zagon, I. S., and Goodman, S. R. (1987) Brain spectrin(240/235) and brain spectrin(240/235E): differential expression during mouse brain development. *J. Neurosci.*, **7**, 864–74.

Ritchie, K. and Lovestone, S. (2002) The dementias. *Lancet*, **360**, 1759–66.

Rivas, R. J. and Hatten, M. E. (1995) Motility and cytoskeletal organization of migrating cerebellar granule neurons. *J. Neurosci.*, **15**, 981–9.

Rosenmund, C. and Westbrook, G. L. (1993) Calcium-induced actin depolymerization reduces NMDA channel activity. *Neuron*, **10**, 805–14.

Rubtsov, A. M. and Lopina, O. D. (2000) Ankyrins. *FEBS Lett.*, **482**, 1–5.

Sanchez, C., Diaz-Nido, J., and Avila, J. (2000) Phosphorylation of microtubule-associated protein 2 (MAP2) and its relevance for the regulation of the neuronal cytoskeleton function. *Prog. Neurobiol.*, **61**, 133–68.

Sasaki, S., Shionoya, A., Ishida, M., Gambello, M. J., Yingling, J., Wynshaw-Boris, A., *et al.* (2000) A LIS1/NUDEL/cytoplasmic dynein heavy chain complex in the developing and adult nervous system. *Neuron*, **28**, 681–96.

Scott, E. K. and Luo, L. (2001) How do dendrites take their shape? *Nat. Neurosci.*, **4**, 359–65.

Shahani, N. and Brandt, R. (2002) Functions and malfunctions of the tau proteins. *Cell Mol. Life Sci.*, **59**, 1668–80.

Sharp, D. J., Yu, W., and Baas, P. W. (1995) Transport of dendritic microtubules establishes their nonuniform polarity orientation. *J. Cell Biol.*, **130**, 93–103.

Shea, T. B. and Flanagan, L. A. (2001) Kinesin, dynein and neurofilament transport. *Trends Neurosci.*, **24**, 644–8.

Shiina, N., Gotoh, Y., and Nishida, E. (1995) Microtubule severing-activity in M phase. *Trends Cell Biol.*, **5**, 283–6.

Smith, S. J. (1988) Neuronal cytomechanics: the actin-based motility of growth cones. *Science*, **242**, 708–15.

Stamer, K., Vogel, R., Thies, E., Mandelkow, E., and Mandelkow, E. M. (2002) Tau blocks traffic of organelles, neurofilaments, and APP vesicles in neurons and enhances oxidative stress. *J. Cell Biol.*, **156**, 1051–63.

Steward, O. and Banker, G. A. (1992) Getting the message from the gene to the synapse: sorting and intracellular transport of RNA in neurons. *Trends Neurosci.*, **15**, 180–6.

Stossel, T. P. (1993) On the crawling of animal cells. *Science*, **260**, 1086–94.

Suetsugu, S., Hattori, M., Miki, H., Tezuka, T., Yamamoto, T., Mikoshiba, K., *et al.* (2002) Sustained activation of N-WASP through phosphorylation is essential for neurite extension. *Dev. Cell*, **3**, 645–58.

Sullivan, K. F. (1988) Structure and utilization of tubulin isotypes. *Annu. Rev. Cell Biol.*, **4**, 687–716.

Takei, Y., Teng, J., Harada, A., and Hirokawa, N. (2000) Defects in axonal elongation and neuronal migration in mice with disrupted tau and map1b genes. *J. Cell Biol.*, **150**, 989–1000.

Terada, S. and Hirokawa, N. (2000) Moving on to the cargo problem of microtubule-dependent motors in neurons. *Curr. Opin. Neurobiol.*, **10**, 566–73.

Tsai, M. Y., Morfini, G., Szebenyi, G., and Brady, S. T. (2000) Release of kinesin from vesicles by hsc70 and regulation of fast axonal transport. *Mol. Biol. Cell*, **11**, 2161–73.

Tsukita, S., Kobayashi, T., and Matsumoto, G. (1986) Subaxolemmal cytoskeleton in squid giant axon. II. Morphological identification of microtubule- and microfilament-associated domains of axolemma. *J. Cell Biol.*, **102**, 1710–25.

Tucker, R. P. (1990) The roles of microtubule-associated proteins in brain morphogenesis: a review. *Brain Res. Brain Res. Rev.*, **15**, 101–20.

Tytell, M., Black, M. M., Garner, J. A., and Lasek, R. J. (1981) Axonal transport: each major rate component reflects the movement of distinct macromolecular complexes. *Science*, **214**, 179–81.

Ulloa, L., Díez-Guerra, F. J., Avila, J., and Díaz-Nido, J. (1994) Localization of differentially phosphorylated isoforms of microtubule-associated protein 1B in cultured rat hippocampal neurons. *Neuroscience*, **61**, 211–23.

Verhey, K. J., Meyer, D., Deehan, R., Blenis, J., Schnapp, B. J., Rapoport, T. A., *et al.* (2001) Cargo of kinesin identified as JIP scaffolding proteins and associated signaling molecules. *J. Cell Biol.*, **152**, 959–70.

Wang, L., Ho, C. L., Sun, D., Liem, R. K., and Brown, A. (2000) Rapid movement of axonal neurofilaments interrupted by prolonged pauses. *Nat. Cell Biol.*, **2**, 137–41.

Wegner, A. (1985) Subtleties of actin assembly. *Nature*, **313**, 97–8.

Wittmann, C. W., Wszolek, M. F., Shulman, J. M., Salvaterra, P. M., Lewis, J., Hutton, M., *et al.* (2001) Tauopathy in Drosophila: neurodegeneration without neurofibrillary tangles. *Science*, **293**, 711–14.

Xia, C. H., Roberts, E. A., Her, L. S., Liu, X., Williams, D. S., Cleveland, D. W., *et al.* (2003) Abnormal neurofilament transport caused by targeted disruption of neuronal kinesin heavy chain KIF5A. *J. Cell Biol.*, **161**, 55–66.

Yano, H., Lee, F. S., Kong, H., Chuang, J., Arevalo, J., Perez, P., *et al.* (2001) Association of Trk neurotrophin receptors with components of the cytoplasmic dynein motor. *J. Neurosci.*, **21**, RC125.

Ye, G. J., Vaughan, K. T., Vallee, R. B., and Roizman, B. (2000) The herpes simplex virus 1 U(L)34 protein interacts with a cytoplasmic dynein intermediate chain and targets nuclear membrane. *J. Virol.*, **74**, 1355–63.

Zakharenko, S. and Popov, S. (1998) Dynamics of axonal microtubules regulate the topology of new membrane insertion into the growing neurites. *J. Cell Biol.*, **143**, 1077–86.

Zhao, C., Takita, J., Tanaka, Y., Setou, M., Nakagawa, T., Takeda, S., *et al.* (2001) Charcot-Marie-Tooth disease type 2A caused by mutation in a microtubule motor KIF1Bbeta. *Cell*, **105**, 587–97.

Chapter 14

Neuronal plasticity

Brian J. Morris

Neurones possess the ability to alter certain aspects of their biochemical and morphological character in response to changes in their local environment or their level of activity. This plasticity presumably allows them to adapt and survive in their altered circumstances, or to assume a different functional role. Neurones are far from being unique in this respect, and many, perhaps the majority, of different cell types in the body show some degree of plasticity. However, it can be argued that neurones have brought this particular part of the repertoire of cellular response to its highest level of sophistication. The plastic changes observed in neurones following a stimulus can be short or long-lasting, subtle or dramatic. A large number of different mechanisms can be invoked, and the changes that are induced can occur in isolation or be part of a complex, coordinated response.

In mature neurones, attention has focussed on the plasticity that underlies the processes of learning and memory, in particular the ability of specific synapses to alter the efficiency of their neurotransmission. A number of experimental models have been developed and characterized across a wide range of species, and it has become clear that a number of features are common to some or all of the paradigms.

14.1 Experimental models of neuronal plasticity

14.1.1 Hippocampal long-term potentiation

Long-term potentiation (LTP) was first observed in the hippocampal formation, a brain region with a role in the processes of learning and memory. While it has since become clear that LTP can be detected in many different brain regions, the likelihood remains that the phenomenon provides a basis for a learning-like change in the properties of networks of neurones (Bliss and Collingridge 1993). Activity-dependent changes in the efficiency of synaptic transmission are observed in particular pathways: typically a brief high-frequency burst of stimulation results in a potentiation of transmission that can last for many days *in vivo*. During this time, a number of temporal components can be identified, each caused by the activation of distinct intracellular mechanisms (Bliss and Collingridge 1993).

At the synapses of two of the major hippocampal pathways — the perforant path/dentate gyrus synapses, and the Schaffer collateral/CA1 synapses — the induction of

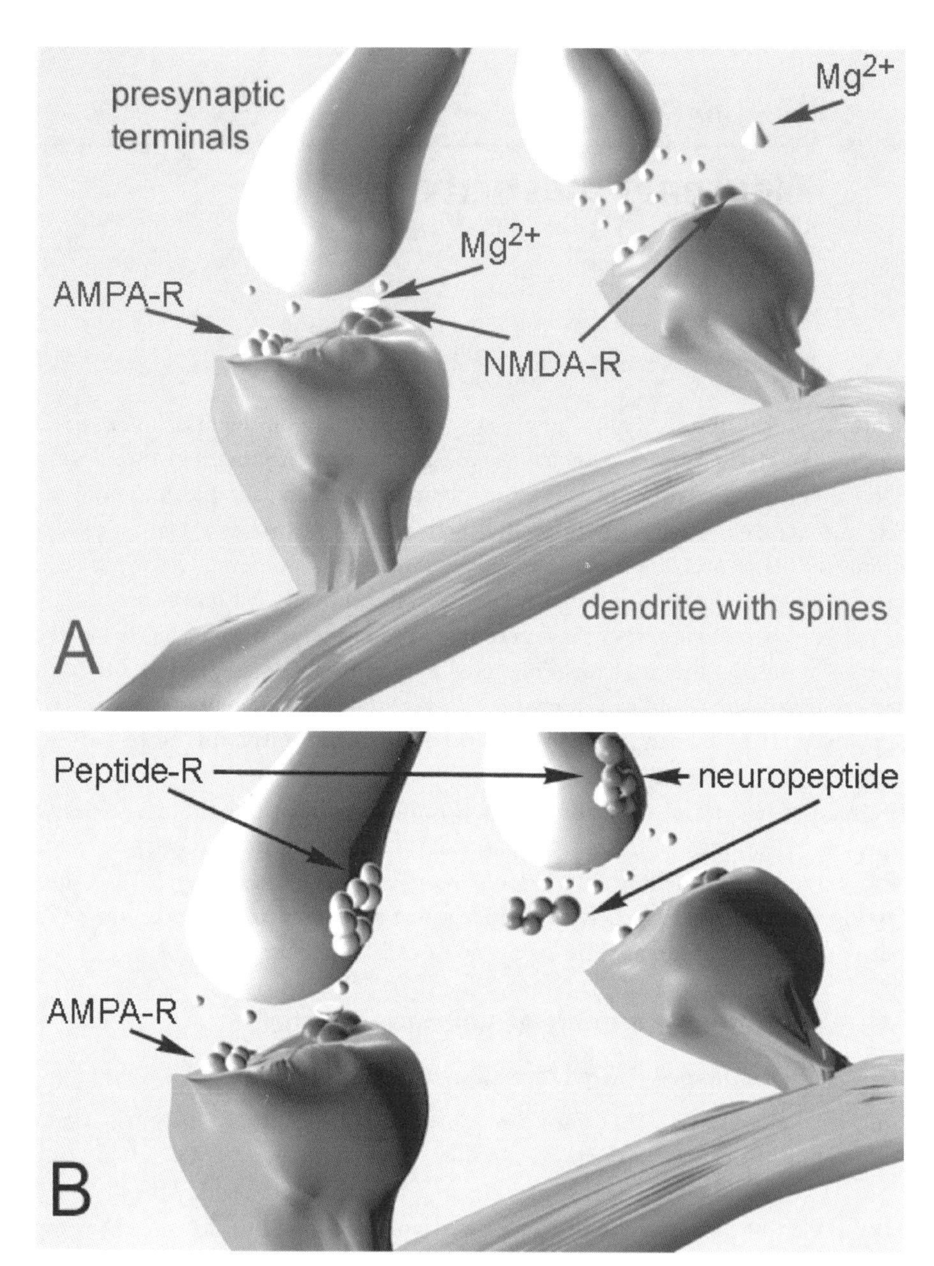

Fig. 14.1 Schematic diagram of two cellular mechanisms where high-frequency firing of afferent fibres leads to a sustained enhancement in synaptic efficiency. (A) Under normal (low) rates of firing, the Mg^{2+} block of the NMDA receptor (NMDA-R) prevents any Ca^{2+} influx into the post-synaptic spine. Synaptic transmission is mediated by Na^{+} influx through AMPA receptors (AMPA-R). Higher firing rates produce sufficient depolarization to relieve the Mg^{2+} block, and the ensuing Ca^{2+} influx through the NMDA receptor activates calcium-dependent enzymes in the spine to alter synaptic properties. The induction of LTP in the CA1 region of the hippocampus is thought to follow such a mechanism.

LTP is dependent on activation of the NMDA class of glutamate receptor (see Chapter 8). This receptor has the unique property of allowing Ca^{2+} influx in a manner that is subject to a voltage-dependent block by Mg^{2+} ions. If high-frequency afferent stimulation results in sufficient release of glutamate from the presynaptic terminals, there will be enough post-synaptic depolarization (mediated by non-NMDA glutamate receptors) to remove the Mg^{2+} blockade, and the NMDA receptor will allow influx of Ca^{2+} ions (Fig. 14.1A).

At another major hippocampal synapse — the mossy fibre/CA3 synapse — induction of LTP is not dependent on activation of NMDA receptors, but rather on the kainate class of glutamate receptor (Bortolotto *et al.* 1999). It also appears that metabotropic glutamate receptors (Conquet *et al.* 1994) and opioid receptors (Morris and Johnston 1995) are involved. Indeed, peptide neurotransmitters would be particularly convenient mediators of synaptic plasticity, since there is a great deal of evidence, from both the peripheral and central nervous systems, that peptides which coexist with conventional neurotransmitters may only be released by high-frequency nerve activity. It is easy to imagine opioid peptides, which are present in the mossy fibres, playing a role in the LTP that follows high-frequency mossy fibre firing (Fig. 14.1B). At this synapse, the intracellular pathways transducing the stimulus for plasticity may involve primarily cAMP elevations, rather than Ca^{2+} (Huang *et al.* 1994).

The mechanisms involved in hippocampal LTP are unlikely to be unique to that brain region or experimental paradigm. Indeed, there is a great deal of evidence that similar mechanisms operate in many of the other forms of neuronal plasticity that have been studied, such as LTP in the basal ganglia (Kombian and Malenka 1994), long-term depression (LTD) in the hippocampus, cerebellum, or basal ganglia (Nakazawa *et al.* 1993; Kombian and Malenka 1994), or the sensitisation to excitatory stimuli that occurs during limbic system 'kindling'.

14.1.2 Limbic system kindling

In limbic system kindling, electrodes are chronically implanted in regions of the limbic system such as the hippocampus or amygdala. A daily or twice-daily stimulus is then applied, at a constant level which initially has no overt behavioural effect. Over a period of few weeks, this same stimulus starts to produce a behavioural response, eventually leading to a generalized seizure. Once this sensitization phenomenon, or kindling, has occurred, the same stimulation will continue to produce a seizure whenever

Fig. 14.1 (B) Neuropeptides (for example opioid peptides), stored in a separate population of vesicles, coexist with glutamate in the afferent fibres. Under normal (low) rates of firing, only glutamate is released, and AMPA (or kainate) receptors mediate synaptic transmission. Higher rates of firing result in peptide release as well, and the post-synaptic (or pre-synaptic) peptide receptors (Peptide-R) then activate G-proteins (G) and alter the activity of cAMP-dependent enzymes to affect synaptic properties. The induction of LTP in the CA3 region of the hippocampus is thought partially to follow such a mechanism.

it is given, even if no stimulation has been given for a number of days or weeks. The sensitivity of the kindling procedure to blockade by antagonists of the NMDA receptor, along with the ability of NMDA and related agonists to precipitate seizure activity, both *in vivo* and *in vitro*, has provided strong evidence that NMDA receptors are involved at some stage in the plastic process.

The kindling procedure is widely used as an experimental model for epilepsy, and the biochemical and morphological changes observed in the hippocampi from kindled animals are similar to those detected in the brains of patients with temporal lobe epilepsy.

14.1.3 Cerebellar long-term depression

Purkinje cells in the cerebellum receive synaptic input from both climbing fibres and parallel fibres. Simultaneous stimulation of both inputs results in a long-lasting depression of transmission at the parallel fibre synapses, and the phenomenon has been suggested to form the basis of cerebellar motor learning and memory. In this case, NMDA receptors are not involved: rather, it is the AMPA class of glutamate receptor, most likely in association with metabotropic receptors (Conquet *et al.* 1994) and nitric oxide release (Nakazawa *et al.* 1993), that provides the initial stimulus.

14.1.4 Invertebrate models

The marine snail *Aplysia* has been exploited to study the cellular mechanisms underlying plastic phenomena because of the relatively large size of its neurones and the simplicity of the nervous system which they form. A simple reflex, where a gill is withdrawn following a stimulus to its body, is enhanced by an unconditioned stimulus to another part of the body. Serotonin released from an interneurone by the unconditioned stimulus raises cAMP levels and PKA activity in the sensory neurone involved in the reflex. Morphological changes, including presynaptic varicosity outgrowth, also seem to play a role in the consolidation of the synaptic plasticity (Glanzman *et al.* 1990).

Studies of *Drosophila melanogaster* (fruitfly) with learning deficits in simple behavioural tasks (i.e. avoiding flying towards odours associated with electric shocks) have identified various gene mutations which are presumably affecting memory processes and the associated synaptic plasticity (Nighorn *et al.* 1991; Tully *et al.* 1994).

14.2 Temporal phases of synaptic plasticity

A particular combination of neurochemical events at the postsynaptic membrane, probably combined with specific events at the presynaptic terminal, is sufficient to trigger the plastic response. The key components of the triggering events have been relatively well characterized in recent years. A separate series of neurochemical events are then required to sustain the plastic response. It has gradually become clear that there are three sequential but distinct components to the maintenance of synaptic plasticity (Fig. 14.2). The initial phase (phase I), lasting from shortly after the initiating

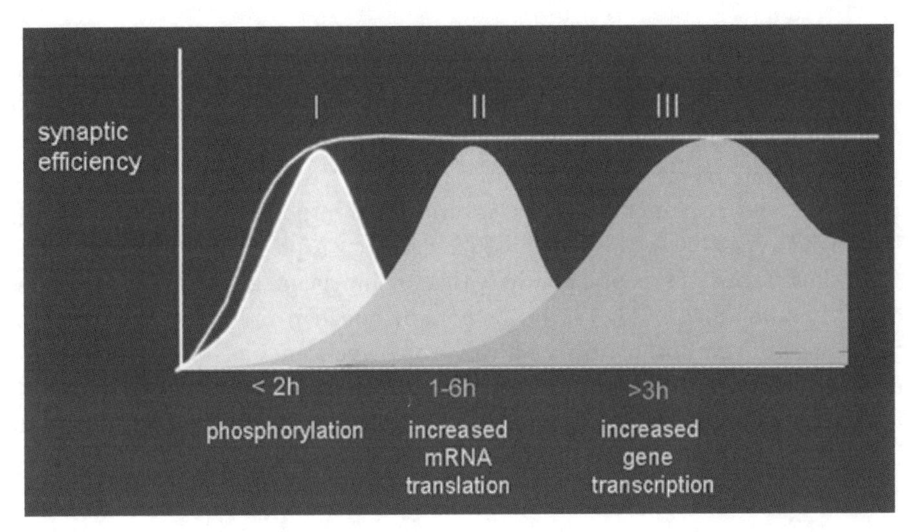

Fig. 14.2 Temporal phases of LTP. While the exact duration and timing of the different temporal components varies according to the experimental model, the same general pattern seems to be evident in all cases.

stimulus for up to 3 hours or so, according to the experimental model, is dependent on covalent modification of existing proteins (i.e. phosphorylation, nitrosylation). The intermediate phase (phase II) is dependent on the synthesis of new protein via mRNA translation, but is not dependent on gene transcription. This phase is apparent between around 2 and 8 hours after the triggering stimulus, according to the experimental model. In contrast, the slowest but most sustained phase (phase III) is dependent not only on the synthesis of new protein, but also on gene transcription and the synthesis of new mRNA.

The evidence suggests that these three phases can be observed in all the different models of neuronal plasticity, whether derived from aplysia (Barzilai *et al.* 1989, Ghirardi *et al.* 1995), helix (Schilhab and Christoffersen 1996), hermissenda (Crow *et al.* 1999), lamprey (Parker and Grillner 1999), drosophila (Xia *et al.* 1998) or hippocampal LTD (Manahan-Vaughan *et al.* 2000) as well as LTP in the amygdala (Bailey *et al.* 1999), and in the dentate gyrus (Otani *et al.* 1989) CA1 (Frey *et al.* 1996*a*; Nguyen and Kandel 1997) and CA3 (Huang *et al.* 1994) regions of the hippocampal formation. Hence these three sequential and mechanistically distinct phases are likely to represent a general feature of sustained plasticity.

14.3 Ca^{2+} as the trigger

The involvement of NMDA receptors in many forms of neuronal plasticity suggests that Ca^{2+} influx may be the initial stimulus that activates the intracellular processes contributing to neuronal plasticity. This is consistent with the fact that many of the intracellular mechanisms thought to be important for neuronal plasticity are dependent on elevations in intracellular Ca^{2+} (see below). However, the spatial and temporal pattern of Ca^{2+} influx may be critical for determining the changes that occur. Influx of Ca^{2+} through voltage-gated channels in hippocampal neurones, for example, produces a different temporal profile of raised intracellular Ca^{2+}, compared to NMDA receptor activation, and a distinct series of transcriptional responses (Bading *et al.* 1993; Hardingham *et al.* 1999). This is discussed in detail in Chapter 10.

For those forms of synaptic plasticity that are independent of NMDA receptors, cAMP elevation may be a key factor. In both mammalian and invertebrate models, those forms of neuronal plasticity which are not dependent on NMDA receptors are frequently observed to involve adenylyl cyclase (Alberini *et al.* 1994; Huang *et al.* 1994). However, since cAMP is also important for NMDA receptor-dependent plasticity, local increases in cAMP, resulting from activation of Ca^{2+}-dependent adenyl cyclase, may be a crucial downstream mediator of the plasticity resulting from Ca^{2+} influx.

14.4 **Rapid, transient plasticity**

We have what appears to be a relatively comprehensive understanding of the signalling events that contribute to the induction of synaptic plasticity. A number of neurotransmitter receptors, structural molecules, and signalling intermediates have been linked to synaptic plasticity, and an overview of these is provided in Table 14.1. In particular, various protein kinases are involved at an early stage, although, in general, the effects of phosphorylation are relatively short-lived, and the original properties of the protein are later restored by dephosphorylation (through the action of protein phosphatases).

Table 14.1 Molecules implicated in early phases of LTP — (either activation observed during early LTP, or LTP compromised by specific antagonists or gene knockout). Those from this group where altered synthesis is also observed after LTP induction are indicated. Space limitations preclude citing of individual references, but many excellent reviews of the mechanisms of early LTP are available

Molecule	Synthesis elevated after LTP	Molecule	Synthesis elevated after LTP
Receptors		α adrenoceptors	
NMDA R1	✓	β adrenoceptors	
NMDA R2s		5HT Rs	
AMPA GluRs	✓	Cannabinoid CB1 Rs	
Kainate Rs		Dopamine Rs	
mGluRs	✓	Opioid Rs	
mAChRs		trkB	

Table 14.1 (continued)

Molecule	Synthesis elevated after LTP
Channels	
IP3/ryanodine Rs	✓
L-type channels	
cGMP-gated channels	
Acid-sensing ion channel	
Structural	
NCAM	✓
L1 CAM	✓
integrins	
syndecan-3	
tenascins	✓
HNK1	
neuroplastin	
telencephalin	
PSD-95	
tropomodulin 2	
actin	
Presynaptic	
RIM1	
Rab3a	
Complexins	
Signalling	
Adenyl cyclase	
Guanyl cyclase	
cAMP PDE	✓
PKA	
PKG	
PKC	✓
CamKII	✓
CamKIV	
calcineurin	
NOS	
MEK	
RasGAPs	
PYK2 (CAKβ)	
Src	
Fyn	
SGK	✓
Kit	
PI3 kinase	
mTOR	
Lim kinase	
proteasome	
Intercellular mediators	
NGF	✓
BDNF	✓
NT4	
Arachidonic acid	
PAF	
MHC complex	
Transcription	
CREB	
C/EBP	
CBP	
Zif268 (egr1)	✓

Determination of the molecular structures of neurotransmitter receptors has, in all cases, revealed potential sites for phosphorylation (see Chapters 8, 9). In many studies using G-protein-coupled receptors, it has become clear that receptor phosphorylation is associated with a decreased responsiveness, providing a likely mechanism for the well-characterized phenomenon of desensitization (Chapter 9). Here then is a clear example of how activation of protein kinases can give rise to an altered neuronal sensitivity that outlasts the original stimulus. It is now clear that ionotropic glutamate receptors can show increased responses after phosphorylation by protein kinases, and this is thought to be one of the major mechanisms for rapid synaptic plasticity at glutamatergic synapses. Phosphorylation of cellular substrates at the synapses can also recruit hidden receptors, sequestered intracellularly, to the synaptic membrane (see below) (Liao *et al.* 1999). In addition, phosphorylation of cytoskeletal components is known to result in plasticity in neuronal morphology (see Chapter 13). The rapid rearrangements of synaptic architecture that have been detected during hippocampal LTP may therefore also be driven by the activation of kinases.

In many cases, not unique to neurones, the kinases function in cascades, where the phosphorylation of the several kinases in sequence broadens and prolongs the functional consequences of the original stimulus. Such cascades can involve serine/threonine kinases, tyrosine kinases, or both. The intracellular control of the phosphorylation state of vast numbers of different proteins must therefore be seen as a dynamic and extraordinarily complex regulatory process. These aspects of neuronal function are covered in detail in Chapters 11 and 12, but some specific details should be considered with regard to neuronal plasticity.

While they have widespread effects in a number of different cell types, five serine/threonine kinases have been particularly linked to hippocampal LTP (Soderling and Derkach 2000): cAMP-dependent protein kinase (PKA), cGMP-dependent protein kinase (PKG), mitogen-activated protein kinase (MAP kinase, aka extracellular signal-regulated kinase or ERK), protein kinase C (PKC), and calcium/calmodulin-dependent protein kinase II (CamKII) — the latter being found only in the central nervous system.

Mutant mice lacking functional type I (Ca^{2+}-stimulated) adenylyl cyclase show deficits in memory tests and in hippocampal LTP (Wu *et al.* 1992), suggesting that the convergence of Ca^{2+} and cAMP signalling by this enzyme has an important function in hippocampal plasticity. The induction of hippocampal LTP can also be prevented by selectively inhibiting the action of PKA, PKG, PKC, or CamKII (Bliss and Collingridge 1993), or ERK (English and Sweatt 1997). These studies have been more or less confirmed by the generation of null recombinant mice lacking the corresponding kinases (Silva *et al.* 1992; Abeliovich *et al.* 1994; Huang *et al.* 1995). This suggests that all of these kinases play a role both in sustaining the earliest phases of LTP, and also in triggering the changes which give rise to the slower, more sustained phases. The fact that LTP can be more or less completely blocked by inhibition of an individual kinase suggests that they each fulfil some critical role in synaptic potentiation.

The molecular structure of CamKII makes it ideally suited as a switch to convert transient ion fluxes into more enduring changes in neuronal function. Activation of CamKII by Ca^{2+}/calmodulin, following NMDA receptor activation, results in autophosphorylation of the enzyme, and the consequent generation of a Ca^{2+}-independent form that can maintain the phosphorylating activity in the absence of any further stimulation (Fukuraga *et al.* 1995). The high concentrations of CamKII in the synaptic area suggest that the subsequent effects should be dramatic. Indeed, there is considerable evidence that CamKII is vital, not only for hippocampal plasticity, but also for long-term change in other brain regions (Gordon *et al.* 1996; Mayford *et al.* 1996; Frankland *et al.* 2001).

Activation of PKA is also essential for synaptic facilitation in *Aplysia* neurones, while in drosophila, the mutants *dunce* and *rutabaga*, isolated in a behavioural screen for associative learning deficits, are deficient in the function of a cAMP-dependent phosphodiesterase and a Ca^{2+}/calmodulin-activated adenylyl cyclase, respectively (Nighorn *et al.* 1991; Levin *et al.* 1992). The actions of adenylyl cyclase and PKA therefore assume a central role in a wide variety of different models of synaptic plasticity.

The target proteins for plasticity-associated phosphorylation in these models have yet to be conclusively identified, but, at least in the mammalian hippocampus, the glutamate receptor subunits are attractive candidates, since phosphorylation by CamKII, PKC or PKA, and possibly PKG, is known to increase their responsiveness (Dev and Morris 1994; Dev and Henley 1998; Koles *et al.* 2001; MacDonald *et al.* 2001).

Tyrosine kinase activation (Chapter 11) is also critical for the expression of hippocampal LTP. Activation of the kinases src, fyn, and pyk2 (cakβ) is necessary for hippocampal LTP (Grant *et al.* 1992; Lu *et al.* 1998; Huang *et al.* 2001). It is thought that tyrosine phosphorylation of NMDA receptor subunits by src, following src activation via PYK2 (which in turn may have been activated by PKC), may contribute to a large part of this effect.

It follows from this clear role of protein kinases in initiating plastic changes in neuronal function that protein phosphatases will be similarly important in regulating the changes that occur. Evidence has been obtained that neuronal phosphatases are essential for normal LTP to occur, and also that the induction of hippocampal LTD may be primarily due to the activation of neuronal phosphatases, with consequent opposite effects on synaptic function to the effects of kinase activation. However, PKA is also required for hippocampal LTD (Brandon *et al.* 1995).

One key aspect of early-phase synaptic plasticity, dependent on phosphorylation events, is the emergence of previously 'silent synapses'. Thus, following NMDA receptor stimulation, previously inactive synaptic responses appear, mediated by AMPA receptor activation. It has become clear that AMPA receptors are continuously recycled through the early endosomal compartment, following dynamin-dependent internalization via clathrin-coated pits (Man *et al.* 2000). Stimulation of NMDA receptors can increase the rate of AMPA receptor reinsertion into the synaptic membrane, and this, combined with morphological alterations in the structure of the synapse, increases the post-synaptic response to subsequent stimulation. It is thought that CamKIIα plays a major role in the AMPA receptor reinsertion process, and may also alter NMDA receptor distribution under the synapse (Gardoni *et al.* 2001).

Other protein modifications apart from phosphorylation are also likely to play a role in rapid neuronal plasticity. Nitric oxide has been suggested to play a role in hippocampal LTP (Böhme *et al.* 1991). Direct nitrosylation or indirect ADP-ribosylation of proteins by nitric oxide can alter their functional properties (Stammler 1994), and altered ADP-ribosylation of hippocampal proteins has in fact been detected following induction of LTP (Duman *et al.* 1993). Also, protein glycosylation may be regulated by neuronal activity. The neuronal cell adhesion molecule N-CAM is extensively modified by addition of sialic acid polymers, and the properties of the N-CAM molecules are affected by the degree of polysialylation (Doherty *et al.* 1995). The extent of polysialylation is reported to vary during development and, in the hippocampus, during learning (Doyle *et al.* 1992; Doherty *et al.* 1995).

It is generally accepted that a component of hippocampal LTP is expressed presynaptically — that is, the increased efficiency of synaptic transmission is partially due to enhanced neurotransmitter release from the presynaptic terminal (A in Fig. 14.3). This is likely to result from the action of a retrograde messenger — possibly nitric oxide, endogenous cannabinoids, and/or arachidonic acid — released from the post-synaptic dendrite and acting on the presynaptic terminal. Assuming this to be correct, then it can be assumed that the mechanism leading to increased transmitter release involves phosphorylation or some other covalent modification of presynaptic target proteins (Herrero *et al.* 1992; Meffert *et al.* 1994).

14.5 **Slower, sustained plasticity**

In contrast to these earlier phases of neuronal plasticity, which rely heavily on modifications to existing proteins, there is abundant evidence that the later, most sustained phases of LTP are dependent on the synthesis of new proteins (Bliss and Collingridge 1993).

For phase II, the intermediate phase of synaptic plasticity which is not dependent on gene transcription and the synthesis of new mRNA, this is likely to involve enhanced translation of existing mRNAs. This could reflect either increased efficiency of the translational apparatus, or else increased stability of particular mRNAs. Since, for perhaps the majority of cellular proteins, mRNA availability is the rate-limiting step in protein synthesis (Hargrove and Schmidt 1989; Morris 1993; Jacobson and Peltz 1996) increased mRNA stability, and resulting elevated levels of the particular mRNA species, is predicted to increase protein levels. Indeed, it has been shown that the levels of CamKIIα mRNA and MAP2 mRNA increase after the induction of LTP (Thomas *et al.* 1994*b*; Roberts *et al.* 1996, 1998*a*), and that plasticity-related stimuli increase the stability of these mRNAs in hippocampal neurones, and this then increases the levels of the corresponding proteins (Morris 1997).

It is striking that both CamKIIα mRNA and MAP2 mRNA are members of the small group of mRNAs that are not restricted to the neuronal cell body, but are also found in high levels in neuronal dendrites, in the region of the synapses. Hence the stability of these mRNAs can be modulated locally in the region of synaptic stimulation without the necessity of communicating with the cell body (Kang and Schuman 1996; Huber *et al.* 2000). The possibility that this is a mechanism of widespread significance is

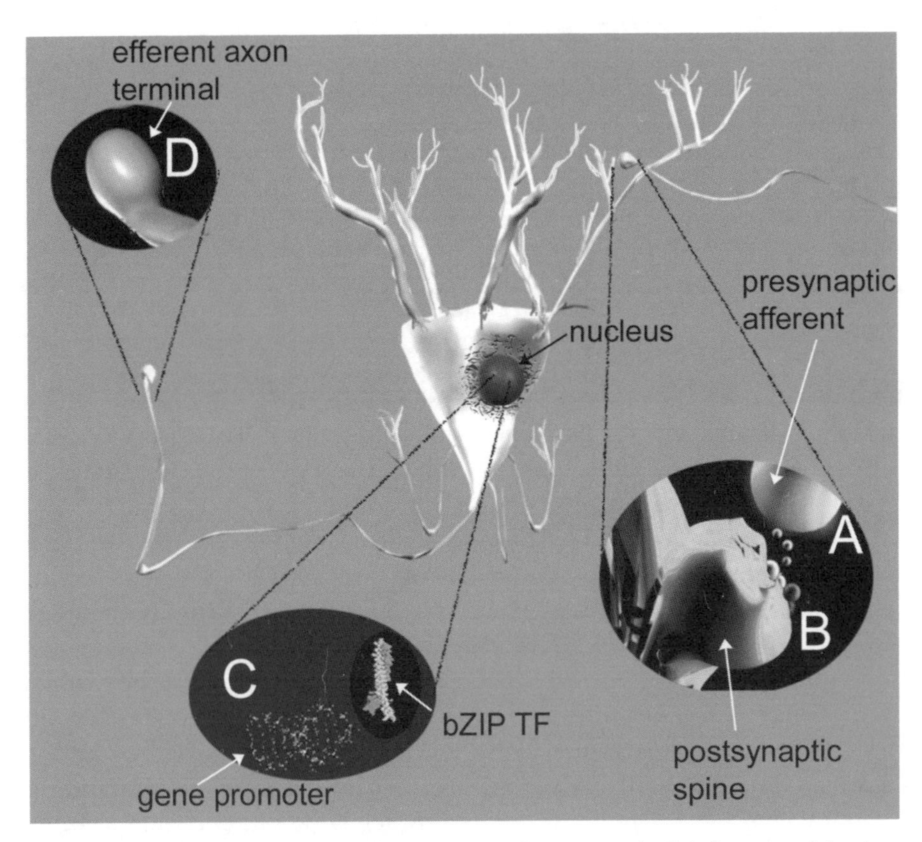

Fig. 14.3 Cellular sites where neuronal plasticity may be expressed. High-frequency firing in an afferent fibre on the right of the figure induces a number of sustained changes in the presynaptic terminal and the post-synaptic neurone. (A) — a retrograde messenger from the post-synaptic cell enhances the amount of neurotransmitter released from the presynaptic terminal by subsequent action potentials. (B) — transient activation of post-synaptic protein kinases alters the sensitivity of the synapse to subsequent stimulation, partly by phosphorylating receptor proteins. Also, more slowly, morphological changes in the structure of the dendrite occur in the region of the potentiated synapse — most likely an increase in the number and/or shape of the dendritic spines (Moser *et al.* 1994). Changes in the activity of second messenger systems occur — maybe restricted to the vicinity of the potentiated synapse, maybe extending through a significant proportion of the cytoplasm. In some cases, intracellular messengers are specifically translocated to the neuronal nucleus (C) where activation of TFs takes place, leading to an altered pattern of gene expression. (D) An increase in the efficiency of the next synapse 'in series' ('domino' plasticity) can occur, due to elevated expression of presynaptic terminal proteins in the neurone post-synaptic to the original stimulus.

supported by evidence that CamKIIα mRNA and MAP2 mRNA levels are increased in other brain regions in association with plasticity (Tighilet *et al.* 1998; Woolf *et al.* 1999; Xue *et al.* 2001).

For phase III, the slowest phase of synaptic plasticity, increased transcription of specific genes underlies the ultimate elevation in protein levels. Considerable attention has focussed not only on the late-response genes which sustain the increased responsiveness, but also on the transcription factors which bind to their specific targets in the genome and switch on (or off) the late-response genes.

14.5.1 Transcription factor families

A group of transcription factor (TF) genes show increased transcription relatively rapidly (15–45 minutes) after the plasticity-inducing stimulus. These genes are classified as immediate-early genes, in that their increased expression is fast, and is not dependent on the synthesis of other proteins.

A large number of TFs have been identified, and many of these have been shown to be expressed in neurones. These various TFs can be divided into a number of families with closely related structures. In relation to neuronal plasticity, much interest has centred on a group of TFs that contain within their protein sequence a region of basic amino acids and also a region containing regularly spaced leucine residues. There is evidence that this latter structure forms a 'leucine zipper' to dimerize with another TF of the same structural family, and hence the group of TFs have become known as the bZIP family. While the leucine zipper allows the formation of functional homo- or heterodimers, the basic region is the part of the molecule that interacts with the promoter region of the target DNA. The dimer adopts a three-dimensional conformation reminiscent of a pair of scissors, with the basic region (the blades of the scissors) gripping the DNA (Glover and Harrison 1995; Chen *et al.* 1998).

This family includes the widely studied c-fos gene, along with other related genes (fos B, fos-related antigen) and the jun proteins (c-jun, junB, junD). A major target of protein dimers formed from these TFs is the AP1 site on genomic DNA, which has the typical sequence TGAGTCA. However, a closely related sequence, TGACGTCA, is known as the cAMP-response element (CRE), and is involved in the transcriptional activation of many genes in response to elevated intracellular cAMP levels. The so-called CRE-binding protein (CREB) can be a component of the dimers that can bind to the CRE, and also has a bZIP structure (see Chapters 3 and 10 for further details). A number of 'activating transcription factors' (ATFs) with homology to CREB also belong to this family. One bZIP protein, known commonly as C/EBP, may have a particularly important role in neuronal plasticity (see below).

Another family of TFs adopt a three-dimensional structure with finger-like processes, and contain zinc. Of these 'zinc finger' TFs, zif/268 (Milbrandt 1987), also known as egr1, NGF-IA, and Krox24, has been widely studied, and binds to a target sequence of the form GCGGGGGCG. Other members of this family include Krox20 and egr3.

Other TF families with a possible but as yet incompletely explored role in neuronal plasticity include those TFs with a basic-helix-loop-helix structure (bHLH), those with a so-called POU domain, which recognize a DNA sequence of the form $(A/T)_{4\text{-}5}T(A/T)TGCAT$, and the rel family related to the inflammatory cell TF NF-κB.

14.5.2 Induction of transcription factors

NF-κB is a transcription factor that is constitutively present in the cytoplasm in many neurones throughout the CNS. NF-κB is present in hippocampal neurones, and can be activated by NMDA receptor stimulation. While definitive evidence for its participation in hippocampal plasticity has not yet been produced, NF-κB is activated in association with long-term memory in the crab (Freudenthal and Romano 2000), and it is possible that NF-κB is involved in many forms of synaptic plasticity. Proteins of the CREB family are also constitutively present in a latent form in neuronal cytoplasm, and are essential for long-term memory in *Drosphila* (Yin *et al.* 1994). CREB can be activated via phosphorylation by various kinases, including PKA and CamKII (see also Chapters 3 and 10), and CREB phosphorylation has been demonstrated to occur in hippocampal neurones after induction of LTP (Moore *et al.* 1996). Activation of CREB family members has been demonstrated in hippocampal LTP (Schulz *et al.* 1999), and a potentially important role for these proteins is supported by evidence that mice with a targeted disruption of the CREB gene lacked the late phase of LTP, and were severely compromised in several tests of associative learning (Bourtchuladze *et al.* 1994; Silva *et al.* 1998). Furthermore, CREB-related proteins, in particular C/EBP, are required for the expression of long-term facilitation in Aplysia neurones (Alberini *et al.* 1994). Evidence is also emerging that activation of the CREB-binding protein (CBP) may also be important for the full expression of plasticity (Hu *et al.* 1999) — CBP can be activated by PKA, CamKII, or CamKIV (see Chapter 10). The complexity of the transcriptional control is emphasized by the finding that ATF4 (CREB2) appears to function as a constitutive repressor of synaptic facilitation in Aplysia neurones (Bartsch *et al.* 1995), possibly by forming an inactive dimer with C/EBP or CREB. It is clear that the presence of CREB, CREB-related proteins, and NF-κB, in neurones under basal conditions potentially allows their target genes to be induced rapidly after the activating stimulus.

Other TFs need to be synthesized via gene induction before they can affect their target genes. An increased level of the mRNA encoding junB is observed within a few minutes of induction of LTP in the perforant path synapses onto the hippocampal dentate gyrus (Cole *et al.* 1989; Wisden *et al.* 1990). The evidence for other members of the fos/jun family is less clear, but a supramaximal LTP-inducing stimulus, which it has been suggested produces a longer-lasting LTP, results also in increased c-fos expression (Jeffrey *et al.* 1990). Elevated expression of c-jun has also been observed during hippocampal LTP. The c-fos knockout mouse has failed to confirm an important role for fos in hippocampal plasticity.

The closest correlation between TF expression and LTP induction in the hippocampus has been observed for zif/268. High-frequency stimulation of perforant path synapses onto dentate granule cells, and of Schaffer collaterals onto CA1 pyramidal cells, results in dramatic increases in zif/268 mRNA levels in the post-synaptic neurone (Cole *et al.* 1989; Wisden *et al.* 1990; Roberts *et al.* 1996). Furthermore, zif/268 is also induced in the rat hippocampus following spatial learning (Fordyce *et al.* 1994), in the rat visual cortex by light adaptation (Worley *et al.* 1991), in the monkey temporal lobe by visual associative learning (Okuno and Y. 1996; Miyashita *et al.* 1998), in the dorsal horn of the rat spinal cord by primary afferent stimulation (Herdegen *et al.* 1990), in the mouse accessory olfactory bulb following odor presentation (Brennan *et al.* 1999), and in the forebrain of songbirds by song presentation (Mello *et al.* 1992). Since these are all conditions resulting in a sustained change in neuronal activity, the evidence is at least suggestive that zif/268 is playing a critical role in the long-term plasticity responses. This is reinforced by the observation that the duration of LTP is correlated with the degree of zif/268 induction (Richardson *et al.* 1992). However, even then, there are situations when LTP is clearly expressed without any induction of zif/268, and also cases where zif/268 is induced in the absence of any LTP (Wisden *et al.* 1990; Johnston and Morris 1994*e*). In the former case, this may be related to the fact that other TFs of this family are also likely to be involved in hippocampal plasticity. For example, egr3 is also induced by LTP (Yamagata *et al.* 1994*a*). Nevertheless, the key role of zif/268 in plasticity has been reinforced by the recent report on hippocampal function in zif/268 knockout mice (Jones *et al.* 2001). These mice exhibit normal early-phase LTP, but late-phase LTP is completely absent. Similarly, short-term memory is unimpaired, while long-term memory is severely compromised.

A working hypothesis of synaptic plasticity, which incorporates many of the above observations (Alberini *et al.* 1994), proposes that brief activation of protein kinases results in a local phosphorylation of target proteins (B in Fig. 14.3). Stronger stimulation results in the translocation of the activated kinase to the nucleus, where it phosphorylates TFs such as CREB. Upon phosphorylation, these TFs then activate other TF genes, such as zif/268 or c-fos, and after transcription and translation have occurred, the pattern of downstream gene expression can be altered (C in Fig. 14.3) (see Chapters 3 and 10 for a detailed description of the potential signalling pathways involved).

14.5.3 **Induction of other immediate-early genes**

Apart from these putative TFs, a number of other genes are also reported to be rapidly induced in hippocampal neurones following high-frequency stimulation (Table 14.2). These include the growth factor β-activin (Andreasson and Worley 1995), the spectrin-like 'arc', the protease tissue plasminogen activator (tPA — (Qian *et al.* 1993)), and the prostaglandin synthetic enzyme cyclooxygenase II (cox2 — (Yamagata *et al.* 1993)). Each of these genes also appears to be induced during limbic system kindling, which may strengthen the idea that they play some significant role in the plasticity process. For β-activin and cox-2, direct evidence that these proteins are necessary for synaptic plasticity has yet to be obtained, although it may be relevant that the tPA knockout mouse is hyporesponsive to seizure stimuli (Tsirka *et al.* 1995) — implying that tPA may increase the excitability of hippocampal neurones. In addition, evidence suggests that there are deficits in LTP induction in the CA1 and CA3 regions of the hippocampus from tPA knockout mice (Frey *et al.* 1996*b*; Huang *et al.* 1996). Since tPA is also induced in association with cerebellar LTP (Seeds *et al.* 1995), it is possible that tPA plays a generally important role in different forms of neuronal plasticity.

Apart from a relatively small subset of these genes, including zif/268 and tPA, the extent to which the association of these various molecules with plasticity is unique to the hippocampus, or even to particular groups of neurones within the hippocampus, remains to be explored. Nevertheless, evidence is accumulating that zif/268, CamKII, and tPA may play some quite fundamental role in the plasticity process.

14.5.4 Induction of late-response genes

The mRNA encoding an IEG is typically induced within 15–45 minutes of the initial stimulus, and then in most cases the level of mRNA expression has returned to normal within a few hours. A number of genes are induced (or suppressed) with a slower time course, the mRNA levels being altered perhaps 2–3 hours after the stimulus, and remaining affected for up to 48 hours. The lag period before induction/suppression presumably reflects the action of induced TFs, which need to be synthesized and modified before returning to the nucleus to alter the transcription of their target genes. These 'late-response' genes belong to various functional classes, including protein kinases, peptide neurotransmitters, growth factors, and proteases (Table 14.2).

Table 14.2 Molecules showing elevated expression during the later phases of LTP — (either at the mRNA or protein level). Corresponding references are given for further details. Those from this group where inhibitor or gene-targetting studies have suggested that the molecule is actively involved in the plasticity process are indicated by shading

Gene	Function	Experimental model	Direction of change in expression	Time	Reference
c-fos	transcription factor	LTP learning	increased	<2 hours	Jeffrey *et al.*, 1990; Worley *et al.*, 1993
c-jun	transcription factor	LTP	increased	<2 hours	Demmer *et al.*, 1993; Worley *et al.*, 1993
junB	transcription factor	LTP	increased	<2 hours	Cole *et al.*, 1989; Wisden *et al.*, 1990
zif/268 (egr1)	**transcription factor**	**LTP learning**	**increased**	**<2 hours**	**Cole *et al.*, 1989; Wisden *et al.*, 1990; Roberts *et al.*, 1996**
egr3	transcription factor	LTP	increased	<4 hours	Yamagata *et al.*, 1994*a*
CHOP	transcription factor	LTP	increased		Matsuo *et al.*, 2000

Table 14.2 (continued)

Gene	Function	Experimental model	Direction of change in expression	Time	Reference
IP3-receptor2	**Signalling receptor**	**learning**	**increased**	**1–3 hours**	**Luo *et al.*, 2001**
rheb	GTPase	LTP learning	increased	2 hours	Yamagata *et al.*, 1994*b*
Reps1	GTPase-related protein	learning	increased	1–3 hours	Luo *et al.*, 2001
CamKII-α	**protein kinase subunit**	**LTP learning**	**increased/ decreased**	**2 hours 24 hours**	**Mackler *et al.*, 1992; Thomas *et al.*, 1994*a*; Johnston and Morris, 1995; Roberts *et al.*, 1996**
PKC-Mζ	protein kinase fragment	LTP	increased	<2 hours	Osten *et al.*, 1996
PKC-β	**protein kinase**	**LTP**	**unchanged decreased**	**48–72 hours**	**Thomas *et al.*, 1994*a* Meberg *et al.*, 1993**
PKC-γ	protein kinase	LTP	increased decreased	2–24 hours	Thomas *et al.*, 1994*a* Meberg *et al.*, 1993
ERK-2	**protein kinase**	**LTP**	**increased**	**24 hours**	**Thomas *et al.*, 1994*a***
Map Kinase phosphatase	phosphatase	LTP	increased		Qian *et al.*, 1994
raf-B	protein kinase	LTP	increased	24 hours	Thomas *et al.*, 1994*a*
EDPK	protein kinase	learning	increased	1–3 hours	Luo *et al.*, 2001
Pim-1	protein kinase	LTP	increased	<2 hours	Konietzko *et al.*, 1999
Pim-3	protein kinase	LTP	increased	<2 hours	Feldman *et al.*, 1998
SnK	protein kinase	LTP	increased	<2 hours	Kauselmann *et al.*, 1999
FmK	protein kinase	LTP	increased	<2 hours	Kauselmann *et al.*, 1999
SGK	**protein kinase**	**learning**	**increased**	**>24 hours**	**Tsai *et al.*, 2002**
PDE4B3	phosphodiesterase	LTP	increased	2–6 hours	Ahmed and Frey, 2003
NCS1	Ca^{2+} binding protein	LTP	increased	1–3 hours	Genin *et al.*, 2001
activin	growth factor	LTP	increased	<2 hours	Andreasson and Worley,1995
NGF	growth factor	LTP	increased unchanged	4 hours	Castren *et al.*, 1993; Patterson *et al.*, 1992

Table 14.2 (continued)

Gene	Function	Experimental model	Direction of change in expression	Time	Reference
BDNF	**growth factor**	**LTP**	**increased**	**4 hours**	**Patterson *et al.*, 1992;** Castren *et al.*, 1993
NT3	growth factor	LTP	increased decreased	4 hours	Patterson *et al.*, 1992; Castren *et al.*, 1993
IGF1	growth factor	learning	increased	24 hours	Cavallaro *et al.*, 2001
IGF-2-R	IGF-1 receptor	learning	increased	1–3 hours	Luo *et al.*, 2001
cpg1	growth factor?	LTP	increased	6 hours	Nedivi *et al.*, 1993
Neu differentiation factor	adhesion-related growth factor	LTP	increased	<2 hours	Eilam *et al.*, 1998
N-cadherin	**cell adhesion molecule**	**LTP**	**increased**	**<2 hours**	**Bozdagi *et al.*, 2000**
arcadlin	cell adhesion molecule	LTP	increased		Yamagata *et al.*, 1999
Integrin-associated protein	**cell adhesion molecule**	**learning**	**increased**	**3 hours**	**Huang *et al.*, 1998**
NCAM	cell adhesion molecule	LTP	increased	24 hours	Schuster *et al.*, 1998
tenascin-R	**extracellular matrix**	**LTP**	**increased**	**>4 hours**	**Nakic *et al.*, 1998**
arc	cytoskeletal protein?	LTP	increased	<2 hours	Link *et al.*, 1995; Lyford *et al.*, 1995
arcadlin	adhesion molecule	LTP	increased		Yamagata *et al.*, 1999
actin	**cytoskeletal protein**	**LTP**	**increased**	**<2 hours**	**Fukazawa *et al.*, 2003**
MAP2	dendritic cytoskeletal protein	LTP	increased	24 hours	Johnston and Morris,1994*b*; Roberts *et al.*, 1996; Roberts *et al.*, 1998*a*
synaptopodin	dendritic cytoskeletal protein	LTP	increased	1–3 hours	Yamazaki *et al.*, 2001
Narp	scaffolding molecule ?	LTP	increased	<2 hours	Tsui *et al.*, 1996
Homer 1a/ Vesl	scaffolding molecule	LTP	increased	<2 hours	Kato *et al.*, 1997
LIRF / HZFw1	unknown	LTP	increased	<2 hours	Matsuo *et al.*, 2001
syntaxin	Presynaptic vesicle-associated protein	LTP	increased	2–5 hours	Smirnova *et al.*, 1993

Table 14.2 (continued)

Gene	Function	Experimental model	Direction of change in expression	Time	Reference
SNAP-25	Presynaptic vesicle-associated protein	LTP	increased	2 hours	Roberts *et al.*, 1998*b*
synapsin	Presynaptic vesicle-associated protein	LTP	increased	>3 hours	Lynch *et al.*, 1994
GAP43	presynaptic, growth-associated protein	LTP	decreased	72 hours	Meberg *et al.*, 1993
proenke-phalin	**peptide neurotransmitter**	**LTP**	**increased**	**24 hours**	**Morris *et al.*, 1988; Johnston and Morris, 1994*a* Roberts *et al.*, 1997*b***
prody-norphin	**peptide neurotransmitter**	**LTP**	**decreased**	**24 hours**	**Morris *et al.*, 1988; Johnston and Morris, 1994*a***
tPA	protease	LTP	increased	< 2 hours	Qian *et al.*, 1993;
neuropsin	protease?	LTP	decreased	4–24 hrs	Chen *et al.*, 1995
TIMP	protease inhibitor	LTP	increased	6 hours	Nedivi *et al.*, 1993
cox2	prostaglandin synthetic enzyme	LTP	increased	<2 hours	Yamagata *et al.*,1993
mGluR5	receptor	LTP	increased	24–48 hours	Manahan-Vaughan *et al.*, 2003
AMPA Receptor *R1/R2*	**receptor subunits**	**LTP**	**increased**	**2 hours**	**Nayak *et al.*, 1998**
NMDA receptor R1	**receptor subunit**	**LTP**	**increased**	**48 hours**	**Thomas *et al.*, 1994*b***

Considering the function of these different late-response genes, it is clear that a number of genes are affected where the protein will be expressed in the presynaptic terminal of the neurone involved. Such genes include the vesicle-associated proteins syntaxin and synapsin, thought to play a role in exocytosis, and the peptide neurotransmitters (Lynch *et al.* 1994). The post-synaptic induction of those genes with a presynaptic function has some interesting functional implications. Induction of the presynaptic proteins and the peptide transmitters suggests a kind of 'domino' plasticity (Morris and Johnston 1995), where potentiation of a synapse in the hippocampus results, via these changes in gene expression, in potentiation, some hours or days later, of the next synapse 'in series' (D in Fig. 14.3). When the first edition of this book was published, one of the unanswered questions posed at the end of this chapter was whether domino plasticity, postulated on the basis of these changes in gene expression, would actually be shown to occur. In the intervening time, the postulated phenomenon has been shown to be a reality (Yeckel and Berger 1998; Kleschevnikov and Routtenberg 2003).

Evidence is accumulating to suggest that growth factors have an important acute function in modulating hippocampal excitability (Lo 1995). However, a number of growth factor genes are modulated at the transcriptional level following the induction of hippocampal LTP, raising the possibility that the subsequent changes in protein levels may exert a long-lasting effect on neurotransmission. Mice lacking a functional BDNF gene show an impaired ability to express hippocampal LTP (Korte *et al.* 1995), and reduced sensitivity to limbic system kindling (Kokala *et al.* 1995). In addition, it has recently been suggested that nerve growth factor is involved in the morphological changes in the hippocampus associated with kindling and temporal lobe epilepsy (Van der Zee *et al.* 1995). The growth factor-related protein Narp is thought to play a role in AMPA receptor clustering (O'Brien *et al.* 1999).

It is clear from Table 14.2 that increased expression of various protein kinase genes has been detected following induction of LTP. Considering the importance of phosphorylation reactions for the most rapid phase of synaptic plasticity (Section 14.3 above), altered expression of these genes is likely to have a prolonged effect on the rapid responses to subsequent stimulation, and possibly on the later responses as well. It is notable that a series of components of a single well-defined signalling pathway — RasGRF, RafB, and ERK2 — are all induced after LTP. Increased activity of this pathway could hence be considered as a likely mediator of sustained plasticity. Indeed, induction of RafB expression has also been observed during spatial learning (Richter-Levin *et al.* 1998). In addition, despite the increased expression of CamKIIα mRNA during phase II of hippocampal plasticity, markedly reduced expression of the CamKIIα gene is observed 24 hours after hippocampal NMDA receptor stimulation (Johnston and Morris 1995), and mice carrying an inactivating mutation in this gene show spontaneous epileptiform seizures (Butler *et al.* 1995). This provides strong evidence that a reduction in CamKIIα gene expression following plastic change in the hippocampus leads to a prolonged elevation in hippocampal excitability. The mechanistic explanation for this remains unclear.

A picture emerges of functional plasticity that lasts for days, and includes a component affecting the likelihood of further plastic change. Such alterations are probably consolidated by morphological changes, involving the altered conformation of synapses (B in Fig. 14.3) (Moser *et al.* 1994). The altered expression of proteases and their inhibitors (such as tPA, neuropsin, and TIMP), of molecules modulating cell connection and adhesion (such as the cadherins and NCAMs) and of genes regulating cytoskeletal structure (such as MAP2) (Table 14.2) may be a part of this structural remodelling process. In particular, NCAM expression can clearly have a direct influence on synaptic efficiency by modifying synaptic architecture (Dityatev *et al.* 2000).

A small group of genes has been identified, where the time course of induction following the generation of hippocampal LTP is exceptionally slow. It is reported that the levels of the mRNA encoding GAP43 — a protein thought to be involved in the function of the presynaptic terminal — are decreased 3 days after the stimulus, but not earlier (Meberg *et al.* 1993). Similarly, the levels of mRNA encoding certain splice variants of the NR1 subunit of the NMDA receptor are first found to be increased 48 hours after LTP induction (Thomas *et al.* 1994*a*).

14.5.5 **Relationship between early and late responses**

It has proved difficult to link the induction of the above TFs with any of these changes in late-response gene expression. Experiments using transfected cell lines had suggested a possible causal relationship between c-fos induction and increased transcription of the proenkephalin gene, but this has proved not to be the case in CNS neurones (Konradi *et al.* 1993; Johnston and Morris 1994*a*). Induction of zif/268 also does not seem to be involved in regulating hippocampal proenkephalin, prodynorphin or CamKII expression (Johnston and Morris 1994*e*). The best evidence to date linking TF and late response gene induction suggests that zif268 may induce synapsin transcription (Thiel *et al.* 1994), that c-fos may regulate NGF expression (Hengerer *et al.* 1990), and that NF-κB regulates induction of NCAM expression (Simpson and Morris 2000). The possible role of the various bZIP family members remains an important target for investigation. However, it is very likely that the pattern of late gene induction can vary according to the particular pattern of afferent activity, and that this in turn is determined by the complement of transcription factors that are activated (Bading *et al.* 1997; Dolmetsch *et al.* 1997; Itoh *et al.* 1997).

It has generally been assumed that increased synaptic efficiency following afferent activity is restricted to the activated synapse, and does not spread to potentiate neighbouring synapses. Although recent evidence suggests that this may be something of a generalization, and that there can be some limited spread of the plasticity to nearby synapses (Engert and Bonhoeffer 1997), it is nevertheless clear that only a small, spatially-restricted number of a neurone's afferent synapses will be affected during a phenomenon such as LTP. In phase II of LTP, those dendritic mRNAs that are in the immediate vicinity of the activated synapses can be induced to support increased protein synthesis, and there is no conceptual problem in maintaining the input specificity of the potentiation. However, for phase III, gene transcription must take place in the cell nucleus. The question then arises as to how the newly synthesized proteins are transported out specifically to that small region of the neurone's complex dendritic tree where the plastic response has been initiated. Evidence has been obtained, in both invertebrate and mammalian models, that the activated synapse is 'tagged', so that proteins undergoing transport outward along the dendrites can be captured at the site of stimulation (Frey and Morris 1997). The precise mechanisms, and the molecular identity of the tag(s), remain to be identified.

14.6 Unanswered questions

Apart from identifying the synaptic 'tag', and also the various TFs or combinations of TFs that produce the altered pattern of late-response gene activation observed in hippocampal LTP (Table 14.2), a number of other crucially important issues remain unresolved.

1) To what extent are these long-term changes common to other forms of neuronal plasticity apart from hippocampal LTP? Are there other genes affected uniquely in other paradigms or in other brain areas?
2) What other genes are affected in the hippocampus? Clearly the data summarized in Table 14.2 are unlikely to be complete.
3) How important are the various genes recorded in Table 14.2 for the processes of hippocampal plasticity? Experiments are still largely at the stage of recording the phenomena, rather than directly manipulating the expression of these genes individually and recording the effect on neuronal function.
4) The concept of LTP in the hippocampus is built around the potentiation of transmission at a single synapse. Activating gene transcription in the nucleus of the neurone would appear to endanger that exquisite anatomical specificity. Is Phase III of LTP as input-specific as Phase I and II ?
5) How important are various possible post-transcriptional processes (i.e. modulation of mRNA stability/ribosomal availability or efficiency) in regulating gene expression during Phase II of functional plasticity?
6) To what extent are phases II and III of synaptic plasticity dependent on phase I, and phase III dependent on phase II ?

An understanding of the molecular mechanisms that combine to generate neuronal plasticity is going to be vital for all aspects of neurobiology. Progress so far has been rapid, and already many of the ingredients have been identified. The challenge is to discover the ones that are missing from the list, and to find out exactly how they are all mixed together to produce the final product.

References

Abeliovich, A., Chen, C., Goda, Y., Silva, A., Stevens, C. F., and Tonegawa, S. (1994) Modified hippocampal LTP in PKC-γ-deficient mice. *Cell*, **75**, 1253–62.

Ahmed, T. and Frey, U. (2003) Expression of the specific type IV phosphodiesterase gene PDE4B3 during different phases of long-term potentiation in single hippocampal slices of rats in vitro. *Neuroscience*, **117**, 627–38.

Alberini, C. M., Ghiradi, M., Metz, R., and Kandel, E. R. (1994) C/EBP is an immediate-early gene required for the consolidation of long-term facilitation in Aplysia. *Cell*, **75**, 1099–114.

Andreasson, K. and Worley, P. (1995) Induction of β-activin expression by synaptic activity and during neocortical development. *Neuroscience*, **69**, 781–96.

Bading, H., Ginty, D. D., and Greenberg, M. E. (1993) Regulation of gene expression in hippocampal neurons by distinct calcium signalling pathways. *Science*, **260**, 181–5.

Bading, H., Hardingham, G. E., Johnson, C. M., and Chawla, S. (1997) Gene regulation by nuclear and cytoplasmic calcium signals. *Biochem. Biophys. Res. Commun.*, **236**, 541–3.

Bailey, D. J., Sun, W., Thompson, R. F., Kim, J. J., and Helmstetter, F. (1999) Acquisition of fear conditioning in rats requires the synthesis of mRNA in the amygdala. *Behav. Neurosci.*, **113**, 276–82.

Bartsch, D., Ghiradi, M., Skehel, P. A., Karl, K. A., Herder, S. P., Chen, M., *et al.* (1995) Aplysia CREB2 represses long-term facilitation relief of repression converts transient facilitation into long-term functional and structural change. *Cell*, **83**, 979–92.

Barzilai, A., Kennedy, T. E., Sweatt, J. D., and Kandel, E. R. (1989) 5HT modulates protein synthesis and the expression of specific proteins during long-term facilitation in Aplysia. *Neuron*, **2**, 1577–86.

Bliss, T. V. P. and Collingridge, G. L. (1993) A synaptic model of memory, long-term potentiation in the hippocampus. *Nature*, **361**, 31–8.

Böhme, G. A., Bon, C., Stutzmann J.-M., Doble, A., and Blanchard J-C. (1991) Possible involvement of nitric oxide in long-term potentiation. *Eur. J. Pharmacol.*, **199**, 379–81.

Bortolotto, Z. A., Clarke, V. R. J., Delany, C. M., Parry, M. C., Smolders, I., Vignes, M., *et al.* (1999) Kainate receptors are involved in synaptic plasticity. *Nature*, **402**, 297–301.

Bourtchuladze, R., Frenguelli, B., Blendy, J., Cioffi, D., Schutz, G., and Silva, A. J. (1994) Deficient long-term memory in mice with a targeted mutation of the cAMP response element binding protein. *Cell*, **79**, 59–68.

Bozdagi, O., Shan, W., Tanaka, H., Benson, D. L., and Huntley, G. W. (2000) Increasing numbers of synaptic puncta during late-phase LTP: N-cadherin is synthesized, recruited to synaptic sites, and required for potentiation. *Neuron*, **28**, 245–59.

Brandon, E. P., Zhuo, M., Huang, Y. Y., Qi, M., Gerhold, K. A., Burton, K. A., *et al.* (1995) Hippocampal long-term depression and depotentiation are defective in mice carrying a targeted disruption of the gene encoding the ri-beta subunit of camp-dependent protein-kinase. *Proc. Natl. Acad. Sci. USA*, **92**, 8851–5.

Brennan, P. A., Schellinck, H. M., and Keverne, E. B. (1999) Patterns of expression of the immediate-early gene **egr**-1 in the accessory olfactory bulb of female mice exposed to pheromonal constituents of male urine. *Neuroscience*, **90**, 1463–70.

Butler, L. S., Silva, A. J., Abeliovich, A., Watanabe, Y., Tonegawa, S., and McNamara, J. O. (1995) Limbic epilepsy in transgenic mice carrying a Ca^{2+}/calmodulin-dependent kinase II alpha-subunit mutation. *Proc. Natl. Acad. Sci. (USA)*, **92**, 6852–5.

Castren, E., Pitkanen, M., Sirvio, J., Parsadanian, A., Lindhom, D., Thoenen, H., *et al.* (1993) The induction of LTP increases BDNF and NGF mRNA but decreases NT-3 mRNA in the dentate gyrus. *Neuroreport*, **4**, 895–8.

Cavallaro, S., Schreurs, B. G., Zhao, W., D'Agata, V., and Alkon, D. (2001) Gene expression profiles during long-term memory consolidation. *Eur. J. Neurosci.*, **13**, 1809–15.

Chen, Z., Yoshida, S., Kato, K., Momota, Y., Suzuki, J., Tanaka, T., *et al.* (1995) Expression and activity-dependent changes of a novel limbic-serine protease gene in the hippocampus. *J. Neurosci.*, **15**, 5088–97.

Chen, L., Glover, J. N. M., Hogan, P. G., Rao, A., and Harrison, S. (1998) Structure of the DNA binding domains from NFAT, Fos and Jun bound specifically to DNA. *Nature*, **392**, 42–8.

Cole, A. J., Saffen, D. W., Baraban, J. M., and Worley, P. F. (1989) Rapid increase of an immediate-early gene mRNA in hippocampal neurons by NMDA receptor activation. *Nature*, **340**, 474–6.

Conquet, F., Bashir, Z., Davies, C., Daniel, H., Ferraguti, F., Bordi, F., *et al.* (1994) Motor deficit and impairment of synaptic plasticity in mice lacking mGluR1. *Nature*, **372**, 237–43.

Crow, T., Xue-Bian, J. J., and Siddiqi, V. (1999) Protein synthesis-dependent and mRNA synthesis-independent intermediate phase of memory in Hermissenda. *J. Neurophysiol.*, **82**, 495–500.

Demmer, J., Dragunow, M., Lawlor, P. A., Mason, S. E., Leah, J. D., Abraham, W. C., *et al.* (1993) Differential expression of immediate-early genes after hippocampal long-term potentiation in awake rats. *Mol. Brain Res.*, **17**, 279–86.

Dev, K. K. and Henley, J. M. (1998) The regulation of AMPA receptor-binding sites. *Mol. Neurobiol.*, **17**, 33–58.

Dev, K. K. and Morris, B. J. (1994) Modulation of alpha-amino-3-hydroxy-5-methylisoxazole-4-propionic acid (AMPA) binding sites by nitric oxide. *J. Neurochem.*, **63**, 946–52.

Dityatev, A., Dityateva, G., and Schachner, M. (2000) Synaptic strength as a function of post- versus presynaptic expression of the neural cell adhesion molecule NCAM. *Neuron*, **26**, 207–17

Doherty, P., Fazeli, M. S., Walsh, F. S. (1995) The neural cell adhesion molecule and synaptic plasticity. *J. Neurobiol.*, **26**, 437–46.

Dolmetsch, R. E., Lewis, R. S., Goodnow, C. C., and Healy, J. I. (1997) Differential activation of transcription factors induced by Ca^{2+} response. *Nature*, **386**, 855–858.

Doyle, E., Nolan, P. M., Bell, R., and Regan, C. M. (1992) Hippocampal-NCAM 180 transiently increases sialylation during the acquisition and consolidation of a passive-avoidance response in the adult-rat. *J. Neurosci. Res.*, **31**, 513–23.

Duman, R. S., Terwilliger, R. Z., and Nestler, E. J. (1993) Alterations in nitric oxide-stimulated endogenous ADP-ribosylation associated with long-term potentiation in rat hippocampus. *J. Neurochem.*, **61**, 1542–5.

Eilam, R., Pinkas-Kramarski, R., Ratzkin, B. J., Segal, M., and Yarden, Y. (1998) Activity-dependent regulation of Neu differentiation factor/neuregulin expression in rat brain. *Proc. Natl. Acad. Sci. USA*, **95**, 1888–93.

Engert, F. and Bonhoeffer, T. (1997) Synapse specificity of long-term potentiation breaks down at short distances. *Nature*, **388**, 279–84.

English, J. D. and Sweatt, J. D. (1997) A requirement for the MAP kinase cascade in hippocampal LTP. *J. Biol. Chem.*, **272**, 19103–6.

Feldman, J. D., Vician, L., Crispino, M., Tocco, G., Marcheselli, V. L., Bazan, N. G., *et al.* (1998) KID-1, a protein kinase induced by depolarization in brain. *J. Biol. Chem.*, **273**, 16535–3.

Fordyce, D. E., Bhat, R. V., Baraban, J., and Wehner, J. M. (1994) Genetic and activity-dependent regulation of **zif**268 expression, Association with spatial learning. *Hippocampus*, **4**, 559–68.

Frankland, P. W., O'Brien, C., Ohno, M., Kirkwood, A., and Silva, A. J. (2001) alpha-CaMKII-dependent plasticity in the cortex is required for permanent memory. *Nature*, **411**, 309–13.

Freudenthal, R. and Romano, A. (2000) Participation of Rel/NF-kappaB transcription factors in long-term memory in the crab *Chasmagnathus. Brain Res.*, **855**, 274–81.

Frey, U., Frey, S., Schollmeier, F., and Krug, M. (1996*a*) Influence of Actinomycin-D, a RNA-synthesis inhibitor, on long-term potentiation in rat hippocampal-neurons in-vivo and in-vitro. *J. Physiol. London*, **490**, 703–11.

Frey, U., Mueller, M., and Kuhl, D. (1996*b*) A different form of long-lasting potentiation revealed in tissue-plasminogen activator mutant mice. *J. Neurosci.*, **16**, 2057–63.

Frey, U., and Morris, R. G. M. (1997) Synaptic tagging and long-term potentiation. *Nature*, **385**, 533–6.

Fukazawa, Y., Saitoh, Y., Ozawa, F., Ohta, Y., Mizuno, K., and Inokuchi, K. (2003) Hippocampal LTP is accompanied by enhanced F-actin content within the dendritic spine that is essential for late LTP maintenance in vitro. *Neuron*, **38**, 447–60.

Fukuraga, K., Muller, D., and Miyamoto, E. (1995) Increased phosphorylation of Ca/calmodulin-dependent protein kinase II and its endogenous substrates in the induction of LTP. *J. Biol. Chem.*, **270**, 6119–24.

Gardoni, F., Schrama, L. H., Kamal, A., Gispen, W. H., Cattabeni, F., and Di Luca, M. (2001) Hippocampal synaptic plasticity involves competition between CamKII and PSD95 for binding to the NR2A subunit of the NMDA receptor. *J. Neurosci.*, **21**, 1501–9.

Genin, A., Davis, S., Meziane, H., Doyere, V., Jeromin, A., Roder, J., *et al.* (2001) Regulated expression of the neuronal calcium sensor-1 gene during long-term potentiation in the dentate gyrus in vivo. *Neuroscience*, **106**, 571–7.

Ghirardi, M., Montarolo, P. G., and Kandel, E. R. (1995) A novel intermediate stage in the transition between short- and long-term facilitation in the sensory to motor neuron synapse of aplysia. *Neuron*, **14**, 413–20.

Glanzman, D. L., Kandel, E. R., and Scacher, S. (1990) Target-dependent structural changes accompanying long-term synaptic facilitation in aplysia neurones. *Science*, **249**, 799–802.

Glover J. N. M. and Harrison, S. C. (1995) Crystal-structure of the heterodimeric bzip transcription factor c-fos-c-jun bound to DNA. *Nature*, **373**, 257–61.

Gordon, J. A., Cioffi, D., Silva, A. J., and Stryker, M. P. (1996) Deficient plasticity in the primary visual cortex of alpha- calcium/calmodulin-dependent protein kinase II mutant mice. *Neuron*, **17**, 491–9.

Grant S. G. N, O'Dell, T. J., Karl, K. A., Stein, P. L., Soriano, P., and Kandel, E. R. (1992) Impaired long-term potentiation, spatial learning, and hippocampal development in fyn mutant mice. *Science*, **258**, 1903–8.

Hardingham, G. E., Chawla, S., Cruzalegui, F. H., and Bading, H. (1999) Control of recruitment and transcription-activating function of CBP determines gene regulation by NMDA receptors and L-type calcium channels. *Neuron*, **22**, 789–98.

Hargrove, J. L. and Schmidt, F. H. (1989) The role of messenger-RNA and protein stability in gene-expression. *Faseb J.*, **3**, 2360–70.

Hengerer, B., Lindhom, D., Heumann, R., Ruether, U., Wagner, E. F., and Thoenen, H. (1990) Lesion-induced increase in nerve growth factor mRNA is mediated by c-fos. *Proc. Natl. Acad. Sci. (USA)*, **87**, 3899–910.

Herdegen, T., Walker, T., Leah, J. D., Bravo, R., and Zimmermann, M. (1990) The Krox24 protein, a new transcription regulating factor, expression in the rat CNS following somatosensory stimulation. *Neurosci. Lett.*, **120**, 21–24.

Herrero, I., Mirasportugal, M. T., and Sanchezprieto, J. (1992) Positive feedback of glutamate exocytosis by metabotropic presynaptic receptor stimulation. *Nature*, **360**, 163–6.

Hu, S. C., Chrivia, J., and Ghosh, A. (1999) Regulation of CBP-mediated transcription by neuronal calcium signaling. *Neuron*, **22**, 799–808.

Huang, A. M., Wang, H. L. Tang, Y. P., and Lee, E. H. Y. (1998) Expression of Integrin-associated protein gene associated with memory formation in rats. *J. Neurosci.*, **18,** 4305–13.

Huang, Y. Q., Lu, W. Y., Ali, D. W., Pelkey, K. A., Pitcher, G. M., Lu, Y. M., *et al.* (2001) CAK beta/Pyk2 kinase is a signaling link for induction of long-term potentiation in CA1 hippocampus. *Neuron*, **29**, 485–96.

Huang, Y.-Y., Bach, M. E., Lipp, H.-P., Zhuo, M., Wolfer, D. P., Hawkins, R. D., *et al.* (1996) Mice lacking the gene encoding tissue-type plasminogen activator show a selective interference with late-phase LTP in both Schaffer collateral and mossy fibre pathways. *Proc. Natl. Acad. Sci. (USA)*, **93**, 8699–704.

Huang, Y.-Y., Kandel, E. R., Varscahvsky, L., Brandon, E. P., Qi, M., Idzerda, R. J., *et al.* (1995) A genetic test of the effects of mutations in PKA on mossy fibre LTP and its relation to spatial and contextual learning. *Cell*, **83**, 1211–22.

Huang, Y.-Y., Li, X-C., and Kandel, E. R. (1994) cAMP contributes to mossy fibre LTP by initiating both a covalently mediated early phase and a macromolecular synthesis-dependent late phase. *Cell*, **79**, 69–79.

Huber, K. M., Kayser, M. S., and Bear, M. F. (2000) Role for rapid dendritic protein synthesis in hippocampal mGluR- dependent long-term depression. *Science*, **288**, 1254–6.

Itoh, K., Ozaki, M., Stevens, B., and Fields, R. D. (1997) Activity-dependent regulation of N-cadherin in DRG neurons: differential regulation of N-cadherin, NCAM, and L1 by distinct patterns of action potentials. *J. Neurobiol.*, **33**, 735–48

Jacobson, A. and Peltz, S. W. (1996) Interrelationships of the pathways of mRNA decay and translation in eukaryotic cells. *Annu. Rev. Biochem.*, **65**, 693–739.

Jeffrey, K. J., Abraham, W. C., Dragunow, M., and Mason, S. E. (1990) Induction of fos-like immunoreactivity and the maintenance of long-term potentiation in the dentate gyrus of unanaesthetised rats. *Mol. Brain Res.*, **8**, 267–84.

Johnston, H. M. and Morris, B. J. (1994*a*) Increased c-fos expression does not cause the elevated proenkephalin mRNA levels in the dentate gyrus following NMDA receptor stimulation. *Mol. Brain Res.*, **25**, 147–50.

Johnston, H. M., Morris, B. J. (1994*b*) Nitric oxide alters proenkephalin and prodynorphin gene expression in hippocampal granule cells. *Neuroscience*, **61**, 435–9.

Johnston, H. M. and Morris, B. J. (1994*c*) NMDA and nitric oxide increase MAP2 gene expression in hippocampal granule cells. *J. Neurochem.*, **63**, 379–82.

Johnston, H. M. and Morris, B. J. (1994*d*) Selective regulation of dendritic MAP2 mRNA levels in hippocampal granule cells by nitric oxide. *Neurosci. Lett.*, **177**, 5–10.

Johnston, H. M. and Morris, B. J. (1994*e*) zif/268 does not mediate increases in proenkephalin mRNA levels after NMDA receptor stimulation. *Neuroreport*, **5**, 1498–500.

Johnston, H. M. and Morris, B. J. (1995) *N*-Methyl-D-Aspartate and nitric oxide regulate the expression of calcium/calmodulin-dependent protein kinase II in the hippocampal dentate gyrus. *Mol. Brain Res.*, **31**, 141–50.

Jones, M. W., Errington, M. L., French, P. J., Fine, A., Bliss, T. V. P., Garel, S., *et al.* (2001) A requirement for the immediate early gene Zif268 in the expression of late LTP and long-term memories. *Nat. Neurosci.*, **4**, 289–96.

Kang, H. and Schuman, E. M. (1996) A requirement for local protein synthesis in neurotrophin-induced hippocampal synaptic plasticity. *Science*, **273**, 1402–6.

Kato, A., Ozawa, F., Saitoh, Y., Hirai, K., and Inokuchi, K. (1997) Vesl, a gene encoding VASP/Ena family related protein, is upregulated during seizure, long-term potentiation and synaptogenesis. *FEBS Lett.*, **412**, 183–9.

Kauselmann, G., Weiler, M., Wulff, P., Jessberger, S., Konietzko, U., Scafidi, J., *et al.* (1999) The polo-like protein kinases Fnk and Snk associate with a Ca^{2+}- and integrin-binding protein and are regulated dynamically with synaptic plasticity. *EMBO J.*, **18**, 5528–39.

Kleschevnikov, A. M. and Routtenberg, A. (2003) Long-term potentiation recruits a trisynaptic excitatory associative network within the mouse dentate gyrus. *Eur. J. Neurosci.*, **17**, 2690–9.

Kokala, M., Ernfors, P., Kokala, Z., Elmer, E., Jaenisch, R., and Lindvall, O. (1995) Suppressed epileptogenesis in BDNF mutant mice. *Exp. Neurol.*, **133**, 215–24.

Koles, L., Wirkner, K., and Illes, P. (2001) Modulation of ionotropic glutamate receptor channels. *Neurochem. Res.*, **26**, 925–32.

Kombian, S. B. and Malenka, R. C. (1994) Simultaneous LTP of non-NMDA receptor-mediated and LTD of NMDA-receptor-mediated responses in the nucleus accumbens. *Nature*, **368**, 242–6.

Konietzko, U., Kauselmann, G., Scafidi, J., Staubli, U., Mikkers, H., Berns, A., *et al.* (1999) Pim kinase expression is induced by LTP stimulation and required for the consolidation of enduring LTP. *EMBO J.*, **18**, 3359–69.

Konradi, C., Kobierski, L., Nguyen, T. V., Heckers, S., and Hyman, S. E. (1993) The cAMP-response element binding protein interacts, but fos does not interact, with the proenkephalin enhancer in rat striatum. *Proc. Natl. Acad. Sci. (USA)*, **90**, 7005–9.

Korte, M., Carroll, P., Wlf, E., Brem, G., Thoenen, H., and Bonhoeffer, T. (1995) Hippocampal long-term potentiation is impaired in mice lacking BDNF. *Proc. Natl. Acad. Sci. (USA)*, **92**, 8856–60.

Levin, L. R., Han, P.-L., Hwaung, P. M., Feinstein, P. G, and Davis, R. L. (1992) The drosophila learning and memory gene rutabaga encodes a Ca/calmodulin responsive adenylyl cyclase. *Cell*, **68**, 469–79.

Link, W., Konietzo, U., Kauselmann, G., Krug, M., Schwanke, B., Frey, U., *et al.* (1995) Somatodendritic expression of an immediate early gene is regulated by synaptic activity. *Proc. Natl. Acad. Sci. USA*, **92**, 5734–8.

Lo, D. L.(1995) Neurotrophic factors and synaptic plasticity. *Neuron*, **15**, 979–81.

Lu, Y. M., Roder, J. C., Davidow, J., and Salter, M. W. (1998) Src activation in the induction of long-term potentiation in CA1 hippocampal neurons. *Science*, **279**, 1363–7.

Luo, Y., Long, J. M., Spangler, E. L., Longo, D. L., Ingram, D. K., and Weng, N.-P. (2001) Identification of maze-learning asociated-genes in rat hippocampus by cDNA microarray. *J. Mol. Neurosci.*, **17**, 397–404.

Lyford, G. L., Yamagata, K., Kaufmann, W. E., Barnes, C. A., Sanders, L. K., Copeland, N. G., *et al.* (1995) Arc, a growth factor and activity regulated gene, encodes a novel cytoskeleton-associated protein that is enriched in neuronal dendrites. *Neuron*, **14**, 433–45.

Lynch, M. A., Voss, K. I., Rodriguez, J., and Bliss, T. V. P. (1994) Increase in synaptic vesicle proteins accompanies long-term potentiation in the dentate gyrus. *Neuroscience*, **60**, 1–4.

MacDonald, J. F., Kotecha, S. A., Lu, W. Y., and Jackson, M. F. (2001) Convergence of PKC-dependent kinase signal cascades on NMDA receptors. *Curr. Drug Targets*, **2**, 299–312.

Mackler, S. A., Brooks, B. P., and Eberwine, J. H. (1992) Stimulus-induced coordinate changes in mRNA abundance in single post-synaptic hippocampal neurones. *Neuron*, **9**, 539–51.

Man, H. Y., Ju, W., Ahmadian, G., and Wang, Y. (2000) Intracellular trafficking of AMPA receptors in synaptic plasticity. *Cell. Mol. Life Sci.*, **57**, 1526–34.

Manahan-Vaughan, D., Kulla, A., and Frey, J. U. (2000) Requirement of translation but not transcription for the maintenance of long-term depression in the CA1 region of freely moving rats. *J. Neurosci.*, **20**, 8572–6.

Manahan-Vaughan, D., Ngomba, R. T., Storto, M., Kulla, A., Catania, M. V., Chiechio, S., *et al.* (2003) An increased expression of the mGlu5 receptor protein following LTP induction at the perforant path-dentate gyrus synapse in freely moving rats. *Neuropharmacology*, **44**, 17–25.

Matsuo, R., Asada, A., Fujitani, K., and Inokuchi, K. (2001) LIRF, a gene induced during hippocampal long-term potentiation as an immediate-early gene, encodes a novel RING finger protein. *Biochem. Biophys. Res. Commun.*, **289**, 479–84.

Matsuo, R., Murayama, A., Saitoh, Y., Sakaki, Y., and Inokuchi, K. (2000) Identification and cataloging of genes induced by long-lasting long-term potentiation in awake rats. *J. Neurochem.*, **74**, 2239–49.

Mayford, M., Bach, M. E., Huang, Y.-Y., Wang, L., Hawkins, R. D., and Kandel, E. R. (1996) Control of memory formation through regulated expression of a CamKII transgene. *Science*, **274**, 1678–83.

Meberg, P. J., Barnes, C. A., McNaughton, B. L., and Routtenberg, A. (1993) Protein kinase C and F1/GAP43 gene expression in hippocampus inversely related to synaptic enhancement lasting 3 days. *Proc. Natl. Acad. Sci. (USA)*, **90**, 12050–4.

Meffert, M. K., Premack, B. A., and Schulman, H. (1994) Nitric-oxide stimulates Ca^{2+}-independent synaptic vesicle release. *Neuron*, **12**, 1235–44.

Mello, C., Vicario, D. S., and Clayton, D. (1992) Song presentation induces gene expression in the songbird forebrain. *Proc. Natl. Acad. Sci. (USA)*, **89**, 6818–22.

Milbrandt, J. (1987) A nerve-growth factor-induced gene encodes a possible transcription regulatory factor. *Science*, **238**, 797–81.

Miyashita, Y., Kameyama, M., Hasegawa, I., and Fukushima, T. (1998) Consolidation of visual associative long-term memory in the temporal cortex of primates. *Neurobiology of Learning* and *Memory*, **70**, 197–211.

Moore, A. N., Waxham, M. N., and Dash, P. K. (1996) Neuronal activity increases the phosphorylation of the transcription factor cAMP response element-binding protein (CREB) in rat hippocampus and cortex. *J. Biol. Chem.*, **271**, 14214–20.

Morris, B. J. (1993) Control of receptor sensitivity at the mRNA level. *Mol. Neurobiol.*, **7**, 1–17.

Morris, B. J. (1997) Stabilization of dendritic mRNAs by nitric oxide allows localized, activity-dependent enhancement of hippocampal protein synthesis. *Eur. J. Neurosci.*, **9**, 2334–9.

Morris, B. J., Feasey, K. J., ten Bruggencate, G., Herz, A., and Höllt, V. (1988) Electrical stimulation in vivo increases the expression of proenkephalin mRNA and decreases the expression of prodynorphin mRNA in rat hippocampal granule cells. *Proc. Natl. Acad. Sci. (USA)*, **85**, 3226–30.

Morris, B. J. and Johnston, H. M. (1995) A role for hippocampal opioids in long-term functional plasticity. *Trends Neurosci.*, **18**, 350–5.

Moser, M-B., Trommald, M., and Andersen, P. (1994) An increase in dendritic spine density on hippocampal CA1 pyramidal cells following spatial learning in adult rats suggests the formation of new synapses. *Proc. Natl. Acad. Sci. (USA)*, **91**, 12673–5.

Nakazawa, K., Karachot, L., Nakabeppu, Y., and Yamamori, T. (1993) The conjunctive stimuli that cause long-term desensitisation also predominantly induce c-fos and junB in cerebellar Purkinje cells. *Neuroreport*, **4**, 1275–8.

Nakic, M., Manahan-Vaughan, D., Reymann, K. G., and Schachner, M. (1998) Long-term potentiation in vivo increases rat hippocampal tenascin-C expression. *J. Neurobiol.*, **37**, 393–404.

Nayak, A., Zastrow, D. J., Lickteig, R., Zahniser, N. R., and Browning, M. D. (1998) Maintenance of late-phase LTP is accompanied by PKA-dependent increase in AMPA receptors synthesis. *Nature*, **394**, 680–3.

Nedivi, E., Hevroni, D., Naot, D., Israeli, D., and Citri, Y. (1993) Numerous candidate plasticity-related genes revealed by differential cDNA cloning. *Nature*, **363**, 718–22.

Nguyen, P. V. and Kandel, E. R. (1997) Brief theta-burst stimulation induces a transcription-dependent late phase of LTP requiring cAMP in area CA1 of the mouse hippocampus. *Learning and Memory*, **4**, 230–43.

Nighorn, A., Healy, M. J., and Davis, R. L. (1991) The cyclicAMP phosphodiesterase encoded by the drosophila Dunce gene is concentrated in the mushroom body neurophil. *Neuron*, **6**, 455–467.

O'Brien, R. J., Xu, D., Petralia, R. S., Steward, O., Huganir, R. L., and Worley, P. (1999) Synaptic clustering of AMPA receptors by the extracellular immediate — early gene product Narp. *Neuron*, **23**, 309–23.

Okuno, H., and Y, M. (1996) Expression of the transcription factor **Zif**268 in the temporal cortex of monkeys during visual paired associate learning. *Eur. J. Neurosci.*, **8**, 2118–28.

Osten, P., Valsamis, L., Harris, A., and Sacktor, T. (1996) Protein synthesis-dependent formation of protein kinase Mzeta in long-term potentiation. *J. Neurosci.*, **16**, 2444–51.

Otani, S., Marshall, C. J., Tate, W. P., Goddard, G. V., and Abraham, W. C. (1989) Maintenance of long-term potentiation in rat dentate gyrus requires protein synthesis but not messenger RNA synthesis immediately post-tetanisation. *Neuroscience*, **28**, 519–30.

Parker, D. and Grillner, S. (1999) Long-lasting substance-P-mediated modulation of NMDA-induced rhythmic activity in the lamprey locomotor network involves separate RNA- and protein-synthesis-dependent stages. *Eur. J. Neurosci.*, **11**, 1515–22.

Patterson, S. L., Grover, L. M., Schwartzkroin, P. A., and Bothwell, M. (1992) Neurotrophin expression in rat hippocampal slices: a stimulus paradigm inducing LTP in CA1 evokes increases in BDNF and NT3 mRNAs. *Neuron*, **9**, 1081–8.

Qian, Z., Gilbert, M., and Kandel, E. (1994) Temporal and spatial regulation pf the expression of BAD2, a MAP kinase phosphatase, during seizure, kindling and LTP. *Learning and Memory 1*, 180–8.

Qian, Z., Gilbert, M. E., Colicos, M. A., Kandel, E. R., and Kuhl, D. (1993) Tissue plasminogen activator is induced as an immediate-early gene during seizure, kindling and long-term potentiation. *Nature*, **361**, 453–6.

Reed, R. R., Liao, D., Zhang, X., O'Brien, R., Ehlers, M. D., and Huganir, R. O. (1999) Regulation of morphological postsynaptic silent synapses in developing hippocampal neurons. *Nature, Neuroscience*, **2**, 37–43.

Richardson, C. L., Tate, W. P., Mason, S. P., Lawlow, P. A., Dragunow, M., and Abraham, W. C. (1992) Correlation betwen the induction of an immediate-early gene, zif/268, and long-term potentiation. *Brain Res.*, **580**, 147–54.

Richter-Levin, G., Thomas, K., Hunt, S. P., and Bliss, T. V. P. (1998) Dissociation between genes activated in long-term potentiation and in spatial learning in the rat. *Neurosci. Lett.*, **251**, 41–4.

Roberts, L. A., Higgins, M., O'Shaughnessy, C. T., Stone, T. W., and Morris, B. J. (1996) Changes in hippocampal gene expression associated with the induction of long-term potentiation. *Mol. Brain Res.*, **42**, 123–7.

Roberts, L. A., Large, C. H., Higgins, M. H., Stone, T. W., O'Shaughnessy, C. T., and Morris, B. J. (1998*a*) Increased expression of dendritic mRNAs following the induction of long-term potentiation. *Mol. Brain Res.*, **56**, 38–44.

Roberts, L. A., Large, C. H., O'Shaughnessy, C. T., and Morris, B. J. (1997*b*) Long-term potentiation in perforant path/granule cell synapses is associated with a post-synaptic induction of proenkephalin gene expression. *Neurosci. Lett.*, **227**, 205–8.

Roberts, L. A., Morris, B. J., and O'Shaughnessy, C. T. (1998*b*) Involvement of two isoforms of SNAP-25 in the expression of long-term. *Neuroreport*, **9**, 33–6.

Schilhab, T. S. S. and Christoffersen, G. R. J. (1996) Role of protein synthesis in the transition from synaptic short-term to long-term depression in neurons of helix pomatia. *Neuroscience*, **73**, 999–1007.

Schulz, S., Siemer, H., Krug, M., and Hollt, V. (1999) Direct evidence for biphasic cAMP responsive element-binding protein phosphorylation during long-term potentiation in the rat dentate gyrus in vivo. *J. Neurosci.*, **19**, 5683–92.

Schuster, T., Krug, M., Hassan, H., and Schachner, M. (1998) Increase in proportion of hippocampal spine synapses expressing neural cell adhesion molecule NCAM180 following long-term potentiation. *J. Neurobiol.*, **37**, 359–72.

Seeds, N. W., Williams, B. L., and Bickford, P. C. (1995) Tissue plasminogen activator induction in purkinje neurones after cerebellar motor learning. *Science*, **270**, 1992–5.

Silva, A. J., Kogan, J. H., Frankland, P. W., and Kida, S. (1998) CREB and memory. *Annu. Rev. Neurosci.*, 21.

Silva, A. J., Stevens, C. F., Tonegawa, S., and Wang, Y. (1992) Deficient hippocampal long-term potentiation in CamKIImutant mice. *Science*, **257**, 201–6.

Simpson, C. S. and Morris, B. J. (2000) Regulation of neuronal cell adhesion molecule (NCAM) expression by NF-kB. *J. Biol. Chem.*, **275**, 16879–84.

Smirnova, T., Laroche, S., Errington, M. L., Hicks, A., Bliss, T. V. P., and Mallet, J. (1993) Trans-synaptic expression of a presynaptic glutamate receptor during hippocampal long-term potentiation. *Science*, **262**, 433–7.

Soderling, T. R. and Derkach, V. A. (2000) Postsynaptic protein phosphorylation and LTP. *Trends Neurosci.* **23**, 75–80.

Stammler, J. S. (1994) Redox signalling: nitrosylation and related target interactions of nitric oxide. *Cell*, **78**, 931–6.

Thiel, G., Schoch, S., and Petersohn, D. (1994) Regulation of synapsin I gene expression by the zinc finger transcription factor zif/268/egr 1. *J. Biol. Chem.*, **269**, 15294–8.

Thomas, K. L., Davis, S., Laroche, S., and Hunt, S. P. (1994*a*) Regulation of the expression of NR1 NMDA glutamate receptor subunits during hippocampal LTP. *Neuroreport*, **6**, 119–23.

Thomas, K. L., Laroche, S., Errington, M. L., Bliss, T. V. P., and Hunt, S. P. (1994*b*) Spatial and temporal changes in signal transduction pathways during LTP. *Neuron*, **13**, 737–45.

Tighilet, B., Hashikawa, T., and Jones, E. G. (1998) Cell- and lamina-specific expression and activity-dependent regulation of type II calcium/calmodulin-dependent protein kinase isoforms in monkey visual cortex. *J. Neurosci.*, **18**, 2129–46.

Tsai, K. J., Chen, S. K., Ma, Y. L., Hsu, W. L., and Lee, E. H. Y. (2002) Sgk, a primary glucocorticoid-induced gene, facilitates memory consolidation of spatial learning in rats. *Proc. Natl. Acad. Sci. (USA)*, **99**, 3990–5.

Tsirka, S. E., Gualandris, A., Amaral, D. G., and Strickland, S. (1995) Excitotoxin-induced neuronal degeneration and seizure are mediated by tissue-plasminogen activator. *Nature*, **377**, 340–3.

Tsui, C. C., Copeland, N. G., Gilbert, D. J., Jenkins, N. A., Barnes, C., and Worley, P. F. (1996) Narp, a novel member of the pentraxin family, promotes neurite outgrowth and is dynamically regulated by neuronal activity. *J. Neurosci.*, **16**, 2463–78.

Tully, T., Preat, T., Boynton, S. C., and Del Vecchio, M. (1994) Genetic dissection of consolidated memory in Drosophila. *Cell*, **79**, 35–47.

Van der Zee, C. E. E. M., Rashid, K., Le, K., Moore, K. A., Stanisz, J., Diamond, J., *et al.* (1995) Intraventricular administration of antibodies to nerve growth factor retards kindling and blocks mossy fibre sprouting in adult rats. *J. Neurosci.*, **15**, 5316–23.

Wisden, W., Errington, M. L., Williams, S., Dunnett, S. B., Waters, C., Hitchcock, D., *et al.* (1990) Differential expression of immediate-early genes in the hippocampus and spinal cord. *Neuron*, **4**, 603–14.

Woolf, N. J., Zinnerman, M. D., and Johnson, G. V. W. (1999) Hippocampal microtubule-associated protein-2 alterations with contextual memory. *Brain Res.*, **821**, 241–9.

Worley, P. F., Bhat, R. V., Baraban, J. M., Erickson, C. A., McNaughton, B. L., and Barnes, C. A. (1993) Thresholds for synaptic activation of transcription factors in hippocampus — correlation with long-term enhancement. *J. Neurosci.*, **13**, 4776–86.

Worley, P. F., Christy, B. A., Nakabeppu, Y., Bhat, R. V., Cole, A. J., and Baraban, J. M. (1991) Constitutive expression of zif268 in neocortex is regulated by synaptic activity. *Proc. Natl. Acad. Sci. (USA)*, **88**, 5106–10.

Wu, Z.-L., Thomas, S. A., Villacres, E. C., Xia, Z., Simmons, M. L., Chavkin, C., *et al.* (1992) Altered behaviour and long-term potentiation in type I adenyly cyclase mutant mice. *Proc. Natl. Acad. Sci., (USA)*, **92**, 220–4.

Xia, S. Z., Feng, C. H., and Guo, A. K. (1998) Multiple-phase model of memory consolidation confirmed by behavioral and pharmacological analyses of operant conditioning in Drosophila. *Pharmacol. Biochem. Behav.*, **60**, 809–16.

Xue, J., Li, G., Laabich, A., and Cooper, N. G. F. (2001) Visual-mediated regulation of retinal CaMKII and its GluR1 substrate is age-dependent. *Mol. Brain Res.*, **93**, 95–104.

Yamagata, K., Andreasson, K. I., Kaufmann, W. E., Barnes, C. A., and Worley, P. (1993) Expression of a mitogen-inducible cyclooxygenase in brain neurons: regulation by synaptic activity. *Neuron*, **11**, 371–9.

Yamagata, K., Andreasson, K. I., Sugiura, H., Maru, E., Dominique, M., Irie, Y., *et al.* (1999) Arcadlin is a neural activity-regulated cadherin involved in long term potentiation. *J. Biol. Chem.*, **274**, 19473–9.

Yamagata, K., Kaufmann, W. E., Lanahan, A., Papavlou, M., Barnes, C. A., Andreasson, K. I., *et al.* (1994*a*) Egr3/pilot, a zinc finger transcription factor, is rapidly regulated by synaptic activity in brain neurons and colocalises with egr1/zif/268. *Learning and memory*, **1**, 141–52.

Yamagata, K., Sanders, L. K., Kaufmann, W. E., Yee, W., Barnes, C. A., Nathans, D., *et al.* (1994*b*) rheb, a growth factor and synaptic activity-regulated gene, encodes a novel ras-related protein. *J. Biol. Chem.*, **269**, 16333–9.

Yamazaki, M., Matsuo, R., Fukazawa, Y., Ozawa, F., and Inokuchi, K. (2001) Regulated expression of an actin-associated protein, synaptopodin, during long-term potentiation. *J. Neurochem.*, **79**, 192–9.

Yeckel, M. F. and Berger, T. (1998) Spatial distribution of potentiated synapses in hippocampus, Dependence on cellular mechanisms and network properties. *J. Neurosci.*, **18**, 438–50.

Yin, J. C. P., Wallach, J. S., Del Vecchio, M., Wilder, E. L., Zhou, H., Quinn, W. G., *et al.* (1994) Induction of a dominant negative CREB transgene specifically blocks long-term memory in drosophila. *Cell*, **79**, 49–58.

Chapter 15

Genetic basis of human neuronal diseases

Mark E. S. Bailey

15.1 Introduction

Since the first edition of this book in 1997, molecular neurogenetics has truly come of age. The rate at which publications in this field are emerging surpasses the ability of even the most dedicated to keep pace. Of 3850 entries in McKusick's OMIM (Online Mendelian Inheritance in Man; http://www.ncbi.nlm.nih.gov/Omim/) database that include neurological/neuronal phenotype references, more than 600 are listed as having identified gene lesions. Detailed knowledge of the human and mouse genomes has brought many disorders of the brain and central and peripheral nervous systems within the scope of genetic and clinical research groups worldwide, and technology and methodology for studying the genetic factors involved is largely mature and widely available. Single gene disorders have been succumbing at the rate of several every month. Even the complex disorders, in which both environmental and genetic predisposing factors are acting, are beginning to yield to the barrage of statistical and molecular techniques currently available. Progress in the realm of bioinformatics will undoubtedly speed the process of identifying predisposing genes further over the next few years.

The identification of such genes has the primary benefit of directing the attention of biologists and clinicians of all leanings to the biochemical and cellular pathways affected by the genetic lesion in their efforts to understand the aetiological factors operating in the pathogenic process. Attention thus directed leads to the secondary, but no less important, benefit of increased understanding of normal cellular processes — an illustration of the dictum that one good way to understand a biological process is to disrupt a component and study the consequences. This approach does not always work in a straightforward manner. The gene may encode a novel protein whose function takes many more years to discover. The 'gene knockout' in an animal model may not accurately reproduce the disorder, or the mouse (currently the most commonly used model) may not be a suitable model organism (Watase and Zoghbi 2003). The effect of the gene lesion may not relate in a straightforward manner to the pathology of the disorder or to measurable aspects of cellular dysfunction. On the whole, however, identification of the genes responsible for disorders in which neurons are affected has had a profoundly enhancing effect on understanding of disease processes and often of neuronal functioning.

The typical procedure for identifying causal and predisposing genes has altered somewhat recently, largely because of the availability of the bulk of the human genome sequence and its preliminary annotation (see the following websites with 'genome browsers': http://www.ncbi.nlm.nih.gov/genome/guide/human/ and http://www.ensembl.org/Homo_sapiens/). Additional factors have been the construction of accurate, high-density linkage maps (Kong *et al.* 2002) and their integration with the physical (clone-based) maps and DNA sequence. Such linkage maps are indispensable in the mapping of genes for simple genetic disorders. Other useful developments include the ongoing construction of 'linkage disequilibrium' (LD) maps (Dawson *et al.* 2002). The finding of LD in a human disorder is often indicative of an ancient 'founder' mutation that has not (yet) been eliminated by strong natural selection — indeed in some cases disease-associated alleles may be under positive selection for other phenotypes. LD has been, and will increasingly be, useful in the mapping and identification of predisposing and protective genes in complex traits and disorders (Ardlie *et al.* 2002; Gabriel *et al.* 2002).

Commonly used strategies for mapping and identifying genes for traits and disorders are illustrated in Fig. 15.1. For straightforward, Mendelian, single gene disorders that are strongly familial, classical positional cloning strategies centred around a 'genome scan' are still the method of choice. Delineation of a critical region within which a disease gene must lie is usually followed by identification of 'positional–functional' candidate genes (those for which a rationale for their involvement can be evinced). These are then screened for mutations in affected individuals from families showing linkage to that locus, either by sequencing of PCR-amplified genomic DNA, or using a range of sensitive methods for identifying the presence of sequence variants in PCR products (dHPLC, DIGE, heteroduplex analysis, SSCP, and many others). If the positional–functional candidates do not exhibit mutations, other positional candidate genes are screened. Once a mutation has been identified, the usual biochemical, cell biological, and physiological approaches to explaining how the mutation leads to the disease process and pathology are invoked. Animal models, usually involving gene-targeted deletions or 'knock-ins' of the human mutation in mice, may be developed to assist in this functional characterization.

Complex disorders, in which multiple genes operating in various combinations act together to influence the risk of developing the disorder, still pose a problem for classical linkage analysis approaches. Alternative methods for disorders that are less often strongly familial have been focused around searching for regions of the genome that tend to be shared more often than expected by affected pairs of relatives (e.g. affected sib-pairs, ASPs). Once identified, these regions may then be subjected to a form of analysis that depends either on LD or on allelic association, comparing the frequency of a gene variant in disease and normal control populations. Both approaches are fraught with difficulties and have enjoyed limited success. In many cases, however,

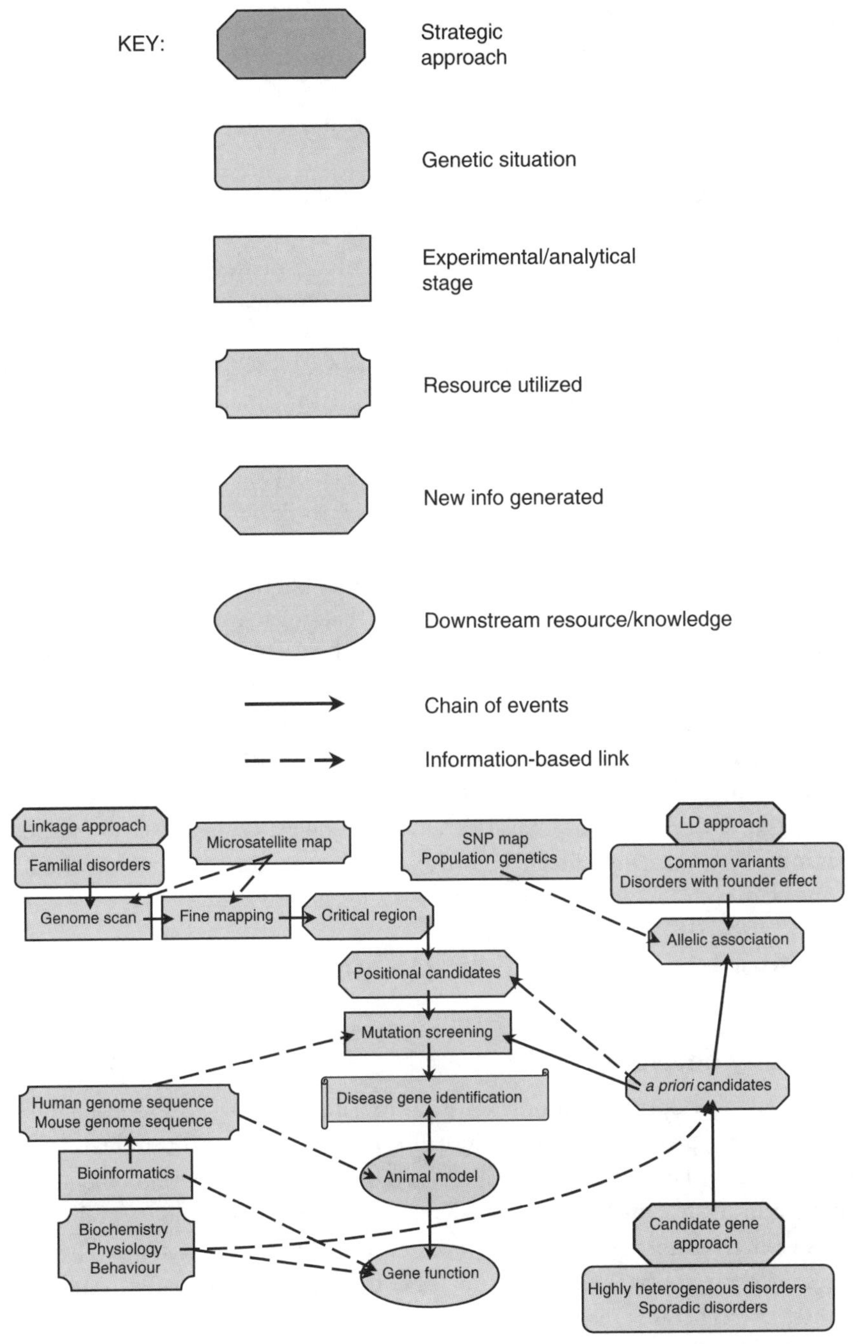

Fig. 15.1 Overview of strategies used to identify disease genes. Summary of approaches used to identify and characterize human disease genes in the post-genome era, focusing on 'post-genome' resources and knowledge base.

ideas about the aetiology of the disorder can direct geneticists to functional candidate genes, which are then tested in association studies, or screened directly for mutations or predisposing variants using the same methods as described above. This strategy is somewhat dangerous, since many sensible candidates often present themselves, but, again, it has met with some success.

In this chapter, I have summarized what is known about the genes underlying a range of human diseases where the causative or associated gene is expressed in neurons and there is a clear primary effect on developing or mature neurons, and have illustrated how subsequent studies have illuminated both the pathogenic process and normal neuronal functioning. Tables giving details of all neuronal disorders with known genes in categories covered in this chapter are contained in Tables 15.1–15.14. I have not tried to cover everything — even taking redundancy of gene function into account, there may be more than 5000 genetic disorders affecting the nervous system — and several categories are covered lightly if at all, including the metabolic storage disorders, disorders involving neurons whose cell bodies or major processes lie outside the brain and spinal cord, disorders involving primarily tumours of neuronal or neuroblast origin, and disorders in which the primary effect is on glial or other cell function and secondarily on neurons. The primary literature is now so vast that, for reasons of limited space, in many places I have cited excellent reviews rather than multiple research papers. The topics covered have been organized according to a necessarily simplistic view of the functional ontogeny of the nervous system, from early neurogenesis, migration, and pattern formation, through neuronal maturation/differentiation and mature neuronal function, to neurodegeneration and cell death. In the tables, 'OMIM Number' refers to the entry for each disease in McKusick's 'Online Mendelian Inheritance in Man' database (http://www.ncbi.nlm.nih.gov/Omim/), in which a substantial amount of clinical and molecular information is summarized and referenced.

15.2 Disorders of neural induction, neurogenesis, and neuronal migration

15.2.1 Early development

A number of developmental disorders are due to failure of neuroectodermal induction and subsequent processes in rostral morphogenesis. Disorders in this category for which genes have been identified are illustrated in Table 15.1. A common theme is that mutations in genes acting early in these developmental pathways tend to produce more pervasive effects, those acting later to produce more limited phenotypes. There is an interesting contrast between two types of genes that have pervasive effects. The first set has these effects because they lie upstream (and hence are expressed earlier in development) in a regulatory cascade. For genes in the other set, the explanation has to do with the fact that they perform different roles at different times in development. For example, many homeodomain transcription factors, when mutated, display a phenotype consistent with defective functioning both in their early developmental role and in their subsequent, more spatially limited, maintenance role.

Neural tube defects (NTDs), such as spina bifida, and conditions such as anencephaly are thought to be related aetiologically. Genetic involvement has been expected from the increased recurrence risk of NTDs in families of affected individuals. An emerging picture is that disruption of biochemical pathways involving folate and methionine metabolism presents an increased risk of NTDs. A common polymorphism in the *MTHFR* gene, designated 665C->T (previously known as 677C->T), appears to confer increased susceptibility to spina bifida. The polymorphism is associated with an apparently conservative amino acid change (A222V) that in fact renders the enzyme heat labile and presumably leads to functional deficiency. Foetuses that are homozygous for the T allele at this polymorphism seem to be at greater risk of NTDs and of failing to develop to term. Boosting of folate levels in the mother of such a foetus is thought to protect the enzyme against this lability, thus reducing both these risks. The encoded enzyme, 5,10-methylenetetrahydrofolate reductase (MTHFR), is involved in methionine metabolism and it is not clear at present why other mutations, resulting in MTHFR deficiency, cause a different range of disorders often characterized by microcephaly (reduced brain size) and retarded psychomotor development, amongst other signs. Polymorphisms in other genes involved in folate provision and utilization, for example those encoding 5,10-methylenetetrahydrofolate dehydrogenase (MTHFD1), methionine synthase (MTR), and methionine synthase reductase (MTRR), also have alleles clearly associated with increased risk of NTDs. Interestingly, several recent studies (Brody *et al.* 2002; Doolin *et al.* 2002) have shown that the risk can come from the maternal genotype at these loci, rather than from that of the foetus, emphasizing the role that maternal factors and the uterine environment play in human development.

Other relatively early processes in neural development, such as regional organogenesis, are also disrupted in human genetic disorders. In benign hereditary chorea (BCH), an early-onset movement disorder mimicking Huntington disease (see Section 15.4.3) in some respects, mutations have been found in *TITF1*, whose product is a transcription factor involved in specification of several tissues and organs occupying ventral locations, including structures lying ventrally in the forebrain, especially the basal ganglia (Breedveld *et al.* 2002). The disrupted development of the basal ganglia in BCH, a dominant phenotype, is presumed to arise from haploinsufficiency i.e. drastic loss of function mutations in one allele result in affected individuals carrying only a single functional allele that cannot provide enough of the gene product for normal function.

The gene encoding another homeodomain-containing transcription factor, *HESX1*, is mutated in septo-optic dysplasia (SOD), a recessive disorder characterized by developmental failure of several midline brain structures as well as other organs, including the pituitary. HESX1 is a paired-like class homeodomain protein expressed in various parts of the presumptive prosencephalon at the induction stage, spreading to the telencephalon and finally ventral diencephalon. As referred to above, HESX1 seems to have an additional role later in development, being expressed in midline structures and in the pituitary gland. The point mutations found in SOD appear to disrupt important residues involved in DNA binding of the homeodomain (Dattani *et al.* 1998), but in

Table 15.1 Disorders of early neurogenesis, CNS patterning and neuronal migration

Disorder type	Disorder	Symbol	OMIM	Gene(s)	Symbol	Location
Neural-tube defects	Neural-tube defect, folate-sensitive	NTDFS	601634	5,10-methylenetetrahydrofolate reductase	*MTHFR*	1p36.3
				5,10-methylenetetrahydrofolate dehydrogenase	*MTHFD1*	14q24
Brain morphogenesis	Holoprosencephaly	HPE2	157170	Sine oculis homeobox gene homologue 3	*SIX3*	2p21
		HPE3	142945	Sonic hedgehog	*SHH*	7q36
		HPE4	142946	TG interacting factor	*TGIF*	18p11.3
		HPE5	603073	Zinc finger protein of cerebellum 2	*ZIC2*	13q32
		HPE7	601309	Patched, Drosophila, homologue of	*PTCH*	9q22.3
	Forebrain midline defects		187395	Teratocarcinoma-derived growth factor 1	*TDGF1*	3p23-p21
	Smith–Lemli–Opitz syndrome	SLOS	270400	7-dehydrocholesterol reductase (7-DHCR)	*DHCR7*	11q12-q13
	Chorea, hereditary benign	BCH	118700	Thyroid transcription factor 1	*TITF1*	14q13
	Septo-optic dysplasia	SOD	182230	Homeobox gene expressed in ES cells	*HESX1*	3p21.2-p21.1
	Sanjad–Sakati syndrome and autosomal recessive Kenney–Caffey syndrome	HRD/AR-KCS	241410 244460	Tubulin-specific chaperone E	*TBCE*	1q43-q44
	Aniridia, type II	AN2	106210	Paired and homeodomain protein 6	*PAX6*	11p13
	Schizencephaly		269160	Empty spiracles homeobox gene homologue 2	*EMX2*	10q26.1

Growth/proliferation/ progenitor apoptosis	Microcephaly, primary, autosomal recessive	MCPH1 MCPH5	251200	Microcephalin Abnormal spindle, microcephaly-associated	*MCPH1* *ASPM*	8p23 1q31
	Amish microcephaly	MCPHA	607196	Deoxynucleotide carrier	*SLC25A19*	17q25
	Opitz syndrome	OS	300000	Midline-1	*MID1*	Xp22.3
	Hoyeraal–Hreidarsson syndrome	HHS	300240	Dyskerin	*DKC1*	Xq28
Neuronal migration	Kallman syndrome	KAL1 KAL2	308700 147950	Anosmin-1 Fibroblast growth factor receptor 1	*KAL1* *FGFR1*	Xp22.3 8p11.2-p11.1
	Hirschsprung disease	HSCR	235730	Smad interacting protein-1	*SMADIP1*	2q22
	Miller–Dieker lissencephaly/ Lissencephaly 1 [Subcortical band heterotopia]	MDS LIS1 [SBH]	247200 and 607432	Platelet-activating factor acetylhydrolase, isoform 1b, α subunit	*PAFAH1B1* *[LIS1]*	17p13.3
	Lissencephaly, X-linked, with ambiguous genitalia	XLAG	300215	Aristaless homologue, X chr.	*ARX*	Xp22.13
	Lissencephaly, X-linked	LISX	300067	Doublecortin	*DCX*	Xq22.3-q23
	Lissencephaly syndrome, Norman–Roberts type [Lissencephaly with cerebellar hypoplasia]	LCH	257320	Reelin	*RELN*	7q22
	Periventricular heterotopia	PH	300049	Filamin A	*FLNA*	Xq28
	Walker–Warburg syndrome	WWS	236670	Protein *O*-mannosyltransferase 1	*POMT1*	9q34.1
	Fukuyama congenital muscular dystrophy	FCMD	253800	Fukutin	*FCMD*	9q31
	Muscle–eye–brain disease	MEB	253280	Protein *O*-mannose β-1,2-*N*-acetylglucosaminyltransferase	*POMGNT1*	1p34-p33

Names and symbols for disorders are taken from the OMIM entry describing them. Names enclosed in [ ] are commonly used synonyms. Disorder symbols are those given in OMIM alongside the name, and correspond to disorder locus, not necessarily to the acronym for the disorder in common use. Locations are given as chromosome with cytogenetic band localization on the p (short) or q (long) arm.

some cases may also cause a phenotype with variable 'penetrance' (probability of the phenotype being displayed, given that the mutation is present) in the heterozygous state (in the presence of one copy of the mutation only; this mutation thus exerts partial dominance at the genetic level). This suggests that the levels of protein required for normal development are pitched precisely on the edge of the 'haplosufficient' threshold and any mild perturbation of levels of functional protein may result in that threshold being crossed. In these cases it is possible that modifier loci exist, and that one's genotype at those other loci may affect the penetrance of the normally recessive mutations in *HESX1*.

Transcriptional specification of particular neuronal populations is not the only organogenesis-stage process disrupted by human mutations. Another well-known set of disorders, holoprosencephaly (HPE) and other forebrain midline defects, is now known to be caused by disruption of intra- and extracellular signalling processes. HPE is characterized by a failure of the developing forebrain to separate into two halves laterally (normally creating the cerebral hemispheres) and consequent lack of proper formation of the ventral forebrain and midline structures of the head and face. With both sporadic and familial forms, HPE is known to be complex, and to date six genes have been reported, *SHH* encoding sonic hedgehog, *SIX3*, *TGIF*, *ZIC2*, *PTCH*, and *TDGF1*, all of whose products are involved in morphogenesis of presumptive forebrain at the midline roofplate. Several signalling pathways are involved, including the nodal/TGF-β pathway, as well as several downstream transcriptional activation/repression pathways, and the interactions between all the components are still somewhat murky, although understanding is accumulating of a hitherto unsuspected convergence of the apparently disparate genetic causes on common downstream pathways (Brown *et al.* 2001b; Muenke and Beachy 2000). In an excellent review, Ming and Muenke (2002) raise the intriguing possibility that digenic inheritance, the requirement for genetic 'hits' at more than one locus, may explain both the small proportion of HPE individuals born with mutations in the known genes and the tendency for the known mutations to be of widely variable penetrance — the mutations are not haploinsufficient or acting in a dominant gain-of-function manner at the molecular level, but may appear to have dominant inheritance in some cases when a mutation in another gene whose product interacts with the first is co-inherited or becomes mutated somatically. Another congenital brain malformation disorder, Smith–Lemli–Opitz syndrome (SLOS) is caused by mutations in *DHCR7*, encoding a sterol synthesis enzyme. Links between SHH signalling function and cholesterol modifications are highlighted by the fact that SLOS mutations in some cases involve an HPE phenotype.

15.2.2 **Neuronal progenitor proliferation**

Microcephaly, defined in terms of reduced overall brain size, is associated with many different disorders of the brain. In some cases, there is no association with mental retardation or incapacity. One such condition is primary autosomal recessive microcephaly (MCPH). This disorder has recently been shown to be caused by null (nonsense) mutations in a novel gene given the name microcephalin, *MCPH1*, located on chr.8p23 (Jackson *et al.* 2002). *MCPH1* is normally expressed at high levels in developing forebrain, particularly in neural progenitors in the periventricular zone, and lack of the gene product is assumed to restrict the ability of these cells to proliferate, thus leading to fewer cortical cells than normal. The predicted protein product has a region of homology with the BRCA1-C-terminal domain (BRCT), and this domain is a feature of cell cycle regulatory proteins that respond to DNA repair. The neuronal proliferation failure may be a result either of increased cell cycle times or of apoptotic progenitor loss. It is known that another locus, MCPH5, carries mutations in the *ASPM* gene (Bond *et al.* 2002), whose product is related to a *Drosophila* protein involved in normal spindle function. Together, these findings provide evidence that brain size (and cortical thickness) in man is related to a large extent to cell division rates, and their balance with rates of apoptosis, during cortical development. Other developmental growth failure disorders are also likely to exhibit disrupted spindle function. One such may be Sanjad–Sakati syndrome, involving mutations in the gene encoding the tubulin chaperone protein, TBCE. Another is X chromosome-linked Opitz syndrome, in which *MID1*, encoding a ubiquitin ligase that is normally involved in targeting a MAP (microtubule-associated protein) phosphatase for degradation, is mutated. Mutations in this gene result in a build up of hypophosphorylated MAPs in the affected cells (Trockenbacher *et al.* 2001). Such disorders involving microtubule function are numerous and are characterized by generalized or region-specific growth failure as well as the consequences, often, of disrupted cell polarity and Golgi/late endosome compartment dysfunction. Other disorders in which microcephaly is a feature may result from disruptions to mitochondrial function during neurogenesis, as in Amish microcephaly (MCPHA), in which a mitochondrial deoxynucleotide carrier gene, *SLC25A19*, is mutated (Rosenberg *et al.* 2002).

15.2.3 **Neuronal migration**

The neuronal migration phase of neural development is also a rich source of genetic disorders. Migrations are of two types, one involving long-distance travel along pathways laid down in the developing brain, and the other involving more local movement of neuronal cell bodies within spaces already occupied by their neuritic processes. A disorder of long-distance migration is Kallman syndrome, caused by mutations in either the X-linked gene, *KAL1*, enoding anosmin, or the autosomal gene, *FGFR1*, encoding a receptor for fibroblast growth factor family members (Dodé *et al.* 2003). It is thought that the extracellular matrix protein anosmin and FGFR1 interact functionally, and particular loss-of-function (LOF) mutations in either gene (possibly including both haploinsufficient mutations and secondarily LOF dominant negative mutations with effects on receptor dimerization in the latter) are associated with disrupted development of the olfactory regions and migration of multiple neuronal types to the forebrain, including that of prospective GnRH neurons to the hypothalamus. Although *KAL1* is located in the pseudoautosomal region of chr.Xp, males only have one functional copy on their X chromosome and in females the gene partially escapes X-inactivation, leading to an increased functional gene dosage in females, which is thought to account for the preponderance of affected males in the X-linked form of the disorder. Another migration disorder, XLAG, results from particular mutations in *ARX*, an X-linked gene that also causes lissencephaly (see below). In XLAG, there is a primary migration defect of interneurons from the ganglionic eminence to their normal position in the cortical layer structure, resulting in the aberrant distribution of these neurons in the cortex (Kitamura *et al.* 2002).

Short range migration defects are seen in a range of disorders characterized by lissencephaly, an absence of the usual highly folded structure of the cortex. Several different phenotypes are known and identified genes, often with partially overlapping phenotypic ranges, include *PAFAH1B1 [LIS1]*, *ARX*, *DCX*, *RELN*, and *FLNA*. The gene products of all of these are involved in regulation and basic functioning of neuronal migration from the ventricular zone to the other layers comprising the cortical plate, and the mutations causing the human disorders are largely thought to be loss-of-function, leading to reductions in amounts of functional protein produced. In the lissencephaly disorders, either the ability of neurons to migrate along radial glia, or their ability to undergo subsequent nuclear/somal translocation towards the pial surface, is disrupted, leading to inverted, bunched, and hypoplastic cortical layer structures as well as to the existence of extra cells in the subventricular regions. ARX, in addition to its function in interneuron migration (see above), is probably involved in proliferation of neurons throughout the ventricular zone, which is reduced in some forms of X-linked lissencephaly.

The *LIS1* gene product appears to interact with both the intracellular dynein motor machinery and microtubules, but its function is still unclear, although several possible models exist (Feng and Walsh 2001; Gupta *et al.* 2002). Lack of this protein probably causes the pathology mainly through disruption of nuclear/somal translocation, and results in inverted cortical layer structures as the affected neurons are unable to 'pass' earlier migrating and more peripherally localized neurons. In a separate, but phenotypically overlapping and similar X-linked lissencephaly, doublecortin, the product of the *DCX* gene, is also probably involved in microtubule function, but mainly functions towards the leading edge of neurite outgrowth in endocytotic and membrane-synthetic vesicle trafficking (Friocourt *et al.* 2003). The *FLNA* gene product, filamin, is an actin network-organizing protein, and mutations probably lead to disruption of neuronal process leading edge motility and signalling functions. Reelin, the product of the *RELN* gene, is an extracellular protein secreted by Cajal–Retzius neurons in the outer marginal zone. It binds to ApoE and VLDL receptors on the cell membranes of migrating neuronal processes and is thereby probably involved in signalling gradients during neuronal migration towards the pial surface (Jossin *et al.* 2003).

Migration of cortical neurons and development of other regions, especially cerebellum, are also affected in Fukuyama congenital muscular dystrophy (FCMD), characterized by micropolygyria, in Walker–Warburg syndrome, characterized by cobblestone lissencephaly, and in muscle–eye–brain disease. In all of these, levels of properly glycosylated α-dystroglycan (DAG1) are reduced. This protein is essential for proper functioning of the extracellular matrix and basal lamina formation, and its absence probably affects functioning of cells at or near the pial surface of the cortex, including the Cajal–Retzius neurons. Fukutin, the *FCMD* gene product, is an extracellular matrix protein (Hayashi *et al.* 2001; Michele *et al.* 2002), whereas *O*-mannosyltransferase and protein *O*-mannose β-1,2-*N*-acetylglucosaminyltransferase, the glycosylation enzymes encoded by *POMT1* and *POMGNT1*, deficient in Walker–Warburg syndrome and muscle–eye–brain disease, respectively, are associated with the Golgi apparatus (Beltrán-Valero de Bernabé *et al.* 2002; Michele *et al.* 2002). Mutations in both probably affect the higher-order glycosylation, and thus structural integrity and/or functioning, of DAG1. Little is known about the integration of the various signalling and effector pathways of cortical development that are disrupted in the lissencephalies and other neuronal migration disorders, but the prospects for rapid attainment of understanding are good.

15.3 Disorders of neuronal terminal differentiation and maturation, and of mature neuronal function

A number of disorders exist in which brain malformations are not overt or major, and in which neurons have originated and migrated effectively, but have not executed their full programme of differentiation and maturation (see Table 15.2). In some cases, the affected neurons have undergone apoptosis as a result of their failure to differentiate, or failure to respond to growth factors and survival signals, and are secondarily absent, whereas in others, no evidence for neuronal loss is visible.

15.3.1 Disorders of neuronal maturation

Rett syndrome (RS) is an example of the latter. Characterized by a reduction in postnatal brain growth rate, developmental regression beginning at 9–18 months and followed by relative stagnation, loss of skills and a number of associated features, such as seizures and breathing difficulties, approximately 80% of classic RS cases, almost all of whom are female, are caused by mutations in *MECP2*, which encodes a member of the methyl-CpG-binding domain family. With an incidence of 1/10 000 or more, RS is the second most common disorder associated with severe intellectual disability in females, after Down syndrome. Around 200 different missense, nonsense, splicing, small insertion/deletion, and large deletion/null mutations (see http://homepages.ed.ac.uk/skirmis/ for tables and graphics of mutation locations and frequency) have been found so far in the coding portion of the gene, which is located in Xq28 and is subject to X-inactivation. It is therefore assumed that complete loss of function of MECP2 causes the disorder, and this is backed up by gene knockout animal models (Shahbazian *et al.* 2002), although the mechanism by which some of the apparently 'milder' missense mutations exert their effect is far from clear. The immediate downstream effects of this loss of function are still not known, however, despite intense interest in the pathogenic process and much scrutiny of the wild-type protein (Nan and Bird 2001; Shahbazian and Zoghbi 2002). This is partly because the purpose and functioning of methylation of the genome in mammals is not well understood, and partly because it is difficult to understand how what appears to be a redundant system for regulating transcription at the level of chromatin structure translates into the pervasive neuronal maturation defect seen in RS. Recent studies have suggested two ways in which MECP2, tethered and targeted to methylated regions of the genome, may carry out its known transcriptional repression effector function. Firstly, repression is known to occur via the recruitment of histone deacetylase complexes, and a story of complex interactions between the histone deacetylase and histone methylase machinery is emerging. These latter set up the 'histone code' of modifications, largely to histones H3 and H4, that are now known to determine the transcriptional state of chromatin, and theories abound as to how these states are maintained and altered in programmes, such that genes are transcribed at the correct temporal and spatial locations in the body (Kouzarides 2002; Lachner and Jenuwein 2002; Jaenisch and Bird 2003). A picture is emerging of how MECP2 and other MBD proteins work, in the context of rapid recent developments in understanding of chromatin-level regulation. Secondly, MECP2 is able, both in the presence and in the absence of DNA methylation, to nucleate the condensation of nucleosomal chromatin into a novel, very tightly compacted conformation (Georgel *et al.* 2003). The interplay between DNA methylation, histone modifications and chromatin conformational states, and the transcriptional and DNA replication machineries is on the point of being understood.

Meanwhile, we do not have a clear idea of the normal targets of MECP2 repression in the genome. Candidate genomic entities include genes whose DNA methylation state is normally regulated during postnatal brain development, genes with genomic imprinting mechanisms based on allele-specific methylation, genes on the X chromosome that normally undergo X-inactivation (both the latter unlikely, given what is known about the biology of these systems), and genes transcribed by RNA polymerases other than RNA Pol II, which transcribes most protein-coding genes. Disruption of RNA polymerase III transcription is an attractive candidate, given that no consistent changes in brain protein-coding gene mRNA levels have been observed in the various *Mecp2* knockout mouse models analysed thus far (Tudor *et al.* 2002). Most RNA Pol III-transcribed genes are involved in the construction of the protein translation machinery or in regulatory aspects of translation, and a generalized neuronal translation defect, or perhaps disruption of the special dendritic translation machinery, could explain several of the cellular features of RS. Neurons in many regions, but especially the frontal cortex, have underdeveloped dendritic arborization, small dendritic fields, and fewer dendritic spines, pointing to a generalized lack of neural connectivity without neuronal loss; they also appear to have smaller somata in some studies. Reduced translation rates at a critical point in neuronal maturation might cause this phenotype, but several other potential mechanisms exist, including defective mRNA and protein trafficking via microtubule-based transport systems.

A disruptive effect on RNA Pol III function would also, however, fit in with one persistent idea of the function of methylation in the genome, the 'genome defence' hypothesis, in which methylation-based repressor systems are thought to have evolved to protect cells against the ravages of 'selfish' transposable elements, which exist in thousands or millions of copies in the genome and in many cases contain transcriptional promoter sites. Lack of functional MECP2 in RS might enable these promoter elements to escape their repressive chromatin environment and act as a sink for transcription factors, which would then be too dilute to carry out their normal function elsewhere in the genome. A major unanswered question is whether the cellular phenotype in RS is cell-autonomous or not. If it is, approximately half the neurons in a patch of cortex, brainstem, or other involved region should be normal, given the random X-inactivation known to exist in most RS patients and the dominant nature of the disorder (patients are thus heterozygous for normal and mutant alleles). Whether a mosaic pattern of affected and normal cells exists in cortical sections has not been adequately explored.

Rett syndrome shares some features, at least during the regression phase, with another disorder that may result from ineffective neuronal maturation, autism. Autism is a complex disorder, with most cases thought to be polygenic or multifactorial, and the genes involved in these cases are almost completely unknown, although several predisposing loci have been mapped (Folstein and Rosen-Sheidley 2001). In rare cases, single gene disorders may give rise to multiple cases of autism within families, and two genes have been identified recently, the related genes *NLGN3* and *NLGN4*, encoding two members of the neuroligin family (Jamain *et al.* 2003). Both genes are located on the X chromosome and the two presumed loss of function mutations are recessive, so that only males have the disorder in the two affected families. X-linked genes have been implicated in predisposition to autism in view of the overall 4 : 1 bias towards males observed in the disorder, but the genes implicated in this study were not thought to be common monogenic causes in autistic individuals. Neuroligins are cell adhesion molecules expressed on the outer surface of dendrites and are thought to mediate physical and functional connectivity at synapses between presynaptic and postsynaptic neurons. The neuroligins contain EF hand domains, which confer Ca^{2+} sensitivity, and the point mutation observed in one family may disrupt the Ca^{2+}-dependent interaction between neuroligin and its known presynaptic binding partner, a neurexin. Autism is thought to result from defective network construction in certain regions of the frontal cortex, and it is presumed that the neuroligin mutations will disrupt synapse formation in these regions. Further insights are eagerly awaited in view of the intense public interest in autism and its causes.

15.3.2 Disorders affecting neuronal maturation and/or mature function

A large number of other monogenic forms of mental retardation (MR) are caused by mutations in X-linked genes, often affecting only one or a few known patients and thus contributing to causes of MR in the population only to a very limited extent (Chelly and Mandel 2001). These comprise a range of non-syndromic, isolated MR disorders (with loci designated 'MRX') and several syndromic forms. The genes mutated in the MRX disorders are varied, but several common themes are emerging, all of which are commensurate with defects in neuronal maturation or mature neuronal function, which can be difficult to tease apart (the disorders covered in the rest of this section may involve neuronal maturation or mature function or both). Neurotransmission defects are involved in some, such as MRX41/MRX48, caused by mutations in *GDI1*, which encodes a Rab-GTPase recycling protein important in vesicle trafficking and neurotransmitter release (D'Adamo *et al.* 1998). Intracellular signalling defects are apparent in MRX30/MRX60/MRX46, involving genes acting in the Rho-GTPase pathway. These are effector proteins responding to several signal transduction mechanisms involving growth factor and repulsive cue receptors, and mostly acting through regulation of neuronal morphogenesis and cellular shape through reorganization of the cytoskeletal actin network. The involvement of PAK/PAK3 kinases in these pathways (Allen *et al.* 1998) implicates defects in formation of focal complexes and lamellipodia during neurite outgrowth. This process is probably also disrupted in the presence of mutations (in two MRX families) in *TM4SF2*, which encodes tetraspanin, a member of a family of plasma membrane-localized, 4-transmembrane (TM) region proteins that interact with β-integrins and thereby mediate communication between the extracellular environment and the actin cytoskeleton.

A number of poorly understood mechanisms are implicated in other MRX disorders. In several individuals, the angiotensin II receptor 2 gene, *AGTR2*, is mutated (Vervoort *et al.* 2002). The role of this 7TM, G-protein-coupled receptor in postnatal and adult brain is not clear, but it has been suggested to have roles in triggering apoptotic cell death programmes, perhaps as part of the 'pruning' process that occurs during network development, as well as a possible role in regulation of neurotransmission efficiency. Another example, MRX34, involves mutations in the gene encoding a member of an IL1 receptor-associated protein family, *IL1RAPL* (Carrié *et al.* 1999). The role of this protein is entirely unknown (but assumed not to be related to immune system functions typical of the IL system proteins). Its expression in hippocampus points to learning and memory defects underlying the phenotype in the few known families with deletions encompassing part or all of the gene.

Many disorders of the brain involve, not surprisingly, genes encoding enzymes involved in synthesizing the important molecules of the neuronal plasma membrane, particularly the complex lipid derivatives, glycolipids, and glycoproteins (see below). Several disorders of membrane protein glycosylation exist and genes have been identified in 10 of these so far (listed in Table 15.2). These disorders usually involve severe brain phenotypes as well as effects on other systems, since glycosylated proteins are extremely diverse and widely expressed, but they are included in this section because of their obvious involvement in maturation of neurons, since many proteins involved in neurite outgrowth and in synaptic development are glycoproteins, and many of the complex membrane lipids are linked to oligosaccharides with roles in structural membrane integrity, as well as in membrane protein targeting and anchoring, functioning of protein complexes, and signalling processes. Almost every stage of glycolipid synthesis and processing is represented in the disorders of glycosylation, as well as several stages of glycoprotein oligosaccharide processing, both in the N-linked and the O-linked oligosaccharide pathways (Jaeken and Matthijs 2001). An unusual example of metabolically-based MR, because of the absence of large scale neurodegeneration and the relatively pure cognitive phenotype, was found to be responsible for the MRX63 locus, and involves mutations in

Table 15.2 Disorders of neuronal differentiation and maturation

Disorder category	Disorder type	Disorder	Symbol	OMIM	Gene(s)	Symbol	Location
X-linked mental retardation (XLMR)		Autism, X-linked		300425	Neuroligin 3 Neuroligin 4	*NLGN3* *NLGN4*	Xq13 Xp22.3
	Syndromic	Rett syndrome	RTT	312750	Methyl-CpG-binding protein-2	*MECP2*	Xq28
		X-linked myoclonic epilepsy with spasticity and intellectual disability and MRX54	XMESID and MRX54	300382	Aristaless homologue, X chr.	*ARX*	Xp22.13
		Mental retardation, aphasia, shuffling gait, and adducted thumbs [MASA syndrome] [CRASH syndrome]	MASA and CRASH	303350	L1 cell adhesion molecule	*L1CAM*	Xq28
		X-linked mental retardation with epilepsy, rostral ventricular enlargement enlargement and cerebellar hypoplasia	MRX60	300127	Oligophrenin 1	*OPHN1*	Xq12
	Non-syndromic	X-linked, non-specific mental retardation	MRX41, MRX48	300104	Rab GDP-dissociation inhibitor α; α-GDI	*GDI1*	Xq28
			MRX34	300426	IL1 receptor accessory protein-like	*IL1RAPL*	Xp22.1-p21.3
			MRX30	300142	p21-activated kinase 3 [oligophrenin 3]	*PAK3* *[OPN3]*	Xq22

Table 15.2 (continued) Disorders of neuronal differentiation and maturation

Disorder category	Disorder type	Disorder	Symbol	OMIM	Gene(s)	Symbol	Location
			MRX46	300346	Rac/Cdc42 guanine exchange factor 6	*ARHGEF6*	Xq26
			MRX	300096	Tetraspanin	*TM4SF2*	Xp11.4
			MRX	300034	Angiotensin II receptor 2	*AGTR2*	Xq22-q23
			MRX63	300387	Fatty acid CoA ligase, long chain 4	*FACL4*	Xq22.3
Disorders of glycosylation	Congenital disorder of lipid-linked oligosaccharide assembly and transfer	Disorders of *N*-glycan assembly	CDGIa	212065	Phosphomannomutase 2	*PMM2*	16p13.3-p13.2
			CDGIc	603147	ALG6*, *S. cerevisiae*, homologue of	*ALG6*	1p22.3
			CDGId	601110	ALG3, *S. cerevisiae*, homologue of	*ALG3*	3
			CDGIe	603503	Dolichyl-phosphate mannosyltransferase 1, catalytic subunit	*DPM1*	20q13.1
			CDGIf	604041	Mannose-P-dolichol utilization defect 1	*MPDU1*	17p13.1-p12
			CDGIg	607143	ALG12, *S. cerevisiae*, homologue of	*ALG12*	22
		Disorders of *N*-glycan processing	CDGIIa	212066	GlcNAc-transferase II	*MGAT2*	14q21
			CDGIIb	606056	Glucosidase I	*GCS1*	2p12-p13
			CDGIId	607091	β-1,4-galactosyltransferase	*B4GALT*	9p13

Table 15.2 (continued) Disorders of neuronal differentiation and maturation

Disorder category	Disorder type	Disorder	Symbol	OMIM	Gene(s)	Symbol	Location
		Disorders of *N*- and *O*-glycan processing	CDGIIc	266265	GDP-fucose transporter1	*FUCT1*	11
Other disorders		Smith–Magenis syndrome	SMS	182290	Retinoic acid induced 1	*RAI1*	17p11.2
		Fibrosis of extraocular muscles, congenital 2	FEOM2	602078	Aristaless homeobox, *Drosophila*, homologue of	*ARIX*	11q13.3-q13.4
		Mental retardation, autosomal recessive		249500	Neurotrypsin	*PRSS12*	4q24-q25
		Dyggve–Melchior–Clausen dysplasia	DMC	223800		*FLJ90130*	18q12-q21.1
		Agenesis of the corpus callosum, with neuronopathy	ACCPN	218000	Solute carrier family 12, member 6 [K^+–Cl^- cotransporter, KCC3]	*SLC12A6*	15q13-q14
		Segawa syndrome, autosomal recessive [Dopa-responsive dystonia, autosomal recessive]		605407	Tyrosine hydroxylase	*TH*	11p15.5
		Segawa syndrome, autosomal dominant/ Dystonia, progressive, with diurnal variation [Dopa-responsive dystonia, autosomal dominant]	DRD	128230	GTP cyclohydrolase I	*GCH1*	14q22.1-q22.2
		Specific language impairment	SLI	602081	Forkhead box P2	*FOXP2*	7q31
		Mucolipidosis type IV	MLIV	252650	Mucolipin 1	*MCOLN1*	19p13.3-p13.2
		Griscelli syndrome, type 1	GS1	214450	Myosin VA	*MYO5A*	15q21

*Encodes dolichyl-P-Glc : $Man_9GlcNAc_2$-pyrophosphate-dolichyl glucosyltransferase.

FACL4, which encodes a fatty acid-CoA ligase whose function is in the synthesis of arachidonate lipid derivatives (Meloni *et al.* 2002). Arachidonate is pro-apoptotic when present at increased levels, so loss-of-function mutations in *FACL4* could result in precocious cell death through failure to remove excess arachidonate by ligation. MRI findings in patients with mutations in this gene were normal, however, so alternative explanations for the phenotype that do not implicate apoptosis may be required.

An important and well-known glycoprotein illustrates a further class of disorders specifically involving cell–cell adhesion. The *L1CAM* gene is mutated in several overlapping disorders, together designated CRASH syndrome (corpus callosum hypoplasia, retardation, adducted thumbs, spastic paraplegia, and hydrocephalus), the components of which may co-occur in various combinations (e.g. MASA syndrome — mental retardation, aphasia, shuffling gait, and adducted thumbs, without hydrocephalus), but all involving more or less specific effects on neuronal and brain development. This example illustrates the tendency of clinicians, including clinical geneticists, to be 'splitters' and of molecular geneticists, with a more gene-centric view, to be 'lumpers' (Fransen *et al.* 1997), although many other examples of genes giving rise to variable and diverse mutant phenotypes exist.

One such gene is *ARX*, which encodes a homologue of the *Drosophila* protein *aristaless*. One mutant phenotype of *ARX* has already been referred to above (in relation to lissencephaly). Other phenotypes include XMESID (X-linked myoclonic epilepsy with spasticity and intellectual disability) and phenotypes that include other seizure types, isolated MR, dystonia, ataxia, and autistic features, and combinations thereof (Scheffer *et al.* 2002; Strømme *et al.* 2002). A widespread role for this paired-class (PAX-related) homeodomain protein is clear, but its regulatory target genes as a transcription factor are not known in detail. The gene encoding a protein in the same family, ARIX, is mutated in a disorder of ocular movement control, FEOM2 (Engle and Leigh 2002). This protein seems to be expressed in many catecholaminergic neurons and the specificity of the phenotype, resulting from maldevelopment of neurons in particular cranial nuclei involved in innervation of ocular muscles, is not understood.

Two forms of dopa-responsive dystonia (DRD, or Segawa syndrome) are the result of mutations in genes involved in dopamine synthesis, *TH* (encoding tyrosine hydroxylase) and *GCH1* (encoding GTP cyclohydrolase I, which produces a co-factor for TH). A large literature documents the molecular biology of this system (Nemeth 2002). The disorders, which share some of the features of Parkinson's disease (PD; see below) and are markedly responsive to treatment with the dopamine mimic, L-DOPA, are characterized by pathological features in the substantia nigra pars compacta (SNc) without overt cell loss (unlike PD), and evidence for nerve terminal defects in the striatum, the target of SNc innervation. Maturation of dopamine neurons is not possible in the absence of their main neurotransmitter. The molecular mechanism of dominance in the form of DRD caused by loss-of-function mutations in *GCH1* is not clear, but may involve dominant negative effects on the normal allele product, since in some cases enzyme activity is reduced to less than 50%.

15.3.3 Disorders of mature neuronal function

15.3.3.1 Channelopathies

'Channelopathy' refers to a set of disorders whose causes are related to ion channel dysfunction. There are more than 400 known ion channels, many of which are expressed in neurons, amongst them the important mediators of membrane potential generation and regulation, action potential generation, responses to neurotransmitters and other signals, other forms of neuromodulation, and physical processes such as cell volume regulation, as covered in other chapters in this book. The disorders comprising the set of channelopathies are predictably varied, but, again, some interesting patterns are emerging (Hübner and Jentsch 2002; Gargus 2003).

The most obvious pattern concerns the association of 'episodic' or recurrent, but not continuous phenotypes with mutations in ion channel genes. This can be seen in Table 15.3, in which a range of more or less monogenic epilepsies, other seizure disorders, episodic ataxias, migraines, and MR-related disorders are listed along with the gene symbols and chromosomal localizations of the causative genes. Most of these genes encode ion channels, and most of those are members of gene families covered in other chapters, so their molecular biology will not be dealt with in detail here. Several interesting things can be noted, however. Firstly, there is some, but only limited, overlap between the neuronal channelopathy genes and the neuromuscular/cardiac channelopathy genes (not shown). This is largely because the affected proteins are members of large protein families in which expression of paralogous genes is tissue-specific to a significant degree.

Secondly, most of the disorders for which genes have been identified represent rare monogenic forms of disorders that are common in the population. Extensive genetic heterogeneity is the rule, both at the locus (different genes causing the same disorder), and allelic (multiple mutations within each gene) levels, but so far the genes involved in these disorders have generally not been found to be associated with the common, non-familial forms of the same disorders. This is particularly the case, for example, with the idiopathic generalized epilepsies (IGEs), in which multiple syndromic mixtures of seizure types are represented both within individuals and in different family members. Mutations in *GABRA1* have been found (Cossette *et al.* 2002) in one family with juvenile myoclonic epilepsy (JME), and in another family, *CLCN2* mutations have been found (Haug *et al.* 2003), but neither gene is mutated in many other families nor in sporadic cases of JME, which make up perhaps 1.5% of all epilepsy patients, despite the *CLCN2* gene occurring within one of the three largest linkage peaks in a genome scan study involving more than 130 IGE families (Sander *et al.* 2000).

There is a notable and substantial degree of overlap between the different syndromes in terms of the causative genes, and patterns are visible here too. Different examples of the same epilepsy syndrome (e.g. ADNFLE, BFNC, GEFS+, episodic ataxia/EA) are caused by mutations in paralogous (duplicated more or less recently in evolution) gene family members (e.g. McLellan *et al.* 2003). In contrast, different epilepsy syndromes

Table 15.3 Disorders of mature neuronal function — epilepsies and other channelopathies

	Gene																	
	CHRNA4 **20q13**	***CHRNB2*** **1q21**	***KCNQ2*** **20q13**	***KCNQ3*** **8q24**	***SCN1A*** **2q24**	***SCN2A*** **2q24**	***SCN1B*** **19q13.1**	***GABRG2*** **5q34**	***GABRA1*** **5q34**	***CLCN2*** **3q26**	***CACNB4*** **2q22-q23**	***CACNA1A*** **19p13**	***ATP1A2*** **1q23**	***KCNA1*** **12p13**	***GLRA1*** **5q32**	***GLRB*** **4q31**	***ARX*** **Xp22**	***LGI1*** **10q24**
Disorder																		
Autosomal dominant nocturnal frontal lobe epilepsy (ADNFLE) OMIM 600513	ENFL1	ENFL3																
Benign familial neonatal convulsions (BFNC) OMIM 121200			EBN1	EBN2														
Generalized epilepsy with febrile seizures + (GEFS+) OMIM 600235, 604233					GEFS+2	GEFS+2	GEFS+1	GEFS+3										
Severe myoclonic epilepsy of infancy (SMEI) OMIM 607208					SMEI			SMEI										
Epilepsy, benign neonatal-infantile OMIM 607745						EBN-I												
Idiopathic generalized epilepsy (IGE) OMIM 606904, 609149, 600669, 600131, 607628									EJM	EJM, ECA, EGMA	EJM							
Generalized epilepsy/episodic ataxia/praxis-induced seizures OMIM 600669											EGI/EA							
Epilepsy/episodic ataxia/LD/Episodic ataxia OMIM 108500												EA2						

Table 15.3 (continued) Disorders of mature neuronal function — epilepsies and other channelopathies

	Gene																	
	CHRNA4 **20q13**	***CHRNB2*** **1q21**	***KCNQ2*** **20q13**	***KCNQ3*** **8q24**	***SCN1A*** **2q24**	***SCN2A*** **2q24**	***SCN1B*** **19q13.1**	***GABRG2*** **5q34**	***GABRA1*** **5q34**	***CLCN2*** **3q26**	***CACNB4*** **2q22-q23**	***CACNA1A*** **19p13**	***ATP1A2*** **1q23**	***KCNA1*** **12p13**	***GLRA1*** **5q32**	***GLRB*** **4q31**	***ARX*** **Xp22**	***LGI1*** **10q24**
Familial hemiplegic migraine/progressive ataxia OMIM 141500												FHM1						
Familial hemiplegic migraine 2 OMIM 602481														FHM2				
Episodic ataxia I (with absence epilepsy) OMIM 160120														EA1				
Hyperekplexia/spastic paraparesis OMIM 149400															STHE	STHE		
X-linked myoclonic epilepsy with generalized spasticity and intellectual disability (XMESID) OMIM																	XMESID1	
X-linked infantile spasms [West syndrome] (ISSX) OMIM 308350																	ISSX1	
Autosomal dominant lateral temporal lobe epilepsy (EPT) a.k.a autosomal dominant partial epilepsy with auditory features (ADPEAF) OMIM 600512																		EPTI

can be caused by different mutations (in some cases even the same mutation) in a particular gene. For example, mutations in *SCN1A*, encoding a sodium channel α subunit occur in a generalized epilepsy/febrile seizures mixed syndrome, GEFS+ and in severe myoclonic epilepsy of infancy, SMEI (Escayg *et al.* 2000, Claes *et al.* 2001). Finally, different epilepsies or other episodic disorders can be caused by mutations in functionally related genes, for example, *GABRG2* in GEFS+ (Baulac *et al.* 2001; Harkin *et al.* 2002) and *GABRA1* in JME — the $GABA_A$ receptor subunits encoded by these two genes co-assemble in the receptor subtype with the widest distribution and highest levels of expression in the brain.

In contrast, very similar disorders can be caused by mutations in unrelated genes (e.g. EA1 and EA2). Many of the channelopathy disorders are genetically dominant, and this is found to be explained most often by the alteration of channel gating properties in the mutated forms, either opening-probability/inactivation profiles or ion flux properties, which change membrane voltages and currents even in the presence of the wild-type gene product. Why these disorders are largely episodic in nature is not clear in most cases, since the proteins mainly act in neurons that are continuously active rather than temporally active. In many cases, further study of the functional properties of the channels in the wild-type and mutated forms may reveal the neurophysiological conditions under which the seizure/ataxia/other phenotype becomes visible (e.g. Jentsch 2000). In addition, the identification of rare or common variants associated with the sporadic forms of these disorders is awaited with eagerness, and a number of studies have provided hints in this direction (e.g. Wilkie *et al.* 2002). The answers will provide not only further understanding of the basic biology of functional neural connectivity, but will also point the way to the development of better treatments with fewer side effects. Furthermore, they will act as a paradigmatic demonstration of whether the 'common disorder–common gene variant' hypothesis of complex disorders turns out to be true (Pritchard and Cox 2002; Smith and Lusis 2002), or whether increasing degrees of heterogeneity resulting from rare variants/mutations is the rule (Gargus 2003).

15.3.3.2 Fragile X and myotonic dystrophy

Table 15.4 lists a number of other disorders thought to be mainly associated with disruption of mature neuron function, but without extensive neurodegeneration. Elsewhere in this chapter, examples of neurodegenerative disorders caused by 'nucleotide expansion' or 'dynamic' mutations are discussed, the so-called 'polyglutamine/polyalanine disorders' (see Section 15.4.3). Together with these, fragile X syndrome types 1 and 2, myotonic dystrophy and Friedreich ataxia comprise the set of known triplet (trinucleotide) repeat disorders. Once considered together because of the likely common pathways in the mechanism of mutation in each case (in most, the characteristic repeat tract length changes in post-meiotic/mitotic germ and somatic cells that constitute the mutation are thought to be a result of single or double-strand break DNA repair mechanisms acting on gaps caused by hairpin 'snapback' structures

Table 15.4 Disorders of mature neuronal function — other disorders

Disorder type	Disorder	Symbol	OMIM	Gene(s)	Symbol	Location
Fragile X disorders	Fragile X syndrome, type A	FRAXA	309550	Fragile site mental retardation 1	*FMR1*	Xq27.3
	Mental retardation, X-linked, associated with fragile site, FRAXE	FRAXE	309548	Fragile site mental retardation 2	*FMR2*	Xq28
	Myotonic dystrophy	DM1	160900	DM protein kinase; myotonin	*DMPK*	19q13.3
		DM2	602668	Zinc finger protein 9	*ZNF9*	3q22
	Angelman syndrome	AS	105830	E6-AP ubiquitin-protein ligase	*UBE3A*	15q11-q13
	Obesity		155541	Melanocortin-4 receptor	*MC4R*	18q22
			601665			
	Coffin–Lowry syndrome	CLS	303600	Ribosomal protein S6 kinase, 90KD, 3; RSK2	*RPS6KA3*	Xp22.2
	Narcolepsy		161400	Hypocretin [orexin]	*HCRT*	17q21
Torsion dystonias	Torsion dystonia 1, autosomal dominant	DYT1	128100	Torsin-A	*DYT1*	9q34
	Myoclonic dystonia	DYT11	159900 126450	ε-sarcoglycan Dopamine receptor, D2	*SGCE* *DRD2*	7q21 11q23

at break sites (see for an example, Kovtun and McMurray 2001). In terms of clinical phenotype and pathogenic mechanisms acting, however, they form quite separate classes.

Myotonic dystrophy (DM) is generally known as a muscular disorder, but although the brain-specific aspects of the phenotype are understudied, they are present, largely manifesting as alterations in affect and cognitive function (Meola *et al.* 1999). Both major types of the disorder, DM1and DM2, are caused by microsatellite repeat expansion mutations. In patients with the adult-onset and congenital forms of DM1, a CTG repeat unit that is normally polymorphic in the population (5–~35 tandem copies of the repeat) is expanded to >50 repeats (and up to >3000 copies). The repeat is located on chromosome 19q in the 3′-untranslated region (3′-UTR) of a protein kinase gene, *DMPK*, where, as CUG in the primary transcript, it has the effect of disrupting the normal splice pattern of the transcribed mRNA, which may lead to part of the muscle phenotype. The primary pathogenic effect, however, is probably related to the presence of the expanded CUG repeat tract in the primary transcription product, creating a 'toxic RNA' effect that acts via a dominant, gain-of-(novel)-function mechanism. Expanded transcripts are retained in the nucleus and form 'foci', in which RNA binding proteins such as muscleblind (MBNL) and CUG-bp are sequestered (Miller *et al.* 2000; Fardaei *et al.* 2001). This is thought to result in dilution of these and perhaps other proteins and consequently altered function of the gene products normally regulated by them.

Although the toxic RNA idea was speculative and controversial for some time, strong support for it came from the identification (Liquori *et al.* 2001) of a second expansion, of a CCTG repeat unit, in the *ZNF9* gene on chromosome 3q in patients with DM2 (also known formerly as proximal myotonic myopathy, PROMM). The function of this gene does not appear to be directly related to muscle physiology, thus lending weight to the toxic RNA idea as being independent of the function of the gene in which the expansion occurs. In addition, no mutations other than (C)CTG expansions have been identified in any cases of DM. Both muscle and brain abnormalities, including abnormal expression of protein tau (see Section 15.4.2 below), are seen in a transgenic mouse model of DM in which the human DM region with an expanded CTG repeat has been inserted into the mouse genome (Seznec *et al.* 2001). A third pathogenic mechanism operating in DM1 (but not DM2) is haploinsufficiency of a homeobox gene of the *Drosophila sine oculis* family, *SIX5*, that lies immediately downstream of *DMPK*. The unstable CTG repeat lies within the regulatory regions of this gene, and disruption of *Six5* in mouse knockout models results primarily in cataract formation (Klesert *et al.* 2000), a universal feature of DM1 that is consistent with the expression pattern of *SIX5* (Winchester *et al.* 1999).

Fragile X syndrome, FRAXA, is an X-linked mental retardation syndrome caused by expansion mutations of an unstable CGG trinucleotide repeat located in the 5′-UTR of the *FMR1* gene in Xq27.3. In the general population, the CGG repeat is polymorphic, with repeat numbers ranging from 7 to ~60. Towards the upper end of this distribution, the length of 'perfect' (uninterrupted) repeat units is sufficient to result in instability of the repeat in the germline, and occasional mutations involving expansions into the 'premutation' range (~60–230 repeats) have occurred, providing a heterogeneous pool of chromosomes in the population with unstable repeats that give rise to the full mutation (230–>1300 repeats) only when transmitted through the female germline (reviewed in Jin and Warren 2000). This provides the explanation for the preponderance of affected males, who have mothers with premutations. Affected females with full mutation-length repeats can occur if the X-inactivation ratio is skewed in an inclement fashion, indicating that the mutation acts in a partially genetically dominant manner. The phenotype results from reduced expression of *FMR1* caused by hypermethylation of the CGG repeat, which sits close to the promoter of the gene and just downstream of a normally unmethylated CpG island, in full mutation expanded alleles only. The hypermethylation is associated with transcriptional repression of *FMR1* via a chromatin-based mechanism involving deacetylation and K9 methylation of histone H3 (Coffee *et al.* 2002; see Section 15.3.1).

There has been considerable progress in understanding the molecular basis for the phenotype in fragile X syndrome in the past few years. The protein encoded by the gene, FMRP, has several RNA binding motifs, two KH boxes and an RGG box. A number of brain mRNAs that bind to FMRP have been identified, many of which possess sequences that form a 'G quartet' secondary structure element, and the translation rates of the proteins made from these mRNAs is known in some cases to be reduced (Darnell *et al.* 2001; Brown *et al.* 2001*a*). This translational repression may represent the main function of FMRP (Laggerbauer *et al.* 2001; Mazrui *et al.* 2002), but it also participates in transport of the same bound mRNAs from the nucleus to the dendritic translation machinery and associated cytoplasm (reviews: Job and Eberwine 2001, Connolly this volume Chapter 4), indicating that FMRP participates in a complex mechanism for provision of a pool of (a particular subset of) mRNAs for translation at appropriate locations and times. The transport complex has several protein members, some of which directly interact with FMRP, and FMRP can also be found in association with ribosomes. This may explain the severe brain phenotype (other clinical features are also present) in fragile X patients, as well as the fact that neuropathological studies have demonstrated that a downstream effect of lack of FMRP is an increase in the length and density of dendritic spines on many neurons. It is easy to see how deregulated translation of some of the mRNAs complexed with FMRP could, in its absence, lead to aberrant spine morphology and function, a feature of several brain disorders (reviewed in Hering and Sheng 2001).

15.3.3.3 Angelman syndrome

Genomic imprinting is a growing area of interest to those interested in human disease pathology, and a number of human genetic disorders are associated with disruption of this process. Imprinted genes are expressed, in the appropriate tissues, only from the chromosome that has originated from one parent — which parent, that is, depends on the gene and the mechanism of establishing and maintaining the imprint, which in almost all known cases involves the differential methylation of a short CpG-rich region, the differentially methylated region (DMR), which in turn is associated with spreading chromatin-level repression of neighbouring genes. In the common, but inaccurate, parlance, maternally imprinted genes are expressed from the maternally derived chromosome, but not from the paternal chromosome. Several dozen imprinted regions exist in the human and mouse genomes (mostly overlapping in gene content). Angelman syndrome, an imprinting disorder resulting from a disruption of maternal allele-specific expression, serves as a good example of the successful identification of the affected gene in one of these regions, on human chromosome 15q, while its corresponding neuronal disorder caused by disruption of paternal allele-specific expression, Prader–Willi syndrome, has so far proved resistant to all such attempts to narrow down the culprit.

Angelman syndrome (AS) is characterized by severe cognitive impairment, seizures, ataxia, and inappropriate laughter, but there are no recognized pathological features in the brains of affected individuals, so it is assumed to be a disorder of neuronal function, rather than developmental or degenerative. It can be caused by several mechanisms (see Nicholls and Knepper 2001), the majority of cases being associated with deletions of various sizes in chromosome 15q11-q13. While some of the bigger deletions disrupt the function of more than one gene in the region, thus implying that in these cases, the severe type of AS that results can be considered a 'contiguous gene syndrome', the majority of cases can be explained by a disorder of imprinting, since AS is only observed when inherited via an unaffected mother. The 'converse' disorder, Prader–Willi syndrome, which has quite separate phenotypic characteristics, results from deletions of largely the same region when inherited via an unaffected father. Some cases of AS are due to uniparental disomy (both alleles coming from one parent), corroborating the idea that it is a disorder of imprinting. The shortest regions of overlap (SRO) of deletions causing Prader–Willi and Angelman syndromes are non-overlapping and occur within and ~35 kb upstream of the promoter of the *SNURF/SNRPN* gene respectively. Short stretches of DNA from both these regions combined are necessary and sufficient to confer allele-specific transcriptional regulation on the whole imprinted region and constitute a bipartite 'imprinting box' (Shemer *et al.* 2000). It is now clear that the results of mutations within these regions are quite different in the two disorders, although the primary mechanism for establishing allele-specific transcriptional competence of neighbouring genes is the same.

Maternal expression of one or two genes in the region is controlled by one of the two SRO/DMRs, and the major imprinted gene affected in AS, *UBE3A*, was finally identified clearly in non-deletion AS cases through the discovery of several point mutations causing loss-of-function of the gene product, an E6-AP ubiquitin protein ligase (see Fang *et al.* 1999). The allele-specificity of expression of this gene only occurs in neurons in the brain (Yamasaki *et al.* 2003), although the gene is widely expressed in other tissues from both alleles, and this specificity is achieved via allele-specific methylation of the upstream AS SRO/DMR and associated chromatin-level gene repression, as in almost all known cases of imprinting. The mechanism by which the repression leads to allele-specific expression of *UBE3A* is almost unique, however. *UBE3A* is transcribed in the opposite orientation to the neighbouring, largely paternally-expressed, genes, but expression of *UBE3A* primary transcript is possible from both alleles, despite methylation of the DMR on the 'expressed' maternal allele. The explanation for this unexpected finding was elucidated with the discovery that the transcript made from the *SNURF/SNRPN* promoter on the opposing DNA strand is continuous through a number of genes encoding small RNAs, and also through *UBE3A* itself, and is then processed in such a way as to create an 'antisense' RNA capable of repressing *UBE3A* (Rougeulle *et al.* 1998). This antisense transcript, which is assumed to interfere with the functioning of *UBE3A* via an undetermined mechanism, is only expressed from the allele with the unmethylated DMR in neurons (Yamasaki *et al.* 2003), thus explaining the tissue-specific imprinted status of *UBE3A*.

15.4 Disorders of apoptotic neuronal cell death and neurodegeneration

15.4.1 Overview of genetic disorders leading to neurodegeneration

The largest group of disorders of CNS neurons, by a considerable margin, includes those that are characterized in the pathology mainly by the occurrence of neurodegeneration. In these disorders, neurons are lost focally, regionally, or throughout the CNS as a result of activation of apoptotic pathways (see Shiels and Davies, this volume Chapter 16). A comprehensive list of degenerative disorders for which the genes responsible have been identified is presented in Tables 15.5–15.14, but for reasons of space, I have limited discussion here to two main classes, the 'polyglutamine disorders' and the 'conformational neuronopathies'. Types of disorder not covered in detail, but representing most of the known examples of neurodegenerative disorders, include disorders of transcriptional regulation, chromatin-level regulation, and DNA replication and repair, such as spinal muscular atrophy (see Lefebvre *et al.* 1998; Monani *et al.* 2000), Rubinstein–Taybi syndrome, α-thalassaemia with mental retardation (ATRX), ataxia telangiectasia, and the DNA excision repair disorders including xeroderma pigmentosum and many others (Table 15.10). These illustrate the importance of genome regulation and maintenance in normal cellular function, particularly in finely-balanced cells like neurons.

Another group not covered in detail here are the 'metabolic storage disorders'. Falling into two main categories, there are the lysosomal storage disorders, mainly resulting from recessive, loss-of-function mutations in genes encoding hydrolase enzymes (Table 15.5) or in genes encoding proteins with functions in transport, targeting and processing of lysosomal proteins (Table 15.6); and the peroxisomal disorders, such as adrenoleukodystrophy and Zellweger syndrome (Table 15.7). These are largely

Table 15.5 Disorders of neuronal degeneration and death 1 — lysosomal storage disorders, hydrolase enzyme deficiencies

Disorder type	Disorder	Symbol	OMIM	Gene(s)	Symbol	Location
NCL Lipofuscinosis	Neuronal ceroid lipofuscinosis	CLN1	256730	Palmitoyl-protein thioesterase 1	*PPT1*	1p32
		CLN2	204500	Ceroid-lipofuscinosis, neuronal 2, late infantile	*TPP1*	11p15
		CLN3	204200	Ceroid-lipofuscinosis, neuronal 3, juvenile [Batten disease]	*CLN3*	16p12
		CLN5	256731	Ceroid-lipofuscinosis, neuronal 5	*CLN5*	13q22
		CLN6	601780	Ceroid-lipofuscinosis, neuronal 6; Linclin	*CLN6*	15q21-q23
	Progressive epilepsy with mental retardation [Northern epilepsy]	CLN8/EPMR	600143	CLN8	*CLN8*	8pter-p22
Gangliosidosis	GM2 gangliosidosis type 1 [Tay–Sachs disease]	GM2-I [TSD]	272800	β-hexosaminidase A	*HEXA*	15q23-q24
	GM2 gangliosidosis type 2 [Sandhoff disease]	GM2-II	268800	β-hexosaminidase B	*HEXB*	5q13
	GM2 gangliosidosis type AB [Tay–Sachs disease, AB variant]	GM2-III	272750	GM2-activator	*GM2A*	5q32-q33.1
	GM1 gangliosidosis type 1	GM1-1	230500	β-galactosidase 1	*GLB1*	3p21.33
Glucocerebrosidosis	Gaucher disease types II/III	GDII/GDIII	230900 231000	Acid β-glucosidase	*GBA*	1q21
Sulphatide lipidosis	Metachromatic leukodystrophy	MLD	250100 249900	Arylsulphatase A Prosaposin	*ARSA* *PSAP*	22q13.31-qter 10q22.1
Sphingomyelinosis	Niemann–Pick disease, type A [Sphingomyelin lipidosis]	NPA	257200	Sphingomyelin phosphodiesterase 1, acid lysosomal	*SMPD1*	11p15.3

Sialidosis	Mucolipidosis I; Neuraminidase deficiency	MLI	256550	Neuraminidase 1	*NEU1*	6p21.3
	Neuraminidase deficiency with β-galactosidase deficiency [Galactosialidosis]	GSL	256540	Protective protein for β-galactosidase (galactosialidosis) [cathepsin A]	*PPGB*	20q13.1
Mucolipidosis	Mucolipidosis II; I-cell disease	ML II	252500	*N*-acetylglucosamine-1-phosphotransferase	*GNPTA*	4q21-q23
	Mucolipidosis III; pseudo-Hurler polydystrophy	ML IIIC	252605	*N*-acetylglucosamine-1-phosphotransferase, γ subunit	*GNPTAG*	16p13.3
Mucopoly-saccharidosis	MPS type I [Hurler and Scheie syndrome]	MPS I	252800	α-L-iduronidase	*IDUA*	4p16.3
	MPS type II [Hunter syndrome]	MPS II	309900	Iduronate-2-sulphatase	*IDS*	Xq28
	MPS type IIIB [Sanfilippo syndrome B]	MPS IIIB	252920	α-*N*-acetylglucosaminidase	*NAGLU*	17q21
	MPS type VII [Sly syndrome]	MPS VII	253220	β-glucuronidase	*GUSB*	7q21.11
Glycosidosis	β-mannosidosis	MANB1	248510	β-mannosidase A	*MANBA*	4q22-q25
	Fucosidosis		230000	α-L-fucosidase 1	*FUCA1*	1p34
	Aspartylglucosaminuria	AGU	208400	Aspartylglucosaminidase	*AGA*	4q32-q33
	Schindler disease		104170	α-galactosidase B	*GALB*	22q11

Table 15.6 Disorders of neuronal degeneration and death 2 — lysosomal storage disorders, transport defects

Disorder type	Disorder	Symbol	OMIM	Gene(s)	Symbol	Location
Sphingolipidosis	Niemann–Pick disease, type C	NPC	257220	Niemann–Pick disease, type C1	*NPC1*	18q11-q12
		NPC2	601015	Niemann–Pick disease, type C2 [Epididymal secretory protein 1]	*NPC2*	14q24.3
Mucolipidosis	Mucolipidosis type IV	MLIV	252650	Mucolipin 1	*MCOLN1*	19p13.3-p13.2
Sialidosis	Infantile sialic acid storage disorder	ISSD/SD	269920	Sialin	*SLC17A5*	6q14-q15
	Salla disease		604369			

Table 15.7 (continued) Disorders of neuronal degeneration and death 3 — peroxisomal disorders

Gene Location	*ABCD1* Xq28	*PEX1* 7q21-q22	*PXR1* 12p13	*PEX10* 1p36.32	*PEX13* 2p15	*PXMP3* 8q21	*PEX3* 6q23-q24	*PEX6* 6p21	*PEX12* 17q21.1	*ABCD3* 1p22-p21	*PHYH* 10pter-p11	*PEX7* 6q22-q24
Disorder												
Adrenoleukodystrophy (ALD) OMIM 300100	+											
Zellweger syndrome (ZWS1) OMIM 214100		+	+	+	+	+	+	+	+	+		
Adrenoleukodystrophy, autosomal neonatal form (NALD) OMIM 202370		+	+	+	+							
Refsum disease, infantile form (IRD) OMIM 266510		+				+						
Refsum disease [Hereditary motor and sensory neuropathy IV (HMSN IV)] OMIM 266500											+	+
Chondrodysplasia punctata, autosomal recessive (RCDP) OMIM 215100												+

+ indicates that the disorder can be caused by mutations in the indicated gene.

Table 15.8 Disorders of neuronal degeneration and death 4 — other metabolic disorders, recessive, loss of function

Disorder type	Disorder	Symbol	OMIM	Gene(s)	Symbol	Location
Polyglucosan storage	Lafora disease [progressive myoclonus epilepsy]	LD/EPM1	254780	EPM region transcript 2A; laforin	*EPM2A*	6q24
Ceramidosis	Farber lipogranulomatosis		228000	*N*-acylsphingosine amidohydrolase	*ASAH*	8p22-p21.3
Sterol synthesis	Mevalonicaciduria		251170	Mevalonate kinase	*MVK*	12q24
	Phenylketonuria	PKU1	261600	Phenylalanine hydroxylase	*PAH*	12q24.1
	2-methyl-3-hydroxybutyryl-CoA dehydrogenase deficiency		300256	2-methyl-3-hydroxybutyryl-CoA dehydrogenase	*HADH2*	Xp11.2
	Citrullinaemia	CTLN1	215700	Argininosuccinate synthetase	*ASS*	9q34
	Lesch–Nyhan Syndrome	LNS	308000 300322	Hypoxanthine-guanine phosphoribosyl transferase	*HPRT*	Xq26-q27.2
	Maple syrup urine disease	MSUD	248600	Branched-chain α-ketoacid dehydrogenase E1a/E2 subunits	*BCKDE1A*	19q13.2
				Dyhydrolipoamide branched-chain transacylase	*DBT*	1p31
	Guanidinoacetate methyltransferase deficiency		601240	Guanidinoacetate methyltransferase	*GAMT*	19p13.3
	Canavan disease		271900	Aspartoacylase	*ASPA*	17pter-p13
	Pantothenate kinase-associated neurodegeneration [Hallervorden–Spatz disease]	PKAN	234200	Pantothenate kinase 2	PANK2	20p13-p12.3
	Sulphocysteinuria		272300	Sulphite oxidase	*SUOX*	12q13.13
	3-methylglutaconicaciduria, type III [Optic atrophy + syndrome]	MGA III [OPA3]	258501	Optic atrophy 3	OPA3	19q13.2-q13.3

Table 15.9 Disorders of neuronal degeneration and cell death 5 — dominant disorders, with gain of function or toxic proteins

Disorder Category	Disorder type	Disorder	Symbol	OMIM	Gene(s)	Symbol	Location
Triplet repeat disorders	Polyglutamine expansion	Basal ganglia disease, adult-onset		606159	Ferritin, light chain	*FTL*	19q13.3
		Spinobulbar muscular atrophy	SBMA	313200	Androgen receptor	*AR*	Xq13-q21
		Huntington disease	HD	143100	Huntingtin	*HTT*	4p16.3
		Dentatorubral-pallidoluysian atrophy	DRPLA	125370	Atrophin-1	*DRPLA*	12p13.31
		Spinocerebellar ataxia, type 1	SCA1	164400	Ataxin-1	*ATX1*	6p23
		Spinocerebellar ataxia, type 2	SCA2	183090	Ataxin-2	*SCA2*	12q24.21
		Spinocerebellar ataxia, type 3 [Machado-Joseph disease]	SCA3/ MJD1	109150	Ataxin-3	*MJD1*	14q32.1
		Spinocerebellar ataxia, type 6	SCA6	183086	Calcium channel, voltage-gated, α1A subunit	*CACNA1A*	19p13
		Spinocerebellar ataxia, type 7	SCA7	164500	Ataxin-7	*SCA7*	3p21.1-p12
		Spinocerebellar ataxia, type 12	SCA12	604326	Protein phosphatase 2, regulatory subunit B, β PR55β	*PPP2R2B*	5q31-q33
		Spinocerebellar ataxia, type 17	SCA17	607136	TATA box-binding protein	*TBP*	6q27
	Polyleucine/ Polyalanine expansion	Huntington disease-like 2	HDL2	606438	Junctophilin-3	*JPH3*	16q24.3
		Partington X-linked mental retardation syndrome	PRTS	309510	Aristaless homologue, X chr.	*ARX*	Xp21.3-p21.1
		Oculopharyngeal muscular dystrophy	OPMD	164300	Poly(A)-binding protein, nuclear 1	*PABPN1*	14q11.2-q13
		Congenital central hypoventilation syndrome [Ondine's curse]	CCHS	209880	Paired mesoderm homeobox 2B	*PHOX2B*	4p12
Other repeat disorders	Pentanucleotide expansion	Spinocerebellar ataxia, type 10	SCA10	603516	Ataxin-10	*SCA10*	22q13-qter

Table 15.10 Disorders of neuronal degeneration and cell death 6 — disorders of DNA repair, transcriptional regulation (chromatin disorders) and RNA processing

Disorder	Symbol	OMIM	Gene(s)	Symbol	Location
Rubinstein–Taybi syndrome	RSTS	180849	CREB-binding protein	*CREBBP*	16p13.3
Ataxia-telangiectasia	AT	208900	Ataxia-telangiectasia mutated	*ATM*	11q22.3
Ataxia-telangiectasia-like disorder	ATLD	604391	Meiotic recombination 11, *S. cerevisiae*, homologue of, A	*MRE11A*	11q21
Seckel syndrome	SCKL	210600	Ataxia-telangiectasia and Rad3-related	*ATR*	3q22-q24
Cockayne syndrome, type B Cerebrooculofacioskeletal syndrome	CSB COFS	133540 278760	Excision-repair cross complementing rodent repair deficiency, complementing group 6	*ERCC6* *ERCC2*	10q11 19q13.2-q13.3
Trichothiodystrophy	TTD		Excision-repair, complementing defective, in Chinese Hamster Ovary cells, 2	*ERCC4*	16p13.3-p13.13
Xeroderma pigmentosum, type F	XPF		Excision repair, complementing defective, in Chinese Hamster, 4		
α-thalassaemia/mental retardation syndrome, X-linked	ATRX	301040	Helicase 2, X-linked	*ATRX*	Xq13
Spinal muscular atrophy I	SMA1	253300	Survival motor neuron 1	*SMN1*	5q12.2-q13.3
Spinal muscular atrophy with respiratory distress 1	SMARD1	604320	Immunoglobulin μ-binding protein 2	*IGHMBP2*	11q13.2-q13.4
Ataxia-oculomotor apraxia syndrome	AOA	208920	Aprataxin	*APTX*	9p13.3
Spinocerebellar ataxia, autosomal recessive, with axonal neuropathy	SCAN1	607250	Tyrosyl-DNA phosphodiesterase 1	*TDP1*	14q31-q32
Börjeson–Forssman–Lehmann syndrome	BFLS	301900	Plant homeodomain-like finger 6	*PHF6*	Xq26-q27

Table 15.11 Disorders of neuronal degeneration and cell death 7 — disorders of cellular signaling and intracellular transport and processing, non-storage

Disorder type	Disorder	Symbol	OMIM	Gene(s)	Symbol	Location
Spastic paraplegia	Spastic paraplegia 4, autosomal dominant	SPG4	182601	Spastin	*SPG4*	2p22-p21
	Spastic paraplegia, hereditary, type 3A	SPG3A	182600	Atlastin	*SPG3A*	14q11-q21
	Spastic paraplegia, hereditary, type 10	SPG10	604187	Kinesin heavy chain	*KIF5A*	12q13
	Spastic paraplegia 20, autosomal recessive [Troyer syndrome]	SPG20 [TRS]	275900	Spartin	*SPG20*	13q12.3
	Diabetes insipidus, neurohypophyseal type		125700	Arginine vasopressin	*AVP*	20p13
	Wolfram syndrome [Diabetes insipidus and mellitus with optic atrophy and deafness]	WFS [DIDMOAD]	222300	Wolframin	*WFS1*	4p16.1
	Choreoacanthocytosis	CHAC	200150	Chorein	*CHAC*	9q21
	Spastic ataxia of Charlevoix–Saguenay, autosomal recessive	ARSACS	270550	Sacsin	*SACS*	13q11
	Amyotrophic lateral sclerosis [motor neurone disease]	ALS1	105400	Superoxide dismutase-1	*SOD1*	21q22.1
	Amyotrophic lateral sclerosis, juvenile Primary lateral sclerosis, juvenile onset; Infantile-onset ascending hereditary spastic paralysis	ALS2 JPLS; IAHSP	205100	Alsin	*ALS2*	2q33
	Giant axonal neuropathy	GAN	256850	Gigaxonin	*GAN*	16q24.1
	Cerebral ataxia, autosomal dominant		601515	Fibroblast growth factor 14	*FGF14*	13q34
	Spinocerebellar ataxia 14	SCA14	605361	Protein kinase C, γ subunit	*PRKCG*	19q13.4
	Myoclonic epilepsy of Unverricht and Lundborg [Progressive myoclonus epilepsy]	EPM1	254800	Cystatin B	*CSTB*	21q22.3

Table 15.12 Disorders of neuronal degeneration and cell death 8 — mitochondrial DNA and respiratory chain disorders

	Gene																					
	TIMM8A Xq22	MTTK	MTTL1	MTTQ	MTND6	MTND4	MTND1	NDUFV1	NDUFS4	NDUFS8	NDUFS1	NDUFS2	NDUFS7	SCO2 22q13	SCO1 17p13-p12	COX10 17p12-p11.2	SURF1 9q34	MTATP6	PDHA1 Xp22.2-p22.1	SDHA 5p15	BCS1L 2q33	MTND5
Disorder																						
Mohr–Tranebjaerg syndrome (MTS) [Dystonia-deafness syndrome] OMIM 304700	+																					
Opticoacoustic nerve atrophy with dementia [Jensen syndrome] OMIM 311150	+																					
Myoclonic epilepsy with ragged-red fibres (MERRF) OMIM 545000		+	+																			
Mitochondrial myopathy, encephalopathy, lactic acidosis, and stroke-like episodes (MELAS) OMIM 540000			+	+	+																	
Dystonia, familial, with visual failure and striatal lucencies [Leber optic atrophy and dystonia] OMIM 500001					+	+																
Leber optic atrophy (LHON) OMIM 535000					+	+	+															+

Table 15.12 (continued) Disorders of neuronal degeneration and cell death 8 — mitochondrial DNA and respiratory chain disorders

	Gene																					
	TIMM8A Xq22	MTTK	MTTL1	MTTQ	MTND6	MTND4	MTND1	NDUFV1	NDUFS4	NDUFS8	NDUFS1	NDUFS2	NDUFS7	SCO2 22q13	SCO1 17p13-p12	COX10 17p12-p11.2	SURF1 9q34	MTATP6	PDHA1 Xp22.2-p22.1	SDHA 5p15	BCS1L 2q33	MTND5
Leigh syndrome, LS OMIM 256000								+	+	+							+	+	+	+	+	+
Complex I, mitochondrial respiratory chain, deficiency of [NADH:Q(1) oxidoreductase deficiency] OMIM 252010								+	+	+	+	+	+									
Complex IV, mitochondrial respiratory chain, deficiency of [cytochrome c oxidase deficiency] OMIM 220110														+	+	+	+					
Cardioencephalomyopathy, fatal infantile, due to cytochrome c oxidase deficiency OMIM 604377														+								
Neuropathy, ataxia, and retinitis pigmentosa (NARP syndrome) OMIM 551500																		+				
Pyruvate dehydrogenase deficiency [ataxia, intermittent, with abnormal pyruvate metabolism] OMIM 312170																				+		

+ indicates that the disorder can be caused by mutations in the indicated gene.

characterized by the accumulation of metabolites, usually the substrates for the enzymes encoded by the mutant genes, which are toxic in large quantities and lead to organellar dysfunction and swelling/vacuolation, followed by cell death. Other storage disorders involving accumulation of metabolites within the cytoplasm, such as Lafora disease, also exist (Table 15.8).

Other classes of degenerative disorder occur as a result of mitochondrial dysfunction (Tables 15.12 and 15.13), including disorders of mitochondrial iron homeostasis, like Friedreich ataxia (see Patel and Isaya 2001), and of respiratory chain dysfunction (see Smeitink *et al.* 2001), which lead to cell death through specific activation, or through energy stress or oxidative stress. Then there are the disorders that result from dysregulation or failure of intracellular signalling and transport mechanisms, such as several of the spastic paraplegias (Crosby and Proukakis 2002), motor neuron disease (amyotrophic lateral sclerosis; see Raoul *et al.* 2002 for intriguing new evidence about the apoptotic mechanism activated in ALS), some cerebellar ataxias (e.g. SCA14; Chen *et al.* 2003), and many others (Table 15.11). Justice cannot be done here to the many and varied mechanisms by which genetic lesions lead to activation of apoptotic pathways in these disorders.

15.4.2 Conformational neuronopathies: Alzheimer´s disease, Tauopathies, Serpinopathies, Prion disorders

A number of important disorders have the common characteristic that the neuronal pathology is associated with intra- or extracellular protein aggregates. These disorders, which include Alzheimer´s disease, Pick disease, Parkinson´s disease, and the prion diseases, might be termed 'conformational neuronopathies' as a major step in the pathology is the accumulation of whole proteins, or naturally processed protein fragments, synthesized in the neurons that come to adopt a destabilized and non-functional conformation through refolding. Table 15.14 lists some of these disorders.

Alzheimer disease (AD) is characterized by progressive memory loss (dementia) caused by widespread neurodegenerative loss of neurons throughout the brain. Associated with intracellular (neurofibrillary tangles), extracellular (amyloid plaques), and neuronal degeneration-linked (neuritic plaques) pathological signs, AD is considered by many to be a normal consequence of ageing processes, and deserving of its clinical disorder designation only when age of onset is early enough (i.e. presenile dementia). The large majority of cases of AD are 'sporadic', or at least are not thought to be the result of a single gene segregating in Mendelian fashion within the family, but progress in understanding the pathogenic mechanisms operating came from the identification, during the early and mid 1990s, of genes harbouring mutations in rare familial forms of the disorder. Mutations in these genes, encoding amyloid precursor protein (APP), and the presenilins 1 and 2, account for a tiny fraction of AD patients. The mutations are generally dominantly acting and of high penetrance with early onset, and all lead to increased production of 'toxic' protein fragments from APP (a variety of fragments cleaved at several sites e.g. Aβ1-42, Aβ1-40, Aβ11-42 etc.), the

Table 15.13 Disorders of neuronal degeneration and cell death 9 — other mitochondrial disorders

Disorder type	Disorder	Symbol	OMIM	Gene(s)	Symbol	Location
Triplet repeat disorder	Friedreich ataxia	FRDA	229300	Frataxin	*FRDA*	9q13
	Anaemia, sideroblastic, and spinocerebellar ataxia	ASAT	301310	ATP-binding cassette, subfamily B, member 7	*ABCB7*	Xq13.1-q13.3
	Spastic paraplegia 7, autosomal recessive	SPG7	607259	Paraplegin	*SPG7*	16q24.3
	Spastic paraplegia 13, autosomal dominant	SPG13	605280	Chaperonin	*HSP60*	2q24

Table 15.14 Disorders with complex inheritance, polygenic/multifactorial

Disorder type	Disorder	Symbol	OMIM	Gene(s)	Symbol	Location
	Major affective disorder 1 [Bipolar affective disorder]	MAFD1	125480	Several (disruption or association)	*G72/G30*	13q32-q33
	Schizophrenia		181500	Several (disruption or association)		
Amyloidopathy	Alzheimer disease, familial	AD1	104300	Amyloid precursor protein	*APP*	21q21.2
		AD3	104311	Presenilin 1	*PSEN1*	14q24.3
		AD4	600759	Presenilin 2	*PSEN2*	1q42.3
	Alzheimer disease, sporadic	AD2	104310	Apolipoprotein E	*APOE*	19q13.2
		FAP	176300	Angiotensinogen converting enzyme, ACE	*DCP1*	17q23
	Familial amyloid polyneuropathy			Transthyretin	*TTR*	18q11.2-q12.1
Tauopathy	Pick disease (PiD);		172700	Microtubule-associated protein Tau	*MAPT*	17q21
	Progressive supranuclear palsy;	PSP	601104			
	Frontotemporal dementia;	FTLD/ FTDP-17	600274			
	Pallidopontonigral degeneration	PPND	168610			
Prion disease	Creutzfeldt-Jakob disease	CJD	123400	Prion protein	*PRNP*	20p12
	Gerstmann-Straussler disease	GSD	137440			
	Familial fatal insomnia	FFI	600072			
	Dementia, Lewy body	DLB	127750			
Serpinopathy	Encephalopathy, familial, with neuroserpin inclusion bodies	FENIB	604218	Protease inhibitor 12 (Serpin1)	*PI12*	3q26
Synucleinopathy	Parkinson disease (PD)	PARK1	168600/1	α-synuclein	*SNCA*	4q21-q22
		PARK2	600116	Parkin	*PARK2*	6q25.2-q27
		PARK5	191342	Ubiquitin carboxyl-terminal esterase	*UCHL1*	4p14
				Nuclear receptor, subfamily 4, Group A, member 2	*NR4A2*	2q22-q23
					SNCAIP	5q23.1-q23.3
		PARK7	168600	Synphilin-1	*DJ-1*	1p36
			606324	Oncogene DJ-1		

primary constituents of amyloid plaques. Generation of these fragments from APP occurs via programmed proteolytic degradation of APP inserted into membranes. The identity of the β-secretase, which participates directly in the release of these fragments, was discovered in 1999 by several groups (Sinha *et al.* 1999; Vassar *et al.* 1999; Yan *et al.* 1999), and was important because of the therapeutic potential of inhibiting that pathway. The identity of the γ-secretase, which carries out the final processing step leading to deposition of Aβ, was a mystery, but it is now thought that the presenilins play a fundamental role in this activity, as well as their role in membrane trafficking functions in cells (Kimberly *et al.* 2003).

The identity of predisposing and protective genes for common, non-familial forms of AD are proving difficult to elucidate, despite some large mapping studies. Genotype at the *APOE* locus on chromosome 19q has been known for some time to have an influence on AD risk, and is now believed to influence age of onset more than predisposition *per se* (Meyer *et al.* 1998); the biology of ApoE and its role in neuronal functioning are discussed in many good reviews (e.g. Herz and Beffert 2000; Mahley and Rall 2000). A *PSEN1* promoter polymorphism that influences levels of expression of presenilin 1 also seems to be associated with risk (Theuns *et al.* 2003), as does the presence of certain haplotypes of *DCP1*, which encodes the blood pressure modifier, angiotensin converting enzyme, ACE (Kehoe *et al.* 2003).

The neurofibrillary tangles seen in AD are associated with aberrantly phosphorylated forms of the microtubule associated protein tau, and recently, several related disorders have been shown to be associated with conformational aberrations of this protein. The 'tauopathies' (Lee and Trojanowski 1999) are caused directly by mutations in the tau encoding gene, *MAPT*, and include Pick's disease and other cases of frontotemporal dementia/pallidopontonigral degeneration with or without supranuclear palsy (Goedert *et al.* 1998; Pickering-Brown *et al.* 2000; Tsuboi *et al.* 2002). The serpin family of serine protease inhibitors is associated with the 'serpinopathies' (Lomas and Carrell 2002) and at least one disorder, familial encephalopathy with neuroserpin inclusion bodies, has been associated with mutations in a member of this gene family, *PI12* (Davis *et al.* 1999). Prion diseases, for example familial forms of Creutzfeld–Jakob disease (CJD), were associated with mutations in *PRNP* on chromosome 20 several years ago, but recent studies have begun to shed light on some of the puzzling aspects of these degenerative disorders, including the role of immune system factors such as complement, and the factors determining strain-specific infectivity of the conformationally aberrant protein-based agent (Johnson and Gibbs 1998; Wadsworth *et al.* 1999; Klein *et al.* 2001). Last but not least, success in understanding the aetiology of Parkinson disease (PD) has been striking over the last few years, with the identification of six genes harbouring mutations in different familial forms of the common degenerative disorder characterized by degeneration of dopaminergic neurons in the substantia nigra pars compacta (see Table 15.14). The roles of α-synuclein (Polymeropoulos *et al.* 1997; Sharon *et al.* 2003), which is deposited in an aberrant conformational state in the Lewy bodies typically present in PD brain (hence the term 'synucleinopathy' for this and related disorders), and the ubiquitin protein ligase, parkin, in protecting against α-synuclein deposition (Chung *et al.* 2001; Petrucelli *et al.* 2002; West *et al.* 2002), and a number of other advances in understanding, particularly of the wider role and functioning in neurons of the ubiquitin-proteasome protein processing apparatus, have occurred in recent years (reviewed in Goedert 2001; McNaught *et al.* 2001; Dekker *et al.* 2003).

15.4.3 Polyglutamine and polyalanine tract disorders

Similar in many ways to the disorders discussed in the last section, the 'polyglutamine disorders' are characterized by aggregation and deposition of proteolytically cleaved fragments of proteins. These disorders, comprising Huntington disease, Kennedy disease (spinal and bulbar muscular atrophy), and most of the spinocerebellar ataxias (generally characterized by loss of purkinje cells), amongst others (see Table 15.9 for a full list), all originate in the expansion of a trinucleotide repeat of the (CAG)n type (the mechanism is probably similar to that operating in myotonic dystrophy and fragile X syndrome). Despite the significant similarities in the phenotypes of at least some of these disorders, there appears to be nothing common to all the genes involved, many of which were novel when identified, their chromosomal locations, their sequences, their detailed expression patterns etc., save the fact that the CAG repeat is located within coding exons in each gene, and is therefore translated into a run of glutamine residues ('polyglutamine') in the cognate protein. Such runs are commonly found in transcription factors and other nuclear proteins, but not all the proteins encoded by the polyglutamine disorder genes are found in the nucleus, nor act in this fashion. In another reading frame the repeat encodes a polyalanine tract, and expansions of these repeats can give rise to broadly similar disorders such as Partington syndrome (yet another phenotype associated with mutations in *ARX*) and Ondine's curse. An expansion of a related repeat gives rise to a disorder that is almost identical to Huntington disease and is associated with a poluleucine/polyalanine tract expansion in the junctophilin 3 gene on chromosome 16q (Holmes *et al.* 2001).

The pathogenic mechanism has been hard to tease out. Several mouse models have been created (reviewed in Rubinsztein 2002), the best of which involve 'knock-ins', i.e. creating a mouse line with its own expanded repeat in the homologous gene to that mutated in the human disorder, or transgenic insertion of non-essential genes unrelated to the disease gene, but containing expanded, polyglutamine-encoding repeats, because a major requirement is to model the dominant mode of action of these mutations. After several years and many studies, it has become near-dogma that the polypeptide aggregates, composed of polyglutamine-containing fragments of protein cleaved by the caspase/proteasomal apparatus after ubiquitination, that are observed mainly in neuronal nuclei, are pathogenic (e.g. Fan *et al.* 2001; Li *et al.* 2000; Yang *et al.* 2002). However, other studies have provided evidence that these aggregates, or neuronal intranuclear inclusions (NIIs), may in fact be protective against the toxic characteristics of the expanded polyglutamine tracts or the fragments containing them (e.g. Verhoef *et al.* 2002; Taylor *et al.* 2003).

Alternatively, there is evidence that the pathology does not depend on the NIIs, signs appearing in subtle form before the NIIs and other aggregates are visible. Disruption of several cellular processes have been implicated, including cAMP-responsive element binding protein (CREB)-dependent transcription (e.g. Shimohata *et al.* 2000). Other interesting features of these disorders, such as the differences in their cell-type specificity, are currently being teased out through the use of the mouse models and comparative studies using *post mortem* HD brain samples. A unifying feature of the 'CNG'-type triplet repeat disorders may turn out to be a result of the mechanism underlying the dynamic nature of the mutations. Expansions resulting from DNA repair processes in post-mitotic neurons (and cell-type specific differences showing up in the details of this process) can probably account for the cell-type specificity, the late onset of the overt phenotype, the progressive quality of the disorders, and the correlation of severity/age-at-onset with inherited repeat size of these disorders — cellular pathology may originate only after the repeat has reached a certain threshold length during the continual process of somatic mutation through repair-induced expansion (Kennedy and Shelbourne 2000). Several good reviews can be consulted for progress in tying all these fascinating aspects of the polyglutamine disorders together (e.g. Cummings and Zoghbi 2000; Ross 2002).

15.5 **Conclusion and prospects for understanding complex disorders**

As is evident from the size of the tables illustrating the current list of disorders of neurons with identified genes, recent progress in understanding the genetic basis of human neuronal disease has been staggering. Understanding of the aetiology and causes of the common disorders not covered in this chapter, for example schizophrenia (e.g. Millar *et al.* 2000; Jacquet *et al.* 2002; Shifman *et al.* 2002; Stefansson *et al.* 2003), bipolar affective disorder, panic disorder (e.g. Gratacòs *et al.* 2001), and other anxiety-related traits, and understanding of the existence of, and basis for, 'normal' behavioural tendencies and traits like speech and language efficiency (e.g. Lai *et al.* 2001), is a revolution in waiting. Gradually these aspects of our biology are succumbing to genomic mapping, association analyses, natural gene deletion/translocation analysis etc., and the promise of the post-genomic era, pharmacogenomics, and rational drug design, may be just around the corner.

Acknowledgements

I would like to express my thanks to Tracey Neilson, who helped with typing part of the manuscript, and to colleagues in the Division of Molecular Genetics of the Institute of Biomedical and Life Sciences in the University of Glasgow for helpful discussions.

References

Allen, K. M., Gleeson, J. G., Bagrodia, S., Partington, M. W., MacMillan, J. C., Cerione, R. A., *et al.* (1998) *PAK3* mutation in nonsyndromic X-linked mental retardation. *Nat. Genet.*, **20**, 25–30.

Ardlie, K. G., Kruglyak, L., and Seielstad, M. (2002) Patterns of linkage disequilibrium in the human genome. *Nat. Rev. Genet.*, **3**, 299–309.

Baulac, S., Huberfeld, G., Gourfinkel-An, I., Mitropoulou, G., Beranger, A., Prud'homme, J-F., *et al.* (2001) First genetic evidence of $GABA_A$ receptor dysfunction in epilepsy: a mutation in the 2-subunit gene. *Nat. Genet.*, **28**, 46–8.

Beltrán-Valero de Bernabé, D., Currier, S., Steinbrecher, A., Celli, J., van Beusekom, E., van der Zwaag, B., *et al.* (2002) Mutations in O-mannosyltransferase gene *POMT1* give rise to the severe neuronal migration disorder Walker-Warburg syndrome. *Am. J. Hum. Genet.*, **71**, 1033–43.

Bond, J., Roberts, E., Mochida, G. H., Hampshire, D. J., Scott, S., Askham, J. M., *et al.* (2002) *ASPM* is a major determinant of cerebral cortical size. *Nat. Genet.*, **32**, 316–20.

Breedveld, G. J., van Dongen, J. W. F., Danesino, C., Guala, A., Percy, A. K., Dure, L. S., *et al.* (2002) Mutations in *TITF1* are associated with benign hereditary chorea. *Hum. Mol. Genet.*, **11**, 971–9.

Brody, L. C., Conley, M., Cox, C., Kirke, P. N., McKeever, M. P., Mills, J. L., *et al.* (2002) A polymorphism, R653Q, in the trifunctional enzyme methylenetetrahydrofolate dehydrogenase/methylenetetrahydrofolate cyclohydrolase/formyltetrahydrofolate synthetase is a maternal genetic risk factor for neural tube defects: report of the Birth Defects Research Group. *Am. J. Hum. Genet.*, **71**, 1207–15.

Brown, V., Jin, P., Ceman, S., Darnell, J. C., O'Donnell, W. T., Tenenbaum, S. A., *et al.* (2001*a*) Microarray identification of FMRP-associated brain mRNAs and altered mRNA translational profiles in fragile X syndrome. *Cell*, **107**, 477–87.

Brown, L. Y., Odent, S., David, V., Blayau, M., Dubourg, C., Apacik, C., *et al.* (2001*b*) Holoprosencephaly due to mutations in *ZIC2*: alanine tract expansion mutations may be caused by parental somatic recombination. *Hum. Mol. Genet.*, **10**, 791–6.

Carrié, A., Jun, L., Bienvenu, T., Vinet, M-C., McDonell, N., Couvert, P., *et al.* (1999) A new member of the IL-1 receptor family highly expressed in hippocampus and involved in X-linked mental retardation. *Nat. Genet.*, **23**, 25–31.

Chelly, J. and Mandel, J-L. (2001) Monogenic causes of X-linked mental retardation. *Nat. Rev. Genet.*, **2**, 669–80.

Chen, D-H., Brkanac, Z., Verlinde, C. L. M. J., Tan, X-J., Bylenok, L., Nochlin, D., *et al.* (2003) Missense mutations in the regulatory domain of PKCγ: a new mechanism for dominant nonepisodic cerebellar ataxia. *Am. J. Hum. Genet.*, **72**, 839–49.

Chung, K. K. K., Zhang, Y., Lim, K. H., Tanaka, Y., Huang, H., Gao, J., *et al.* (2001) Parkin ubiquitinates the α-synuclein-interacting protein, synphilin-1: implications for Lewy-body formation in Parkinson disease. *Nat. Med.*, **7**, 1144–50.

Claes, L., Del-Favero, J., Ceulemans, B., Lagae, L., Van Broeckhoven, C., and De Jonghe, P. (2001) De novo mutations in the sodium-channel gene *SCN1A* cause severe myoclonic epilepsy of infancy. *Am. J. Hum. Genet.*, **68**, 1327–32.

Coffee, B., Zhang, F., Ceman, S., Warren, S. T., and Reines, D. (2002) Histone modifications depict an aberrantly heterochromatinized *FMR1* gene in fragile X syndrome. *Am. J. Hum. Genet.*, **71**, 923–32.

Cossette, P., Liu, L., Brisebois, H., Dong, H., Lortie, A., Vanasse, M., *et al.* (2002) Mutation of *GABRA1* in an autosomal dominant form of juvenile myoclonic epilepsy. *Nat. Genet.*, **31**, 184–9.

Crosby, A. H. and Proukakis, C. (2002) Is the transportation highway the right road for hereditary spastic paraplegia? *Am. J. Hum. Genet.*, **71**, 1009–16.

Cummings, C. J. and Zoghbi, H. Y. (2000) Trinucleotide repeats: mechanisms and pathophysiology. *Ann. Rev. Genomics Hum. Genet.*, **1**, 281–328.

D'Adamo, P., Menegon, A., Lo Nigro, C., Grasso, M., Gulisano, M., Tamanini, F., *et al.* (1998) Mutations in *GDI1* are responsible for X-linked non-specific mental retardation. *Nat. Genet.*, **19**, 134–9.

Darnell, J. C., Jensen, K. B., Jin, P., Brown, V., Warren, S. T., and Darnell, R. B. (2001) Fragile X mental retardation protein targets G quartet mRNAs important for neuronal function. *Cell*, **107**, 489–99.

Dattani, M. T., Martinez-Barbera, J.-P., Thomas, P. Q., Brickman, J. M., Gupta, R., Mårtensson, I.-L., *et al.* (1998) Mutations in the homeobox gene *HESX1/Hesx1* associated with septo-optic dysplasia in human and mouse. *Nat. Genet.*, **19**, 125–33.

Davis, R. L., Shrimpton, A. E., Holohan, P. D., Bradshaw, C., Feiglin, D., Collins, G. H., *et al.* (1999) Familial dementia caused by polymerization of mutant neuroserpin. *Nature*, **401**, 376–9.

Dawson, E., Abecasis, G. R., Bumpstead, S., Chen, Y., Hunt, S., and Beare, D. M. (2002) A first-generation linkage disequilibrium map of human chromosome 22. *Nature*, **418**, 544–8.

Dekker, M. C. J., Bonifati, V., and van Duijn, C. M. (2003) Parkinson's disease: piecing together a genetic jigsaw. *Brain*, **126**, 1722–33.

Des Portes, V., Pinard, J. M., Billuart, P., Vinet, M. C., Koulakoff, A., Carrié, A., *et al.* (1998) A novel CNS gene required for neuronal migration and involved in X-linked subcortical laminar heterotopia and lissencephaly syndrome. *Cell*, **92**, 51–61.

Dodé, C., Levilliers, J., Dupont, J.-M., De Paepe, A., Le Dû, N., Soussi-Yanicostas, N., *et al.* (2003) Loss-of-function mutations in *FGFR1* cause autosomal dominant Kallman syndrome. *Nat. Genet.*, **33**, 1–3.

Doolin, M-T., Barbaux, S., McDonnell, M., Hoess, K., Whitehead, A. S., and Mitchell, L. E. (2002) Maternal genetic effects, exerted by genes involved in homocysteine remethylation, influence the risk of spina bifida. *Am. J. Hum. Genet.*, **71**, 1222–6.

Engle, E. C. and Leigh, R. J. (2002) Genes, brainstem development, and eye movements. *Neurology*, **59**, 304–5.

Escayg, A., MacDonald, B. T., Meisler, M. H., Baulac, S., Huberfield, G., An-Gourfinkel, I., *et al.* (2000) Mutations of *SCN1A*, encoding a neuronal sodium channel, in two families with GEFS+2. *Nat. Genet.*, **24**, 343–5.

Fan, X., Dion, P., Laganiere, J., Brais, B., and Rouleau, G. A. (2001) Oligomerization of polyalanine expanded PABPN1 facilitates nuclear protein aggregation that is associated with cell death. *Hum. Mol. Genet.*, **10**, 2341–51.

Fang, P., Lev-Lehman, E., Tsai, T-F., Matsuura, T., Benton, C. S., Sutcliffe, J. S., *et al.* (1999) The spectrum of mutations in *UBE3A* causing Angelman syndrome. *Hum. Mol. Genet.*, **8**, 129–35.

Fardaei, M., Larkin, K., Brook, J. D., and Hamshere, M. G. (2001) *In vivo* co-localisation of MBNL protein with *DMPK* expanded-repeat transcripts. *Nucleic Acids Res.*, **29**, 2766–71.

Feng, Y. and Walsh, C. A. (2001) Protein-protein interactions, cytoskeletal regulation and neuronal migration. *Nat. Rev. Neurosci.*, **2**, 408–16.

Folstein, S. E. and Rosen-Sheidley, B. (2001) Genetics of autism: complex aetiology for a heterogeneous disorder. *Nat. Rev. Genet.*, **2**, 943–55.

Fox, J. W., Lamperti, E. D., Eksioglu, Y. Z., Hong, S. E., Feng, Y., Graham, D. A., *et al.* (1998) Mutations in *filamin 1* prevent migration of cerebral cortical neurons in human periventricular heterotopia. *Neuron*, **21**, 1315–25.

Fransen, E., Van Camp, G., Vits, L., and Willems, P. J. (1997) L1-associated diseases: clinical geneticists divide, molecular geneticists unite. *Hum. Mol. Genet.*, **6**, 1625–32.

Friocourt, G., Koulakoff, A., Chafey, P., Boucher, D., Fauchereau, F., Chelly, J., *et al.* (2003) Doublecortin functions at the extremities of growing neuronal processes. *Cerebral Cortex*, **13**, 620–6.

Gabriel, S. B., Schaffner, S. F., Nguyen, H., Moore, J. M., Roy, J., and Blumenstiel, B. (2002) The structure of haplotype blocks in the human genome. *Science*, **296**, 2225–9.

Gargus, J. J. (2003) Unraveling monogenic channelopathies and their implications for complex polygenic disease. *Am. J. Hum. Genet.*, **72**, 785–803.

Georgel, P. T., Horowitz-Scherer, R. A., Adkins, N., Woodcock, C. L., Wade, P. A., and Hansen, J. C. (2003) Chromatin compaction by human MeCP2: Assembly of novel secondary chromatin structures in the absence of DNA methylation. *J. Biol. Chem.*, **278**, 32181–8.

Gleeson, J. G., Allen, K. M., Fox, J. W., Lamperti, E. D., Berkovic, S., Scheffer, I., *et al.* (1998) *Doublecortin*, a brain-specific gene mutated in human X-linked lissencephaly and double cortex syndrome, encodes a putative signaling protein. *Cell*, **92**, 63–72.

Goedert, M. (2001) Alpha-synuclein and neurodegenerative diseases. *Nat. Rev. Neurosci.*, **2**, 492–501.

Goedert, M., Crowther, R. A., and Spillantini, M. G. (1998) Tau mutations cause frontotemporal dementias. *Neuron*, **21**, 955–8.

Gratacòs, M., Nadal, M., Martín-Santos, R., Pujana, M. A., Gago, J., Peral, B., *et al.* (2001) A polymorphic genomic duplication on human chromosome 15 is a susceptibility factor for panic and phobic disorders. *Cell*, **106**, 367–79.

Gupta, A., Tsai, L-H., and Wynshaw-Boris, A. (2002) Life is a journey: a genetic look at neocortical development. *Nat. Rev. Genet.*, **3**, 342–55.

Harkin, L. A., Bowser, D. N., Dibbens, L. M., Singh, R., Phillips, F., Wallace, R. H., *et al.* (2002) Truncation of the $GABA_A$-receptor γ2 subunit in a family with generalized epilepsy with febrile seizures plus. *Am. J. Hum. Genet.*, **70**, 530–6.

Haug, K., Warnstedt, M., Alekov, A. K., Sander, T., Ramirez, A., Poser, B., *et al.* (2003) Mutations in *CLCN2* encoding a voltage-gated chloride channel are associated with idiopathic generalized epilepsies. *Nat. Genet.*, **33**, 527–32.

Hayashi, Y. K., Ogawa, M., Tagawa, K., Noguchi, S., Isihara, T., Nonaka, I., *et al.* (2001) Selective deficiency of α-dystroglycan in Fukuyama-type congenital muscular dystrophy. *Neurology*, **57**, 115–21.

Hering, H. and Sheng, M. (2001) Dendritic spines: structure, dynamics and regulation. *Nat. Rev. Neurosci.*, **2**, 880–8.

Herz, J. and Beffert, U. (2000) Apolipoprotein E receptors: linking brain development and Alzheimer's disease. *Nat. Rev. Neurosci.*, **1**, 51–8.

Holmes, S. E., O'Hearn, E., Rosenblatt, A., Callahan, C., Hwang, H. S., Ingersoll-Ashworth, R. G., *et al.* (2001) A repeat expansion in the gene encoding junctophilin-3 is associated with Huntington disease-like 2. *Nat. Genet.*, **29**, 377–8.

Hong, S. E., Shugart, Y. Y., Huang, D. T., Al Shahwan, S., Grant, P. E., Hourihane, J. O'. B., *et al.* (2000) Autosomal recessive lissencephaly with cerebellar hypoplasia is associated with human *RELN* mutations. *Nat. Genet.*, **26**, 93–6.

Hübner, C. A. and Jentsch, T. J. (2002) Ion channel diseases. *Hum. Mol. Genet.*, **11**, 2435–45.

Jackson, A. P., Eastwood, H., Bell, S. M., Adu, J., Toomes, C., Carr, I. M., *et al.* (2002) Identification of microcephalin, a protein implicated in determining the size of the human brain. *Am. J. Hum. Genet.*, **71**, 136–42.

Jacquet, H., Raux, G., Thibaut, F., Hecketsweiler, B., Houy, E., Demilly, C., *et al.* (2002) *PRODH* mutations and hyperprolinemia in a subset of schizophrenic patients. *Hum. Mol. Genet.*, **11**, 2243–9.

Jaeken, J. and Matthijs, G. (2001) Congenital disorders of glycosylation. *Ann. Rev. Genomics Hum. Genet.*, **2**, 129–51.

Jaenisch, R. and Bird, A. (2003) Epigenetic regulation of gene expression: how the genome integrates intrinsic and environmental signals. *Nat. Genet. Supplement*, **33**, 245–54.

Jamain, S., Quach, H., Betancur, C., Råstam, M., Colineaux, C., Gillberg, I. C., *et al.* (2003) Mutations of the X-linked genes encoding neuroligins NLGN3 and NLGN4 are associated with autism. *Nat. Genet.*, **34**, 27–9.

Jentsch, T. J. (2000) Neuronal KCNQ potassium channels: physiology and role in disease. *Nat. Rev. Neurosci.*, **1**, 21–30.

Jin, P. and Warren, S. T. (2000) Understanding the molecular basis of fragile X syndrome. *Hum. Mol. Genet.*, **9**, 901–8.

Job, C. and Eberwine, J. (2001) Localization and translation of mRNA in dendrites and axons. *Nat. Rev. Neurosci.*, **2**, 889–98.

Johnson, R. T. and Gibbs, C. J. (1998) Medical progress: Creutzfeldt-Jakob disease and related transmissible spongiform encephalopathies. *N. Engl. J. Med.*, **339**, 1994–2004.

Jossin, Y., Bar, I., Ignatova, N., Tissir, F., Lambert de Rouvroit, C., and Goffinet, A. M. (2003) The reelin signaling pathway: some recent developments. *Cerebral Cortex*, **13**, 627–33.

Kehoe, P. G., Katzov, H., Feuk, L., Bennet, A. M., Johansson, B., Wiman, B., *et al.* (2003) Haplotypes extending across *ACE* are associated with Alzheimer's disease. *Hum. Mol. Genet.*, **12**, 859–67.

Kennedy, L. and Shelbourne, P. F. (2000) Dramatic mutation instability in HD mouse striatum: does polyglutamine load contribute to cell-specific vulnerability in Huntington's disease? *Hum. Mol. Genet.*, **9**, 2539–44.

Kimberly, W. T., LaVoie, M. J., Ostaszewski, B. L., Ye, W., Wolfe, M. S., and Selkoe, D. J. (2003) γ-secretase is a membrane protein complex comprised of presenilin, nicastrin, Aph-1 and Pen-2. *Proc. Natl. Acad. Sci. USA*, **100**, 6382–7.

Kitamura, K., Yanazawa, M., Sugiyama, N., Miura, H., Iizuka-Kogo, A., Kusaka, M., *et al.* (2002) Mutation of *ARX* causes abnormal development of forebrain and testes in mice and X-linked lissencephaly with abnormal genitalia in humans. *Nat. Genet.*, **32**, 359–69.

Klein, M. A., Kaesar, P. S., Schwarz, P., Weyd, H., Xenarios, I., Zinkernagel, R. M., *et al.* (2001) Complement facilitates early prion pathogenesis. *Nat. Med.*, **7**, 488–92.

Klesert, T. R., Cho, D. H., Clark, J. I., Maylie, J., Adelman, J., Snider, L., *et al.* (2000) Mice deficient in Six5 develop cataracts: implications for myotonic dystrophy. *Nat. Genet.*, **25**, 105–9.

Kong, A., Gudbjartsson, D. F., Sainz, J., Jonsdottir, G. M., Gudjonsson, S. A., Richardsson, B., *et al.* (2002) A high-resolution recombination map of the human genome. *Nat. Genet.*, **31**, 241–7.

Kouzarides, T. (2002) Histone methylation in transcriptional control. *Curr. Opin. Genet. Dev.*, **12**, 198–209.

Kovtun, I. V. and McMurray, C. T. (2001) Trinucleotide expansion in haploid germ cells by gap repair. *Nat. Genet.*, **27**, 407–11.

Lachner, M. and Jenuwein, T. (2002) The many faces of histone lysine methylation. *Curr. Opin. Cell Biol.*, **14**, 286–9.

Laggerbauer, B., Ostareck, D., Keidel, E-M., Ostarek-Lederer, A., and Fischer, U. (2001) Evidence that fragile X mental retardation protein is a negative regulator of translation. *Hum. Mol. Genet.*, **10**, 329–38.

Lai, C. S. L., Fisher, S. E., Hurst, J. A., Vargha-Khadem, F., and Monaco, A. P. (2001) A forkhead-domain gene is mutated in a severe speech and language disorder. *Nature*, **413**, 519–23.

Lee, V. M-Y. and Trojanowski, J. Q. (1999) Neurodegenerative tauopathies: human disease and transgenic mouse models. *Neuron*, **24**, 507–10.

Lefebvre, S., Bürglen, L., Frézal, J., Munnich, A., and Melki, J. (1998) The role of the *SMN* gene in proximal spinal muscular atrophy. *Hum. Mol. Genet.*, **7**, 1531–6.

Li, H., Li, S-H., Johnston, H., Shelbourne, P. F., and Li, X-J. (2000) Amino-terminal fragments of mutant huntingtin show selective accumulation in striatal neurons and synaptic toxicity. *Nat. Genet.*, **25**, 385–9.

Liquori, C. L., Ricker, K., Moseley, M. L., Jacobsen, J. F., Kress, W., Naylor, S. L., *et al.* (2001) Myotonic dystrophy type 2 caused by a CCTG expansion in intron 1 of *ZNF9*. *Science*, **293**, 864–7.

Lomas, D. A. and Carrell, R. W. (2002) Serpinopathies and the conformational dementias. *Nat. Rev. Genet.*, **3**, 759–68.

Mahley, R. W. and Rall, S. C. (2000) Apolipoprotein E: far more than a lipid transport protein. *Ann. Rev. Genomics Hum. Genet.*, **1**, 507–37.

Mazrui, R., Huot, M-E., Tremblay, S., Filion, C., Labelle, Y., and Khandjian, E. W. (2002) Trapping of messenger RNA by Fragile X Mental Retardation protein into cytoplasmic granules induces translation repression. *Hum. Mol. Genet.*, **11**, 3007–17.

McLellan, A., Phillips, H. A., Rittey, C., Kirkpatrick, M., Mulley, J. C., Goudie, D., *et al.* (2003) Phenotypic comparison of two Scottish families with mutations in different genes causing autosomal dominant nocturnal frontal lobe epilepsy. *Epilepsia*, **44**, 613–7.

McNaught, K. S., Olanow, C. W., Halliwell, B., Isacson, O., and Jenner, P. (2001) Failure of the ubiquitin-proteasome system in Parkinson's disease. *Nat. Rev. Neurosci.*, **2**, 589–94.

Meloni, I., Muscettola, M., Raynaud, M., Longo, I., Bruttini, M., Moizard, M-P., *et al.* (2002) *FACL4*, encoding fatty acid-CoA ligase 4, is mutated in nonspecific X-linked mental retardation. *Nat. Genet.*, **30**, 436–40.

Meola, G., Sansone, V., Perani, D., Colleluori, A., Cappa, S., Cotelli, M., *et al.* (1999) Reduced cerebral blood flow and impaired visual-spatial function in proximal myotonic myopathy. *Neurology*, **53**, 1042–50.

Meyer, M. R., Tschantz, J. T., Norton, M. C., Welsh-Bohmer, Steffens, D. C., Wyse, B. W., *et al.* (1998) *APOE* genotype predicts when — not — whether — one is predisposed to develop Alzheimer disease. *Nat. Genet.*, **19**, 321–2.

Michele, D. E., Barresi, R., Kanagawa, M., Saito, F., Cohn, R. D., and Satz, J. S. (2002) Post-translational disruption of dystroglycan-ligand interactions in congenital muscular dystrophies. *Nature*, **418**, 417–22.

Millar, J. K., Wilson-Annan, J. C., Anderson, S., Christie, S., Taylor, M. S., Semple, C. A. M., *et al.* (2000) Disruption of two novel genes by a translocation co-segregating with schizophrenia. *Hum. Mol. Genet.*, **9**, 1415–23.

Miller, J. W., Urbinati, C. R., Teng-umnuay, P., Stenberg, M. G., Byrne, B. J., Thornton, C. A., *et al.* (2000) Recruitment of human muscleblind proteins to $(CUG)_n$ expansions associated with myotonic dystrophy. *EMBO J.*, **19**, 4439–48.

Ming, J. E. and Muenke, M. (2002) Multiple hits during early embryonic development: digenic diseases and holoprosencephaly. *Am. J. Hum. Genet.*, **71**, 1017–32.

Monani, U. R., Coovert, D. D., and Burghes, A. H. M. (2000) Animal models of spinal muscular atrophy. *Hum. Mol. Genet.*, **9**, 2451–7.

Monuki, E. S. and Walsh, C. A. (2001) Mechanism of cerebral cortical patterning in mice and humans. *Nat. Neurosci. suppl.*, **4**, 1199–206.

Muenke, M. and Beachy, P. A. (2000) Genetics of ventral forebrain development and holoprosencephaly. *Curr. Opin. Genet. Dev.*, **10**, 262–9.

Nan, X. and Bird, A. (2001) The biological functions of the methyl-CpG-binding protein MeCP2 and its implications in Rett syndrome. *Brain Dev.*, **23**, S32–7.

Nemeth, A. H. (2002) The genetics of primary dystonias and related disorders. *Brain*, **125**, 695–721.

Nicholls, R. D. and Knepper, J. L. (2001) Genome organization, function, and imprinting in Prader-Willi and Angelman Syndromes. *Ann. Rev. Genomics Hum. Genet.*, **2**, 153–75.

Online Mendelian Inheritance in Man, OMIM^{TM} (2000) McKusick-Nathans Institute for Genetic Medicine, Johns Hopkins University (Baltimore, MD) and National Center for Biotechnology Information, National Library of Medicine (Bethesda, MD). World Wide Web URL: http://www.ncbi.nlm.nih.gov/omim/

Patel, P. I. and Isaya, G. (2001) Friedreich ataxia: from GAA triplet-repeat expansion to frataxin deficiency. *Am. J. Hum. Genet.*, **69**, 15–24.

Petrucelli, L., O'Farrell, C., Lockhart, P. J., Baptista, M., Kehoe, K., Vink, L., *et al.* (2002) Parkin protects against the toxicity associated with mutant α-synuclein: proteasome dysfunction selectively affects catecholaminergic neurons. *Neuron*, **36**, 1007–19.

Pickering-Brown, S., Baker, M., Yen, S-H., Liu, W-K., Hasegawa, M., and Cairns, N. (2000) Pick's disease is associated with mutations in the *tau* gene. *Ann. Neurol.*, **48**, 859–67.

Pilz, D. T., Kuc, J., Matsumoto, N., Bodurtha, J., Bernadi, B., Tassinari, C. A., *et al.* (1999) Subcortical band heterotopia in rare affected males can be caused by missense mutations in *DCX* (*XLIS*) or *LIS1*. *Hum. Mol. Genet.*, **8**, 1757–60.

Polymeropoulos, M. H., Lavedan, C., Leroy, E., Ide, S. E., Dehejia, A., Dutra, A., *et al.* (1997) Mutation in the α-synuclein gene identified in families with Parkinson's disease. *Science*, **276**, 2045–7.

Pritchard, J. K. and Cox, N. J. (2002) The allelic structure of human disease genes: common disease – common variant…or not? *Hum. Mol. Genet.*, **11**, 2417–23.

Raoul, C., Estévez, A. G., Nishimune, H., Cleveland, D. W., deLapeyrière, O., Henderson, C. E., *et al.* (2002) Motoneuron death triggered by a specific pathway downstream of Fas: potentiation by ALS-linked SOD1 mutations. *Neuron*, **35**, 1067–83.

Rosenberg, M. J., Agarwala, R., Bouffard, G., Davis, J., Fiermonte, G., Hilliard, M. S., *et al.* (2002) Mutant deoxynucleotide carrier is associated with congenital microcephaly. *Nat. Genet.*, **32**, 175–9.

Ross, C. A. (2002) Polyglutamine pathogenesis: Emergence of unifying mechanisms for Huntington's disease and related disorders. *Neuron*, **35**, 819–22.

Rougeulle, C., Cardoso, C., Fontes, M., Colleaux, L., and Lalande, M. (1998) An imprinted antisense RNA overlaps *UBE3A* and a second maternally expressed transcript. *Nat. Genet.*, **19**, 15–16.

Rubinsztein, D. C. (2002) Lessons from animal models of Huntington's disease. *Trends Genet.*, **18**, 202–9.

Sander, T., Schulz, H., Saar, K., Gennaro, E., Riggio, M. C., Bianchi, A., *et al.* (2000) Genome search for susceptibility loci of common idiopathic generalized epilepsies. *Hum. Mol. Genet.*, **9**, 1465–72.

Scheffer, I. E., Wallace, R. H., Phillips, F. L., Hewson, P., Reardon, K., Parasivam, G., *et al.* (2002) X-linked myoclonic epilepsy with spasticity and intellectual disability. *Neurology*, **59**, 348–56.

Seznec, H., Agbulut, O., Sergeant, N., Savouret, C., Ghestem, A., Tabti, N., *et al.* (2001) Mice transgenic for the human myotonic dystrophy region with expanded CTG repeats display muscular and brain abnormalities. *Hum. Mol. Genet.*, **10**, 2717–26.

Shahbazian, M. D., Young, J. I., Yuva-Paylor, L. A., Spencer, C. M., Antalffy, B. A., Noebels, J. L., *et al.* (2002) Mice with truncated MeCP2 recapitulate many Rett syndrome features and display hyperacetylation of histone H3. *Neuron*, **35**, 243–54.

Shahbazian, M. D. and Zoghbi, H. Y. (2002) Rett syndrome and MECP2: linking epigenetics and neuronal function. *Am. J. Hum. Genet.*, **71**, 1259–72.

Sharon, R., Bar-Joseph, I., Frosch, M. P., Walsh, D. M., Hamilton, J. A., and Selkoe, D. J. (2003) The formation of highly soluble oligomers of α-synuclein is regulated by fatty acids and enhanced in Parkinson's disease. *Neuron*, **37**, 583–95.

Shemer, R., Hershko, A. Y., Perk, J., Mostoslavsky, R., Tsuberi, B.-Z., Cedar, H., *et al.* (2000) The imprinting box of the Prader-Willi/Angelman syndrome domain. *Nat. Genet.*, **26**, 440–3.

Shifman, S., Bronstein, M., Sternfield, M., Pisanté-Shalom, A., Lev-Lehman, E., Weizman, A., *et al.* (2002) A highly significant association between a COMT haplotype and schizophrenia. *Am. J. Hum. Genet.*, **71**, 1296–1302.

Shimohata, T., Nakajima, T., Yamada, M., Uchida, C., Onodera, O., Naruse, S., *et al.* (2000) Expanded polyglutamine stretches interact with $TAF_{II}130$, interfering with CREB-dependent transcription. *Nat. Genet.*, **26**, 29–36.

Sinha, S., Anderson, J. P., Barbour, R., Basi, G. S., Caccavello, R., Davis, D., *et al.* (1999) Purification and cloning of amyloid precursor protein β-secretase from human brain. *Nature*, **402**, 537–40.

Smeitink, J., van den Heuvel, L., and DiMauro, S. (2001) The Genetics and Pathology of oxidative phosphorylation. *Nat. Rev. Genet.*, **2**, 342–52.

Smith, D. J. and Lusis, A. J. (2002) The allelic structure of common disease. *Hum. Mol. Genet.*, **11**, 2455–61.

Stefansson, H., Sarginson, J., Kong, A., Yates, P., Steinthorsdottir, V., Gudfinnsson, E., *et al.* (2003) Association of neuregulin 1 with schizophrenia confirmed in a Scottish population. *Am. J. Hum. Genet.*, **72**, 83–7.

Strømme, P., Mangelsdorf, M. E., Scheffer, I. E., and Gécz, J. (2002) Infantile spasms, dystonia, and other X-linked phenotypes caused by mutations in Aristaless related homeobox gene, *ARX. Brain Dev.*, **24**, 266–8.

Taylor, J. P., Tanaka, F., Robitschek, J., Sandoval, C. M., Taye, A., Markovic-Plese, *et al.* (2003) Aggresomes protect cells by enhancing the degradation of toxic polyglutamine-containing protein. *Hum. Mol. Genet.*, **12**, 749–57.

Theuns, J., Remacle, J., Killick, R., Corsmit, E., Vennekens, K., Huylebroeck, D., *et al.* (2003) Alzheimer-associated C allele of the promoter polymorphism –22C>T causes a critical neuron-specific decrease of presenilin 1 expression. *Hum. Mol. Genet.*, **12**, 869–77.

Trockenbacher, A., Suckow, V., Foerster, J., Winter, J., Krauß, S., Ropers, H.-H., *et al.* (2001) *MID1*, mutated in Opitz syndrome, encodes an ubiquitin ligase that targets phosphatase 2A for degradation. *Nat. Genet.*, **29**, 287–94.

Tsuboi, Y., Uitti, R. J., Delisle M-B, Ferreira, J. J., Brefel-Courbon, C., Rascol, O., *et al.* (2002) Clinical features and disease haplotypes of individuals with the N279K tau gene mutation: a comparison of the pallidopontonigral degeneration kindred and a French family. *Neurology*, **59**, 943–50.

Tudor, M., Akbarian, S., Chen, R. Z., and Jaenisch, J. (2002) Transcriptional profiling of a mouse model for Rett syndrome reveals subtle transcriptional changes in the brain. *Proc. Natl. Acad. Sci. USA*, **99**, 15536–41.

van der Put, N. M. J., Steegers-Theunissen, R. P. M., Frosst, P., Trijbels, F. J. M., Eskes, T. K. A. B., van den Heuvel, L. P., *et al.* (1995) Mutated methylenetetrahydrofolate reductase as a risk factor for spina bifida. *Lancet*, **346**, 1070–1.

Vassar, R., Bennett, B. D., Babu-Khan, S., Kahn, S., Mendiaz, E. A., Denis, P., *et al.* (1999) β-secretase cleavage of Alzheimer's disease amyloid precursor protein by the transmembrane aspartic protease BACE. *Science*, **286**, 735–41.

Verhoef, L. G. G. C., Lindsten, K., Masucci, M. G., and Dantuma, N. P. (2002) Aggregate formation inhibits proteasomal degradation of polyglutamine proteins. *Hum. Mol. Genet.*, **11**, 2689–700.

Vervoort, V. S., Beachem, M. A., Edwards, P. S., Ladd, S., Miller, K. E., de Mollerat, X., *et al.* (2002) *AGTR2* mutations in X-linked mental retardation. *Science*, **296**, 2401–3.

Wadsworth, J. D. F., Hill, A. F., Joiner, S., Jackson, G. S., Clarke, A. R., and Collinge, J. (1999) Strain-specific prion-protein conformation determined by metal ions. *Nat. Cell Biol.*, **1**, 55–9.

Wakamatsu, N., Yamada, Y., Yamada, K., Ono, T., Nomura, N., Taniguchi, H., *et al.* (2001) Mutations in *SIP1*, encoding Smad interacting protein-1, cause a form of Hirschsprung disease. *Nat. Genet.*, **27**, 369–70.

Walsh, C. A. (1999) Genetic malformations of the human cerebral cortex. *Neuron*, **23**, 19–29.

Walsh, C. A. and Goffinet, A. M. (2000) Potential mechanisms of mutations that affect neuronal migration in man and mouse. *Curr. Opin. Genet. Dev.*, **10**, 270–4.

Watase, K. and Zoghbi, H. Y. (2003) Modelling brain diseases in mice: the challenges of design and analysis. *Nat. Rev. Genet.*, **4**, 296–307.

West, A. B., Maraganore, D., Crook, J., Lesnick, T., Lockhart, P. J., Wilkes, K. M., *et al.* (2002) Functional association of the *parkin* gene promoter with idiopathic Parkinson's disease. *Hum. Mol. Genet.*, **11**, 2787–92.

Wilkie, H., Osei-Lah, A., Chioza, B., Nashef, L., McCormick, D., Asherson, P., *et al.* (2002) Association of μ-opioid receptor subunit gene and idiopathic generalized epilepsy. *Neurology*, **59**, 724–8.

Winchester, C. L., Ferrier, R. K., Sermoni, A., Clark, B. J., and Johnson, K. J. (1999) Characterization of the expression of *DMPK* and *SIX5* in the human eye and implications for pathogenesis in myotonic dystrophy. *Hum. Mol. Genet.* **8**, 481–92.

Yamasaki, K., Joh, K., Ohta, T., Masuzaki, H., Ishimaru, T., Mukai, T., *et al.* (2003) Neurons but not glial cells show reciprocal imprinting of *Ube3a*. *Hum. Mol. Genet.*, **12**, 837–47.

Yan, R., Bienkowski, M. J., Shuck, M. E., Miao, H., Tory, M. C., Pauley, A. M., *et al.* (1999) Membrane-anchored aspartyl protease with Alzheimer's disease β-secretase activity. *Nature*, **402**, 533–6.

Yang, W., Dunlap, J. R., Andrews, R. B., and Wetzel, R. (2002) Aggregated polyglutamine peptides delivered to nuclei are toxic to mammalian cells. *Hum. Mol. Genet.*, **11**, 2905–17.

Chapter 16

Ageing and the death of neurones

Paul G. Shiels and R. Wayne Davies

16.1 Introduction

The causes of neuronal death are legion, including external environmental stresses such as mechanical damage, infectious disease, toxins, and hypoxia, and internal stresses such as free radical toxicity and genetic variation in protein synthesis, structure, and function. Some of these factors cause acute stress, while others cause chronic stress. All such causes are superimposed on a progressive inability to deal with stress as chronological age increases.

All organisms have evolved ways of handling these stresses. Stress responses need to be viewed from two perspectives, those of cellular mechanism and of organismal biology. Within the cells of each tissue, specific molecular mechanisms exist that sense damage, assess the level of stress and determine the appropriate response; either to try to repair the damage or signal the end of the cell and a need for its replacement. It is important to understand these mechanisms, how they vary between cell types, and which version exists in human neurones. However, these cellular mechanisms are adapted to the particular life strategy that each organism has adopted during evolution. The human life strategy differs considerably from that of a mouse, let alone a fruit fly or a nematode worm. Clearly, the reproductive strategy and longevity of an organism requires adaptations at the cellular and molecular level to make the strategy successful. This aspect is often ignored when considering how neurones age and die, but is key to understanding how an ostensibly post-mitotic cell type responds to damage. Surprisingly, it remains to be factored into current models of neurodegeneration.

The most celebrated model of neuronal death has been the 'cumulative damage hypothesis' (Coyle and Puttfarken 1993). This proposes that individual neurones accumulate fibrils and oxidative damage with time, until a damage threshold level is reached and the cell can no longer cope with the biochemical stress and dies. At the level of the organ, this predicts that neuronal death with time can be mathematically represented as a sigmoidal response curve. This model also predicts that the probability of loss of healthy neurones is proportionally dependent on the percentage of damaged neurones at any point in time. Whilst seemingly intuitive, this model has not withstood recent critical scrutiny. A new model, based upon a catastrophic loss of function as the result of random neuronal loss, has been proposed to explain the observation that neuronal loss, in man, occurs essentially at the same rate over time, and independent of disease state progression (Clarke *et al.* 2000). Neither of these models considers ageing, at both the level of the cell and of the organ, as a factor to be equated into how neurones die or how this might relate to pathogenesis.

The maintenance of brain function will depend on the continued correct functioning of the organ's support systems (glial cells, blood supply), on maintenance of individual neurone function and connectivity, and on the replacement of neurones and their connections from stem cells. The extent to which the latter occurs is currently unclear. Both differentiated neurones and stem cells will be subject to ageing processes. Most of our knowledge of these processes derives from studies in non-neural tissues. Senescence has been implicated in the pathophysiology of kidney disease (Halloran *et al.* 1999; Paul 1999), cardiovascular disease (Fukino *et al.* 2002), neurodegeneration (Calabrese *et al.* 2001) and the reduced survival of animals derived by nuclear transplantation (cloning) (Ogonuki *et al.* 2002). Here we consider what is known about the key molecular systems involved in detecting, signalling, and counteracting the damage associated with ageing, how the known systems may function in neurones and neuronal stem cells, and their relevance to neuronal death.

16.2 **Concepts in ageing research**

Ageing as defined by Gompertz (1825), constitutes an exponential increase in the likelihood of mortality with time. Chronological ageing manifests as declining organ mass, cellular function, and integrity with time elapsed since birth. It reflects the lifestyle of the organism, life history and the cumulative burden of oxidative insult at the molecular level (Shiels 1999).

Senescence can be applied as a concept at the cellular level or at the physiological (organ/tissue) level. Cellular senescence is studied largely in tissue culture, and the understanding gained experimentally in culture may not be entirely transferable to cells *in situ* in the context of an organ. Organ senescence has additional components of intrinsic ageing and of effects due to the interdependence of cell types within the organ. Figure 16.1 presents these concepts diagrammatically.

Cellular senescence in culture is characterized by a growth arrest phenotype as a consequence of stress. Typically, it is due to oxidative damage to DNA (both genomic and mitochondrial) in the form of breaks, which lead to destabilization of telomere–nucleoprotein complexes (Shiels 1999; Shiels *et al.* 1999*a*, *b*), and to damage to other cellular components in the form of protein oxidation, lipid peroxidation, and a consequential elevation in cellular responses to stress (Linskens *et al.* 1995). Significantly, these changes occur independently of the chronological age of the organism at the time of tissue isolation for cell culture. However, cell types can and do vary in their senescence characteristics. It is pertinent to note that such a stress response phenomenon is not limited to replicative cells. If the senescence phenotype is assessed in terms of a standard group of molecular changes (rather than growth arrest), it can also be seen to occur in post-mitotic cells, such as hepatocytes or neurones. Essentially, this will be determined by the level of stress to which the cell is exposed. Sub-lethal stress will result in 'stasis' (STimulation And Stress Induced Senescence; Section 16.5), while lethal levels will result in apoptosis or necrosis, with consequent effect on brain physiology in the case of neurones. This is illustrated in Fig. 16.1 and discussed further below.

Senescence at the level of the organ/tissue will be reflected in declining physiological capacity and a decreased ability to withstand insult. This will be a direct reflection of

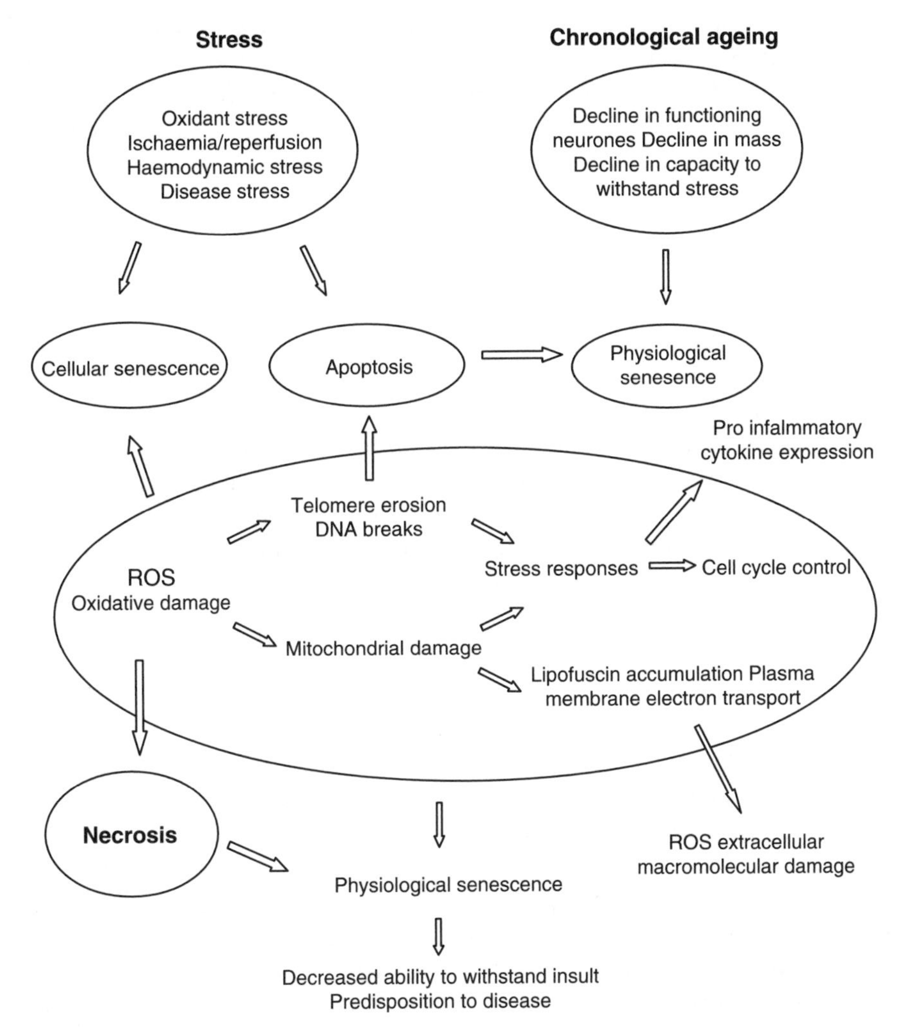

Fig. 16.1 Paths to Senescence. Cellular stresses can lead to cell loss through apoptosis or carriage of the damage if it is sub-lethal. Both processes will accelerate physiological senescence. Damage carried as wear and tear will lead to accelerated cellular senescence or to stasis and can lead to pro-inflammatory changes. These overlapping processes will weaken cells so as to predispose them to disease when subject to further stress (e.g. oxidative damage, hydrodynamic stress, trophic stress).

chronological ageing in combination with cellular mechanisms of senescence, influencing both organ function and structure. This is not envisaged as a uniform process, but the degree to which structure and function are affected will be a consequence of the numbers and location of senescent cells within the organ, any decline in mass with chronological age and critically, any cell loss due to insult. Each organ has multiple cell types, including cells in blood and lymph vessels that supply and drain it, that are subject to differential stresses and exist in different molecular states of differentiation. Cellular stresses will lead to cell loss through apoptosis, or to the organ carrying damaged and sub-functional cells if the damage is sub-lethal. Early failure of one cell type will both impose more demand on the remaining functional cells of that type, and have knock-on effects on other cell types as well, leading to accelerated senescence and predisposing to inflammatory changes. An organ will only be as good as its least senescence resistant cell type.

16.3 **Damage, repair and disposal**

If a cell were not exposed to physical or chemical damage, or if it were able to efficiently sense and repair all damage rapidly and completely, there is no reason why it should not continue functioning indefinitely in a hospitable environment. Unfortunately, both the external environment and the cell's own metabolism generate damage. The internal damage is widely thought to be due to oxidative stress caused by various free radical species that can be generated by elements of the normal biochemistry of the cell, such as complex III of the electron transport chain in mitochondria, flavoproteins, cytochrome p450 and other oxidases (for a review with a neuronal focus see: Ischiropoulos and Beckman 2003). The electron flow through these processes is tightly coupled to avoid partial reduction of free oxygen, but any interference, such as uncoupling of complex 1 or genetic defects in cytochrome oxidase in mitochondria, can contribute to neuronal degeneration (Swerdlow *et al.* 1996; Davis *et al.* 1997). Other enzymes of neuronal relevance that generate free radicals are nitric oxide synthase (NOS) and prostaglandin H synthase. Inflammatory cells also contribute significantly to the free radicals within an organ. The major free radicals present under physiological conditions are reactive oxygen species (ROS), such as superoxide (O_2^{***}), hydroxyl radicals (OH^{**}), and alkoxy radicals (RO^{**}), and reactive nitrogen species such as $ONOO^{**}$, NO^{**}, and NO_2^{**}. Hydrogen peroxide (H_2O_2) is also produced. Neither superoxide nor hydrogen peroxide is extremely toxic, and it has been argued (Beckman 1994) that the peroxynitrite radical ($ONOO^{**}$), which can be rapidly formed when superoxide and NO are in the same cellular compartment, and reacts strongly with proteins to form nitrotyrosine, is most likely to mediate neurotoxicity. Peroxynitrite has been shown to inactivate the dopamine transporter in a neuronal cell line in culture (Park *et al.* 2002); this is a correlate of MPTP killing of dopaminergic neurons. However, the actual source of the free radicals supposed to effect damage remains unclear.

These radicals are used by living systems for many biologically useful purposes, including signalling, enzyme function, phagocytic killing, and regulation (Halliwell 1999). Unfortunately, they can oxidatively damage all major macromolecules of the cell. Cells vary widely in their activity and in their use of glycolysis, mitochondrial, or other electron transport for energy generation, which as we shall see below can be crucial for the level of oxidative damage generated. Certain cells have specialized biochemistry that makes them more vulnerable. A neuronal example is that of dopaminergic neurones, because the degradation pathway of excess dopamine gives rise to *o*-quinones, which are neurotoxic because of the formation of the dopamine *o*-hydroquinone free radical, which reacts with proteins to make quinoproteins (Smythies and Galzigna 1998). It is no coincidence that these neurones are the first to die in the face of a variety of stresses.

Cells have several defence mechanisms against free radical damage. In the nervous system there is a need for extracellular antioxidants, because H_2O_2 is produced (e.g. by NOS) and is freely diffusible. The major extracellular antioxidant is ascorbate (Grünewald 1993), which is released with some neurotransmitters (e.g. glutamate). Carnosine is also packaged in synaptic vesicles with transmitter and co-released, while glutathione is released by astrocytes. Some neurotransmitters, such as dopamine, have antioxidant activity at synapses where they are released. Within the cell there are a number of antioxidant defence systems, ranging from low molecular weight agents such as polyphenols, tocopherol, carotenoids, glutathione, bilirubin, ubiquinol, melatonin, and lipoic acid, to proteins, including both iron and copper sequestration proteins (transferrin, lactoferrin, ferritin, haemopexin, caeruloplasmin) and enzymes such as superoxide dismutase, catalase, peroxiredoxins, thioredoxins, glutaredoxins, and glutathione peroxidases, that convert free radicals into harmless chemicals or inactivate them (Halliwell 1999). If damage nevertheless occurs, the cell can either repair the damage, or dispose the damaged molecule. Thus DNA damage sensing and DNA repair mechanisms are crucial, as is their efficient application both in the nucleus and mitochondrion. Currently, we know very little about the DNA repair system of neurons. The mitochondrial DNA of neurons is able to replicate. On the one hand, this means that mitochondrial DNA can be repaired during replication, but on the other hand there is a danger of conversion to the mutant form and propagation of mutant DNA molecules in the population of mitochondria in the cell. Damage to nuclear DNA can only be detected and repaired by mismatch recognition and repair systems in neurons. There are many unanswered questions in this area.

Our knowledge of protein damage recognition and disposal mechanisms also comes from non-neuronal cells, and is patchy. Proteins that are damaged and unable to fold correctly are thought to be recognized by chaperones and tagged by ubiquitin for disposal through the proteasome system. Proteins that are damaged after folding can be removed by the ubiquitination–proteasome system, or via regular cycling through endosomal compartments, and shunting of damaged molecules into lysosomes for degradation. It is not clear whether these processes are random or partially selective.

Whichever it is, proteins that cycle faster will be less likely to be a source of problems for the cell when damaged, as they will be cleared and replaced. Proteins that remain in place for long periods, such as some components of the postsynaptic density including NMDA receptors, will be more vulnerable, and may require special damage-sensing systems. At the level of the endoplasmic reticulum (ER), a molecular response to stress caused by the local accumulation of unfolded or misfolded proteins, which can lead to apoptosis, has been partially characterized. There are two components to this stress response. ER stress sensor proteins called IREs (Kaufman 1999) mediate ER stress signals, recruiting TRAF2 (TNF receptor-associated factor 2), which in turn recruits and activates the proximal components of the c-Jun N-terminal kinase (JNK) pathway (Urano *et al.* 2000). ER stress also leads to the activation of genes of the unfolded protein response (UPR), which either improves protein folding or results in cell death. Components of the UPR are ATF6, BiP, and caspase 12. ATF6 resides at the ER membrane, is processed on induction of ER stress, translocates to the nucleus and activates transcription of the gene encoding the Grp78/BiP chaperone protein. If the response proves unable to repair the cause of the stress, an apoptotic pathway is activated, with caspase 12, which is located in the outer ER component (Nakagawa *et al.* 2000). Experimentally, this pathway can be induced by tunicamycin, which inhibits N-linked glycosylation.

Clearly, the protein degradation systems of the cell — proteasomes and lysosomes and associated molecules — will be crucial to long-term survival, and we would expect neurones to have active disposal systems. Indeed, partial failure of the ubiquitin tagging system for protein removal results in the eventual death of the most vulnerable neuronal type, dopaminergic neurons. For example, mutations in the parkin gene, encoding an E3 ubiquitin ligase, cause very early onset of Parkinson's disease, while mutations in the gene encoding UCH-L1 (ubiquitin carboxy-terminal hydrolase) also give rise to Parkinson's disease. Damaged lipids are also processed in lysosomes. Lysosomes have been reported to be particularly susceptible to free radical oxidative stress (Bahr and Bendiske 2002), and loss of lysosomal as well as proteasomal processing capacity is likely to be a factor in the onset of diseases of ageing involving protein aggregation (Section 16.10 and Bailey, this volume Chapter 15) and ageing itself.

A further disposal mechanism of the cell is the sequestration of damaged cellular components, dangerous chemical constituents, or excess proteins that are difficult to process, in disposal organelles or cellular inclusion bodies. The quinone by-products of dopamine degradation are further converted into the inert neuromelanin polymer, and stored as pigment granules, giving adult human substantia nigra cells their characteristic black colour. It has been suggested (Cheng *et al.* 1996) that dopamine quinone is normally degraded by 5-cysteinylization and 5-glutathionization, and that only in times of severe oxidative stress, when the cysteine and glutathione pool is depleted, is quinone shunted into neuromelanin biosynthesis. It is likely that the characteristic Lewy bodies of Parkinson's disease (and some other syndromes) are the cell's way of sequestering a protein — in this case alpha synuclein – that is difficult to clear from the cell. The neurones that sequester alpha synuclein best may survive longest, which is why all remaining dopaminergic neurons have Lewy bodies on autopsy. This certainly applies to triplet repeat diseases such as Huntington's disease and spinocerebellar ataxia 1 (SCA1), where there is clear evidence that the formation of inclusion bodies containing the polyQ-proteins is not required for the disease process (Saudou *et al.* 1998; Klement *et al.* 1998; review Kim and Tanzi 1998). Nevertheless, the occurrence of such inclusions is an indication that something out of the usual requires disposal. Clearly, stress from damaged or unwanted proteins is a significant factor in cellular ageing and pathogenesis (review: McNaught and Olanow 2003). One of the indicators of cellular ageing is the accumulation of a substance called lipofuscin, which is derived from mitochondrial membrane material, in lysosomal compartments (Brunk and Terman 2002).

16.4 **Ageing and the MTR trinity**

Until recently it was not clear which elements of molecular cell biology played the key role in assessing the ageing damage process and making decisions on cell fate dependent on that assessment. From an adult neuronal perspective, achieving long-term cell survival would seem to be the primary goal. However, it is important to remember that for many cell types a mechanism for measuring the age of a cell, allowing it to be disposed of before things go badly wrong, is paramount. Thus, when considering the evidence for ageing detection and decision making systems from various cell types, it should be realized that only a subset of these mechanisms may apply in mature neurones.

A number of distinct, but not mutually exclusive components of ageing damage detection and response exist. The key players are the mitochondrion, telomere nucleo-protein complexes, and the rDNA cluster, which form an interactive trinity. Their interactions integrate energy production, protein synthesis, and DNA integrity. As discussed below, the telomere nucleo-protein complexes are part of the sensor system for DNA damage, the mitochondrion generates energy and plays a major part in apoptotic cell death mechanisms, and rDNA controls energy utilization via ribosome production while being in balance with the telomere for protein complex binding. Any dramatic change in the components of this trinity has drastic consequences for the cell. For example, telomere erosion leads to premature cell death, and rDNA hypersynthesis can produce a sink for the telomere nucleo-protein complex, exposing the telomere with the same consequence. Since defects in any of these components will accelerate ageing, they can be regarded as contributing to the ageing process, and several models of ageing have been proposed with one or the other of the trinity on centre stage. Some molecular components of the DNA damage-sensing system are shown in Fig. 16.2.

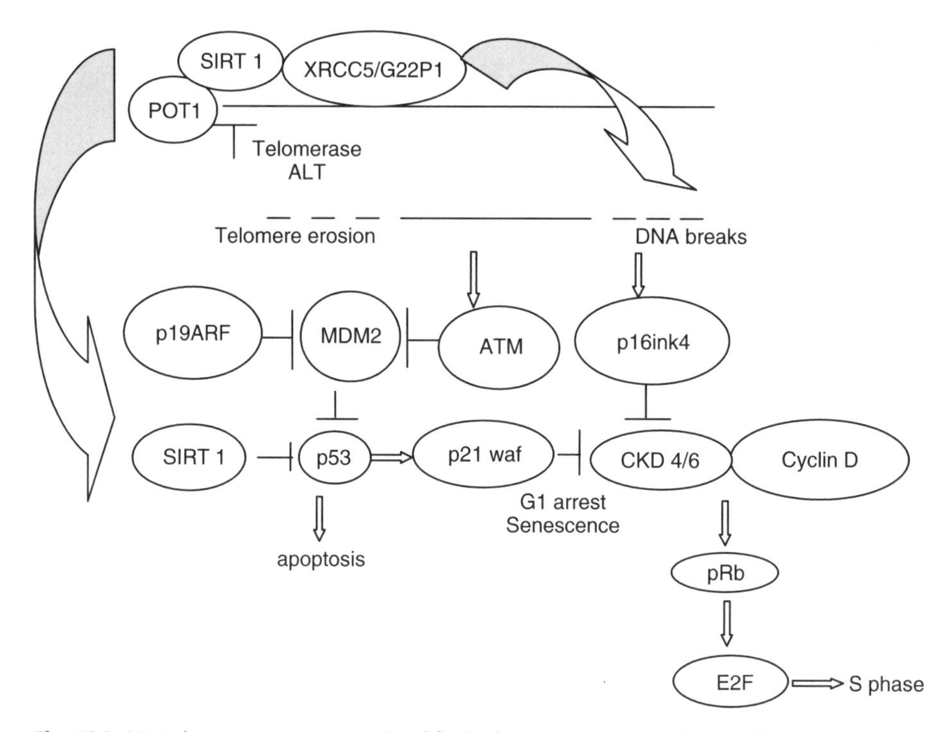

Fig. 16.2 DNA damage responses: a simplified schematic representation of DNA damage response pathways. Telomerase and the alternative lengthening of telomeres (ALT) system maintain telomere length unless inactive. Telomere erosion in the soma or DNA damage will result in repair complexes being recruited to the site of damage. This will result in up-regulation of relevant genes (e.g. XRCC5, G22P1). Heavy damage will perturb telomere structure resulting in up-regulation of its protein components (e.g. hPOT1) in an attempt to maintain structural integrity. SIRT1, acting as a negative regulator of p53 function, will then act as a surrogate sensor of the level of damage. Too much damage and p53 will be activated via decreased SIRT 1 expression setting off an apoptotic cascade. If the damage level is sub-lethal, then the cell will carry it as wear and tear and may enter senescence via p21/p16 activation. The role of free telomere binding proteins (e.g. hPOT 1) as direct effectors of this pathway is still unproven.

16.4.1 **Telomeres as molecular triggers for stress responses**

Telomeres are nucleoprotein complexes found at the ends of eukaryotic chromosomes. In mammals, they comprise a dynamic complex of proteins bound to stretches of a simple DNA repeat $(TTAGGG)_n$ (Moyzis *et al.* 1988). Traditionally, they have been ascribed three roles, namely (i) capping the chromosome to prevent end-to-end attachments; (ii) distinguishing the chromosome end from a damage-related DNA break, thus protecting it from nuclease action; and (iii) ensuring the complete replication of chromosome ends.

The telomeric DNA component of human chromosomes in dividing somatic cells erodes with increasing chronological age. This is due to a combination of incomplete replication of the chromosome ends and nuclease action (Harley *et al.* 1990; Karlseder *et al.* 2002). In germ cells and stem cells, however, this is counteracted by the action of the holo-enzyme telomerase, which facilitates the addition of fresh telomeric repeats onto the chromosome end. In somatic cells, telomere erosion is considered to be an anti-neoplastic mechanism that functions as a mitotic clock, which eventually signals cell death. It is worth noting that rodents, with their short lifespans, are not concerned with avoiding neoplasms, and have both long telomeres and telomerase in somatic cells.

Telomere shortening has been causally implicated in human disease and by general implication, the physiological ageing process in higher animals (Harley *et al.* 1990; Allsopp *et al.* 1992; Oexle and Zwirner 1997). The reason for this is that telomeric proteins form part of a damage-sensing and signalling system. DNA damage checkpoint proteins, which are highly conserved (e.g. Ku70-Ku80, Mre11-Rad50-Nbs1), detect DNA damage including double-strand breaks, and form signalling cascades that inhibit mitosis in dividing cells and can either facilitate DNA repair or apoptosis, dependent on the level of damage (Fig. 16.2). Telomere nucleo-protein complexes store all the DNA repair system proteins and DNA damage checkpoint proteins for the cell apart from those acting on the DNA at any given time (Nugent *et al.* 1998; Zhu *et al.* 2000). Remarkably, these proteins do not recognize telomere ends as double-strand breaks but as special structures to be maintained, and they have also been shown to contribute to telomere maintenance (Khanna and Jackson 2001; Nakamura *et al.* 2002). As discussed below, the status of the telomere nucleo-protein complex is signalled to the mitochondrion. An inability to respond to, or repair damage properly can result in accelerated ageing. Segmental progeric conditions, such as Werners' syndrome, Blooms' syndrome, and Hutchinson–Guilfords' disease — all result from the dysfunction of an individual telomeric protein. In mice, knock out of the telomerase coding sequence resulted in progressive loss of telomeric DNA and progeria (Blasco *et al.* 1997). Reintroduction of telomerase reversed both these effects (Samper *et al.* 2001). Knock out of the double-strand break repair and telomeric-binding protein XRCC5 similarly results in accelerated ageing in mice, highlighting the importance of these structures to cellular and organ integrity.

16.4.2 **Mitochondrial damage and ageing**

Mitochondrial oxidative damage is a strong correlate of ageing. A decrease in mitochondrial respiratory activity and an increase in mitochondrial mutations and fragmentation have been positively associated with increasing age (Harman 1972; Linnane *et al.* 1989, 1990; Hayakawa *et al.* 1992). It seems probable that mitochondrial DNA (mtDNA), located as it is close to a major source of free radicals, will suffer oxidative damage first, but there is no direct evidence of this, and we know even less about mitochondria in neurones. The pronounced age-related accumulation of 8-OhdG in mtDNA relative to nuclear DNA (Hayakawa *et al.* 1992, 1993), and a correlation with a decline in the activity of the mitochondrial electron transfer chain

(Takasawa *et al.* 1993) does support such a hypothesis. Oxidative damage to mtDNA may have pronounced effects, as mitochondria in post-mitotic cells maintain the capacity to replicate (Menzies and Gold 1971), potentially allowing the multiplication of mutant genotypes. Respiratory deficiency and degeneration may also arise as a consequence of mitochondrial electron leak and an inability to degrade damaged mitochondrial macromolecules, which accumulate in lipofuscin (Brunk and Terman 2002). These processes may exacerbate physiological senescence, as cells with mitochondrial genomes carrying even low levels (<1%) of loss of function mutations have been proposed to use plasma membrane electron transport as a compensatory mechanism (deGrey 2002). This generates ROS on the outside of the cell and amplifies the effects of oxidant stress (deGrey 2002). The results of elevated stress resulting from mitochondrial mutations have also been observed clinically in patients with mitochondrial myopathy (Ozawa *et al.* 1995) and in murine models of mitochondrial disease (Esposito *et al.* 1999). Accelerated telomere erosion has also been observed in human mitochondrial diseases (Oexle and Zwirner 1997). However, while many aspects of this hypothesis seem intuitive, evidence in support of it still remains largely circumstantial. It remains unclear whether mitochondrial DNA, proteins, and lipids are really the primary site of early oxidative damage in the cell. Indeed, evidence from caloric restriction studies in yeast (Section 16.4.3) indicates that emphasizing energy production via the mitochondrial electron transport chain is actually beneficial and increases longevity. However, the role of the mitochondrion in apoptosis, once damage becomes severe, is well established.

16.4.3 rDNA, ageing and Sirtuins

Yeast mutants for the *SGS 1* gene, which codes for a RecQ-like helicase, senesce prematurely. The RecQ-like helicase plays a role in recombination-mediated telomere rebuilding. These mutants not only affect telomere maintenance, but also have a dramatic effect on rDNA, demonstrating the link between telomeres and rDNA. What is observed is the accumulation of extrachromosomal rDNA circles (ERCs) in mother cells following successive asymmetric cell divisions (Sinclair *et al.* 1997). Since the rDNA has the same sequences as the telomere for protein complex binding, the accumulation of ERCs results in competition for telomere-binding proteins and deprotection of the telomeres, leading to cell death (Sinclair *et al.* 1997). It is not clear how the release of telomere proteins on telomere erosion leads to ERC formation, but this may be due to nucleosome destabilization, increased transcription, and an enhanced opportunity for recombination to occur within the repeat unit, releasing ERCs. In man, mutation in SGS1 orthologues leads to the segmental progerias, Werners', Blooms' and Rothmund–Thomson syndromes (Gangloff *et al.* 2000). There is currently no evidence for the occurrence of ERCs in man, so the effects of these mutations may relate only to their effect on telomere length maintenance: this is significant since neurones are severely affected in these progerias.

Significantly, nucleolar fragmentation and a disruption of silencing complexes at telomeres accompany the accumulation of ERCs (Sinclair *et al.* 1997). These complexes are composed of Sir (Sirtuin) proteins, one of which, SIR2, regulates ageing in yeast and enhanced life span due to caloric restriction (Lin *et al.* 2000; Luo *et al.* 2001). In man, seven SIR 2 homologues (SIRT 1–7) have been identified. These appear to be NAD-dependent deacetylases. SIRT 1 is a negative regulator of p53, which promotes cell survival under stress (Vaziri *et al.* 2001). DNA damage is known to induce p53 expression and consequentially potentiate growth arrest. SIRT2 is an NAD(+)-dependent tubulin deacetylase. SIRT3 is an NAD-dependent deacetylase localized to mitochondria (Onyango *et al.* 2002). SIRT 7 may have a role in the control of cell proliferation and is found to be up-regulated in thyroid cancer. The functions of these sirtuins require further study, and the roles of the other three sirtuins remain to be defined.

Recent observations in the nematode worm *Caenorhabditis elegans* indicate that strains carrying extra copies of the SIR2 orthologue, sir-2.1, exhibit an extended life span, in keeping with the observations in yeast (Tissenbaum and Guarente 2001). It is reasonable to expect that a similar situation will be found in mammals. If there is ERC formation, or indeed amplification of rDNA in any form with ageing, this would reduce the available SIR2 and lead to reduced life span. ERCs have already been detected in a number of metazoans, though they have not been correlated with ageing (Lumpkin *et al.* 1985). However, a direct correlation in mammals may not be forthcoming readily, given the increased number of sequence homologues for SIR2 and potential multiple functions.

Any direct link is more readily found with respect to caloric restriction (CR), where the NAD dependency of Sir 2 function may link energy production to life span. Interestingly, in yeast CR appears to promote longevity through driving carbon metabolism towards the mitochondrial TCA cycle thus increasing respiration. This results in a reduction in the rate of glycolysis and an increase in the rate of electron transport, with concomitant oxidation of NADH to NAD in mitochondria and the activation of Sir2 (Lin *et al.* 2002). The role of the mitochondrial electron transport chain is pivotal, as its disruption prevents the CR associated longevity. It should be noted that this implies that the mitochondrial electron transport chain (unless perturbed) is not the source of the free radicals that actually do damage, and that the faster and better it functions the longer the cell lives. Thus the simple expectation that cells with high levels of mitochondrial activity, such as many key neurones in the brain, e.g. cortical pyramidal cells, are necessarily more at risk may not be true.

Importantly, CR does not appear to increase resistance to oxidative stress during the replicative life span in yeast. Reactive oxygen species (ROS) do, however, appear to affect survival of stationary phase yeast and post-mitotic dipteran and nematode cells. These observations are relevant to ageing in mammals, where increases in anti-oxidant enzyme levels associated with CR may no longer *per se* be viewed as a direct cause of the longevity, but of CR driving carbon into the TCA, thus increasing respiration. Gene expression profiling in mouse appears to support this scenario (Lee *et al.* 1999).

16.5 Comparative biology of ageing

Contemporary models of ageing have been formulated from *in vitro* observations on replicative cells and extrapolated to the level of the organism. This is not a straightforward process and is complicated by the comparative biology of the organisms used to derive the respective models. Rodent models of ageing, however, may provide a powerful paradigm of how post-mitotic human cells age.

Human replicative senescence is characterized by incremental telomere erosion, which acts as a counting mechanism for determining when a cell should enter growth arrest. Growth arrest is reached after a finite number of cell divisions, a point termed the Hayflick limit (Shay and Wright 2000). This is postulated to minimize the proliferation of mutated cells that may give rise to tumours in a long-lived mammal such as man.

Rodents, however, with a high metabolic rate and associated short lifespan, have a different telomere biology. Their telomeres are typically much longer than man (terminal restriction fragment lengths of 20–150 kb versus 5–20 kb for man) and their somatic cells express the enzyme telomerase. Murine cells do not show replicative senescence in culture. They do, however, undergo growth arrest. This is not based upon telomere shortening, but culture stresses producing a 'stasis' phenotype that mimics replicative senescence, as described for human cells. The term stasis corresponds to Stimulation And Stress Induced Senescence, not replicative senescence. It is a term formulated to reconcile seemingly conflicting human and murine *in vitro* data on growth arrest phenotypes (Wright and Shay 2002). Its translation to a neuronal setting seems appropriate as it does not involve replicative processes. In this respect, rodent models of ageing, remain valid for the study of stress responses and their effect on cell and organ function. Common stress response pathways exist despite the differing telomere biology.

The involvement of neuronal stasis in neurodegenerative processes requires critical testing. Recent data from Atm–/– (ataxia-telangiectasia mutated) and Terc (telomerase RNA component) –/– deficient mice indicates that telomere dysfunction has a critical influence on organ homeostasis and ageing, providing strong *in vivo* support for such a hypothesis (Wong *et al.* 2003). Atm has a central role in signalling the presence of double-strand DNA breaks, and in the co-ordination of the appropriate checkpoint response functions, linked in part to ATM-directed phosphorylation of p53 and many DNA damage response proteins (Kastan and Lim 2000). This shows the central role of DNA damage sensing in controlling ageing damage. Atm–/– mice show only modest neuronal and premature ageing effects, much less than in man. However, mice doubly mutant for Atm and the telomerase RNA component Terc show dramatically accelerated telomere erosion, genomic instability, and progressive neurodegeneration; most significantly, this occurs in conjunction with depleted neural stem cell reserves and impaired cell function (Wong *et al.* 2003). The additional Terc–/– mutation is required to move the phenotype into the range of the human Atm–/– syndrome, reflecting the different somatic expression of telomerase in these species. These mice displayed compromised organ function, accelerated physiological senescence, and premature death.

16.6 **Telomere binding protein-links to mitochondria and rDNA**

The functional triumvirate of the telomere, mitochondrion and rDNA allows the cell to tie in damage responses to energy production and utilization. The common bonds between the three appear to centre on p53.

The telomere acts as a sink for DNA repair proteins, such as XRCC5 and G22P1, damage signalling components, such as the Sirtuins and damage checkpoint regulators such as ATM. In response to genotoxic insult, these proteins un-dock initiating damage signalling and facilitate repair (Fig. 16.2). The degree of protein loss from the telomere may act as an indicator of damage levels, initiating apoptosis, growth arrest, or repair. How this is actually achieved remains unproven and a number of hypotheses exist. These either espouse that (i) critically shortened telomeres are unable to recruit sufficient telomeric proteins to form a functional cap, thus exposing a free broken DNA end to nuclease action, or (ii) it is the shortened telomere repeat stretch *per se*, or a resulting increase in the availability of free telomeric proteins arising through loss of substrate sites, that engenders the necessary signals.

Recent observations indicate that for replicative senescence, the critical factor is the length of the single-strand telomeric overhang (Stewart *et al.* 2003). It is conceivable however, that these hypotheses are not mutually exclusive and that both loss of sequence and availability of free telomeric proteins contribute to damage responses and pathogenesis.

Loss of SIRT 1 from the telomere results in activation of p53 (Langley *et al.* 2002). Human SIR2 deacetylates p53 and antagonizes PML/p53-induced cellular senescence (Vaziri *et al.* 2001). G22P1, the partner of XRCC5, suppresses Bax and prevents apoptosis, unless damage is sufficiently great. As a consequence, these two protein families tie the telomere directly to the mitochondrion and to the initiation of damage responses. It is pertinent to note that XRCC5 knock out mice exhibit neurodegeneration, ataxia, limb rigidity, and tremor as a consequence of accelerated physiological ageing (Karanjawala *et al.* 2002). A further link may be via the activity of SIRT3, which localizes to the mitochondrion, and appears to be a NAD(+)-dependent histone deacetylase analogous to SIRT1 (Onyango *et al.* 2002).

The activities of sirtuins also tie in the regulation of ribosome production to telomere and mitochondrial responses to damage, via p53 activation, as p53 forms part of a complex interacting with ribosomal proteins L5 and 5SrRNA (Marechal *et al.* 1994).

16.7 **How does telomere biology link to apoptosis and necrosis?**

Two principal mechanisms are thought to contribute to neuronal loss, these being apoptosis and necrosis. The molecular mechanisms underpinning these processes are still incompletely understood. Apoptosis occurs as a consequence of genotoxic insult, leading to mitochondrial dysfunction and destabilization of telomere nucleo-protein complexes. The former is well documented in neurobiological literature. The latter, by virtue of a more traditional association with replicative cell biology, has been neglected.

What these processes share in common with senescence is that they use the same damage signalling and sensing mechanisms. In essence they may represent one end of a spectrum of damage responses, dependent upon the degree of damage accrued by the neurone.

16.7.1 Neuronal apoptosis

Apoptosis (Fig. 16.3) is a set of programmed and ATP-dependent cell death pathways, and results in the formation of apoptotic bodies containing the degraded contents of the dead cell, with no resultant inflammatory response. Apoptosis in neurones is typically a p53 dependent process, resulting from cellular stress, inclusive of trophic factor withdrawal and the action of ROS. Typically, these potentiate loss of mitochondrial membrane integrity and thus the release of cytochrome c into the cytosol with

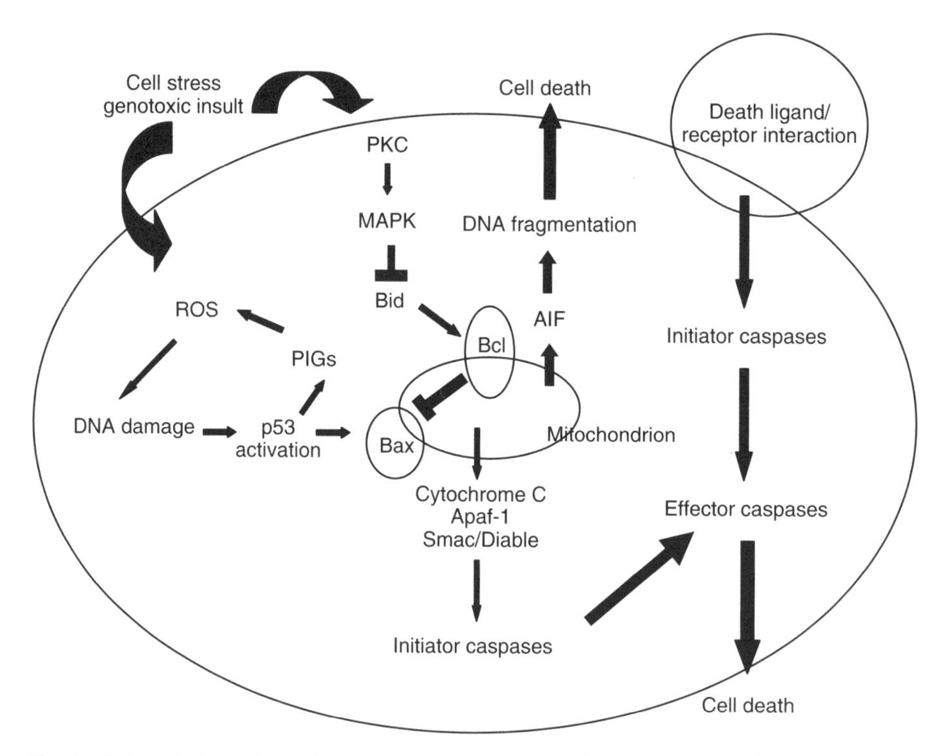

Fig. 16.3 Apoptotic pathways in mammals. Lethal stress initiates a number of possible mechanisms to effect cell death responses. Typically, these are p53 dependent but may be dependent on (cytochrome c mediated), or independent of (AIF mediated) caspase action.

consequent activation of the apoptosome. Cytochrome c release is regulated by the antagonist actions of the pro- and anti-apoptotic gene families, typified by Bax and Bcl-2 respectively. Of these, Bax is upregulated during p53-mediated damage responses, while Bcl-2 is repressed (review: Hengarter 2000).

ROS can trigger the expression of p53-induced genes (PIGs) and multiple signalling cascades, inclusive of PKC MAPKs PLCs, tyrosine kinases, ras and PLC in neurones (Yuan and Yanker 2000). Activation of signalling cascades is complex and may in part determine how the neurone assesses the nature, level, and response to the stress.

A further level of complexity is determined by the composition of the apoptosome. Mammals have at least 14 caspases, two of which appear critical to the functioning of the neuronal apoptosome, namely caspase 9 and caspase 3. Cytochrome c release from the mitochondrion, results in the formation of the cytochrome c-Apaf-1-caspase 9 complex. This induces caspase 9 activation and in turn, caspase 3 activation, resulting in apoptosis. The importance of these two caspases is emphasized by the fact that murine null mutants for either caspase 3 or caspase 9 display extensive neuronal defects as a consequence of improper regulation of neuronal apoptosis during development (Kuida *et al.* 1996, 1998). Significantly, Apaf-1 null mutants show a similar phenotype (Cecconi *et al.* 1998). We note that adult neurones may not necessarily use the same apoptosis system.

Neurones also employ a caspase independent form of apoptosis. Apoptosis-inducing factor (AIF) is a mitochondrial effector of apoptotic cell death. AIF is a flavoprotein of relative molecular mass 57,000, with homology to bacterial oxidoreductases (Susin *et al.* 1999). It normally resides in the inter-membrane space of the mitochondria. Induction of apoptosis results in AIF translocation to the nucleus, where it causes large-scale DNA fragmentation and cell death.

Critically for neurones, commitment to apoptosis is dependent on the nature and level of stress and the ability of the cell to respond to it. The latter will be influenced extensively by cell type, chronological age, and developmental status. Chronologically old cells will be less robust by virtue of cumulative oxidative insult. Cell type may be influential in this respect, as cells with higher metabolic requirements may have an enhanced level of oxidative damage products (but see 16.4.3 above). Differential sensitivity to stress during development, with changes in the degree of damage that triggers apoptosis, is also a well-documented feature of neurones (Pettmann and Henderson 1998; Romero *et al.* 2003).

16.7.2 **Necrosis**

Necrosis (Fig. 16.4) results from extreme adverse stimuli that precipitate significant changes in the cellular microenvironment (e.g. exposure to toxins, ischaemia) leading to cell death. This type of cell death is not as controlled as apoptosis, and does not give rise to apoptotic bodies within the cell. The extent of cell loss due to necrosis is typically large and does not exhibit cell specificity. Significantly, necrosis results in complete disintegration of the affected cells, release of the intracellular contents to the extracellular medium and an associated inflammatory response.

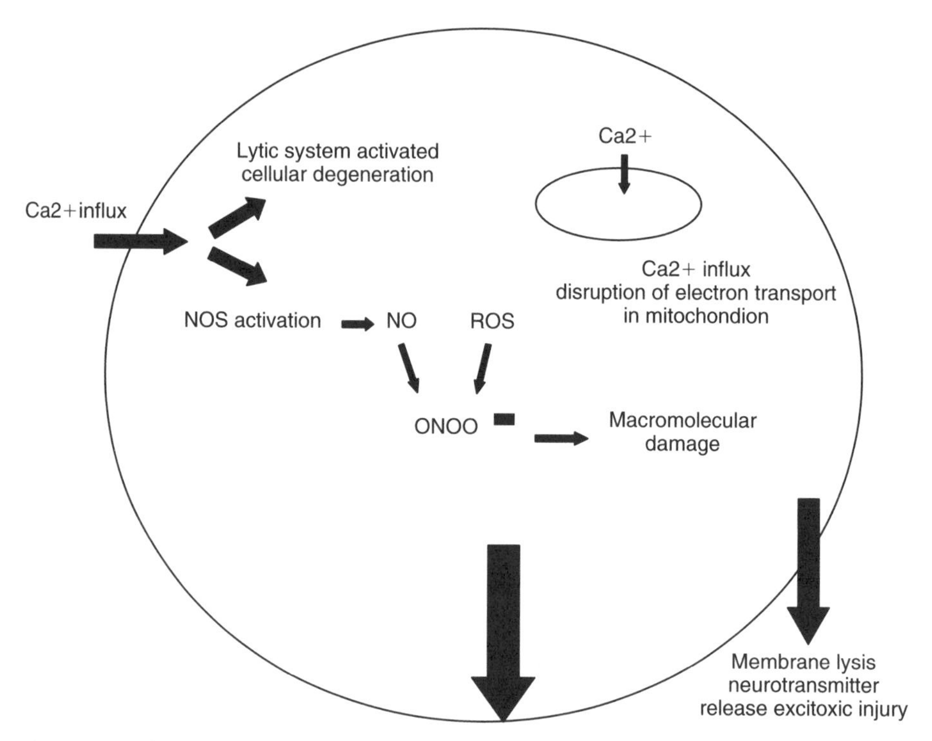

Fig. 16.4 Mechanisms of neuronal necrosis. Necrosis constitutes non-programmed cell death, typified by an inflammatory response and oedema resulting from mitochondrial damage, loss of cellular homeostasis and cell membrane lysis. The release of neurotransmitters, as a result of cell lysis, exacerbates excitotoxic injury in surrounding cells.

It is not known whether there is a link between the telomere DNA damage sensing system and necrosis. It seems likely that a strong stress or a combination of stresses will release the telomere bound-proteins very rapidly, with dramatic effects on the mitochondrion and cellular DNA.

Neuronal necrosis appears to be effected by an increase in cytosolic Ca^{2+}, which activates lytic enzymes, such as calpains, and influxes the mitochondrion, inhibiting the respiratory cycle and the energy production processes (Kim *et al.* 2003) Changes in the mitochondrial status associated with the excessive increase in mitochondrial Ca^{2+} leads to an increase in the production of ROS. Concomitant with these processes is the activation of nitric oxide synthase (NOS) which in turn leads to nitric oxide (NO) production. A consequence of this is that NO and ROS combine to produce peroxinitrite radicals. These are excessively toxic agents that mediate extensive cell damage. The necrosis system is shown diagrammatically in Fig. 16.4.

16.8 How might stasis contribute to pathogenesis?

Given the key actions of cell cycle regulatory proteins in classical damage response pathways and their role as effectors in neuronal death, it would be surprising to find neurones fixed in a terminally differentiated state. A more likely situation would be to find that neurones had the capacity to re-enter the cell cycle, albeit a capacity that did not allow completion of that cycle. Such a situation would facilitate dynamism at telomeric nucleo-protein complexes, and thus damage detection, assessment, and repair involving telomeric proteins.

Precedent for such a hypothesis can be found in a study of hippocampal neurones, which have been postulated to be arrested at cell cycle stage G1 (Nagy *et al.* 1998). Cerebral hypoxia has also been proposed as a trigger for neuronal cell cycle re-entry in Alzheimers disease (AD) and cerebro-vascular disease (Smith *et al.* 1999). In support of this assertion, nuclear cyclin B expression has been detected in these circumstances (Nagy 2000), indicative of the cell cycle G2/M transition point having been reached. A link between DNA damage and AD has previously been reported (Christen 2000), which is in keeping with reports indicating that patients with segmental progerias, due to mutations in genes involved in DNA repair such as WRN, develop AD-like brain pathology (Leverenz *et al.* 1998; Shiels 1999).

Aberrant cell cycle activation has also been implicated in neuronal cell death (Rideout *et al.* 2003). Indeed, proteosomal dysfunction, implicated in Lewy body diseases, can be abrogated through the use of cyclin-dependent kinase (cdk) inhibitors. The possibility remains, however, that rather than the cell cycle activation being aberrant, it is in fact a consequence of stress responses being effected.

A direct prediction of this would be that cytological stress markers would, in the absence of replicative senescence, accumulate independently of cell cycle stage. Furthermore, it may be possible to discriminate between stress responses dependent on the nature of the stress or its level. Sub-lethal stresses, for example, lead to stasis phenomena without loss of telomeric DNA. Lethal stress, resulting in apoptosis, will result in telomere loss and thus a detectable drop in mean terminal restriction fragment lengths (mTRFs). Precedent for this has been established in kidney allograft studies (Joosten *et al.* 2003; Shiels and Jardine 2003). A critical component of this hypothesis is the level of damage that is potentially communicated by the extent of cytochrome c release from stressed mitochondria. Previous observations have already indicated that a discrete level of cytochrome c in the cytosol is required to overcome anti-apoptotic checks and initiate death responses (Scorrano and Korsmeyer 2003). The possibility exists that the threshold level of cytosolic cytochrome c may be cell specific and tied to levels of anti-apoptotic factors in the cell, or the number of mitochondria in the cell. It is worth noting that this might be a factor in the apparent susceptibility of epithelial cells in the kidney to oxidant stress, as these cells have an unusually large mitochondrial complement. By extrapolation, neuronal cell types with high mitochondrial numbers, or low-level anti-apoptotic defences, might similarly show susceptibility to oxidant stress and thus preferentially contribute to pathogenesis. It will be interesting to determine if this is indeed a factor in disease such as Parkinson's.

How such activities might contribute to pathogenesis remains somewhat tenuous and requires experimental testing. An extrapolation to loss, or diminution, of cellular function due to stress, overlapping with cell loss at the level of the organ, is an intuitive link to disease susceptibility. Superimpose this upon genetic background and natural loss of function with chronological age and one has the basis for a progressive neurodegenerative disorder that shows increasing prevalence with age. Similarly, accelerated neurodegenerative conditions associated with segmental progerias affecting DNA repair, are in keeping with this hypothesis.

This scenario would also predict that dysfunction of other genes involved in DNA repair might result in neurodegeneration. In support of this, disruption of XRCC4, DNA ligase IV (Frank *et al.* 2000) and atm, both result in neurodegeneration or neurodevelopmental problems in the mouse (Klocke *et al.* 2002; Wong *et al.* 2003).

16.9 **Neuronal stem cells**

Why do neurodegenerative processes occur when there are stem cell populations present in the brain? Are there insufficient stem cells to effect repair, or do stem cells senesce *in vivo* with increasing chronological age?

Little direct experimental information is available to address these questions, though some data from the atm/terc double knock-out mouse are pertinent; this again involves DNA damage sensing, signalling, and repair machinery. Disruption of both atm and Terc leads to a reduction in the neural stem cell compartment in the adult brain and a decreased ability of the neural stem cells to proliferate (Wong *et al.* 2003). In addition there is an increased rate of apoptosis in such animals in conjunction with accelerated physiological and cellular senescence. These results clearly demonstrate that telomere maintenance is essential to preserve neural stem cell reserves, and indicates that these reserves are indeed subject to ageing processes.

Haemopoetic stem cells (HSCs) also show a limited replicative capacity in both mice and humans, despite being telomerase positive (Allsopp *et al.* 2003). Serial transplantation experiments in mice have demonstrated that sustained telomerase activity, even at elevated levels, is insufficient to maintain extended transplantation capacity. These data are indicative of telomere-independent mechanisms having a potential involvement in stem cell differentiation and subsequent expansion.

One possible explanation for these observations is that HSCs and other adult stem cells (ASCs) may exhibit inherent asymmetric cell kinetics; i.e. the products of a division are not equivalent, one being recognizable as a mother cell that is still a stem cell, the other as a daughter that is already differentiated. The end result of asymmetric cell kinetics in postnatal somatic tissues, is a limitation in replicative potential, hence senescence. Such a kinetic system will enable cell renewal in adult tissues and maintain mass. *In vitro* this may pose a barrier to the expansion of neuronal stem cells, through a dilution of ASCs with serial passaging. *In vivo*, this could imply that there is a limited number of ASCs with a limited expansion potential, as observed for HSCs. Thus with specific neurological insults, degenerative disease may occur in the brain despite the presence of stem cells. This is envisaged as being dependent on the size of the individual's stem cell pool, and the nature and level of insult (mutation/genotoxin) and must be factored in conjunction with increasing chronological age. Interestingly, genes involved in determining the rate of asymmetric cell kinetics include p53 and p21, which have a long-established role in the DNA damage responses and telomere-based senescence processes (Sherley 2002).

16.10 Major genetic triggers of neuronal death

Among the many ways that neuronal death can be caused, genetic defects are particularly instructive to study since they provide defined molecular starting points for the eventual neuronal death process. Moreover, certain mutations have been shown to cause particular patterns of neurodegeneration in man, which are recognized clinically as important diseases. The effects of all these mutations becomes increasingly severe with age, due both to long-term exposure of the cells to the effect of the mutant protein, and to interaction of these effects with the ongoing ageing process. Indeed, family history and age are the most important predictors of the probability of an individual coming to suffer from one of the major neurodegenerative diseases.

Many other mutations in man and experimental animals cause neurodegeneration (Jockusch and Schmitt-John, this volume Chapter 2; Bailey, this volume Chapter 15). Many of them, e.g. the mouse *weaver* mutation, are in ion channels, which is not very instructive as this is a relatively non-specific trigger of cell dysfunction. One interesting case is that a loss of function mutation in the brain-specific signal pathway enzyme PKCγ has been shown to be responsible for a parkinsonian phenotype involving degeneration of dopaminergic neurons in the rat (Craig *et al.* 2001). This is amenable to biochemical study since the molecular consequences of the mutation are defined. Interestingly, work in progress (Wright S, Shiels P, Davies RW, Payne AP, unpublished) indicates that this mutation may also alter cell senescence properties, providing another link between senescence and neurodegeneration.

Here we focus on the two major neurodegenerative disease of ageing, Alzheimer's disease and Parkinson's disease, and briefly consider polyglutamine diseases.

16.10.1 Alzheimer's disease

The most important neurodegenerative disease, in terms of prevalence, is Alzheimer's disease (AD). The onset of this disease is clearly age dependent, although different variants of AD have characteristic probability distributions of onset at particular ages. A great deal of evidence shows that there is likely to be a single primary mediator of toxicity in this disease, which is the beta-amyloid (Aβ) family of peptides. Aβ is derived by proteolytic processing from the amyloid precursor protein (APP), and is a normal component of neurons (Cherny *et al.* 1999). Known mutations that give rise to this disease include triplication of chromosome 21 (Down's Syndrome), mutations affecting proteolytic processing and mutations affecting APP levels (Bailey, this volume Chapter 15). All result in increased levels of Aβ or a proportional increase in the more toxic Aβ42 form. Even mutations in other genes which lead to AD have been shown to mediate their effects via modulation of Aβ production, e.g. presenilin 1 and 2 mutations which affect the γ-secretase APP processing enzyme. It is now widely accepted that all genetic causes that remain to be discovered are likely to act in one way or another via stimulation of abnormal Aβ production, localization or handling.

The mechanism of Aβ toxicity is not understood. Aβ is certainly toxic to cells in culture, with a range of effects, but non-physiologically high levels must be used. In transgenic mice expressing human familial AD mutant forms of APP in the brain, subtle changes in synaptic efficacy are detectable, which have been proposed to lead to stress. However, these transgenic mice do not get AD symptoms other than mild plaque formation, so something more dramatic must happen in the human brain. Aβ may modify protein processing (e.g. of cdk5), and this has been proposed to link to abnormal phosphorylation of tau and cytoskeletal damage. Aβ induces apoptosis via a caspase 12 dependent pathway in cultured cortical neurons (Nakagawa *et al.* 2000): this pathway is also induced by ER stress (Section 16.3). Particularly interesting is the evidence that Aβ generates cytotoxic H_2O_2 through the reduction of Cu^{2+} and Fe^{3+} using O_2 as the substrate (Huang *et al.* 1999*a*, *b*). Aβ also binds and is precipitated by Zn^{2+} (and Cu^{2+}, but there is ten times less of this in the brain extracellular fluid). Thus a current hypothesis (review: Bush and Tanzi 2002) is that Aβ is normally internalized and destroyed by neurones, but that any tardiness in this system is liable to give rise to local precipitation of Aβ by high concentrations (300 μM) of Zn^{2+} ions at or near synapses. Similarly, anything that elevates extracellular Zn^{2+} or Cu^{2+} levels, such as reductions in mitochondrial function affecting Zn^{2+} recycling, or malfunction of Cu^{2+}/Zn^{2+} binding proteins such as ApoE and α2-macroglobulin, would also exacerbate the situation. The amyloid plaques that thus form act as depots for release of the toxic form of Aβ (which would otherwise be cleared), which produces locally high levels of H_2O_2 and thus oxidative stress. This is supported by the marked decrease in Aβ deposition in the brains of mice lacking the synaptic ZnT3 transporter (Lee *et al.* 2002), with concomitant improvement in behavioural and general health parameters. Thus while Aβ may have other actions that contribute to cytotoxicity, it seems likely that one clear mode of toxic action by this peptide is to add to the local ROS free radical load wherever amyloid plaques form. This is therefore very congruent with general ageing processes, and the efficiency of cellular damage sensing and repair mechanisms will determine relative survival under these particularly stressful circumstances. One potential link between general ageing and Alzheimer's disease would seem to be any gradual reduction in mitochondrial function, as this will reduce Zn^{2+} recycling and enhance the risk of plaque formation in those who are susceptible.

16.10.2 **Parkinson's disease**

The second most frequent neurodegenerative disease associated with age is Parkinson's disease (PD). The state of knowledge concerning the molecular triggers of PD is less advanced than for Alzheimer's disease. However, a consensus is emerging that gives alpha synuclein toxicity the central role in at least a proportion of PD cases. Gain of function mutations (Ala30Pro, Ala53Thr) in this protein clearly precipitate the disease in an age-dependent manner, and contribute some cases of familial PD. Strikingly, when wild-type or human mutant alpha synuclein is expressed in all neurons of the Drosophila brain, preferential death of dopaminergic neurones is observed (Feany and Bender 2000), indicating that excess alpha synuclein can be a specific toxin for dopaminergic neurons in unrelated species, and strengthening the hypothesis that overproduction or malfunction of this protein is central to all forms of PD. While juvenile forms of Parkinson's disease can be caused by mutations in parkin, which seemed initially to be a component of a non-specific protein disposal system, evidence is increasing that parkin is responsible for controlling the ubiquitination of a subset of cellular proteins, among them alpha synuclein, thus tying them together. It has been reported that parkin binds alpha synuclein, and that this interaction is disrupted by parkin mutations, while parkin also protects catecholaminergic neurones in primary culture from toxic effects of alpha synuclein and proteasome inhibition (Petrucelli *et al.* 2002: review, Feany and Pallanck 2003). Parkin loss of function mutations do not show neurodegeneration in Drosophila, but do show apoptosis in muscle cells and early mitochondrial dysfunction (Greene *et al.* 2003), indicating that there is a link between proteasome system function and mitochondrial function.

Alpha synuclein is ubiquitously expressed in the brain, and is present in significant amounts in presynaptic axon termini. It has proved to be a difficult protein to assign a clear function to. It has some similarity to 14-3-3 proteins, and has been proposed to have some chaperone activity. Its best known property is a tendency to aggregate, and it has been shown that the familial mutant forms aggregate more rapidly and easily. The relevance of aggregation to toxicity is not established, but it is thought that it may result in the proteasome being unable to degrade the protein efficiently, leading to local accumulations and proteasome stress. Geldanamycin, which is an inhibitor of Hsp90, relieves alpha synuclein toxicity in Drosophila. Hsp90 is an inhibitor of other chaperones, and this result has been interpreted as indicating a need for chaperone function in the correct folding or clearance of alpha synuclein. Indeed, overexpression of the chaperone Hsp70 rescues *Drosophila* dopaminergic neurones from alpha synuclein toxicity (Auluck *et al.* 2002). In differentiated (neuronal) PC12 cells, mutant alpha synuclein expression has been shown to decrease proteasome activity without cell toxicity, and to increase the sensitivity to apoptotic cell death in the presence of sub-lethal levels of proteasome inhibitor (Tanaka *et al.* 2001).

Are there any links between alpha synuclein overexpression and oxidative stress? Alpha synuclein knock-out mice revealed one actual function of the protein: it acts to inhibit the refilling of the rapidly releasable pool of synaptic vesicles at nerve terminals (Abeliovich *et al.* 2000). A separate study of another alpha synuclein knock-out mouse strain (Cabin *et al.* 2002) indicated that alpha synuclein is normally required for the genesis, localization and/or maintenance of some subset of vesicles that make up the reserve or resting pools of presynaptic vesicles. There is evidence that it is a regulator of phospholipase D2 activity (Jenco *et al.* 1998; Ahn *et al.* 2002), which is a key component in signalling for vesicle production at membranes. Another interesting result is that alpha synuclein KO mice are resistant to MPTP (MPP+) toxicity (Dauer *et al.* 2002). MPP+ kills dopaminergic neurones by inhibiting complex 1 of mitochondria. The experimental evidence indicates that resistance in the absence of alpha synuclein may be due to increased vesicular uptake and storage of MPP+, so that by inference alpha synuclein would inhibit dopamine storage; direct action of alpha synuclein at the mitochondrion may also occur. From these and other data it has been hypothesized (Lotharius and Brundin 2002) that mutations in alpha synuclein might lead to impaired dopamine storage in vesicles, which would increase the concentration of cytoplasmic dopamine, thus in turn increasing the oxidative stress on the neurone due to the increased risk of toxic dopamine *o*-hydroquinone radicals. Moreover, mutant alpha synuclein (and very high levels of the wild-type protein) has been shown to enhance dopamine toxicity and increase H_2O_2 induced nitrite production in neuroblastoma cells (Wersinger and Sidhu 2003). It should be noted that normal low levels of alpha synuclein have been reported several times to be neuroprotective, and human alpha synuclein expressed from the tyrosine hydroxylase promoter in transgenic mice confers protection against paraquat toxicity (Manning-Bog *et al.* 2003). Recently it has been shown that oligomerization of alpha synuclein is dependent on oxidation of tyrosine by free radicals, and that oligomerization of mutant alpha synuclein is more sensitive to free radical oxidation (Kang and Kim 2003; Krishnan *et al.* 2003; Norris *et al.* 2003; Yamin *et al.* 2003). To the extent that oligomerization and fibrillogenesis are important to toxicity, this would link ageing to alpha synuclein specific effects.

16.10.3 **Polyglutamine diseases**

Many rarer neurodegenerative diseases of ageing are due to the incorporation of polyglutamine repeats within proteins due to CAG triplet repeat expansion (Bailey, this volume Chapter 15). It is clear that the polyglutamine causes toxicity, not the host protein, but little is known about the mechanism. As for alpha synuclein, polyglutamine neurotoxicity can be modelled in *Drosophila* (Kastemi-Esfaranjani and Benzer 2000). Alpha synuclein and polyglutamine toxicity also have in common that they are ameliorated by parkin (Tsai *et al.* 2003). Proteasome inhibition and resultant protein stress, interference with transcription, protein sequestration (e.g. of normal binding partners of the host protein) into aggregates, and mitochondrial membrane insertion causing an electron leak all have some experimental support. It is particularly interesting that the eventual mode of cell death caused by polyglutamine is not standard apoptosis, since TUNEL staining shows no evidence of DNA fragmentation. The term 'black cell death' is used for this form of cell death that is neither apoptosis nor necrosis. It is possible that this state corresponds to stasis as defined above.

16.11 **Conclusion**

Studying neuronal senescence offers a new perspective on how neurones die. It allows the integration of cellular damage responses with specific neuropathological stresses, and also allows neuropathological processes to be considered in the contexts of organ senescence and chronological ageing. Most importantly, it promises to open up new avenues for therapeutic intervention.

At this time, we do not have much directly acquired knowledge about the molecular mechanisms of senescence in neurones. However, the indications are that at least a subset of the knowledge being acquired from experimental studies on other cell types will be applicable in neurones. The recent data on the atm–/– terc–/– mouse (Section 16.5) shows very clearly that both adult neurones and neuronal stem cells must maintain telomeres, and must sense DNA damage and respond to it correctly using the ATM system. The occurrence of neurodegeneration in XRCC5 knock-out mice emphasizes the importance of DNA repair for neurones. Since terc–/– mice are mutated in the RNA component of telomerase, this enzyme must be involved in telomere maintenance in mice. In man, we know from the progeric diseases, which are due to mutations in a RecQ-like helicase, that recombinational mechanisms play a part at least in telomere maintenance in human neurones. It should be noted that different types of neurones may vary considerably in the efficacy of their DNA damage sensing and repair systems. Evidence for variation in these systems comes from differential amplification of (CAG)n repeats in different neurones of mouse and human brains with increasing age, which is probably due to differences in DNA repair systems (P. Shelbourne, pers.comm.).

In the senescence context, the genotoxic (and environmental) factors that precipitate specific variants of neurodegenerative disease can be viewed as adding extra stresses for the neuron's damage control and response system to deal with. We are beginning to get closer to understanding the molecular mechanisms of these stress factors in some major diseases (Section 16.10) but we have not yet achieved integration with the mechanisms of neuronal senescence. In order to treat these diseases, we must address two questions: how to maintain adult neurones in a functional state for as long as possible despite genotoxic stress, and how to stimulate or enable the replacement of lost neurones from stem cells. It is currently not clear how important adult neuronal stem cells are for most regions of the brain: we do not know the stem cell pool size, nor how much replacement occurs, nor how fast this occurs, nor how the stem cell compartment responds to neuronal dysfunction and loss. Another problem is senescence of stem cells themselves.

In vivo data show that stem cells do undergo senescence, and this may be a factor in the progression of neurodegenerative diseases in adults. This will also apply to stem cells that are introduced artificially in cell replacement therapeutic strategies. Therefore, senescence is a barrier to be overcome for stem cell therapy, although this may sound heretical for a multi-potent cell type that can be propagated indefinitely in cell culture. It should be noted that standard rodent cells can be propagated indefinitely in cell culture because they are telomerase positive, but typically end up in STASIS due to culture shock. With stem cells, p53-dependent asymmetric cell kinetics may also influence this process, again tying together cellular stress responses and proliferation.

Thus senescence needs to be taken into account in any cellular replacement approach to therapy. Expansion of cells *in vitro* to reach requisite numbers for transplantation will, by virtue of culture stress, result in the accumulation of damage and result in premature senescence. This remains a problem for neural allografts and xenografts, and even the use of nuclear transplantation does not escape this problem.

The ability to isolate and propagate neural stem cells and to develop neurones from other stem cell sources does, however, offer substantial hope for the treatment of neurodegeneration. Critically, these cells will also provide an important resource for the testing and development of novel chemical and biological therapeutics. A better understanding of neuronal and stem cell damage responses and associated telomere biology will be crucial in formulating these therapeutic strategies. This knowledge is also fundamental to understanding the neuronal ageing process and its interaction with the molecules that trigger major neurodegenerative diseases.

References

Abeliovich, A., Schmitz, Y., Farinas, I., Choi-Lundberg, D., Ho, W. H., Castillo, P. E., *et al.* (2000) Mice lacking α-synuclein display functional deficits in the nigrostriatal dopamine system. *Neuron*, **25**, 239–52.

Ahn, B. H., Rhim, H., Kim, S. Y., Sung, Y. M., Le, M. Y., Choi, J. Y., *et al.* (2002) Alpha-synuclein interacts with phospholipase D enzymes and inhibits pervanadate induced phospholipase D activation in human embryonic kidney 293 cells. *J. Biol. Chem.*, **277**, 12334–42.

Allsopp, R. C., Vaziri, H., Patterson, C., Futcher, A. B., Greider, C. W., and Harley, C. B. (1992) Telomere length predicts replicative capacity of human fibroblasts. *Proc. Natl. Acad. Sci. USA*, **89**, 10114–18.

Allsopp, R. C., Morin, G. B., Horner, J. W., DePinho, R., Harley, C. B., and Weissman, I. L. (2003) Effect of TERT over-expression on the long-term transplantation capacity of hematopoietic stem cells. *Nat. Med.*, **9**, 369–71.

Auluck, P. K., Chan, H. Y., Trojanowski, J. O., Lee, V. M., and Bonini, N. M. (2002) Chaperone suppression of alpha-synuclein toxicity in a Drosophila model of Parkinson's disease. *Science*, **295**, 865–8.

Bahr, B. A. and Bendiske, J. (2002) The neuropathic contribution of lysosomal dysfunction. *J. Neurochem.*, **83**, 481–9.

Beckman, J. S. (1994) Peroxynitrite versus hydroxyl radical: the role of nitric oxide in superoxide-dependent cerebral injury. In *The Neurobiology of NO and OH* (Church, C.C. and Colton, C. A., ed.). New York Academy of Sciences, New York, New York, USA, pp. 69–75.

Benn, S. C., Perrelet, D., Kato, A. C., Scholz, J., Decosterd, I., Mannion, R. J., *et al.* (2002) Hsp27 upregulation and phosphorylation is required for injured sensory and motor neuron survival. *Neuron*, **36**, 45–56.

Blasco, M. A., Lee H.-W., Hande, P., Samper, E., Lansdorp, P. M., DePinho, R. A., *et al.* (1997) Telomere shortening and tumor formation by mouse cells lacking telomerase RNA. *Cell*, **91**, 25–34.

Brunk, U. T. and Terman, A. (2002) The mitochondrial-lysosomal axis theory of ageing: accumulation of damaged mitochondria as a result of imperfect autophagocytosis. *Eur. J. Biochem.*, **269**, 1996–2002.

Bush, A. I. and Tanzi, R. E. (2002) The galvanization of β–amyloid in Alzheimer's disease. *Proc. Natl. Acad. Sci. USA*, **99**, 7317–19.

Cabin, D. E., Shimazu, K., Murphy, D., Cole, N. B., Gottschalk, W., McIlwain, K. L. *et al.* (2002) Synaptic vesicle depletion correlates with attenuated synaptic responses to prolonged repetitive stimulation in mice lacking α-synuclein. *J. Neurosci.*, **22**, 8797–807.

Calabrese, V., Scapagnini, G., Giuffrida Stella, A. M., Bates, T. E., and Clark, J. B. (2001) Mitochondrial involvement in brain function and dysfunction: relevance to ageing, neurodegenerative disorders and longevity. *Neurochem. Res.*, **26**, 739–64.

Cecconi, F., Alvarez-Bolado, G., Meyer, B. I., Roth, K. A., and Gruss, P. (1998) Apafl CED-4 homolog regulates programmed cell death in mammalian development. *Cell*, **94**, 727–37.

Cheng F-C, Kuo J-S., Chia L-G., and Dryhurst, G. (1996) Elevated 5-S-cysteinyldopamine and reduced homovanillic acid in cerebrospinal fluid: possible markers for and potential insights into the pathoetiology of Parkinson's Disease. *J. Neurol. Transm.*, **103**, 433–46.

Christen, Y. (2000) Oxidative stress and Alzheimer's disease. *Am. J. Clin. Nutr.*, **71**, 621S–629S.

Clarke, G., Collins, R. A., Leavitt, B. R., Andrews, D. F., Hayden, M. R., Lumsden, C. J., *et al.* (2000) A one-hit model of cell death in inherited neuronal degenerations. *Nature*, **406**, 195–9.

Coyle, J. T. and Puttfarken, P. (1993) Oxidative stress, glutamate, and neurodegenerative disorders. *Science*, **262**, 689–95.

Craig, N. J., Duran Alonso, M. B., Hawker, K. L., Shiels, P., Glencorse, T. A., Campbell, J. M., *et al.* (2001) A candidate gene for human neurodegenerative disorders: a rat PKC gamma mutation causes a Parkinsonian syndrome. *Nat. Neurosci.*, **4**, 1061–2.

Dauer, W., Kholodilov, N., Vila, M., Trillat A.-C, Goodchild, R., Larsen, K. E., *et al.* (2002) Resistance of α-synuclein null mice to the parkinsonian neurotoxin MPTP. *Proc. Natl. Acad. Sci. USA*, **99**, 14524–9.

Davis, R. E., Miller, S. W., Herrnstadt, C., Ghosh, S. S., Fahy, E., Shinobu, L. A., *et al.* (1997) Mutations in mitochondrial cytochrome c oxidase genes segregate with late-onset Alzheimer's disease. *Proc. Natl. Acad. Sci. USA*, **94**, 4526–31.

de Grey, A. (2002) The reductive hotspot hypothesis of mammalian ageing: membrane metabolism magnifies mutant mitochondrial mischief. *Eur. J. Biochem.*, **269**, 2003–9.

Esposito, L. A., Melov, S., Panov, A., Cottrell, B. A., and Wallace, D. C. (1999) Mitochondrial disease in mouse results in increased oxidative stress. *Proc. Natl. Acad. Sci. USA*, **96**, 4820–5.

Feany, M. B. and Bendes, W. W. (2000) A Drosophilia model of Parkinson's Disease. *Hum. Mol. Genet.*, **12**, 2457–66.

Feany, M. B. and Pallanck, L. J. (2003) Parkin: a multipurpose neuroprotective agent? *Neuron*, **38**, 13–16.

Frank, K. M., Sharpless, N. E., Gao, Y., Sekiguchi, J. M., Ferguson, D. O., Zhu, C., *et al.* (2000) DNA ligase, I. deficiency in mice leads to defective neurogenesis and embryonic lethality via the p53 pathway. *Mol. Cell*, **5**, 993–1002.

Fukino, K., Suzuki, T., Saito, Y., Shindo, T., Amaki, T., Kurabayashi, M., *et al.* (2002) Regulation of angiogenesis by the ageing suppressor gene klotho. *Biochem. Biophys. Res. Commun.*, **293**, 332–7.

Gangloff, S., Soustelle, C., and Fabre, F. (2000) Homologous recombination is responsible for cell death in the absence of the Sgsl and Srs2 helicases. *Nat. Genet.*, **25**, 192–4.

Gompertz, B. (1825) On the nature of the function expressive of the law of human mortality, and on the mode of determining life contingencies. *Phil. Trans. Roy. Soc.*, **115**, 513–85.

Greene, J. C., Whitworth, A. J., Kuo, I., Andrews, C. A., Feany, M. B., and Pallanck, L. J. (2003) Mitochondrial pathology and apoptotic muscle degeneration in Drosophilia Parbin mutants. *Proc. Natl. Acad. Sci. USA*, **100**, 4078–83.

Grünewald, R. A. (1993) Ascorbic acid in the brain. *Brain Res. Rev.*, **18**, 123–33.

Halliwell, B. (1999) Free Radicals in Biology and Medicine. 3rd edition. Oxford University Press, New York.

Halloran, P. F., Melk, A., and Barth, C. (1999) Rethinking chronic allograft nephropathy: the concept of accelerated senescence. *J. Am. Soc. Nephrol.*, **10**, 167–81.

Harley, C. B., Futcher, A. B., and Greider, C. W. (1990) Telomeres shorten during ageing of human fibroblasts. *Nature*, **345**, 458–60.

Harman, D. (1972) The biologic clock: the mitochondria? *J. Am. Geriat. Soc.*, **20**, 145–7.

Hayakawa, M., Hattori, K., Sugiyama, S., and Ozawa, T. (1992) Age-associated oxygen damage and mutations in mitochondrial DNA in human hearts. *Biochem. Biophys. Res. Commun.*, **189**, 979–85.

Hayakawa, M., Sugiyama, S., Hattori, K., Takasawa, M., and Ozawa, T. (1993) Age-associated damage in mitochondrial DNA in human hearts. *Mol. Cell. Biochem.*, **119**, 95–110.

Hengarter, M. O. (2000) The biochemistry of apoptosis. *Nature*, **407**, 770–6.

Huang, X., Atwood, C. S., Hartshorn, M. A., Multhaup, G., Goldstein, L. E., Scarpa, R. C., *et al.* (1999*a*) The A beta peptide of Alzheimer's disease directly produces hydrogen peroxide through metal ion reduction. *Biochemistry*, **38**, 7609–16.

Huang, X., Cuajungeo, M. P., Atwood, C. S., Hartshorn, M. A., Tyndall, J., Hanson, G. R., *et al.* (1999*b*) Cu(II). potentiation of Alzheimer abeta neurotoxicity: correlation with cell free hydrogen peroxide production and metal reduction. *J. Biol. Chem.*, **274**, 37111–16.

Ischiropoulos H. and Beckman, J. S. (2003) Oxidative stress and nitration in neurodegeneration: cause, effect, or association? *J. Clin. Invest.*, **111**, 163–9.

Jenco, J. M., Rawlingson, A., Daniels, B., and Morris, A. J. (1998) Regulation of phospholipase D2: selective inhibition of mammalian phospholipase D isozymes by α- and β-synucleins. *Biochemistry*, **37**, 4901–409.

Joosten, S. A., van Ham, V., Nolan, C. E., Borrias, M. C., Jardine, A. G., Shiels, P. G., *et al.* (2003) Telomere shortening and cellular senescence in a model of chronic renal allograft rejection. *Am. J. Pathol.*, **162**, 1305–12.

Kang, J. H. and Kim, K. S. (2003) Enhanced oligomerization of the alpha-synuclein mutant by the Cu,Zn superoxide dismutase and hydrogen peroxide system. *Mol. Cells*, **15**, 87–93.

Karanjawala, Z. E., Murphy, N., Hinton, D. R., Hsieh, C. L., and Lieber, M. R. (2002) Oxygen metabolism causes chromosome breaks and is associated with the neuronal apoptosis observed in DNA double-strand break repair mutants. *Curr. Biol.*, **12**, 397–402.

Karlseder, J., Smogorzewska, A., and de Lange, T. (2002) Senescence induced by altered telomere state, not telomere loss. *Science*, **295**, 2446–9.

Kastan, M. B. and Lim, D. S. (2000) The many substrates and functions of ATM. *Nat. Rev. Mol. Cell Biol.*, **1**, 179–86.

Kastemi-Esfaranjani, P. and Benzer, S. (2000) Genetic suppression of polyglutamine toxicity in Drosophila. *Science*, **287**, 1837–40.

Kaufman, R. (1999) Stress signalling from the lumen of the endoplasmic reticulum: co-ordination of gene transcriptional and translational controls. *Genes Dev.*, **13**, 1211–33.

Khanna, K. K. and Jackson, S. P. (2001) DNA Double-strand breaks: signaling, repair and the cancer connection. *Nat. Genet.*, **27**, 247–54.

Kim J-S., He, L., and Lemasters, J. J. (2003) Mitochondrial permeability transition: a common pathway to necrosis and apoptosis. *Biochim. Biophys. Res. Commun.*, **304**, 463–70.

Kim, T.-W. and Tanzi, R. E. (1998) Neuronal intranuclear inclusions in polyglutamine diseases: nuclear weapons or nuclear fallout? *Neuron*, **21**, 657–59.

Klement, I. A., Skinner, P. J., Kaytor, M. D., Yi, H., Hersch, S. M., Clark, H. B., *et al.* (1998) Ataxin-1 nuclear localisation and aggregation in polyglutamine-induced disease in SCA1 transgenic mice. *Cell*, **95**, 41–53.

Klocke, B. J., Latham, C. B., D'Sa, C., and Roth, K. A. (2002) p53 deficiency fails to prevent increased programmed cell death in the Bcl-X(L)-deficient nervous system. *Cell Death Differ.*, **9**, 1063–8.

Krishnan, S., Chi, E. Y., Wood, S. J., Kendrick, B. S., Li, C., Garzon-Rodriguez, W., *et al.* (2003) Oxidative dimer formation is the critical rate-limiting step for Parkinson's disease alpha-synuclein fibrillogenesis. *Biochemistry*, **42**, 829–37.

Kuida, K, Zheng, T. S., Na, S., Kuan, C., Yang, D., Karasuyama, H., *et al.* (1996) Decreased apoptosis in the brain and premature lethality in CPP32-deficient mice. *Nature*, **384**, 368–72.

Kuida, K, Haydar, T. F., Kuan, C. Y., Gu, Y., Taya, C., Karasuyama, H., *et al.* (1998) Reduced apoptosis and cytochrome c-mediated caspase activation in mice lacking caspase 9. *Cell*, **94**, 325–37.

Langley, E., Pearson, M., Faretta, M., Bauer, U. M., Frye, R. A., Minucci, S., *et al.* (2002) Human SIR2 deacetylates p53 and antagonizes PML/p53-induced cellular senescence. *EMBO J.*, **21**, 2383–96.

Lee, C. K., Klopp, R. G., Weindruch, R., and Prolla, T. A. (1999) Gene expression profile of ageing and its retardation by caloric restriction. *Science*, **285**, 1390–3.

Lee J.-Y, Cole, T. B., Palmiter, R. D., Suh, S. W., and Koh, J.-Y. (2002) Contribution by synaptic zinc to the gender-disparate plaque formation in human Swedish mutant APP transgenic mice. *Proc. Natl. Acad. Sci. USA*, **99**, 7705–10.

Leverenz, J. B., Yu, C. E., and Schellenberg, G. D. (1998) Ageing-associated neuropathology in Werner syndrome. *Acta Neuropathol. (Berl.)*, **96**, 421–4.

Lin, S. J., Defossez, P. A., and Guarente, L. (2000) Requirement of NAD and SIR2 for life-span extension by calorie restriction in *Saccharomyces cerevisiae*. *Science*, **289**, 2126–8.

Lin, S. J., Kaeberlein, M., Andalis, A. A., Sturtz, L. A., Defossez, P. A., Culotta, V. C., *et al.* (2002) Calorie restriction extends *Saccharomyces cerevisiae* lifespan by increasing respiration. *Nature*, **418**, 344–8.

Linnane, A. W., Baumer, A., Maxwell, R. J., Preston, H., Zhang, C. F., and Marzuki, S. (1990) Mitochondrial gene mutation: the ageing process and degenerative diseases. *Biochem. Int.*, **22**, 1067–76.

Linnane, A. W., Marzuki, S., Ozawa, T., and Tanaka, M. (1989) Mitochondrial DNA mutations as an important contributor to ageing and degenerative diseases. *Lancet*, **1**, 642–5.

Linskens, M. H., Feng, J., Andrews, W. H., Enlow, B. E., Saati, S. M., Tonkin, L. A., *et al.* (1995) Cataloging altered gene expression in young and senescent cells using enhanced differential display. *Nucleic Acids Res*, **23**, 3244–51.

Lotharius, J. and Brundin, P. (2002) Impaired dopamine storage resulting from alpha-synuclein mutations may contribute to the pathogenesis of Parkinson's disease. *Hum. Mol. Genet.*, **11**, 2395–407.

Lumpkin Jr, C., McGill, J. R., Riabowol, K. T., Moerman, E. J., Reis, R. J., and Goldstein, S. (1985) Extrachromosomal circular DNA and ageing cells. *Adv. Exp. Med. Biol.*, **190**, 479–93.

Luo, J., Nikolaev, A. Y., Imai, S., Chen, D., Su, F., Shiloh, A., *et al.* (2001) Negative control of p53 by Sir2alpha promotes cell survival under stress. *Cell*, **107**, 137–48.

Manning-Bog, A. B., McCormack, A. L., Purisai, M. G., Bolin, A. M., and Di Monte, D. A. (2003) Alpha-synuclein overexpression protects against paraquat-induced neurodegeneration. *J. Neurosci.*, **23**, 3095–9.

Marechal, V., Elenbaas, B., Piette, J., Nicolas, J. C., and Levine, A. J. (1994) The ribosomal L5 protein is associated with mdm-2 and mdm-2-p53 complexes. *Mol. Cell Biol.*, **14**, 7414–20.

McNaught, K. and Olanow, C. W. (2003) Proteolytic stress: a unifying concept for the etiopathogenesis of Parkinson's Disease. *Ann. Neurol.*, 53 Suppl. **3**, 73–84.

Menzies, R. A. and Gold, P. H. (1971) The turnover of mitochondria in a variety of tissues in young adult and aged rats. *J. Biol. Chem.*, **246**, 2425–9.

Moyzis, R. K., Buckingham, J. M., Cram, L. S., Dani, M., Deaven, L. L., Jones, M. D., *et al.* (1988) A highly conserved repetitive DNA sequence, (TTAGGG).n, present at the telomeres of human chromosomes. *Proc. Natl. Acad. Sci. USA*, **85**, 6622–6.

Nagy, Z., Esiri, M. M., and Smith, A. D. (1998) The cell division cycle and the pathophysiology of Alzheimer's disease. *Neuroscience*, **87**, 731–9.

Nagy, Z. (2000) Cell cycle regulatory failure in neurones: causes and consequences. *Neurobiol. Aging*, **21**, 761–9.

Nakagawa, T., Zhu, H., Morishima, N., Li, E., Xu, S., Yankner, B. A., *et al.* (2000) Caspase 12 mediates endoplasmic-reticulum-specific apoptosis and cytotoxicity by amyloid-beta. *Nature*, **403**, 98–103.

Nakamura, T. M., Moser, B. A., and Russell, P. (2002) Telomere binding of checkpoint sensor and DNA repair proteins contributes to maintenance of functional fission yeast telomeres. *Genetics*, **161**, 1437–52.

Norris, E. H., Giasson, B. I., Ischiropoulos, H., and Lee, V. M. (2003) Effects of oxidative and nitrative challenges on alpha-synuclein fibrillogenesis involve mechanisms of protein modification. *J. Biol. Chem.*, May 6 2003 ahead of press.

Nugent, C. I., Bosco, G., Ross, L. O., Evans, S. K., Salinger, A. P., Moore, J. K., *et al.* (1998) Telomere maintenance is dependent on activities required for end repair of double-strand breaks. *Curr. Biol.*, **8**, 657–60.

Oexle, K. and Zwirner A. (1997) Advanced telomere shortening in respiratory chain disorders. *Hum. Mol. Genet.*, **6**, 905–8.

Ogonuki, N., Inoue, K., Yamamoto, Y., Noguchi, Y., Tanemura, K., Suzuki, O., *et al.* (2002) Early death of mice cloned from somatic cells. *Nat. Genet.*, **30**, 253–4.

Onyango, P., Celic, I., McCaffery, J. M., Boeke, J. D., and Feinberg, A. P. (2002) SIRT3, a human SIR2 homologue, is an NAD-dependent deacetylase localized to mitochondria. *Proc. Natl. Acad. Sci. USA*, **99**, 13653–8.

Ozawa, T., Sahashi, K., Nakase, Y., and Chance, B. (1995) Extensive tissue oxygenation associated with mitochondrial DNA mutations. *Biochem. Biophys. Res. Commun.*, **213**, 432–38.

Park, S. U., Ferrer, J. V., Javitch, J. A., and Kuhn, D. M. (2002) Peroxynitrite inactivates the human dopamine transporter by modification of cysteine 342: potential mechanism of neurotoxicity in dopamine neurones. *J. Neurosci.*, **22**, 4399–405.

Paul, L.C. (1999) Chronic allograft nephropathy: an update. *Kid. Int.*, **56**,783–93.

Petrucelli, L., O'Farrell, C., Lockhart, P. J., Baptista, M., Kehoe, K., Vink, L., *et al.* (2002) Parkin protects against the toxicity associated with mutant alpha-synuclein: proteasome dysfunction selectively affects catecholaminergic neurones. *Neuron*, **36**, 1007–19.

Pettmann, B. and Henderson, C. E. (1998) Neuronal cell death. *Neuron*, **20**, 633–47.

Rideout, H. J., Wang, Q., Park, D. S., and Stefanis, L. (2003) Cyclin-dependent kinase activity is required for apoptotic death but not inclusion formation in cortical neurones after proteasomal inhibition. *J. Neurosci.*, **23**, 1237–45.

Romero, A. A., Gross, S. R., Cheng, K. Y., Goldsmith, N. K., and Geller, H. M. (2003) An age-related increase in resistance to DNA damage-induced apoptotic cell death is associated with development of DNA repair mechanisms. *J. Neurochem.*, **84**, 1275–87.

Samper, E., Flores, J. M., and Blasco, M. A. (2001) Mammalian Ku86 protein prevents telomeric fusions independently of the length of TTAGGG repeats and the G-strand overhang. *EMBO Rep.*, **2**, 800–7.

Saudou, F., Finkbeiner, S., Devys, D., and Greenberg, M. E. (1998) Huntingtin acts in the nucleus to induce apoptosis but death does not correlate with the formation of intranuclear inclusions. *Cell*, **95**, 55–66.

Scorrano, L. and Korsmeyer, S. J. (2003) Mechanisms of cytochrome c release by proapoptotic Bcl-2 family members. *Biochem. Biophys. Res. Comm.*, **304**, 437–44.

Shay, J. W. and Wright, W. E. (2000) Hayflick, his limit, and cellular ageing. *Nat. Rev. Mol. Cell Biol.*, **1**, 72–6.

Sherley, J. L. (2002) Asymmetric cell kinetics genes: the key to expansion of adult stem cells in culture. *Scientific World Journal*, **2**, 1906–21.

Shiels, P. G. (1999) Somatic cell nuclear transfer as a tool for investigating ageing processes. *Mammal Gene Ther. Mol. Biol.*, **4**, 11–22.

Shiels, P. and Jardine, A. G. (2003) Dolly, no longer the exception: telomeres and implications for transplantation. *Cloning and Stem Cells*, **5**, 149–52.

Shiels, P. G., Kind, A. J., Campbell, K. H., Waddington, D., Wilmut, I., Colman, A., *et al.* (1999*a*) Analysis of telomere lengths in cloned sheep. *Nature*, **399**, 316–17.

Shiels, P. G., Kind, A. J., Campbell, K. H., Waddington, D., Wilmut, I., Colman, A., *et al.* (1999*b*) Analysis of telomere length in Dolly, a sheep derived by nuclear transfer. *Cloning*, **1**, 1–7.

Sinclair, D. A., Mills, K., and Guarente, L. (1997) Accelerated ageing and nucleolar fragmentation in yeast sgs 1 mutants. *Science*, **277**, 1313–16.

Smith, M. Z., Nagy, Z., and Esiri, M. M. (1999) Cell cycle-related protein expression in vascular dementia and Alzheimer's disease. *Neurosci. Lett.*, **271**, 45–8.

Smythies, J. R. and Galzigna, L. (1998) The oxidative metabolism of catecholamines in the brain: a review. *Biochim. Biophys. Acta*, **1380**, 159–62.

Susin, S. A., Lorenzo, H. K., Al Zamzami, N., Marzo, I., Snow, B. E., Brothers, G. M., *et al.* (1999) Molecular characterization of mitochondrial apoptosis-inducing factor. *Nature*, **397**, 441–6.

Stewart, G. S., Wang, B., Bignell, C. R., Taylor, A. M., and Elledge, S. J. (2003) MDC1 is a mediator of the mammalian DNA damage checkpoint. *Nature*, **421**, 961–6.

Swerdlow, R. H., Parks, J. K., Miller, S. W., Tuttle, J. B., Trimmer, P. A., Sheehan, J. P., *et al.* (1996) Origin and functional consequences of the complex I defect in Parkinson's disease. *Ann. Neurol.*, **40**, 663–71.

Takasawa, M., Hayakawa, M., Sugiyama, S., Hattori, K., Ito, T., and Ozawa, T. (1993) Age-associated damage in mitochondrial function in rat hearts. *Exp. Gerontol.*, **28**, 269–80.

Tanaka, Y., Engelender, S., Igarashi, S., Rao, R. K., Wanner, T., Tanzi, R. E., *et al.* (2001) Inducible expression of mutant alpha-synuclein decreases proteasome activity and increases sensitivity to mitochondria-dependent apoptosis. *Hum. Mol. Genet.*, **10**, 919–26.

Tissenbaum, H. A. and Guarente, L. (2001) Increased dosage of a sir-2 gene extends lifespan in *Caenorhabditis elegans*. *Nature*, **410**, 227–30.

Tsai, Y. C., Fishman, P. S., Thakor, N. V., and Oyler, G. A. (2003) Parkin facilitates the elimination of expanded polyglutamine proteins and leads to preservation of proteasome function. *J. Biol. Chem.*, **278**, 22044–55.

Urano, F., Wang, X., Bertolotti, A., Zhang, Y., Chung, P., Harding, H. P., *et al.* (2000) Coupling of stress in the ER to activation of JNK protein kinases by transmembrane protein kinases IRE1. *Science*, **287**, 664–66.

Vaziri, H., Dessain, S. K., Ng Eaton, E., Imai, S. I., Frye, R. A., Pandita, T. K., *et al.* (2001) hSIR2(SIRT1). functions as an NAD-dependent p53 deacetylase. *Cell*, **107**, 149–59.

Wersinger, C. and Sidhu, A. (2003) Differential cytotoxicity of dopamine and H2O2 in a human neuroblastoma divided cell line transfected with alpha-synuclein and its familial Parkinson's diseases-linked mutants. *Neurosci. Lett.*, **342**, 124–28.

Wong, K. K., Maser, R. S., Bachoo, R. M., Menon, J., Carrasco, D. R., Gu, Y., *et al.* (2003) Telomere dysfunction and Atm deficiency compromises organ homeostasis and accelerates ageing. *Nature*, **421**, 643–8.

Wright, W. E. and Shay, J. W. (2002) Historical claims and current interpretations of replicative ageing. *Nat. Biotechnol.*, **20**, 682–8.

Yamin, G., Glaser, C. B., Uversky, V. N., and Fink, A. L. (2003) Certain metals trigger fibrillation of methionine-oxidized alpha-synuclein. *J. Biol. Chem.,* May 16 ahead of press.

Yuan, J. and Yanker, B. A. (2000) Apoptosis in the nervous system. *Nature*, **407**, 802–9.

Zhu, X. D., Kuster, B., Mann, M., Petrini, J. H., and Lange, T. (2000) Cell-cycle-regulated association of RAD50/MRE11/NBS1 with TRF2 and human telomeres. *Nat. Genet.*, **25**, 347–52.

Index

NB: page numbers in **bold** refer to illustrations and tables